APPLIED
NUMERICAL
LINEAR
ALGEBRA

APPLIED NUMERICAL LINEAR ALGEBRA

James W. Demmel
University of California
Berkeley, California

siam® Society for Industrial and Applied Mathematics

Philadelphia

Library of Congress Cataloging-in-Publication Data

Demmel, James W.
    Applied numerical linear algebra / James W. Demmel.
      p. cm.
    Includes bibliographical references and index.
    ISBN 978-0-898713-89-3 (pbk.)
    1. Algebras, Linear. 2. Numerical calculations. I. Title.
  QA184.D455 1997
  512'.5--dc21   97-17290

MATLAB is a registered trademark of The MathWorks, Inc. For MATLAB product information, please contact The MathWorks, Inc., 3 Apple Hill Drive, Natick, MA 01760-2098 USA, 508-647-7000, fax 508-647-7001, *info@mathworks.com, www.mathworks.com.*

The four images on the cover show an original image of a baby as well as three versions compressed using the singular value decomposition. See Example 3.4 on pages 113–116 for details.

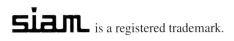

Contents

Preface   ix

1 Introduction   **1**
   1.1   Basic Notation 1
   1.2   Standard Problems of Numerical Linear Algebra 1
   1.3   General Techniques 2
      1.3.1   Matrix Factorizations 3
      1.3.2   Perturbation Theory and Condition Numbers 4
      1.3.3   Effects of Roundoff Error on Algorithms 5
      1.3.4   Analyzing the Speed of Algorithms 5
      1.3.5   Engineering Numerical Software 6
   1.4   Example: Polynomial Evaluation 7
   1.5   Floating Point Arithmetic 9
      1.5.1   Further Details 12
   1.6   Polynomial Evaluation Revisited 15
   1.7   Vector and Matrix Norms 19
   1.8   References and Other Topics for Chapter 1 23
   1.9   Questions for Chapter 1 24

2 Linear Equation Solving   **31**
   2.1   Introduction 31
   2.2   Perturbation Theory 32
      2.2.1   Relative Perturbation Theory 35
   2.3   Gaussian Elimination 38
   2.4   Error Analysis 44
      2.4.1   The Need for Pivoting 45
      2.4.2   Formal Error Analysis of Gaussian Elimination 46
      2.4.3   Estimating Condition Numbers 50
      2.4.4   Practical Error Bounds 54
   2.5   Improving the Accuracy of a Solution 60
      2.5.1   Single Precision Iterative Refinement 62
      2.5.2   Equilibration 62
   2.6   Blocking Algorithms for Higher Performance 63
      2.6.1   Basic Linear Algebra Subroutines (BLAS) 66
      2.6.2   How to Optimize Matrix Multiplication 67
      2.6.3   Reorganizing Gaussian Elimination to Use Level 3 BLAS 72
      2.6.4   More About Parallelism and Other Performance Issues . 75

2.7   Special Linear Systems   76
    2.7.1   Real Symmetric Positive Definite Matrices   76
    2.7.2   Symmetric Indefinite Matrices   79
    2.7.3   Band Matrices   79
    2.7.4   General Sparse Matrices   83
    2.7.5   Dense Matrices Depending on Fewer Than $O(n^2)$ Parameters   90
2.8   References and Other Topics for Chapter 2   93
2.9   Questions for Chapter 2   93

3   Linear Least Squares Problems   **101**
3.1   Introduction   101
3.2   Matrix Factorizations That Solve the Linear Least Squares Problem   105
    3.2.1   Normal Equations   106
    3.2.2   QR Decomposition   107
    3.2.3   Singular Value Decomposition   109
3.3   Perturbation Theory for the Least Squares Problem   117
3.4   Orthogonal Matrices   118
    3.4.1   Householder Transformations   119
    3.4.2   Givens Rotations   121
    3.4.3   Roundoff Error Analysis for Orthogonal Matrices   123
    3.4.4   Why Orthogonal Matrices?   124
3.5   Rank-Deficient Least Squares Problems   125
    3.5.1   Solving Rank-Deficient Least Squares Problems Using the SVD   128
    3.5.2   Solving Rank-Deficient Least Squares Problems Using QR with Pivoting   130
3.6   Performance Comparison of Methods for Solving Least Squares Problems   132
3.7   References and Other Topics for Chapter 3   134
3.8   Questions for Chapter 3   134

4   Nonsymmetric Eigenvalue Problems   **139**
4.1   Introduction   139
4.2   Canonical Forms   140
    4.2.1   Computing Eigenvectors from the Schur Form   148
4.3   Perturbation Theory   148
4.4   Algorithms for the Nonsymmetric Eigenproblem   153
    4.4.1   Power Method   154
    4.4.2   Inverse Iteration   155
    4.4.3   Orthogonal Iteration   156
    4.4.4   QR Iteration   159
    4.4.5   Making QR Iteration Practical   163
    4.4.6   Hessenberg Reduction   164

        4.4.7   Tridiagonal and Bidiagonal Reduction 166
        4.4.8   QR Iteration with Implicit Shifts 167
    4.5   Other Nonsymmetric Eigenvalue Problems 173
        4.5.1   Regular Matrix Pencils and Weierstrass Canonical Form 173
        4.5.2   Singular Matrix Pencils and the Kronecker
                 Canonical Form 180
        4.5.3   Nonlinear Eigenvalue Problems 183
    4.6   Summary 184
    4.7   References and Other Topics for Chapter 4 187
    4.8   Questions for Chapter 4 187

5   The Symmetric Eigenproblem and Singular Value Decompo-
    sition   195
    5.1   Introduction 195
    5.2   Perturbation Theory 197
        5.2.1   Relative Perturbation Theory 207
    5.3   Algorithms for the Symmetric Eigenproblem 210
        5.3.1   Tridiagonal QR Iteration 212
        5.3.2   Rayleigh Quotient Iteration 214
        5.3.3   Divide-and-Conquer 216
        5.3.4   Bisection and Inverse Iteration 228
        5.3.5   Jacobi's Method 232
        5.3.6   Performance Comparison 235
    5.4   Algorithms for the Singular Value Decomposition 237
        5.4.1   QR Iteration and Its Variations for the Bidiagonal SVD 242
        5.4.2   Computing the Bidiagonal SVD to High Relative Accuracy 245
        5.4.3   Jacobi's Method for the SVD 248
    5.5   Differential Equations and Eigenvalue Problems 254
        5.5.1   The Toda Lattice 255
        5.5.2   The Connection to Partial Differential Equations 259
    5.6   References and Other Topics for Chapter 5 260
    5.7   Questions for Chapter 5 260

6   Iterative Methods for Linear Systems   265
    6.1   Introduction 265
    6.2   On-line Help for Iterative Methods 266
    6.3   Poisson's Equation 267
        6.3.1   Poisson's Equation in One Dimension 267
        6.3.2   Poisson's Equation in Two Dimensions 270
        6.3.3   Expressing Poisson's Equation with Kronecker Products 274
    6.4   Summary of Methods for Solving Poisson's Equation 277
    6.5   Basic Iterative Methods 279
        6.5.1   Jacobi's Method 281
        6.5.2   Gauss–Seidel Method 282
        6.5.3   Successive Overrelaxation 283

| | | 6.5.4 | Convergence of Jacobi's, Gauss–Seidel, and SOR(ω) Methods on the Model Problem | 285 |

         6.5.4   Convergence of Jacobi's, Gauss–Seidel, and
SOR(ω) Methods on the Model Problem 285

         6.5.5   Detailed Convergence Criteria for Jacobi's,
Gauss–Seidel, and SOR(ω) Methods 286

         6.5.6   Chebyshev Acceleration and Symmetric SOR (SSOR) . 294

   6.6   Krylov Subspace Methods 299

         6.6.1   Extracting Information about A via Matrix-Vector Multiplication 301

         6.6.2   Solving $Ax = b$ Using the Krylov Subspace $\mathcal{K}_k$ 305

         6.6.3   Conjugate Gradient Method 307

         6.6.4   Convergence Analysis of the Conjugate Gradient Method 312

         6.6.5   Preconditioning 316

         6.6.6   Other Krylov Subspace Algorithms for Solving $Ax = b$. 319

   6.7   Fast Fourier Transform 321

         6.7.1   The Discrete Fourier Transform 323

         6.7.2   Solving the Continuous Model Problem Using Fourier Series 324

         6.7.3   Convolutions 325

         6.7.4   Computing the Fast Fourier Transform 326

   6.8   Block Cyclic Reduction 327

   6.9   Multigrid 331

         6.9.1   Overview of Multigrid on the Two-Dimensional Poisson's Equation 332

         6.9.2   Detailed Description of Multigrid on the One-Dimensional Poisson's Equation 337

   6.10 Domain Decomposition 347

         6.10.1 Nonoverlapping Methods 348

         6.10.2 Overlapping Methods 351

   6.11 References and Other Topics for Chapter 6 356

   6.12 Questions for Chapter 6 356

7   Iterative Methods for Eigenvalue Problems   **361**

   7.1   Introduction 361

   7.2   The Rayleigh–Ritz Method 362

   7.3   The Lanczos Algorithm in Exact Arithmetic 366

   7.4   The Lanczos Algorithm in Floating Point Arithmetic 375

   7.5   The Lanczos Algorithm with Selective Orthogonalization 382

   7.6   Beyond Selective Orthogonalization 383

   7.7   Iterative Algorithms for the Nonsymmetric Eigenproblem . . . 384

   7.8   References and Other Topics for Chapter 7 386

   7.9   Questions for Chapter 7 386

Bibliography   **389**

Index   **409**

Preface

This textbook covers both direct and iterative methods for the solution of linear systems, least squares problems, eigenproblems, and the singular value decomposition. Earlier versions have been used by the author in graduate classes in the Mathematics Department of the University of California at Berkeley since 1990 and at the Courant Institute before then.

In writing this textbook I aspired to meet the following goals:

1. The text should be attractive to first-year graduate students from a variety of engineering and scientific disciplines.

2. It should be self-contained, assuming only a good undergraduate background in linear algebra.

3. The students should learn the mathematical basis of the field, as well as how to build or find good numerical software.

4. Students should acquire practical knowledge for solving real problems efficiently. In particular, they should know what the state-of-the-art techniques are in each area or when to look for them and where to find them, even if I analyze only simpler versions in the text.

5. It should all fit in one semester, since that is what most students have available for this subject.

Indeed, I was motivated to write this book because the available textbooks, while very good, did not meet these goals. Golub and Van Loan's text [121] is too encyclopedic in style, while still omitting some important topics such as multigrid, domain decomposition, and some recent algorithms for eigenvalue problems. Watkins's [252] and Trefethen's and Bau's [243] also omit some state-of-the-art algorithms.

While I believe that these five goals were met, the fifth goal was the hardest to manage, especially as the text grew over time to include recent research results and requests from colleagues for new sections. A reasonable one-semester curriculum based on this book would cover

- Chapter 1, excluding section 1.5.1;

- Chapter 2, excluding sections 2.2.1, 2.4.3, 2.5, 2.6.3, and 2.6.4;

- Chapter 3, excluding sections 3.5 and 3.6;

- Chapter 4, up to and including section 4.4.5;

- Chapter 5, excluding sections 5.2.1, 5.3.5, 5.4 and 5.5;

- Chapter 6, excluding sections 6.3.3, 6.5.5, 6.5.6, 6.6.6, 6.7.2, 6.7.3, 6.7.4, 6.8, 6.9.2, and 6.10; and

- Chapter 7, up to and including section 7.3.

Notable features of this book include

- a class homepage with Matlab source code for examples and homework problems in the text;

- frequent recommendations and pointers to the best software currently available (from LAPACK and elsewhere);

- a discussion of how modern cache-based computer memories impact algorithm design;

- performance comparisons of competing algorithms for least squares and symmetric eigenvalue problems;

- a discussion of a variety of iterative methods, from Jacobi's to multigrid, with detailed performance comparisons for solving Poisson's equation on a square grid;

- detailed discussion and numerical examples for the Lanczos algorithm for the symmetric eigenvalue problem;

- numerical examples drawn from fields ranging from mechanical vibrations to computational geometry;

- sections on "relative perturbation theory" and corresponding high-accuracy algorithms for symmetric eigenvalue problems and the singular value decomposition; and

- dynamical systems interpretations of eigenvalue algorithms.

The URL for the class homepage will be abbreviated to HOMEPAGE throughout the text, standing for http://www.siam.org/books/demmel/ demmel_class. Two other abbreviated URLs will be used as well. PARALLEL_ HOMEPAGE is an abbreviation for http://www.siam.org/books/demmel/ demmel_parallelclass and points to a related on-line class by the author on parallel computing. NETLIB is an abbreviation for http://www.netlib.org.

Homework problems are marked Easy, Medium, or Hard, according to their difficulty. Problems involving significant amounts of programming are marked "programming."

Many people have contributed to this text. Most notably, Zhaojun Bai used this text at Texas A&M and the University of Kentucky, contributed numerous questions, and made many useful suggestions. Alan Edelman (who used this book at MIT), Martin Gutknecht (who used this book at ETH Zurich), Velvel Kahan (who used this book at Berkeley), Richard Lehoucq, Beresford Parlett, and many anonymous referees made detailed comments on various parts of the text. In addition, Alan Edelman and Martin Gutknecht provided hospitable surroundings while this final edition was being prepared. Table 2.2 is taken from the Ph.D. thesis of my former student Xiaoye Li. Mark Adams, Tzu-Yi Chen, Inderjit Dhillon, Jian Xun He, Melody Ivory, Xiaoye Li, Bernd Pfrommer, Huan Ren, and Ken Stanley, along with many other students at Courant, Berkeley, Kentucky, and MIT over the years, helped debug the text. Bob Untiedt and Selene Victor were of great help in typesetting and producing figures. Megan supplied the cover photo. Finally, Kathy Yelick has contributed support over more years than either of us expected this project to take.

James Demmel
Berkeley, California
June 1997

1

Introduction

1.1. Basic Notation

In this course we will refer frequently to *matrices*, *vectors*, and *scalars*. A matrix will be denoted by an upper case letter such as A, and its (i,j)th element will be denoted by a_{ij}. If the matrix is given by an expression such as $A + B$, we will write $(A + B)_{ij}$. In detailed algorithmic descriptions we will sometimes write $A(i,j)$ or use the Matlab^TM [184] notation $A(i : j, k : l)$ to denote the submatrix of A lying in rows i through j and columns k through l. A lower-case letter like x will denote a vector, and its ith element will be written x_i. Vectors will almost always be column vectors, which are the same as matrices with one column. Lower-case Greek letters (and occasionally lower-case letters) will denote scalars. $\mathbb{R}$ will denote the set of real numbers; $\mathbb{R}^n$, the set of n-dimensional real vectors; and $\mathbb{R}^{m \times n}$, the set of m-by-n real matrices. $\mathbb{C}$, $\mathbb{C}^n$, and $\mathbb{C}^{m \times n}$ denote complex numbers, vectors, and matrices, respectively. Occasionally we will use the shorthand $A^{m \times n}$ to indicate that A is an m-by-n matrix. A^T will denote the *transpose* of the matrix A: $(A^T)_{ij} = a_{ji}$. For complex matrices we will also use the *conjugate transpose* A^*: $(A^*)_{ij} = \bar{a}_{ji}$. $\Re z$ and $\Im z$ will denote the real and imaginary parts of the complex number z, respectively. If A is m-by-n, then $|A|$ is the m-by-n matrix of absolute values of entries of A: $(|A|)_{ij} = |a_{ij}|$. Inequalities like $|A| \leq |B|$ are meant componentwise: $|a_{ij}| \leq |b_{ij}|$ for all i and j. We will also use this absolute value notation for vectors: $(|x|)_i = |x_i|$. Ends of proofs will be marked by □, and ends of examples by ⋄. Other notation will be introduced as needed.

1.2. Standard Problems of Numerical Linear Algebra

We will consider the following standard problems:

---

[1] Matlab is a registered trademark of The MathWorks, Inc., 24 Prime Park Way, Natick, MA 01760, USA, tel. 508-647-7000, fax 508-647-7001, info@mathworks.com, http://www.mathworks.com.

- *Linear systems of equations*: Solve $Ax = b$. Here A is a given n-by-n nonsingular real or complex matrix, b is a given column vector with n entries, and x is a column vector with n entries that we wish to compute.

- *Least squares problems*: Compute the x that minimizes $\|Ax - b\|_2$. Here A is m-by-n, b is m-by-1, x is n-by-1, and $\|y\|_2 \equiv \sqrt{\sum_i |y_i|^2}$ is called the *two-norm* of the vector y. If $m > n$ so that we have more equations than unknowns, the system is called *overdetermined*. In this case we cannot generally solve $Ax = b$ exactly. If $m < n$, the system is called *underdetermined*, and we will have infinitely many solutions.

- *Eigenvalue problems*: Given an n-by-n matrix A, find an n-by-1 nonzero vector x and a scalar λ so that $Ax = \lambda x$.

- *Singular value problems*: Given an m-by-n matrix A, find an n-by-1 nonzero vector x and scalar λ so that $A^T A x = \lambda x$. We will see that this special kind of eigenvalue problem is important enough to merit separate consideration and algorithms.

We choose to emphasize these standard problems because they arise so often in engineering and scientific practice. We will illustrate them throughout the book with simple examples drawn from engineering, statistics, and other fields. There are also many variations of these standard problems that we will consider, such as generalized eigenvalue problems $Ax = \lambda Bx$ (section 4.5) and "rank-deficient" least squares problems $\min_x \|Ax - b\|_2$, whose solutions are nonunique because the columns of A are linearly dependent (section 3.5).

We will learn the importance of exploiting any *special structure* our problem may have. For example, solving an n-by-n linear system costs $2/3 n^3$ floating point operations if we use the most general form of Gaussian elimination. If we add the information that the system is symmetric and positive definite, we can save half the work by using another algorithm called Cholesky. If we further know the matrix is *banded* with *semibandwidth* $\sqrt{n}$ (i.e., $a_{ij} = 0$ if $|i-j| > \sqrt{n}$), then we can reduce the cost further to $O(n^2)$ by using band Cholesky. If we say quite explicitly that we are trying to solve Poisson's equation on a square using a 5-point difference approximation, which determines the matrix nearly uniquely, then by using the multigrid algorithm we can reduce the cost to $O(n)$, which is nearly as fast as possible, in the sense that we use just a constant amount of work per solution component (section 6.4).

1.3. General Techniques

There are several general concepts and techniques that we will use repeatedly:

1. matrix factorizations;

2. perturbation theory and condition numbers;

3. effects of roundoff error on algorithms, including properties of floating point arithmetic;

4. analysis of the speed of an algorithm;

5. engineering numerical software.

We discuss each of these briefly below.

1.3.1. Matrix Factorizations

A *factorization* of the matrix A is a representation of A as a product of several "simpler" matrices, which make the problem at hand easier to solve. We give two examples.

EXAMPLE 1.1. Suppose that we want to solve $Ax = b$. If A is a lower triangular matrix,

$$
\begin{bmatrix}
a_{11} & & & \\
a_{21} & a_{22} & & \\
\vdots & \vdots & \ddots & \\
a_{n1} & a_{n2} & \cdots & a_{nn}
\end{bmatrix}
\begin{bmatrix}
x_1 \\
x_2 \\
\vdots \\
x_n
\end{bmatrix}
=
\begin{bmatrix}
b_1 \\
b_2 \\
\vdots \\
b_n
\end{bmatrix}
$$

is easy to solve using *forward substitution*:

for $i = 1$ to n
$\qquad x_i = (b_i - \sum_{k=1}^{i-1} a_{ik}x_k)/a_{ii}$
end for

An analogous idea, *back substitution*, works if A is upper triangular. To use this to solve a general system $Ax = b$ we need the following matrix factorization, which is just a restatement of Gaussian elimination.

THEOREM 1.1. *If the n-by-n matrix A is nonsingular, there exist a permutation matrix P (the identity matrix with its rows permuted), a nonsingular lower triangular matrix L, and a nonsingular upper triangular matrix U such that $A = P \cdot L \cdot U$. To solve $Ax = b$, we solve the equivalent system $PLUx = b$ as follows:*
$$
\begin{aligned}
LUx &= P^{-1}b = P^Tb \quad &\textit{(permute entries of b)}, \\
Ux &= L^{-1}(P^Tb) \quad &\textit{(forward substitution)}, \\
x &= U^{-1}(L^{-1}P^Tb) \quad &\textit{(back substitution)}.
\end{aligned}
$$

We will prove this theorem in section 2.3. ◇

EXAMPLE 1.2. The *Jordan canonical factorization* $A = VJV^{-1}$ exhibits the eigenvalues and eigenvectors of A. Here V is a nonsingular matrix, whose columns include the eigenvectors, and J is the *Jordan canonical form* of A, a special triangular matrix with the eigenvalues of A on its diagonal. We will learn that it is numerically superior to compute the *Schur factorization* $A = UTU^*$, where U is a unitary matrix (i.e., U's columns are orthonormal) and T is upper triangular with A's eigenvalues on its diagonal. The Schur form T can be computed faster and more accurately than the Jordan form J. We discuss the Jordan and Schur factorizations in section 4.2. ◇

1.3.2. Perturbation Theory and Condition Numbers

The answers produced by numerical algorithms are seldom exactly correct. There are two sources of error. First, there may be errors in the input data to the algorithm, caused by prior calculations or perhaps measurement errors. Second, there are errors caused by the algorithm itself, due to approximations made within the algorithm. In order to estimate the errors in the computed answers from both these sources, we need to understand how much the solution of a problem is changed (or *perturbed*) if the input data are slightly perturbed.

EXAMPLE 1.3. Let $f(x)$ be a real-valued differentiable function of a real variable x. We want to compute $f(x)$, but we do not know x exactly. Suppose instead that we are given $x + \delta x$ and a bound on δx. The best that we can do (without more information) is to compute $f(x + \delta x)$ and to try to bound the absolute error $|f(x + \delta x) - f(x)|$. We may use a simple linear approximation to f to get the estimate $f(x + \delta x) \approx f(x) + \delta x f'(x)$, and so the error is $|f(x+\delta x) - f(x)| \approx |\delta x| \cdot |f'(x)|$. We call $|f'(x)|$ the *absolute condition number* of f at x. If $|f'(x)|$ is large enough, then the error may be large even if δx is small; in this case we call f *ill-conditioned* at x. ◇

We say *absolute* condition number because it provides a bound on the absolute error $|f(x + \delta x) - f(x)|$ given a bound on the absolute change $|\delta x|$ in the input. We will also often use the following essentially equivalent expression to bound the error:

$$\frac{|f(x + \delta x) - f(x)|}{|f(x)|} \approx \frac{|\delta x|}{|x|} \cdot \frac{|f'(x)| \cdot |x|}{|f(x)|}.$$

This expression bounds the *relative error* $|f(x + \delta x) - f(x)|/|f(x)|$ as a multiple of the *relative* change $|\delta x|/|x|$ in the input. The multiplier, $|f'(x)| \cdot |x|/|f(x)|$, is called the *relative condition number*, or often just *condition number* for short.

The condition number is all that we need to understand how error in the input data affects the computed answer: we simply multiply the condition number by a bound on the input error to bound the error in the computed solution.

For each problem we consider, we will derive its corresponding condition number.

1.3.3. Effects of Roundoff Error on Algorithms

To continue our analysis of the error caused by the algorithm itself, we need to study the effect of roundoff error in the arithmetic, or simply roundoff for short. We will do so by using a property possessed by most good algorithms: *backward stability*. We define it as follows.

> If $\text{alg}(x)$ is our algorithm for $f(x)$, including the effects of roundoff, we call $\text{alg}(x)$ a *backward stable algorithm* for $f(x)$ if for all x there is a "small" δx such that $\text{alg}(x) = f(x + \delta x)$. δx is called the *backward error*. Informally, we say that we get the exact answer $(f(x + \delta x))$ for a slightly wrong problem $(x + \delta x)$.

This implies that we may bound the error as

$$\text{error} = |\text{alg}(x) - f(x)| = |f(x + \delta x) - f(x)| \approx |f'(x)| \cdot |\delta x|,$$

the product of the absolute condition number $|f'(x)|$ and the magnitude of the backward error $|\delta x|$. Thus, if $\text{alg}(\cdot)$ is backward stable, $|\delta x|$ is always small, so the error will be small unless the absolute condition number is large. Thus, backward stability is a desirable property for an algorithm, and most of the algorithms that we present will be backward stable. Combined with the corresponding condition numbers, we will have error bounds for all our computed solutions.

Proving that an algorithm is backward stable requires knowledge of the roundoff error of the basic floating point operations of the machine and how these errors propagate through an algorithm. This is discussed in section 1.5.

1.3.4. Analyzing the Speed of Algorithms

In choosing an algorithm to solve a problem, one must of course consider its speed (which is also called performance) as well as its backward stability. There are several ways to estimate speed. Given a particular problem instance, a particular implementation of an algorithm, and a particular computer, one can of course simply run the algorithm and see how long it takes. This may be difficult or time consuming, so we often want simpler estimates. Indeed, we typically want to estimate how long a particular algorithm would take *before* implementing it.

The traditional way to estimate the time an algorithm takes is to count the *flops*, or *floating point operations*, that it performs. We will do this for all the algorithms we present. However, this is often a misleading time estimate on modern computer architectures, because it can take significantly more time to move the data inside the computer to the place where it is to be multiplied, say, than it does to actually perform the multiplication. This is especially true on parallel computers but also is true on conventional machines such as workstations and PCs. For example, matrix multiplication on the IBM RS6000/590 workstation can be sped up from 65 Mflops (millions of floating point operations per second) to 240 Mflops, nearly four times faster, by judiciously reordering the operations of the standard algorithm (and using the correct compiler optimizations). We discuss this further in section 2.6.

If an algorithm is *iterative*, i.e., produces a series of approximations converging to the answer rather than stopping after a fixed number of steps, then we must ask how many steps are needed to decrease the error to a tolerable level. To do this, we need to decide if the convergence is *linear* (i.e., the error decreases by a constant factor $0 < c < 1$ at each step so that $|\text{error}_i| \leq c \cdot |\text{error}_{i-1}|$) or faster, such as *quadratic* ($|\text{error}_i| \leq c \cdot |\text{error}_{i-1}|^2$). If two algorithms are both linear, we can ask which has the smaller constant c. Iterative linear equation solvers and their convergence analysis are the subject of Chapter 6.

1.3.5. Engineering Numerical Software

Three main issues in designing or choosing a piece of numerical software are *ease of use*, *reliability*, and *speed*. Most of the algorithms covered in this book have already been carefully programmed with these three issues in mind. If some of this existing software can solve your problem, its ease of use may well outweigh any other considerations such as speed. Indeed, if you need only to solve your problem once or a few times, it is often easier to use general purpose software written by experts than to write your own more specialized program.

There are three programming paradigms for exploiting other experts' software. The first paradigm is the traditional software library, consisting of a collection of subroutines for solving a fixed set of problems, such as solving linear systems, finding eigenvalues, and so on. In particular, we will discuss the LAPACK library [10], a state-of-the-art collection of routines available in Fortran and C. This library, and many others like it, are freely available in the public domain; see NETLIB on the World Wide Web.[2] LAPACK provides reliability and high speed (for example, making careful use of matrix multiplication, as described above) but requires careful attention to data structures and calling sequences on the part of the user. We will provide pointers to such software throughout the text.

The second programming paradigm provides a much easier-to-use environment than libraries like LAPACK, but at the cost of some performance. This paradigm is provided by the commercial system Matlab [184], among others. Matlab provides a simple interactive programming environment where all variables represent matrices (scalars are just 1-by-1 matrices), and most linear algebra operations are available as built-in functions. For example, "$C = A * B$" stores the product of matrices A and B in C, and "$A = \text{inv}(B)$" stores the inverse of matrix B in A. It is easy to quickly prototype algorithms in Matlab and to see how they work. But since Matlab makes a number of algorith-

---

[2]Recall that we abbreviate the URL prefix http://www.netlib.org to NETLIB in the text.

mic decisions automatically for the user, it may perform more slowly than a carefully chosen library routine.

The third programming paradigm is that of *templates*, or recipes for assembling complicated algorithms out of simpler building blocks. Templates are useful when there are a large number of ways to construct an algorithm but no simple rule for choosing the best construction for a particular input problem; therefore, much of the construction must be left to the user. An example of this may be found in *Templates for the Solution of Linear Systems: Building Blocks for Iterative Methods* [24]; a similar set of templates for eigenproblems is currently under construction.

1.4. Example: Polynomial Evaluation

We illustrate the ideas of perturbation theory, condition numbers, backward stability, and roundoff error analysis with the example of *polynomial evaluation*:

$$p(x) = \sum_{i=0}^{d} a_i x^i.$$

Horner's rule for polynomial evaluation is

$$p = a_d$$
for $i = d - 1$ down to 0
$$p = x * p + a_i$$
end for

Let us apply this to $p(x) = (x - 2)^9 = x^9 - 18x^8 + 144x^7 - 672x^6 + 2016x^5 - 4032x^4 + 5376x^3 - 4608x^2 + 2304x - 512$. In the bottom of Figure 1.1, we see that near the zero $x = 2$ the value of $p(x)$ computed by Horner's rule is quite unpredictable and may justifiably be called "noise." The top of Figure 1.1 shows an accurate plot.

To understand the implications of this figure, let us see what would happen if we tried to find a zero of $p(x)$ using a simple zero finder based on Bisection, shown below in Algorithm 1.1.

Bisection starts with an interval $[x_{low}, x_{high}]$ in which $p(x)$ changes sign $(p(x_{low}) \cdot p(x_{high}) < 0)$ so that $p(x)$ must have a zero in the interval. Then the algorithm computes $p(x_{mid})$ at the interval midpoint $x_{mid} = (x_{low} + x_{high})/2$ and asks whether $p(x)$ changes sign in the bottom half interval $[x_{low}, x_{mid}]$ or top half interval $[x_{mid}, x_{high}]$. Either way, we find an interval of half the original length containing a zero of $p(x)$. We can continue bisecting until the interval is as short as desired.

So the decision between choosing the top half interval or bottom half interval depends on the sign of $p(x_{mid})$. Examining the graph of $p(x)$ in the bottom half of Figure 1.1, we see that this sign varies rapidly from plus to minus as

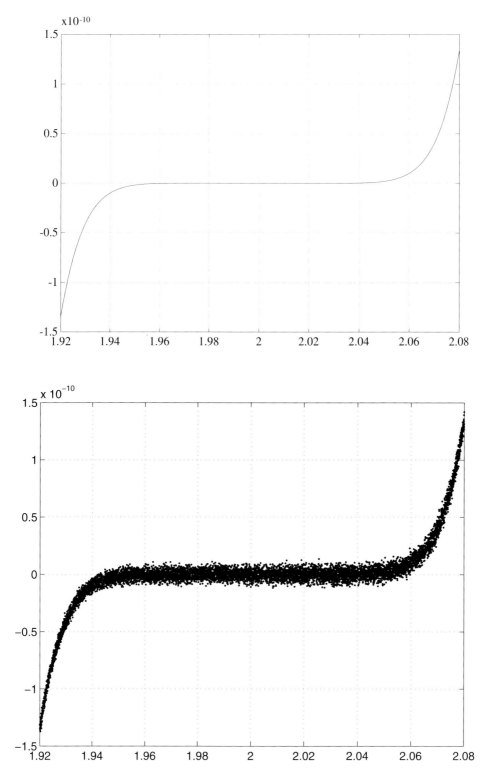

Fig. 1.1. *Plot of $y = (x-2)^9 = x^9 - 18x^8 + 144x^7 - 672x^6 + 2016x^5 - 4032x^4 + 5376x^3 - 4608x^2 + 2304x - 512$ evaluated at 8000 equispaced points, using $y = (x-2)^9$ (top) and using Horner's rule (bottom).*

x varies. So changing x_{low} or x_{high} just slightly could completely change the sequence of sign decisions and also the final interval. Indeed, depending on the initial choices of x_{low} and x_{high}, the algorithm could converge *anywhere* inside the "noisy region" from 1.95 to 2.05 (see Question 1.21).

To explain this fully, we return to properties of floating point arithmetic.

ALGORITHM 1.1. *Finding zeros of $p(x)$ using Bisection.*

```
proc bisect (p, x_low, x_high, tol)
/* find a root of p(x) = 0 in [x_low, x_high]
      assuming p(x_low) · p(x_high) < 0 */
/* stop if zero found to within ±tol */
p_low = p(x_low)
p_high = p(x_high)
while x_high − x_low > 2 · tol
      x_mid = (x_low + x_high)/2
      p_mid = p(x_mid)
      if p_low · p_mid < 0 then /* there is a root in [x_low, x_mid] */
            x_high = x_mid
            p_high = p_mid
      else if p_mid · p_high < 0 then /* there is a root in [x_mid, x_high] */
            x_low = x_mid
            p_low = p_mid
      else /* x_mid is a root */
            x_low = x_mid
            x_high = x_mid
      end if
end while
root = (x_low + x_high)/2
```

1.5. Floating Point Arithmetic

The number -3.1416 may be expressed in *scientific notation* as follows:

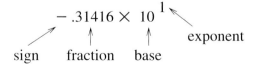

Computers use a similar representation called *floating point*, but generally the base is 2 (with exceptions, such as 16 for IBM 370 and 10 for some spreadsheets and most calculators). For example, $.10101_2 \times 2^3 = 5.25_{10}$.

A floating point number is called *normalized* if the leading digit of the fraction is nonzero. For example, $.10101_2 \times 2^3$ is normalized, but $.010101_2 \times 2^4$ is not. Floating point numbers are usually normalized, which has two advantages:

each nonzero floating point value has a unique representation as a bit string, and in binary the leading 1 in the fraction need not be stored explicitly (because it is always 1), leaving one extra bit for a longer, more accurate fraction.

The most important parameters describing floating point numbers are the base; the number of digits (bits) in the fraction, which determines the precision; and the number of digits (bits) in the exponent, which determines the exponent range and thus the largest and smallest representable numbers. Different floating point arithmetics also differ in how they round computed results, what they do about numbers that are too near zero (underflow) or too big (overflow), whether $\pm\infty$ is allowed, and whether useful nonnumbers (sometimes called NaNs, indefinites, or reserved operands) are provided. We discuss each of these below.

First we consider the precision with which numbers can be represented. For example, $.31416 \times 10^1$ has five decimal digits, so any information less than $.5 \times 10^{-4}$ may have been lost. This means that if x is a real number whose best five-digit approximation is $.31416 \times 10^1$, then the *relative representation error* in $.31416 \times 10^1$ is

$$\frac{|x - .31416 \times 10^1|}{.31416 \times 10^1} \leq \frac{.5 \times 10^{-4}}{.31416 \times 10^1} \approx .16 \times 10^{-4}.$$

The maximum relative representation error in a normalized number occurs for $.10000 \times 10^1$, which is the most accurate five-digit approximation of all numbers in the interval from .999995 to 1.00005. Its relative error is therefore bounded by $.5 \cdot 10^{-4}$. More generally, the *maximum relative representation error* in a floating point arithmetic with p digits and base β is $.5 \times \beta^{1-p}$. This is also half the distance between 1 and the next larger floating point number, $1 + \beta^{1-p}$.

Computers have historically used many different choices of base, number of digits, and range, but fortunately the *IEEE standard for binary arithmetic* is now most common. It is used on Sun, DEC, HP, and IBM workstations and all PCs. IEEE arithmetic includes two kinds of floating point numbers: *single precision* (32 bits long) and *double precision* (64 bits long).

IEEE single precision

If s, e, and $f < 1$ are the 1-bit sign, 8-bit exponent, and 23-bit fraction in the IEEE single precision format, respectively, then the number represented is $(-1)^s \cdot 2^{e-127} \cdot (1 + f)$. The maximum relative representation error is $2^{-24} \approx 6 \cdot 10^{-8}$, and the range of positive normalized numbers is from 2^{-126} (the *underflow threshold*) to $2^{127} \cdot (2 - 2^{-23}) \approx 2^{128}$ (the *overflow threshold*), or about 10^{-38} to 10^{38}. The positions of these floating point numbers on the real

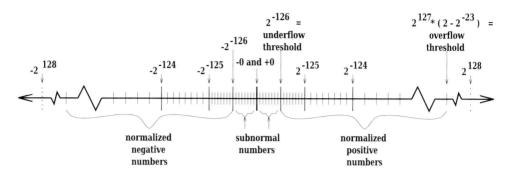

Fig. 1.2. *Real number line with floating point numbers indicated by solid tick marks. The range shown is correct for IEEE single precision, but a 3-bit fraction is assumed for ease of presentation so that there are only $2^3 - 1 = 7$ floating point numbers between consecutive powers of 2, not $2^{23} - 1$. The distance between consecutive tick marks is constant between powers of 2 and doubles/halves across powers of 2 (among the normalized floating point numbers). $+2^{128}$ and -2^{128}, which are one unit in the last place larger in magnitude than the overflow threshold (the largest finite floating point number, $2^{127} \cdot (2 - 2^{-23})$), are shown as dotted tick marks. The figure is symmetric about 0; $+0$ and -0 are distinct floating point bit strings but compare as numerically equal. Division by zero is the only binary operation that gives different results, $+\infty$ and $-\infty$, for different signed zero arguments.*

number line are shown in Figure 1.2 (where we use a 3-bit fraction for ease of presentation).

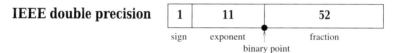

IEEE double precision

If s, e, and $f < 1$ are the 1-bit sign, 11-bit exponent, and 52-bit fraction in IEEE double precision format, respectively, then the number represented is $(-1)^s \cdot 2^{e-1023} \cdot (1 + f)$. The maximum relative representation error is $2^{-53} \approx 10^{-16}$, and the exponent range is 2^{-1022} (the *underflow threshold*) to $2^{1023} \cdot (2 - 2^{-52}) \approx 2^{1024}$ (the *overflow threshold*), or about 10^{-308} to 10^{308}.

When the true value of a computation $a \odot b$ (where $\odot$ is one of the four binary operations $+$, $-$, $*$, and $/$) cannot be represented exactly as a floating point number, it must be approximated by a nearby floating point number before it can be stored in memory or a register. We denote this approximation by $\mathrm{fl}(a \odot b)$. The difference $(a \odot b) - \mathrm{fl}(a \odot b)$ is called the *roundoff error*. If $\mathrm{fl}(a \odot b)$ is a nearest floating point number to $a \odot b$, we say that the arithmetic *rounds correctly* (or just *rounds*). IEEE arithmetic has this attractive property. (IEEE arithmetic breaks ties, when $a \odot b$ is exactly halfway between two adjacent floating point numbers, by choosing $\mathrm{fl}(a \odot b)$ to have its least significant bit zero; this is called *rounding to nearest even*.) When rounding correctly, if $a \odot b$ is within the exponent range (otherwise we get *overflow* or *underflow*), then we can write

$$\text{fl}(a \odot b) = (a \odot b)(1 + \delta),\tag{1.1}$$

where $|\delta|$ is bounded by ε, which is called variously *machine epsilon, machine precision*, or *macheps*. Since we are rounding as accurately as possible, ε is equal to the maximum relative representation error $.5 \cdot \beta^{1-p}$. IEEE arithmetic also guarantees that $\text{fl}(\sqrt{a}) = \sqrt{a}(1 + \delta)$, with $|\delta| \leq \varepsilon$. This is the most common model for roundoff error analysis and the one we will use in this book. A nearly identical formula applies to complex floating point arithmetic; see Question 1.12. However, formula (1.1) does ignore some interesting details.

1.5.1. Further Details

IEEE arithmetic also includes *subnormal numbers*, i.e., unnormalized floating point numbers with the minimum possible exponent. These represent tiny numbers between zero and the smallest normalized floating point number; see Figure 1.2. Their presence means that a difference $\text{fl}(x - y)$ can never be zero because of underflow, yielding the attractive property that the predicate $x = y$ is true if and only if $\text{fl}(x - y) = 0$. To incorporate errors caused by underflow into formula (1.1) one would change it to

$$\text{fl}(a \odot b) = (a \odot b)(1 + \delta) + \eta,$$

where $|\delta| \leq \varepsilon$ as before, and $|\eta|$ is bounded by a tiny number equal to the largest error caused by underflow ($2^{-150} \approx 10^{-45}$ in IEEE single precision and $2^{-1075} \approx 10^{-324}$ in IEEE double precision).

IEEE arithmetic includes the symbols $\pm\infty$ and NaN (*Not a Number*). $\pm\infty$ is returned when an operation overflows, and behaves according to the following arithmetic rules: $x/\pm\infty = 0$ for any finite floating point number x, $x/0 = \pm\infty$ for any nonzero floating point number x, $+\infty + \infty = +\infty$, etc. An NaN is returned by any operation with no well-defined finite or infinite result, such as $\infty - \infty$, $\frac{\infty}{\infty}$, $\frac{0}{0}$, $\sqrt{-1}$, NaN $\odot x$, etc.

Whenever an arithmetic operation is invalid and so produces an NaN, or overflows or divides by zero to produce $\pm\infty$, or underflows, an *exception flag* is set and can later be tested by the user's program. These features permit one to write both more reliable programs (because the program can detect and correct its own exceptions, instead of simply aborting execution) and faster programs (by avoiding "paranoid" programming with many tests and branches to avoid possible but unlikely exceptions). For examples, see Question 1.19, the comments following Lemma 5.3, and [81].

The most expensive error known to have been caused by an improperly handled floating point exception is the crash of the Ariane 5 rocket of the European Space Agency on June 4, 1996. See HOME/ariane5rep.html for details.

Not all machines use IEEE arithmetic or round carefully, although nearly all do. The most important modern exceptions are those machines produced by

Cray Research,[3] although future generations of Cray machines may use IEEE arithmetic.[4] Since the difference between $\mathrm{fl}(a \odot b)$ computed on a Cray machine and $\mathrm{fl}(a \odot b)$ computed on an IEEE machine usually lies in the 14th decimal place or beyond, the reader may wonder whether the difference is important. Indeed, most algorithms in numerical linear algebra are insensitive to details in the way roundoff is handled. But it turns out that some algorithms are easier to design, or more reliable, when rounding is done properly. Here are two examples.

When the Cray C90 subtracts 1 from the next smaller floating point number, it gets -2^{-47}, which is twice the correct answer, -2^{-48}. Getting even tiny differences to high relative accuracy is essential for the correctness of the divide-and-conquer algorithm for finding eigenvalues and eigenvectors of symmetric matrices, currently the fastest algorithm available for the problem. This algorithm requires a rather nonintuitive modification to guarantee correctness on Cray machines (see section 5.3.3).

The Cray machine may also yield an error when computing $\arccos(x/\sqrt{x^2 + y^2})$ because excessive roundoff causes the argument of arccos to be larger than 1. This cannot happen in IEEE arithmetic (see Question 1.17).

To accommodate error analysis on a Cray C90 or other Cray machines we may instead use the model $\mathrm{fl}(a \pm b) = a(1+\delta_1) \pm b(1+\delta_2)$, $\mathrm{fl}(a*b) = (a*b)(1+\delta_3)$, and $\mathrm{fl}(a/b) = (a/b)(1 + \delta_3)$, with $|\delta_i| \leq \varepsilon$, where ε is a small multiple of the maximum relative representation error.

Briefly, we can say that correct rounding and other features of IEEE arithmetic are designed to preserve as many mathematical relationships used to derive formulas as possible. It is easier to design algorithms knowing that (barring overflow or underflow) $\mathrm{fl}(a - b)$ is computed with a small relative error (otherwise divide-and-conquer can fail), and that $-1 \leq c \equiv \mathrm{fl}(x/\sqrt{x^2 + y^2}) \leq 1$ (otherwise $\arccos(c)$ can fail). There are many other such mathematical relationships that one relies on (often unwittingly) to design algorithms. For more details about IEEE arithmetic and its relationship to numerical analysis, see [159, 158, 81].

Given the variability in floating point across machines, how does one write portable software that depends on the arithmetic? For example, iterative algorithms that we will study in later chapters frequently have loops such as

> repeat
>
>   · · ·
>
>   update e
> until "e is negligible compared to f,"

---

[3] We include machines such as the NEC SX-4, which has a "Cray mode" in which it performs arithmetic the same way. We exclude the Cray T3D and T3E, which are parallel computers built from DEC Alpha processors, which use IEEE arithmetic very nearly (underflows are flushed to zero for speed's sake).

[4] Cray Research was purchased by Silicon Graphics in 1996.

where $e \geq 0$ is some error measure, and $f > 0$ is some comparison value (see section 4.4.5 for an example). By negligible we mean "is $e \leq c \cdot \varepsilon \cdot f$?," where $c \geq 1$ is some modest constant, chosen to trade off accuracy and speed of convergence. Since this test requires the machine-dependent constant ε, this test has in the past often been replaced by the *apparently* machine-independent test "is $e + cf = cf$?" The idea here is that adding e to cf and rounding will yield cf again if $e < c\varepsilon f$ or perhaps a little smaller. But this test can fail (by requiring e to be *much* smaller than necessary, or than attainable), depending on the machine and compiler used (see the next paragraph). So the best test indeed uses ε explicitly. It turns out that with sufficient care one can compute ε in a machine-independent way, and software for this is available in the LAPACK subroutines `slamch` (for single precision) and `dlamch` (for double precision). These routines also compute or estimate the overflow threshold (without overflowing!), the underflow threshold, and other parameters. Another portable program that uses these explicit machine parameters is discussed in Question 1.19.

Sometimes one needs higher precision than is available from IEEE single or double precision. For example, higher precision is of use in algorithms such as iterative refinement for improving the accuracy of a computed solution of $Ax = b$ (see section 2.5.1). So IEEE defines another, higher precision called *double extended*. For example, *all* arithmetic operations on an Intel Pentium (or its predecessors going back to the Intel 8086/8087) are performed in 80-bit double extended registers, providing 64-bit fractions and 15-bit exponents. Unfortunately, not all languages and compilers permit one to declare and compute with double-extended precision variables.

Few machines offer anything beyond double-extended arithmetic in hardware, but there are several ways in which more accurate arithmetic may be simulated in software. Some compilers on DEC Vax and DEC Alpha, Sun Sparc, and IBM RS6000 machines permit the user to declare *quadruple precision* (or *real*16* or *double double precision*) variables and to perform computations with them. Since this arithmetic is simulated using shorter precision, it may run several times slower than double. Cray's single precision is similar in precision to IEEE double, and so Cray double precision is about twice IEEE double; it too is simulated in software and runs relatively slowly. There are also algorithms and packages available for simulating much higher precision floating point arithmetic, using either integer arithmetic [20, 21] or the underlying floating point (see Question 1.18) [204, 218].

Finally, we mention *interval arithmetic*, a style of computation that automatically provides guaranteed error bounds. Each variable in an interval computation is represented by a pair of floating point numbers, one a lower bound and one an upper bound. Computation proceeds by rounding in such a way that lower bounds and upper bounds are propagated in a guaranteed fashion. For example, to add the intervals $a = [a_l, a_u]$ and $b = [b_l, b_u]$, one rounds $a_l + b_l$ *down* to the nearest floating point number, c_l, and rounds $a_u + b_u$

up to the nearest floating point number, c_u. This guarantees that the interval $c = [c_l, c_u]$ contains the sum of any pair of variables from a and from b. Unfortunately, if one naively takes a program and converts all floating point variables and operations to interval variables and operations, it is most likely that the intervals computed by the program will quickly grow so wide (such as $[-\infty, +\infty]$) that they provide no useful information at all. (A simple example is to repeatedly compute $x = x - x$ when x is an interval; instead of getting $x = 0$, the width $x_u - x_l$ of x doubles at each subtraction.) It is possible to modify old algorithms or design new ones that do provide useful guaranteed error bounds [4, 140, 162, 190], but these are often several times as expensive as the algorithms discussed in this book. The error bounds that we present in this book are not guaranteed in the same mathematical sense that interval bounds are, but they are reliable enough in almost all situations. (We discuss this in more detail later.) We will not discuss interval arithmetic further in this book.

1.6.   Polynomial Evaluation Revisited

Let us now apply roundoff model (1.1) to evaluating a polynomial with Horner's rule. We take the original program,

$$p = a_d$$
for $i = d - 1$ down to 0
$$\quad p = x \cdot p + a_i$$
end for

Then we add subscripts to the intermediate results so that we have a unique symbol for each one (p_0 is the final result):

$$p_d = a_d$$
for $i = d - 1$ down to 0
$$\quad p_i = x \cdot p_{i+1} + a_i$$
end for

Then we insert a roundoff term $(1 + \delta_i)$ at each floating point operation to get

$$p_d = a_d$$
for $i = d - 1$ down to 0
$$\quad p_i = ((x \cdot p_{i+1})(1 + \delta_i) + a_i)(1 + \delta_i'), \qquad \text{where } |\delta_i|, |\delta_i'| \leq \varepsilon$$
end for

Expanding, we get the following expression for the final computed value of the polynomial:

$$p_0 = \sum_{i=0}^{d-1} \left[(1 + \delta_i') \prod_{j=0}^{i-1} (1 + \delta_j)(1 + \delta_j') \right] a_i x^i + \left[\prod_{j=0}^{d-1} (1 + \delta_j)(1 + \delta_j') \right] a_d x^d .$$

This is messy, a typical result when we try to keep track of every rounding error in an algorithm. We simplify it using the following upper and lower bounds:

$$
\begin{aligned}
(1+\delta_1)\cdots(1+\delta_j) &\leq (1+\varepsilon)^j \leq \frac{1}{1-j\varepsilon} = 1+j\varepsilon + O(\varepsilon^2), \\
(1+\delta_1)\cdots(1+\delta_j) &\geq (1-\varepsilon)^j \geq 1-j\varepsilon.
\end{aligned}
$$

These bounds are correct, provided that $j\varepsilon < 1$. Typically, we make the reasonable assumption that $j\varepsilon \ll 1$ ($j \ll 10^7$ in IEEE single precision) and make the approximations

$$
1 - j\varepsilon \leq (1+\delta_1)\cdots(1+\delta_j) \leq 1+j\varepsilon.
$$

This lets us write

$$
\begin{aligned}
p_0 &= \sum_{i=0}^{d}(1+\bar{\delta}_i)a_i x^i, \quad \text{where } |\bar{\delta}_i| \leq 2d\varepsilon \\
&= \sum_{i=0}^{d}\bar{a}_i x^i
\end{aligned}
$$

So the computed value p_0 of $p(x)$ is the exact value of a slightly different polynomial with coefficients $\bar{a}_i$. This means that evaluating $p(x)$ is "backward stable," and the "backward error" is $2d\varepsilon$ measured as the maximum relative change of any coefficient of $p(x)$.

Using this backward error bound, we bound the error in the computed polynomial:

$$
\begin{aligned}
|p_0 - p(x)| &= \left| \sum_{i=0}^{d}(1+\bar{\delta}_i)a_i x^i - \sum_{i=0}^{d}a_i x^i \right| \\
&= \left| \sum_{i=0}^{d}\bar{\delta}_i a_i x^i \right| \leq \sum_{i=0}^{d}\varepsilon 2d|a_i \cdot x^i| \\
&\leq 2d\varepsilon \sum_{i=0}^{d}|a_i \cdot x^i|.
\end{aligned}
$$

Note that $\sum_i |a_i x^i|$ bounds the largest value that we could compute if there were no cancellation from adding positive and negative numbers, and the error bound is $2d\varepsilon$ times smaller. This is also the case for computing dot products and many other polynomial-like expressions.

By choosing $\bar{\delta}_i = \varepsilon \cdot \text{sign}(a_i x^i)$, we see that the error bound is attainable to within the modest factor $2d$. This means that we may use

$$
\frac{\sum_{i=0}^{d}|a_i x^i|}{|\sum_{i=0}^{d}a_i x^i|}
$$

as the *relative condition number* for polynomial evaluation.

We can easily compute this error bound, at the cost of doubling the number of operations:

> $p = a_d$, $bp = |a_d|$
> for $i = d - 1$ down to 0
> $\quad p = x \cdot p + a_i$
> $\quad bp = |x| \cdot bp + |a_i|$
> end for
> error bound $= bp = 2d \cdot \varepsilon \cdot bp$

so the true value of the polynomial is in the interval $[p - bp, p + bp]$, and the number of guaranteed correct decimal digits is $- \log_{10}(|\frac{bp}{p}|)$. These bounds are plotted in the top of Figure 1.3 for the polynomial discussed earlier, $(x - 2)^9$. (The reader may wonder whether roundoff errors could make this computed error bound inaccurate. This turns out not to be a problem and is left to the reader as an exercise.)

The graph of $- \log_{10}|\frac{bp}{p}|$ in the bottom of Figure 1.3, a lower bound on the number of correct decimal digits, indicates that we expect difficulty computing $p(x)$ to high relative accuracy when $p(x)$ is near 0. What is special about $p(x) = 0$? An arbitrarily small error ε in computing $p(x) = 0$ causes an infinite relative error $\frac{\varepsilon}{p(x)} = \frac{\varepsilon}{0}$. In other words, our relative error bound $2d\varepsilon \sum_{i=0}^{d} |a_i x^i| / |\sum_{i=0}^{d} a_i x^i|$ is infinite.

DEFINITION 1.1. *A problem whose condition number is infinite is called* ill-posed. *Otherwise it is called* well-posed.[5]

There is a simple geometric interpretation of the condition number: it tells us how far $p(x)$ is from a polynomial which is ill-posed.

DEFINITION 1.2. *Let* $p(z) = \sum_{i=0}^{d} a_i z^i$ *and* $q(z) = \sum_{i=0}^{d} b_i z^i$. *Define the relative distance* $d(p, q)$ *from* p *to* q *as the smallest value satisfying* $|a_i - b_i| \leq d(p, q) \cdot |a_i|$ *for* $0 \leq i \leq d$. *(If all* $a_i \neq 0$, *then we can more simply write* $d(p, q) = \max_{0 \leq i \leq d} |\frac{a_i - b_i}{a_i}|$.)

Note that if $a_i = 0$, then b_i must also be zero for $d(p, q)$ to be finite.

---

[5]This definition is slightly nonstandard, because ill-posed problems include those whose solutions are continuous as long as they are nondifferentiable. Examples include multiple roots of polynomials and multiple eigenvalues of matrices (section 4.3). Another way to describe an ill-posed problem is one in which the number of correct digits in the solution is not always within a constant of the number of digits used in the arithmetic in the solution. For example, multiple roots of polynomials tend to lose *half* or more of the precision of the arithmetic.

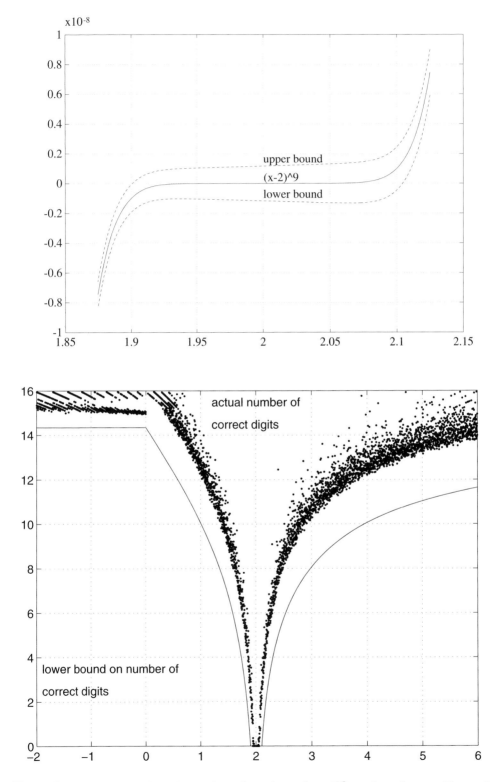

Fig. 1.3. *Plot of error bounds on the value of* $y = (x - 2)^9$ *evaluated using Horner's rule.*

THEOREM 1.2. *Suppose that $p(z) = \sum_{i=0}^{d} a_i z^i$ is not identically zero.*

$$\min\{d(p,q) \text{ such that } q(x) = 0\} = \frac{|\sum_{i=0}^{d} a_i x^i|}{\sum_{i=0}^{d} |a_i x^i|}.$$

In other words, the distance from p to the nearest polynomial q whose condition number at x is infinite, i.e., $q(x) = 0$, is the reciprocal of the condition number of $p(x)$.

Proof. Write $q(z) = \sum b_i z^i = \sum (1 + \varepsilon_i) a_i z^i$ so that $d(p,q) = \max_i |\varepsilon_i|$. Then $q(x) = 0$ implies $|p(x)| = |q(x) - p(x)| = |\sum_{i=0}^{d} \varepsilon_i a_i x^i| \leq \sum_{i=0}^{d} |\varepsilon_i a_i x^i| \leq \max_i |\varepsilon_i| \sum_i |a_i x^i|$, which in turn implies $d(p,q) = \max |\varepsilon_i| \geq |p(x)| / \sum_i |a_i x^i|$. To see that there is a q this close to p, choose

$$\varepsilon_i = \frac{-p(x)}{\sum |a_i x^i|} \cdot \text{sign}(a_i x^i). \quad \square$$

This simple reciprocal relationship between condition number and distance to the nearest ill-posed problem is very common in numerical analysis, and we shall encounter it again later.

At the beginning of the introduction we said that we would use canonical forms of matrices to help solve linear algebra problems. For example, knowing the exact Jordan canonical form makes computing exact eigenvalues trivial. There is an analogous canonical form for polynomials, which makes accurate polynomial evaluation easy: $p(x) = a_d \prod_{i=1}^{d} (x - r_i)$. In other words, we represent the polynomial by its leading coefficient a_d and its roots $r_1, \ldots, r_n$. To evaluate $p(x)$ we use the obvious algorithm

```
p = a_d
for i = 1 to d
    p = p · (x − r_i)
end for
```

It is easy to show the computed $p = p(x) \cdot (1 + \delta)$, where $|\delta| \leq 2d\varepsilon$; i.e., we always get $p(x)$ with high relative accuracy. But we need the roots of the polynomial to do this!

1.7. Vector and Matrix Norms

Norms are used to measure errors in matrix computations, so we need to understand how to compute and manipulate them.

Missing proofs are left as problems at the end of the chapter.

DEFINITION 1.3. *Let $\mathcal{B}$ be a real (complex) linear space $\mathbb{R}^n$ (or $\mathbb{C}^n$). It is normed if there is a function $\| \cdot \| : \mathcal{B} \to \mathbb{R}$, which we call a norm, satisfying all of the following:*

1) $\|x\| \geq 0$, *and* $\|x\| = 0$ *if and only if* $x = 0$ (positive definiteness),
2) $\|\alpha x\| = |\alpha| \cdot \|x\|$ *for any real (or complex) scalar α* (homogeneity),
3) $\|x + y\| \leq \|x\| + \|y\|$ (the triangle inequality).

EXAMPLE 1.4. The most common norms are $\|x\|_p = (\sum_i |x_i|^p)^{1/p}$ for $1 \leq p < \infty$, which we call *p-norms*, as well as $\|x\|_\infty = \max_i |x_i|$, which we call the *$\infty$-norm* or *infinity-norm*. Also, if $\|x\|$ is any norm and C is any nonsingular matrix, then $\|Cx\|$ is also a norm. ◇

We see that there are many norms that we could use to measure errors; it is important to choose an appropriate one. For example, let $x_1 = [1, 2, 3]^T$ in meters and $x_2 = [1.01, 2.01, 2.99]^T$ in meters. Then x_2 is a good approximation to x_1 because the relative error $\frac{\|x_1 - x_2\|_\infty}{\|x_1\|_\infty} \approx .0033$, and $x_3 = [10, 2.01, 2.99]^T$ is a bad approximation because $\frac{\|x_1 - x_3\|_\infty}{\|x_1\|_\infty} = 3$. But suppose the first component is measured in kilometers instead of meters. Then in this norm $\hat{x}_1$ and $\hat{x}_3$ look close:

$$\hat{x}_1 = \begin{bmatrix} .001 \\ 2 \\ 3 \end{bmatrix}, \ \hat{x}_3 = \begin{bmatrix} .01 \\ 2.01 \\ 2.99 \end{bmatrix}, \text{ and } \frac{\|\hat{x}_1 - \hat{x}_3\|_\infty}{\|\hat{x}_1\|_\infty} \approx .0033.$$

To compare $\hat{x}_1$ and $\hat{x}_3$, we should use

$$\|\hat{x}\|_s \equiv \left\| \begin{bmatrix} 1000 & & \\ & 1 & \\ & & 1 \end{bmatrix} \hat{x} \right\|_\infty$$

to make the units the same or so that equally important errors make the norm equally large.

Now we define *inner products*, which are a generalization of the standard dot product $\sum_i x_i y_i$, and arise frequently in linear algebra.

DEFINITION 1.4. *Let $\mathcal{B}$ be a real (complex) linear space.* $\langle \cdot, \cdot \rangle : \mathcal{B} \times \mathcal{B} \to \mathbb{R}(\mathbb{C})$ *is an* inner product *if all of the following apply:*

1) $\langle x, y \rangle = \langle y, x \rangle$ *(or $\overline{\langle y, x \rangle}$)*,
2) $\langle x, y + z \rangle = \langle x, y \rangle + \langle x, z \rangle$,
3) $\langle \alpha x, y \rangle = \alpha \langle x, y \rangle$ *for any real (or complex) scalar α,*
4) $\langle x, x \rangle \geq 0$, *and* $\langle x, x \rangle = 0$ *if and only if* $x = 0$.

EXAMPLE 1.5. Over $\mathbb{R}$, $\langle x, y \rangle = y^T x = \sum_i x_i y_i$, and over $\mathbb{C}$, $\langle x, y \rangle = y^* x = \sum_i x_i \bar{y}_i$ are inner products. (Recall that $y^* = \bar{y}^T$ is the conjugate transpose of y.) ◇

DEFINITION 1.5. x *and* y *are* orthogonal *if* $\langle x, y \rangle = 0$.

The most important property of an inner product is that it satisfies the Cauchy–Schwartz inequality. This can be used in turn to show that $\sqrt{\langle x, x \rangle}$ is a norm, one that we will frequently use.

LEMMA 1.1. Cauchy–Schwartz inequality. $|\langle x, y \rangle| \leq \sqrt{\langle x, x \rangle \cdot \langle y, y \rangle}$.

LEMMA 1.2. $\sqrt{\langle x, x \rangle}$ is a norm.

There is a one-to-one correspondence between inner products and *symmetric (Hermitian) positive definite matrices*, as defined below. These matrices arise frequently in applications.

DEFINITION 1.6. *A real symmetric (complex Hermitian) matrix A is* positive definite *if $x^T A x > 0$ ($x^* A x > 0$) for all $x \neq 0$. We abbreviate symmetric positive definite to s.p.d., and Hermitian positive to h.p.d.*

LEMMA 1.3. *Let $\mathcal{B} = \mathbb{R}^n$ (or $\mathbb{C}^n$) and $\langle \cdot, \cdot \rangle$ be an inner product. Then there is an n-by-n s.p.d. (h.p.d.) matrix A such that $\langle x, y \rangle = y^T A x$ ($y^* A x$). Conversely, if A is s.p.d (h.p.d.), then $y^T A x$ ($y^* A x$) is an inner product.*

The following two lemmas are useful in converting error bounds in terms of one norm to error bounds in terms of another.

LEMMA 1.4. *Let $\| \cdot \|_\alpha$ and $\| \cdot \|_\beta$ be two norms on $\mathbb{R}^n$ (or $\mathbb{C}^n$). There are constants $c_1, c_2 > 0$ such that, for all x, $c_1 \|x\|_\alpha \leq \|x\|_\beta \leq c_2 \|x\|_\alpha$. We also say that norms $\| \cdot \|_\alpha$ and $\| \cdot \|_\beta$ are* equivalent *with respect to constants c_1 and c_2.*

LEMMA 1.5.

$$\begin{array}{rcccl}
\|x\|_2 & \leq & \|x\|_1 & \leq & \sqrt{n}\|x\|_2, \\
\|x\|_\infty & \leq & \|x\|_2 & \leq & \sqrt{n}\|x\|_\infty, \\
\|x\|_\infty & \leq & \|x\|_1 & \leq & n\|x\|_\infty.
\end{array}$$

In addition to vector norms, we will also need *matrix norms* to measure errors in matrices.

DEFINITION 1.7. $\| \cdot \|$ *is a* matrix norm *on m-by-n matrices if it is a vector norm on $m \cdot n$ dimensional space:*

1) $\|A\| \geq 0$ *and* $\|A\| = 0$ *if and only if* $A = 0$,
2) $\|\alpha A\| = |\alpha| \cdot \|A\|$,
3) $\|A + B\| \leq \|A\| + \|B\|$.

EXAMPLE 1.6. $\max_{ij} |a_{ij}|$ *is called the* max norm, *and* $(\sum |a_{ij}|^2)^{1/2} = \|A\|_F$ *is called the Frobenius norm.* $\diamond$

The following definition is useful for bounding the norm of a product of matrices, something we often need to do when deriving error bounds.

DEFINITION 1.8. *Let $\| \cdot \|_{m \times n}$ be a matrix norm on m-by-n matrices, $\| \cdot \|_{n \times p}$ be a matrix norm on n-by-p matrices, and $\| \cdot \|_{m \times p}$ be a matrix norm on m-by-p matrices. These norms are called* mutually consistent *if $\|A \cdot B\|_{m \times p} \leq \|A\|_{m \times n} \cdot \|B\|_{n \times p}$, where A is m-by-n and B is n-by-p.*

DEFINITION 1.9. *Let A be m-by-n, $\| \cdot \|_{\hat{m}}$ be a vector norm on $\mathbb{R}^m$, and $\| \cdot \|_{\hat{n}}$ be a vector norm on $\mathbb{R}^n$. Then*

$$\|A\|_{\hat{m}\hat{n}} \equiv \max_{\substack{x \neq 0 \\ x \in \mathbb{R}^n}} \frac{\|Ax\|_{\hat{m}}}{\|x\|_{\hat{n}}}$$

is called an operator norm *or* induced norm *or* subordinate matrix norm.

The next lemma provides a large source of matrix norms, ones that we will use for bounding errors.

LEMMA 1.6. *An operator norm is a matrix norm.*

Orthogonal and *unitary matrices*, defined next, are essential ingredients of nearly all our algorithms for least squares problems and eigenvalue problems.

DEFINITION 1.10. *A* real square matrix Q *is* orthogonal *if $Q^{-1} = Q^T$. A complex square matrix is* unitary *if $Q^{-1} = Q^*$.*

All rows (or columns) of orthogonal (or unitary) matrices have unit 2-norms and are orthogonal to one another, since $QQ^T = Q^TQ = I$ ($QQ^* = Q^*Q = I$).

The next lemma summarizes the essential properties of the norms and matrices we have introduced so far. We will use these properties later in the book.

LEMMA 1.7.   1. *$\|Ax\| \leq \|A\| \cdot \|x\|$ for a vector norm and its corresponding operator norm, or the vector two-norm and matrix Frobenius norm.*

2. *$\|AB\| \leq \|A\| \cdot \|B\|$ for any operator norm or for the Frobenius norm. In other words, any operator norm (or the Frobenius norm) is mutually consistent with itself.*

3. *The max norm and Frobenius norm are not operator norms.*

4. *$\|QAZ\| = \|A\|$ if Q and Z are orthogonal or unitary for the Frobenius norm and for the operator norm induced by $\| \cdot \|_2$. This is really just the Pythagorean theorem.*

5. *$\|A\|_\infty \equiv \max_{x \neq 0} \frac{\|Ax\|_\infty}{\|x\|_\infty} = \max_i \sum_j |a_{ij}| = $ maximum absolute row sum.*

6. $\|A\|_1 \equiv \max_{x \neq 0} \frac{\|Ax\|_1}{\|x\|_1} = \|A^T\|_\infty = \max_j \sum_i |a_{ij}| = $ *maximum absolute column sum.*

7. $\|A\|_2 \equiv \max_{x \neq 0} \frac{\|Ax\|_2}{\|x\|_2} = \sqrt{\lambda_{\max}(A^*A)}$, *where $\lambda_{\max}$ denotes the largest eigenvalue.*

8. $\|A\|_2 = \|A^T\|_2$.

9. $\|A\|_2 = \max_i |\lambda_i(A)|$ *if A is* normal, *i.e., $AA^* = A^*A$.*

10. *If A is n-by-n, then* $n^{-1/2}\|A\|_2 \leq \|A\|_1 \leq n^{1/2}\|A\|_2$.

11. *If A is n-by-n, then* $n^{-1/2}\|A\|_2 \leq \|A\|_\infty \leq n^{1/2}\|A\|_2$.

12. *If A is n-by-n, then* $n^{-1}\|A\|_\infty \leq \|A\|_1 \leq n\|A\|_\infty$.

13. *If A is n-by-n, then* $\|A\|_1 \leq \|A\|_F \leq n^{1/2}\|A\|_2$.

Proof. We prove part 7 only and leave the rest to Question 1.16.

Since A^*A is Hermitian, there exists an eigendecomposition $A^*A = Q\Lambda Q^*$, with Q a unitary matrix (the columns are eigenvectors), and $\Lambda = \text{diag}(\lambda_1, \ldots, \lambda_n)$, a diagonal matrix containing the eigenvalues, which must all be real. Note that all $\lambda_i \geq 0$ since if one, say λ, were negative, we would take q as its eigenvector and get the contradiction $0 \leq \|Aq\|_2^2 = q^T A^T A q = q^T \lambda q = \lambda\|q\|_2^2 < 0$. Therefore

$$
\begin{aligned}
\|A\|_2 &= \max_{x \neq 0} \frac{\|Ax\|_2}{\|x\|_2} = \max_{x \neq 0} \frac{(x^*A^*Ax)^{1/2}}{\|x\|_2} = \max_{x \neq 0} \frac{(x^*Q\Lambda Q^*x)^{1/2}}{\|x\|_2} \\
&= \max_{x \neq 0} \frac{((Q^*x)^*\Lambda Q^*x)^{1/2}}{\|Q^*x)\|_2} = \max_{y \neq 0} \frac{(y^*\Lambda y)^{1/2}}{\|y\|_2} = \max_{y \neq 0} \sqrt{\frac{\sum \lambda_i y_i^2}{\sum y_i^2}} \\
&\leq \max_{y \neq 0} \sqrt{\lambda_{\max}} \sqrt{\frac{\sum y_i^2}{\sum y_i^2}} = \sqrt{\lambda_{\max}},
\end{aligned}
$$

which is attainable by choosing y to be the appropriate column of the identity matrix. $\square$

1.8.   References and Other Topics for Chapter 1

At the end of each chapter we will list the references most relevant to that chapter. They are also listed alphabetically in the bibliography at the end. In addition we will give pointers to related topics not discussed in the main text.

The most modern comprehensive work in this area is by G. Golub and C. Van Loan [121], which also has an extensive bibliography. A recent undergraduate level or beginning graduate text in this material is by D. Watkins [252]. Another good graduate text is by L. Trefethen and D. Bau [243]. A classic work that is somewhat dated but still an excellent reference is by J. Wilkinson [262]. An older but still excellent book at the same level as Watkins is by G. Stewart [235].

More detailed information on error analysis can be found in the recent book by N. Higham [149]. Older but still good general references are by J. Wilkinson [261] and W. Kahan [157].

"What every computer scientist should know about floating point arithmetic" by D. Goldberg is a good recent survey [119]. IEEE arithmetic is described formally in [11, 12, 159] as well as in the reference manuals published by computer manufacturers. Discussion of error analysis with IEEE arithmetic may be found in [54, 70, 159, 158] and the references cited therein.

A more general discussion of condition numbers and the distance to the nearest ill-posed problem is given by the author in [71] as well as in a series of papers by S. Smale and M. Shub [219, 220, 221, 222]. Vector and matrix norms are discussed at length in [121, sects. 2.2, 2.3].

1.9.   Questions for Chapter 1

QUESTION 1.1. *(Easy; Z. Bai)* Let A be an orthogonal matrix. Show that $\det(A) = \pm 1$. Show that if B also is orthogonal and $\det(A) = -\det(B)$, then $A + B$ is singular.

QUESTION 1.2. *(Easy; Z. Bai)* The *rank* of a matrix is the dimension of the space spanned by its columns. Show that A has rank one if and only if $A = ab^T$ for some column vectors a and b.

QUESTION 1.3. *(Easy; Z. Bai)* Show that if a matrix is orthogonal and triangular, then it is diagonal. What are its diagonal elements?

QUESTION 1.4. *(Easy; Z. Bai)* A matrix is *strictly upper triangular* if it is upper triangular with zero diagonal elements. Show that if A is strictly upper triangular and n-by-n, then $A^n = 0$.

QUESTION 1.5. *(Easy; Z. Bai)* Let $\| \cdot \|$ be a vector norm on $\mathbb{R}^m$ and assume that $C \in \mathbb{R}^{m \times n}$. Show that if $\text{rank}(C) = n$, then $\|x\|_C \equiv \|Cx\|$ is a vector norm.

QUESTION 1.6. *(Easy; Z. Bai)* Show that if $0 \neq s \in \mathbb{R}^n$ and $E \in \mathbb{R}^{n \times n}$, then

$$\left\| E \left(I - \frac{ss^{\mathrm{T}}}{s^T s} \right) \right\|_F^2 = \|E\|_F^2 - \frac{\|Es\|_2^2}{s^T s}.$$

QUESTION 1.7. *(Easy; Z. Bai)* Verify that $\|xy^H\|_F = \|xy^H\|_2 = \|x\|_2 \|y\|_2$ for any $x, y \in \mathbf{C}^n$.

QUESTION 1.8. *(Medium)* One can identify the degree d polynomials $p(x) = \sum_{i=0}^{d} a_i x^i$ with $\mathbb{R}^{d+1}$ via the vector of coefficients. Let x be fixed. Let S_x be the set of polynomials with an infinite relative condition number with respect to evaluating them at x (i.e., they are zero at x). In a few words, describe S_x geometrically as a subset of $\mathbb{R}^{d+1}$. Let $S_x(\kappa)$ be the set of polynomials whose relative condition number is κ or greater. Describe $S_x(\kappa)$ geometrically in a few words. Describe how $S_x(\kappa)$ changes geometrically as $\kappa \to \infty$.

QUESTION 1.9. *(Medium)* Consider the figure below. It plots the function $y = \log(1 + x)/x$ computed in two different ways. Mathematically, y is a smooth function of x near $x = 0$, equaling 1 at 0. But if we compute y using this formula, we get the plots on the left (shown in the ranges $x \in [-1, 1]$ on the top left and $x \in [-10^{-15}, 10^{-15}]$ on the bottom left). This formula is clearly unstable near $x = 0$. On the other hand, if we use the algorithm

$$d = 1 + x$$
$$\text{if } d = 1 \text{ then}$$
$$\quad y = 1$$
$$\text{else}$$
$$\quad y = \log(d)/(d - 1)$$
$$\text{end if}$$

we get the two plots on the right, which are correct near $x = 0$. Explain this phenomenon, proving that the second algorithm must compute an accurate answer in floating point arithmetic. Assume that the log function returns an accurate answer for any argument. (This is true of any reasonable implementation of logarithm.) Assume IEEE floating point arithmetic if that makes your argument easier. (Both algorithms can malfunction on a Cray machine.)

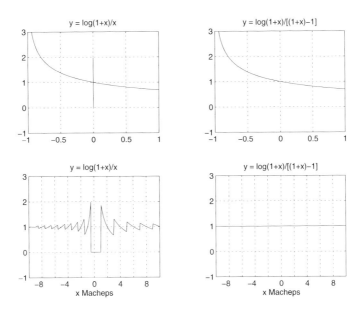

QUESTION 1.10. *(Medium)* Show that, barring overflow or underflow, $\mathrm{fl}(\sum_{i=1}^{d} x_i y_i) = \sum_{i=1}^{d} x_i y_i (1 + \delta_i)$, where $|\delta_i| \leq d\varepsilon$. Use this to prove the following fact. Let $A^{m \times n}$ and $B^{n \times p}$ be matrices, and compute their product in the usual way. Barring overflow or underflow show that $|\mathrm{fl}(A \cdot B) - A \cdot B| \leq n \cdot \varepsilon \cdot |A| \cdot |B|$. Here the absolute value of a matrix $|A|$ means the matrix with entries $(|A|)_{ij} = |a_{ij}|$, and the inequality is meant componentwise.

The result of this question will be used in section 2.4.2, where we analyze the roundoff errors in Gaussian elimination.

QUESTION 1.11. *(Medium)* Let L be a lower triangular matrix and solve $Lx = b$ by forward substitution. Show that barring overflow or underflow, the computed solution $\hat{x}$ satisfies $(L + \delta L)\hat{x} = b$, where $|\delta l_{ij}| \leq n\varepsilon |l_{ij}|$, where ε is the machine precision. This means that forward substitution is backward stable. Argue that backward substitution for solving upper triangular systems satisfies the same bound.

The result of this question will be used in section 2.4.2, where we analyze the roundoff errors in Gaussian elimination.

QUESTION 1.12. *(Medium)* In order to analyze the effects of rounding errors, we have used the following model (see equation (1.1)):

$$\mathrm{fl}(a \odot b) = (a \odot b)(1 + \delta),$$

where $\odot$ is one of the four basic operations $+$, $-$, $*$, and $/$, and $|\delta| \leq \varepsilon$. To show that our analyses also work for *complex* data, we need to prove an analogous formula for the four basic complex operations. Now δ will be a tiny *complex* number bounded in absolute value by a small multiple of ϵ. Prove that this is true for complex addition, subtraction, multiplication, and division. Your algorithm for complex division should successfully compute $a/a \approx 1$, where $|a|$ is either very large (larger than the square root of the overflow threshold) or very small (smaller than the square root of the underflow threshold). Is it true that both the real and imaginary parts of the complex product are always computed to high relative accuracy?

QUESTION 1.13. *(Medium)* Prove Lemma 1.3.

QUESTION 1.14. *(Medium)* Prove Lemma 1.5.

QUESTION 1.15. *(Medium)* Prove Lemma 1.6.

QUESTION 1.16. *(Medium)* Prove all parts except 7 of Lemma 1.7. Hint for part 8: Use the fact that if X and Y are both n-by-n, then XY and YX have the same eigenvalues. Hint for part 9: Use the fact that a matrix is normal if and only if it has a complete set of orthonormal eigenvectors.

QUESTION 1.17. *(Hard; W. Kahan)* We mentioned that on a Cray machine the expression $\arccos(x/\sqrt{x^2 + y^2})$ caused an error, because roundoff caused $(x/\sqrt{x^2 + y^2})$ to exceed 1. Show that this is impossible using IEEE arithmetic, barring overflow or underflow. Hint: You will need to use more than the simple model $fl(a \odot b) = (a \odot b)(1 + \delta)$ with $|\delta|$ small. Think about evaluating $\sqrt{x^2}$, and show that, barring overflow or underflow, $fl(\sqrt{x^2}) = x$ *exactly*; in numerical experiments done by A. Liu, this failed about 5% of the time on a Cray YMP. You might try some numerical experiments and explain them. Extra credit: Prove the same result using correctly rounded *decimal* arithmetic. (The proof is different.) This question is due to W. Kahan, who was inspired by a bug in a Cray program of J. Sethian.

QUESTION 1.18. *(Hard)* Suppose that a and b are normalized IEEE double precision floating point numbers, and consider the following algorithm, running with IEEE arithmetic:

if $(|a| < |b|)$, swap a and b
$s_1 = a + b$
$s_2 = (a - s_1) + b$

Prove the following facts:

1. Barring overflow or underflow, the only roundoff error committed in running the algorithm is computing $s_1 = fl(a + b)$. In other words, both subtractions $s_1 - a$ and $(s_1 - a) - b$ are computed *exactly.*

2. $s_1 + s_2 = a + b$, *exactly.* This means that s_2 is actually the roundoff error committed when rounding the exact value of $a + b$ to get s_1.

Thus, this program in effect simulates *quadruple* precision arithmetic, representing the true sum $a + b$ as the higher-order bits (s_1) and the lower-order bits (s_2).

   Using this and similar tricks in a systematic way, it is possible to efficiently simulate all four basic floating point operations in *arbitrary* precision arithmetic, using only the underlying floating point instructions and no "bit-fiddling" [204]. 128-bit arithmetic is implemented this way on the IBM RS6000 and Cray (but much less efficiently on the Cray, which does not have IEEE arithmetic).

QUESTION 1.19. *(Hard; Programming)* This question illustrates the challenges in engineering highly reliable numerical software. Your job is to write a program to compute the two-norm $s \equiv \|x\|_2 = (\sum_{i=1}^{n} x_i^2)^{1/2}$ given $x_1, \ldots, x_n$. The most obvious (and inadequate) algorithm is

$s = 0$
for $i = 1$ to n
    $s = s + x_i^2$

endfor
$s = \text{sqrt}(s)$

This algorithm is inadequate because it does not have the following desirable properties:

1. It must compute the answer accurately (i.e., nearly all the computed digits must be correct) unless $\|x\|_2$ is (nearly) outside the range of normalized floating point numbers.

2. It must be nearly as fast as the obvious program above in most cases.

3. It must work on any "reasonable" machine, possibly including ones not running IEEE arithmetic. This means it may not cause an error condition, unless $\|x\|_2$ is (nearly) larger than the largest floating point number.

To illustrate the difficulties, note that the obvious algorithm fails when $n = 1$ and x_1 is larger than the square root of the largest floating point number (in which case x_1^2 overflows, and the program returns $+\infty$ in IEEE arithmetic and halts in most non-IEEE arithmetics) or when $n = 1$ and x_1 is smaller than the square root of the smallest normalized floating point number (in which case x_1^2 underflows, possibly to zero, and the algorithm may return zero). Scaling the x_i by dividing them all by $\max_i |x_i|$ does not have property 2), because division is usually many times more expensive than either multiplication or addition. Multiplying by $c = 1/\max_i |x_i|$ risks overflow in computing c, even when $\max_i |x_i| > 0$.

This routine is important enough that it has been standardized as a *Basic Linear Algebra Subroutine*, or *BLAS*, which should be available on all machines [169]. We discuss the BLAS at length in section 2.6.1, and documentation and sample implementations may be found at NETLIB/blas. In particular, see NETLIB/cgi-bin/netlibget.pl/blas/snrm2.f for a sample implementation that has properties 1) and 3) but not 2). These sample implementations are intended to be starting points for implementations specialized to particular architectures (an easier problem than producing a completely portable one, as requested in this problem). Thus, when writing your own numerical software, you should think of computing $\|x\|_2$ as a building block that should be available in a numerical library on each machine.

For another careful implementation of $\|x\|_2$, see [35].

You can extract test code from NETLIB/blas/sblat1 to see if your implementation is correct; all implementations turned in must be thoroughly tested as well as timed, with times compared to the obvious algorithm above on those cases where both run. See how close to satisfying the three conditions you can come; the frequent use of the word "nearly" in conditions (1), (2) and (3) shows where you may compromise in attaining one condition in order to more nearly attain another. In particular, you might want to see how much easier the problem is if you limit yourself to machines running IEEE arithmetic.

Hint: Assume that the values of the overflow and underflow thresholds are available for your algorithm. Portable software for computing these values is available (see NETLIB/cgi-bin/netlibget.pl/lapack/util/slamch.f).

QUESTION 1.20. *(Easy; Medium)* We will use a Matlab program to illustrate how sensitive the roots of polynomial can be to small perturbations in the coefficients. The program is available[6] at HOMEPAGE/Matlab/polyplot.m. Polyplot takes an input polynomial specified by its roots r and then adds random perturbations to the polynomial coefficients, computes the perturbed roots, and plots them. The inputs are

    r = vector of roots of the polynomial,

    e = maximum relative perturbation to make to each coefficient of the polynomial,

    m = number of random polynomials to generate, whose roots are plotted.

1. *(Easy)* The first part of your assignment is to run this program for the following inputs. In all cases choose m high enough that you get a fairly dense plot but don't have to wait too long. m = a few hundred or perhaps 1000 is enough. You may want to change the axes of the plot if the graph is too small or too large.
   - r=(1:10); e = 1e-3, 1e-4, 1e-5, 1e-6, 1e-7, 1e-8,
   - r=(1:20); e = 1e-9, 1e-11, 1e-13, 1e-15,
   - r=[2,4,8,16,..., 1024]; e=1e-1, 1e-2, 1e-3, 1e-4.

   Also try your own example with complex conjugate roots. Which roots are most sensitive?

2. *(Medium)* The second part of your assignment is to modify the program to compute the condition number $c(i)$ for each root. In other words, a relative perturbation of e in each coefficient should change root $r(i)$ by at most about $e*c(i)$. Modify the program to plot circles centered at $r(i)$ with radii $e*c(i)$, and confirm that these circles enclose the perturbed roots (at least when e is small enough that the linearization used to derive the condition number is accurate). You should turn in a few plots with circles and perturbed eigenvalues, and some explanation of what you observe.

3. *(Medium)* In the last part, notice that your formula for $c(i)$ "blows up" if $p'(r(i)) = 0$. This condition means that $r(i)$ is a *multiple root* of $p(x) = 0$. We can still expect some accuracy in the computed value of a multiple

---

root, however, and in this part of the question, we will ask how sensitive a multiple root can be: First, write $p(x) = q(x) \cdot (x - r(i))^m$, where $q(r(i)) \neq 0$ and m is the multiplicity of the root r(i). Then compute the m roots nearest r(i) of the slightly perturbed polynomial $p(x) - q(x)\epsilon$, and show that they differ from r(i) by $|\epsilon|^{1/m}$. So that if $m = 2$, for instance, the root r(i) is perturbed by $\epsilon^{1/2}$, which is much larger than ϵ if $|\epsilon| \ll 1$. Higher values of m yield even larger perturbations. If ϵ is around machine epsilon and represents rounding errors in computing the root, this means an m-tuple root can lose all but $1/m$-th of its significant digits.

QUESTION 1.21. *(Medium)* Apply Algorithm 1.1, Bisection, to find the roots of $p(x) = (x - 2)^9 = 0$, where $p(x)$ is evaluated using Horner's rule. Use the Matlab implementation in HOMEPAGE/Matlab/bisect.m, or else write your own. Confirm that changing the input interval slightly changes the computed root drastically. Modify the algorithm to use the error bound discussed in the text to stop bisecting when the roundoff error in the computed value of $p(x)$ gets so large that its sign cannot be determined.

2

Linear Equation Solving

2.1. Introduction

This chapter discusses perturbation theory, algorithms, and error analysis for solving the linear equation $Ax = b$. The algorithms are all variations on Gaussian elimination. They are called *direct methods*, because in the absence of roundoff error they would give the exact solution of $Ax = b$ after a finite number of steps. In contrast, Chapter 6 discusses *iterative methods*, which compute a sequence $x_0, x_1, x_2, \ldots$ of ever better approximate solutions of $Ax = b$; one stops iterating (computing the next x_{i+1}) when x_i is accurate enough. Depending on the matrix A and the speed with which x_i converges to $x = A^{-1}b$, a direct method or an iterative method may be faster or more accurate. We will discuss the relative merits of direct and iterative methods at length in Chapter 6. For now, we will just say that direct methods are the methods of choice when the user has no special knowledge about the source[7] of matrix A or when a solution is required with guaranteed stability and in a guaranteed amount of time.

The rest of this chapter is organized as follows. Section 2.2 discusses perturbation theory for $Ax = b$; it forms the basis for the practical error bounds in section 2.4. Section 2.3 derives the Gaussian elimination algorithm for dense matrices. Section 2.4 analyzes the errors in Gaussian elimination and presents practical error bounds. Section 2.5 shows how to improve the accuracy of a solution computed by Gaussian elimination, using a simple and inexpensive iterative method. To get high speed from Gaussian elimination and other linear algebra algorithms on contemporary computers, care must be taken to organize the computation to respect the computer memory organization; this is discussed in section 2.6. Finally, section 2.7 discusses faster variations of Gaussian elimination for matrices with special properties commonly arising in practice, such as symmetry $(A = A^T)$ or sparsity (when many entries of A are zero).

---

[7]For example, in Chapter 6 we consider the case when A arises from approximating the solution to a particular differential equation, Poisson's equation.

Sections 2.2.1 and 2.5.1 discuss recent innovations upon which the software in the LAPACK library depends.

There are a variety of open problems, which we shall mention as we go along.

2.2. Perturbation Theory

Suppose $Ax = b$ and $(A + \delta A)\hat{x} = b + \delta b$; our goal is to bound the norm of $\delta x \equiv \hat{x} - x$. Later, $\hat{x}$ will be the computed solution of $Ax = B$. We simply subtract these two equalities and solve for δx: one way to do this is to take

$$
\begin{array}{rcl}
(A + \delta A)(x + \delta x) & = & b + \delta b \\
- \qquad\qquad\qquad [Ax & = & b] \\
\hline
\delta Ax + (A + \delta A)\delta x & = & \delta b
\end{array}
$$

and rearrange to get

$$\delta x = A^{-1}(-\delta A\hat{x} + \delta b). \tag{2.1}$$

Taking norms and using part 1 of Lemma 1.7 as well as the triangle inequality for vector norms, we get

$$\|\delta x\| \leq \|A^{-1}\|(\|\delta A\| \cdot \|\hat{x}\| + \|\delta b\|). \tag{2.2}$$

(We have assumed that the vector norm and matrix norm are consistent, as defined in section 1.7. For example, any vector norm and its induced matrix norm will do.) We can further rearrange this inequality to get

$$\frac{\|\delta x\|}{\|\hat{x}\|} \leq \|A^{-1}\| \cdot \|A\| \cdot \left(\frac{\|\delta A\|}{\|A\|} + \frac{\|\delta b\|}{\|A\| \cdot \|\hat{x}\|} \right). \tag{2.3}$$

The quantity $\kappa(A) = \|A^{-1}\| \cdot \|A\|$ is the *condition number*[8] of the matrix A, because it measures the relative change $\frac{\|\delta x\|}{\|\hat{x}\|}$ in the answer as a multiple of the relative change $\frac{\|\delta A\|}{\|A\|}$ in the data. (To be rigorous, we need to show that inequality (2.2) is an equality for some nonzero choice of δA and δb; otherwise $\kappa(A)$ would only be an upper bound on the condition number. See Question 2.3.) The quantity multiplying $\kappa(A)$ will be small if δA and δb are small, yielding a small upper bound on the relative error $\frac{\|\delta x\|}{\|\hat{x}\|}$.

The upper bound depends on δx (via $\hat{x}$), which makes it seem hard to interpret, but it is actually quite useful in practice, since we know the computed solution $\hat{x}$ and so can straightforwardly evaluate the bound. We can also derive a theoretically more attractive bound that does not depend on δx as follows:

---

[8]More pedantically, it is the condition number with respect to the problem of matrix inversion. The problem of finding the eigenvalues of A, for example, has a different condition number.

LEMMA 2.1. *Let $\| \cdot \|$ satisfy $\|AB\| \leq \|A\| \cdot \|B\|$. Then $\|X\| < 1$ implies that $I - X$ is invertible, $(I - X)^{-1} = \sum_{i=0}^{\infty} X^i$, and $\|(I - X)^{-1}\| \leq \frac{1}{1-\|X\|}$.*

Proof. The sum $\sum_{i=0}^{\infty} X^i$ is said to converge if and only if it converges in each component. We use the fact (from applying Lemma 1.4 to Example 1.6) that for any norm, there is a constant c such that $|x_{jk}| \leq c \cdot \|X\|$. We then get $|(X^i)_{jk}| \leq c \cdot \|X^i\| \leq c \cdot \|X\|^i$, so each component of $\sum X^i$ is dominated by a convergent geometric series $\sum c\|X\|^i = \frac{c}{1-\|X\|}$ and must converge. Therefore $S_n = \sum_{i=0}^{n} X^i$ converges to some S as $n \to \infty$, and $(I - X)S_n = (I - X)(I + X + X^2 + \cdots + X^n) = I - X^{n+1} \to I$ as $n \to \infty$, since $\|X^i\| \leq \|X\|^i \to 0$. Therefore $(I - X)S = I$ and $S = (I - X)^{-1}$. The final bound is $\|(I - X)^{-1}\| = \|\sum_{i=0}^{\infty} X^i\| \leq \sum_{i=0}^{\infty} \|X^i\| \leq \sum_{i=0}^{\infty} \|X\|^i = \frac{1}{1-\|X\|}$. □

Solving our first equation $\delta A x + (A + \delta A)\delta x = \delta b$ for δx yields

$$
\begin{aligned}
\delta x &= (A + \delta A)^{-1}(-\delta A x + \delta b) \\
&= [A(I + A^{-1}\delta A)]^{-1}(-\delta A x + \delta b) \\
&= (I + A^{-1}\delta A)^{-1} A^{-1}(-\delta A x + \delta b).
\end{aligned}
$$

Taking norms, dividing both sides by $\|x\|$, using part 1 of Lemma 1.7 and the triangle inequality, and assuming that δA is small enough so that $\|A^{-1}\delta A\| \leq \|A^{-1}\| \cdot \|\delta A\| < 1$, we get the desired bound:

$$
\begin{aligned}
\frac{\|\delta x\|}{\|x\|} &\leq \|(I + A^{-1}\delta A)^{-1}\| \cdot \|A^{-1}\| \left(\|\delta A\| + \frac{\|\delta b\|}{\|x\|} \right) \\
&\leq \frac{\|A^{-1}\|}{1 - \|A^{-1}\| \cdot \|\delta A\|} \left(\|\delta A\| + \frac{\|\delta b\|}{\|x\|} \right) \quad \text{by Lemma 2.1} \\
&= \frac{\|A^{-1}\| \cdot \|A\|}{1 - \|A^{-1}\| \cdot \|A\| \frac{\|\delta A\|}{\|A\|}} \left(\frac{\|\delta A\|}{\|A\|} + \frac{\|\delta b\|}{\|A\| \cdot \|x\|} \right) \\
&\leq \frac{\kappa(A)}{1 - \kappa(A)\frac{\|\delta A\|}{\|A\|}} \left(\frac{\|\delta A\|}{\|A\|} + \frac{\|\delta b\|}{\|b\|} \right) \qquad (2.4) \\
&\text{since } \|b\| = \|Ax\| \leq \|A\| \cdot \|x\|.
\end{aligned}
$$

This bound expresses the relative error $\frac{\|\delta x\|}{\|x\|}$ in the solution as a multiple of the relative errors $\frac{\|\delta A\|}{\|A\|}$ and $\frac{\|\delta b\|}{\|b\|}$ in the input. The multiplier, $\kappa(A)/(1 - \kappa(A)\frac{\|\delta A\|}{\|A\|})$, is close to the condition number $\kappa(A)$ if $\|\delta A\|$ is small enough.

The next theorem explains more about the assumption that $\|A^{-1}\| \cdot \|\delta A\| = \kappa(A) \cdot \frac{\|\delta A\|}{\|A\|} < 1$: it guarantees that $A + \delta A$ is nonsingular, which we need for δx to exist. It also establishes a geometric characterization of the condition number.

THEOREM 2.1. *Let A be nonsingular. Then*

$$
\min \left\{ \frac{\|\delta A\|_2}{\|A\|_2} : A + \delta A \text{ singular} \right\} = \frac{1}{\|A^{-1}\|_2 \cdot \|A\|_2} = \frac{1}{\kappa(A)}.
$$

Therefore, the distance to the nearest singular matrix (ill-posed problem) $=$ $\frac{1}{\text{condition number}}$.

Proof. It is enough to show $\min \{\|\delta A\|_2 : A + \delta A \text{ singular}\} = \frac{1}{\|A^{-1}\|_2}$.

To show this minimum is at least $\frac{1}{\|A^{-1}\|_2}$, note that if $\|\delta A\|_2 < \frac{1}{\|A^{-1}\|_2}$, then $1 > \|\delta A\|_2 \cdot \|A^{-1}\|_2 \geq \|A^{-1}\delta A\|_2$, so Lemma 2.1 implies that $I + A^{-1}\delta A$ is invertible, and so $A + \delta A$ is invertible.

To show the minimum equals $\frac{1}{\|A^{-1}\|_2}$, we construct a δA of norm $\frac{1}{\|A^{-1}\|_2}$ such that $A + \delta A$ is singular. Note that since $\|A^{-1}\|_2 = \max_{x \neq 0} \frac{\|A^{-1}x\|_2}{\|x\|_2}$, there exists an x such that $\|x\|_2 = 1$ and $\|A^{-1}\|_2 = \|A^{-1}x\|_2 > 0$. Now let $y = \frac{A^{-1}x}{\|A^{-1}x\|_2} = \frac{A^{-1}x}{\|A^{-1}\|_2}$ so $\|y\|_2 = 1$. Let $\delta A = \frac{-xy^T}{\|A^{-1}\|_2}$.

Then

$$\|\delta A\|_2 = \max_{z \neq 0} \frac{\|xy^T z\|_2}{\|A^{-1}\|_2 \, \|z\|_2} = \max_{z \neq 0} \frac{|y^T z|}{\|z\|_2} \frac{\|x\|_2}{\|A^{-1}\|_2} = \frac{1}{\|A^{-1}\|_2},$$

where the maximum is attained when z is any nonzero multiple of y, and $A + \delta A$ is singular because

$$(A + \delta A)y = Ay - \frac{xy^T y}{\|A^{-1}\|_2} = \frac{x}{\|A^{-1}\|_2} - \frac{x}{\|A^{-1}\|_2} = 0 . \quad \square$$

We have now seen that the distance to the nearest ill-posed problem equals the reciprocal of the condition number for two problems: polynomial evaluation and linear equation solving. This reciprocal relationship is quite common in numerical analysis [71].

Here is a slightly different way to do perturbation theory for $Ax = b$; we will need it to derive practical error bounds later in section 2.4.4. If $\hat{x}$ is any vector, we can bound the difference $\delta x \equiv \hat{x} - x = \hat{x} - A^{-1}b$ as follows. We let $r = A\hat{x} - b$ be the *residual* of $\hat{x}$; the residual r is zero if $\hat{x} = x$. This lets us write $\delta x = A^{-1}r$, yielding the bound

$$\|\delta x\| = \|A^{-1}r\| \leq \|A^{-1}\| \cdot \|r\|. \tag{2.5}$$

This simple bound is attractive to use in practice, since r is easy to compute, given an approximate solution $\hat{x}$. Furthermore, there is no apparent need to estimate δA and δb. In fact our two approaches are very closely related, as shown by the next theorem.

THEOREM 2.2. *Let $r = A\hat{x} - b$. Then there exists a δA such that $\|\delta A\| = \frac{\|r\|}{\|\hat{x}\|}$ and $(A + \delta A)\hat{x} = b$. No δA of smaller norm and satisfying $(A + \delta A)\hat{x} = b$ exists. Thus, δA is the smallest possible backward error (measured in norm). This is true for any vector norm and its induced norm (or $\| \cdot \|_2$ for vectors and $\| \cdot \|_F$ for matrices).*

Proof. $(A+\delta A)\hat{x} = b$ if and only if $\delta A \cdot \hat{x} = b - A\hat{x} = -r$, so $\|r\| = \|\delta A \cdot \hat{x}\| \leq$ $\|\delta A\| \cdot \|\hat{x}\|$, implying $\|\delta A\| \geq \frac{\|r\|}{\|\hat{x}\|}$. We complete the proof only for the two-norm and its induced matrix norm. Choose $\delta A = \frac{-r \cdot \hat{x}^T}{\|\hat{x}\|_2^2}$. We can easily verify that $\delta A \cdot \hat{x} = -r$ and $\|\delta A\|_2 = \frac{\|r\|_2}{\|\hat{x}\|_2}$. $\square$

Thus, the smallest $\|\delta A\|$ that could yield an $\hat{x}$ satisfying $(A+\delta A)\hat{x} = b$ and $r = A\hat{x} - b$ is given by Theorem 2.2. Applying error bound (2.2) (with $\delta b = 0$) yields

$$\|\delta x\| \leq \|A^{-1}\| \left(\frac{\|r\|}{\|\hat{x}\|} \cdot \|\hat{x}\| \right) = \|A^{-1}\| \cdot \|r\|,$$

the same bound as (2.5).

All our bounds depend on the ability to estimate the condition number $\|A\| \cdot \|A^{-1}\|$. We return to this problem in section 2.4.3. Condition number estimates are computed by LAPACK routines such as `sgesvx`.

2.2.1. Relative Perturbation Theory

In the last section we showed how to bound the norm of the error $\delta x = \hat{x} - x$ in the approximate solution $\hat{x}$ of $Ax = b$. Our bound on $\|\delta x\|$ was proportional to the condition number $\kappa(A) = \|A\| \cdot \|A^{-1}\|$ times the norms $\|\delta A\|$ and $\|\delta b\|$, where $\hat{x}$ satisfies $(A + \delta A)\hat{x} = b + \delta b$.

In many cases this bound is quite satisfactory, but not always. Our goal in this section is to show when it is too pessimistic and to derive an alternative perturbation theory that provides tighter bounds. We will use this perturbation theory later in section 2.5.1 to justify the error bounds computed by the LAPACK subroutines like `sgesvx`.

This section may be skipped on a first reading.

Here is an example where the error bound of the last section is much too pessimistic.

EXAMPLE 2.1. Let $A = \mathrm{diag}(\gamma, 1)$ (a diagonal matrix with entries $a_{11} = \gamma$ and $a_{22} = 1$) and $b = [\gamma, 1]^T$, where $\gamma > 1$. Then $x = A^{-1}b = [1, 1]^T$. Any reasonable direct method will solve $Ax = b$ very accurately (using two divisions b_i/a_{ii}) to get $\hat{x}$, yet the condition number $\kappa(A) = \gamma$ may be arbitrarily large. Therefore our error bound (2.3) may be arbitrarily large.

The reason that the condition number $\kappa(A)$ leads us to overestimate the error is that bound (2.2), from which it comes, assumes that δA is bounded in norm *but is otherwise arbitrary*; this is needed to prove that bound (2.2) is attainable in Question 2.3. In contrast, the δA corresponding to the actual rounding errors is not arbitrary but has a special structure not captured by its norm alone. We can determine the smallest δA corresponding to $\hat{x}$ for our problem as follows: A simple rounding error analysis shows that $\hat{x}_i = (b_i/a_{ii})/(1+\delta_i)$, where $|\delta_i| \leq \varepsilon$. Thus $(a_{ii} + \delta_i a_{ii})\hat{x}_i = b_i$. We may rewrite this as $(A + \delta A)\hat{x} = b$, where $\delta A = \text{diag}(\delta_1 a_{11}, \delta_2 a_{22})$. Then $\|\delta A\|$ can be as large max$_i |\varepsilon a_{ii}| = \varepsilon\gamma$. Applying error bound (2.3) with $\delta b = 0$ yields

$$\frac{\|\delta x\|_\infty}{\|\hat{x}\|_\infty} \le \gamma \left(\frac{\varepsilon\gamma}{\gamma}\right) = \varepsilon\gamma.$$

In contrast, the actual error satisfies

$$
\begin{aligned}
\|\delta x\|_\infty &= \|\hat{x} - x\|_\infty \\
&= \left\| \begin{bmatrix} (b_1/a_{11})/(1 + \delta_1) - (b_1/a_{11}) \\ (b_2/a_{22})/(1 + \delta_2) - (b_2/a_{22}) \end{bmatrix} \right\|_\infty \\
&= \left\| \begin{bmatrix} -\delta_1/(1 + \delta_1) \\ -\delta_2/(1 + \delta_2) \end{bmatrix} \right\|_\infty \\
&\le \frac{\varepsilon}{1 - \varepsilon}
\end{aligned}
$$

or

$$\frac{\|\delta x\|_\infty}{\|\hat{x}\|_\infty} \le \varepsilon/(1 - \varepsilon)^2,$$

which is about γ times smaller. $\diamond$

For this example, we can describe the structure of the actual δA as follows: $|\delta a_{ij}| \le \epsilon|a_{ij}|$, where ϵ is a tiny number. We write this more succinctly as

$$|\delta A| \le \epsilon|A| \tag{2.6}$$

(see section 1.1 for notation). We also say that δA is a *small componentwise relative perturbation in* A. Since δA can often be made to satisfy bound (2.6) in practice, along with $|\delta b| \le \epsilon|b|$ (see section 2.5.1), we will derive perturbation theory using these bounds on δA and δb.

We begin with equation (2.1):

$$\delta x = A^{-1}(-\delta A\hat{x} + \delta b).$$

Now take absolute values, and repeatedly use the triangle inequality to get

$$
\begin{aligned}
|\delta x| &= |A^{-1}(-\delta A\hat{x} + \delta b)| \\
&\le |A^{-1}|(|\delta A| \cdot |\hat{x}| + |\delta b|) \\
&\le |A^{-1}|(\epsilon|A| \cdot |\hat{x}| + \epsilon|b|) \\
&= \epsilon(|A^{-1}|(|A| \cdot |\hat{x}| + |b|)).
\end{aligned}
$$

Now using any vector norm (like the infinity-, one-, or Frobenius norms), where $\| |z| \| = \|z\|$, we get the bound

$$\|\delta x\| \le \epsilon\||A^{-1}|(|A| \cdot |\hat{x}| + |b|)\|. \tag{2.7}$$

Assuming for the moment that $\delta b = 0$, we can weaken this bound to

$$\|\delta x\| \le \epsilon \| |A^{-1}| \cdot |A| \| \cdot \|\hat{x}\|$$

or

$$\frac{\|\delta x\|}{\|x\|} \le \epsilon \| |A^{-1}| \cdot |A| \|. \tag{2.8}$$

This leads us to define $\kappa_{CR}(A) \equiv \| |A^{-1}| \cdot |A| \|$ as the *componentwise relative condition number of A*, or just *relative condition number* for short. It is sometimes also called the Bauer condition number [26] or Skeel condition number [225, 226, 227]. For a proof that bounds (2.7) and (2.8) are attainable, see Question 2.4.

Recall that Theorem 2.1 related the condition number $\kappa(A)$ to the distance from A to the nearest singular matrix. For a similar interpretation of $\kappa_{CR}(A)$, see [72, 208].

EXAMPLE 2.2. *Consider our earlier example with* $A = \mathrm{diag}(\gamma, 1)$ *and* $b = [\gamma, 1]^T$. *It is easy to confirm that* $\kappa_{CR}(A) = 1$, *since* $|A^{-1}| \cdot |A| = I$. *Indeed*, $\kappa_{CR}(A) = 1$ *for any diagonal matrix* A, *capturing our intuition that a diagonal system of equations should be solvable quite accurately.* $\diamond$

More generally, suppose that D is any nonsingular diagonal matrix and B is an arbitrary nonsingular matrix. Then

$$
\begin{aligned}
\kappa_{CR}(DB) &= \| |(DB)^{-1}| \cdot |(DB)| \| \\
&= \| |B^{-1}D^{-1}| \cdot |DB| \| \\
&= \| |B^{-1}| \cdot |B| \| \\
&= \kappa_{CR}(B).
\end{aligned}
$$

This means that if DB is *badly scaled*, i.e., B is well-conditioned but DB is badly conditioned (because D has widely varying diagonal entries), then we should hope to get an accurate solution of $(DB)x = b$ despite DB's ill-conditioning. This is discussed further in sections 2.4.4, 2.5.1, and 2.5.2.

Finally, as in the last section we provide an error bound using only the residual $r = A\hat{x} - b$:

$$\|\delta x\| = \|A^{-1}r\| \le \| |A^{-1}| \cdot |r| \|, \tag{2.9}$$

where we have used the triangle inequality. In section 2.4.4 we will see that this bound can sometimes be much smaller than the similar bound (2.5), in particular when A is badly scaled. There is also an analogue to Theorem 2.2 [193].

THEOREM 2.3. *The smallest* $\epsilon > 0$ *such that there exist* $|\delta A| \le \epsilon |A|$ *and* $|\delta b| \le \epsilon |b|$ *satisfying* $(A + \delta A)\hat{x} = b + \delta b$ *is called the* componentwise relative backward error. *It may be expressed in terms of the residual* $r = A\hat{x} - b$ *as follows:*

$$\epsilon = \max_i \frac{|r_i|}{(|A| \cdot |\hat{x}| + |b|)_i}.$$

For a proof, see Question 2.5.

LAPACK routines like `sgesvx` compute the componentwise backward relative error ϵ (the LAPACK variable name for ϵ is `BERR`).

2.3.   Gaussian Elimination

The basic algorithm for solving $Ax = b$ is *Gaussian elimination*. To state it, we first need to define a *permutation matrix*.

DEFINITION 2.1. *A permutation matrix P is an identity matrix with permuted rows.*

The most important properties of a permutation matrix are given by the following lemma.

LEMMA 2.2. *Let P, P_1, and P_2 be n-by-n permutation matrices and X be an n-by-n matrix. Then*

1. *PX is the same as X with its rows permuted. XP is the same as X with its columns permuted.*

2. *$P^{-1} = P^T$.*

3. $\det(P) = \pm 1$.

4. *$P_1 \cdot P_2$ is also a permutation matrix.*

For a proof, see Question 2.6.

Now we can state our overall algorithm for solving $Ax = b$.

ALGORITHM 2.1. *Solving $Ax = b$ using Gaussian elimination:*

1. *Factorize A into $A = PLU$, where*

$$\begin{aligned} P &= & \text{permutation matrix,} \\ L &= & \text{unit lower triangular matrix (i.e., with ones on the diagonal),} \\ U &= & \text{nonsingular upper triangular matrix.} \end{aligned}$$

2. *Solve $PLUx = b$ for LUx by permuting the entries of b: $LUx = P^{-1}b = P^T b$.*

3. *Solve $LUx = P^{-1}b$ for Ux by forward substitution: $Ux = L^{-1}(P^{-1}b)$.*

4. *Solve $Ux = L^{-1}(P^{-1}b)$ for x by back substitution: $x = U^{-1}(L^{-1}P^{-1}b)$.*

We will derive the algorithm for factorizing $A = PLU$ in several ways. We begin by showing why the permutation matrix P is necessary.

DEFINITION 2.2. *The* leading *j*-by-*j* principal submatrix *of A is* $A(1:j,1:j)$.

THEOREM 2.4. *The following two statements are equivalent:*

1. *There exists a unique unit lower triangular L and nonsingular upper triangular U such that $A = LU$.*

2. *All leading principal submatrices of A are nonsingular.*

Proof. We first show (1) implies (2). $A = LU$ may also be written

$$
\begin{bmatrix} A_{11} & A_{12} \\ A_{21} & A_{22} \end{bmatrix} = \begin{bmatrix} L_{11} & 0 \\ L_{21} & L_{22} \end{bmatrix} \begin{bmatrix} U_{11} & U_{12} \\ 0 & U_{22} \end{bmatrix}
$$

$$
= \begin{bmatrix} L_{11}U_{11} & L_{11}U_{12} \\ L_{21}U_{11} & L_{21}U_{12} + L_{22}U_{22} \end{bmatrix},
$$

where A_{11} is a *j*-by-*j* leading principal submatrix, as are L_{11} and U_{11}. Therefore $\det A_{11} = \det(L_{11}U_{11}) = \det L_{11} \det U_{11} = 1 \cdot \prod_{k=1}^{j}(U_{11})_{kk} \neq 0$, since L is unit triangular and U is triangular.

We prove that (2) implies (1) by induction on n. It is easy for 1-by-1 matrices: $a = 1 \cdot a$. To prove it for *n*-by-*n* matrices $\tilde{A}$, we need to find unique $(n-1)$-by-$(n-1)$ triangular matrices L and U, unique $(n-1)$-by-1 vectors l and u, and a unique nonzero scalar η such that

$$
\tilde{A} = \begin{bmatrix} A & b \\ c^T & \delta \end{bmatrix} = \begin{bmatrix} L & 0 \\ l^T & 1 \end{bmatrix} \begin{bmatrix} U & u \\ 0 & \eta \end{bmatrix} = \begin{bmatrix} LU & Lu \\ l^T U & l^T u + \eta \end{bmatrix}.
$$

By induction, unique L and U exist such that $A = LU$. Now let $u = L^{-1}b$, $l^T = c^T U^{-1}$, and $\eta = \delta - l^T u$, all of which are unique. The diagonal entries of U are nonzero by induction, and $\eta \neq 0$ since $0 \neq \det(\tilde{A}) = \det(U) \cdot \eta$. ◻

Thus LU factorization without pivoting can fail on (well-conditioned) nonsingular matrices such as the permutation matrix

$$
P = \begin{bmatrix} 0 & 1 & 0 \\ 0 & 0 & 1 \\ 1 & 0 & 0 \end{bmatrix};
$$

the 1-by-1 and 2-by-2 leading principal minors of P are singular. So we need to introduce permutations into Gaussian elimination.

THEOREM 2.5. *If A is nonsingular, then there exist permutations P_1 and P_2, a unit lower triangular matrix L, and a nonsingular upper triangular matrix U such that $P_1 A P_2 = LU$. Only one of P_1 and P_2 is necessary.*

Note: $P_1 A$ reorders the rows of A, $A P_2$ reorders the columns, and $P_1 A P_2$ reorders both.

Proof. As with many matrix factorizations, it suffices to understand block 2-by-2 matrices. More formally, we use induction on the dimension n. It is easy for 1-by-1 matrices: $P_1 = P_2 = L = 1$ and $U = A$. Assume that it is true for dimension $n - 1$. If A is nonsingular, then it has a nonzero entry; choose permutations P_1' and P_2' so that the $(1,1)$ entry of $P_1' A P_2'$ is nonzero. (We need only one of P_1' and P_2' since nonsingularity implies that each row and each column of A has a nonzero entry.)

Now we write the desired factorization and solve for the unknown components:

$$
\begin{aligned}
P_1' A P_2' &= \begin{bmatrix} a_{11} & A_{12} \\ A_{21} & A_{22} \end{bmatrix} = \begin{bmatrix} 1 & 0 \\ L_{21} & I \end{bmatrix} \cdot \begin{bmatrix} u_{11} & U_{12} \\ 0 & \tilde{A}_{22} \end{bmatrix} \\
&= \begin{bmatrix} u_{11} & U_{12} \\ L_{21} u_{11} & L_{21} U_{12} + \tilde{A}_{22} \end{bmatrix},
\end{aligned}
\tag{2.10}
$$

where A_{22} and $\tilde{A}_{22}$ are $(n-1)$-by-$(n-1)$ and L_{21} and U_{12}^T are $(n-1)$-by-1.

Solving for the components of this 2-by-2 block factorization we get $u_{11} = a_{11} \neq 0$, $U_{12} = A_{12}$, and $L_{21} u_{11} = A_{21}$. Since $u_{11} = a_{11} \neq 0$, we can solve for $L_{21} = \frac{A_{21}}{a_{11}}$. Finally, $L_{21} U_{12} + \tilde{A}_{22} = A_{22}$ implies $\tilde{A}_{22} = A_{22} - L_{21} U_{12}$.

We want to apply induction to $\tilde{A}_{22}$, but to do so we need to check that $\det \tilde{A}_{22} \neq 0$: Since $\det P_1' A P_2' = \pm \det A \neq 0$ and also

$$
\det P_1' A P_2' = \det \begin{bmatrix} 1 & 0 \\ L_{21} & I \end{bmatrix} \cdot \det \begin{bmatrix} u_{11} & U_{12} \\ 0 & \tilde{A}_{22} \end{bmatrix} = 1 \cdot (u_{11} \cdot \det \tilde{A}_{22}),
$$

then $\det \tilde{A}_{22}$ must be nonzero.

Therefore, by induction there exist permutations $\tilde{P}_1$ and $\tilde{P}_2$ so that $\tilde{P}_1 \tilde{A}_{22} \tilde{P}_2 = \tilde{L} \tilde{U}$, with $\tilde{L}$ unit lower triangular and $\tilde{U}$ upper triangular and nonsingular. Substituting this in the above 2-by-2 block factorization yields

$$
\begin{aligned}
P_1' A P_2' &= \begin{bmatrix} 1 & 0 \\ L_{21} & I \end{bmatrix} \begin{bmatrix} u_{11} & U_{12} \\ 0 & \tilde{P}_1^T \tilde{L} \tilde{U} \tilde{P}_2^T \end{bmatrix} \\
&= \begin{bmatrix} 1 & 0 \\ L_{21} & I \end{bmatrix} \begin{bmatrix} 1 & 0 \\ 0 & \tilde{P}_1^T \tilde{L} \end{bmatrix} \begin{bmatrix} u_{11} & U_{12} \\ 0 & \tilde{U} \tilde{P}_2^T \end{bmatrix} \\
&= \begin{bmatrix} 1 & 0 \\ L_{21} & \tilde{P}_1^T \tilde{L} \end{bmatrix} \begin{bmatrix} u_{11} & U_{12} \tilde{P}_2 \\ 0 & \tilde{U} \end{bmatrix} \begin{bmatrix} 1 & 0 \\ 0 & \tilde{P}_2^T \end{bmatrix} \\
&= \begin{bmatrix} 1 & 0 \\ 0 & \tilde{P}_1^T \end{bmatrix} \begin{bmatrix} 1 & 0 \\ \tilde{P}_1 L_{21} & \tilde{L} \end{bmatrix} \begin{bmatrix} u_{11} & U_{12} \tilde{P}_2 \\ 0 & \tilde{U} \end{bmatrix} \begin{bmatrix} 1 & 0 \\ 0 & \tilde{P}_2^T \end{bmatrix},
\end{aligned}
$$

so we get the desired factorization of A:

$$
\begin{aligned}
P_1 A P_2 &= \left(\begin{bmatrix} 1 & 0 \\ 0 & \tilde{P}_1 \end{bmatrix} P_1' \right) A \left(P_2' \begin{bmatrix} 1 & 0 \\ 0 & \tilde{P}_2 \end{bmatrix} \right) \\
&= \begin{bmatrix} 1 & 0 \\ \tilde{P}_1 L_{21} & \tilde{L} \end{bmatrix} \begin{bmatrix} u_{11} & U_{12} \tilde{P}_2 \\ 0 & \tilde{U} \end{bmatrix}. \quad \square
\end{aligned}
$$

The next two corollaries state simple ways to choose P_1 and P_2 to guarantee that Gaussian elimination will succeed on a nonsingular matrix.

COROLLARY 2.1. *We can choose $P_2' = I$ and P_1' so that a_{11} is the largest entry in absolute value in its column, which implies $L_{21} = \frac{A_{21}}{a_{11}}$ has entries bounded by 1 in absolute value. More generally, at step i of Gaussian elimination, where we are computing the ith column of L, we reorder rows i through n so that the largest entry in the column is on the diagonal. This is called "Gaussian elimination with partial pivoting," or GEPP for short. GEPP guarantees that all entries of L are bounded by one in absolute value.*

GEPP is the most common way to implement Gaussian elimination in practice. We discuss its numerical stability in the next section. Another more expensive way to choose P_1 and P_2 is given by the next corollary. It is almost never used in practice, although there are rare examples where GEPP fails but the next method succeeds in computing an accurate answer (see Question 2.14). We discuss briefly it in the next section as well.

COROLLARY 2.2. *We can choose P_1' and P_2' so that a_{11} is the largest entry in absolute value in the whole matrix. More generally, at step i of Gaussian elimination, where we are computing the ith column of L, we reorder rows and columns i through n so that the largest entry in this submatrix is on the diagonal. This is called "Gaussian elimination with complete pivoting," or GECP for short.*

The following algorithm embodies Theorem 2.5, performing permutations, computing the first column of L and the first row of U, and updating A_{22} to get $\tilde{A}_{22} = A_{22} - L_{21}U_{12}$. We write the algorithm first in conventional programming language notation and then using Matlab notation.

ALGORITHM 2.2. *LU factorization with pivoting:*

> *for $i = 1$ to $n-1$*
>   *apply permutations so $a_{ii} \neq 0$ (permute L and U too)*
>   */* for example, for GEPP, swap rows j and i of A and of L*
>   *where $|a_{ji}|$ is the largest entry in $|A(i:n,i)|$;*
>   *for GECP, swap rows j and i of A and of L,*
>   *and columns k and i of A and of U,*
>   *where $|a_{jk}|$ is the largest entry in $|A(i:n,i:n)|$ */*
>   */* compute column i of L (L_{21} in (2.10)) */*
>   *for $j = i+1$ to n*
>   *$l_{ji} = a_{ji}/a_{ii}$*
>   *end for*
>   */* compute row i of U (U_{12} in (2.10)) */*
>   *for $j = i$ to n*

$$u_{ij} = a_{ij}$$
$$\text{end for}$$
/* update A_{22} (to get $\tilde{A}_{22} = A_{22} - L_{21}U_{12}$ in (2.10)) */
for $j = i + 1$ to n
    for $k = i + 1$ to n
        $a_{jk} = a_{jk} - l_{ji} * u_{ik}$
    end for
    end for
end for

Note that once column i of A is used to compute column i of L, it is never used again. Similarly, row i of A is never used again after computing row i of U. This lets us overwrite L and U on top of A as they are computed, so we need no extra space to store them; L occupies the (strict) lower triangle of A (the ones on the diagonal of L are not stored explicitly), and U occupies the upper triangle of A. This simplifies the algorithm to the following algorithm.

ALGORITHM 2.3. *LU factorization with pivoting, overwriting L and U on A:*

for $i = 1$ to $n - 1$
    apply permutations (see Algorithm 2.2 for details)
    for $j = i + 1$ to n
        $a_{ji} = a_{ji}/a_{ii}$
    end for
    for $j = i + 1$ to n
        for $k = i + 1$ to n
            $a_{jk} = a_{jk} - a_{ji} * a_{ik}$
        end for
    end for
end for

Using Matlab notation this further reduces to the following algorithm.

ALGORITHM 2.4. *LU factorization with pivoting, overwriting L and U on A:*

for $i = 1$ to $n - 1$
    apply permutations (see Algorithm 2.2 for details)
    $A(i + 1 : n, i) = A(i + 1 : n, i)/A(i, i)$
    $A(i + 1 : n, i + 1 : n) =$
        $A(i + 1 : n, i + 1 : n) - A(i + 1 : n, i) * A(i, i + 1 : n)$
end for

In the last line of the algorithm, $A(i+1 : n, i) * A(i, i+1 : n)$ is the product of an $(n - i)$-by-1 matrix (L_{21}) by a 1-by-$(n - i)$ matrix (U_{12}), which yields an $(n - i)$-by-$(n - i)$ matrix.

We now rederive this algorithm from scratch starting from perhaps the most familiar description of Gaussian elimination: "Take each row and subtract multiples of it from later rows to zero out the entries below the diagonal." Translating this directly into an algorithm yields

for $i = 1$ to $n - 1$ /* for each row i */
    for $j = i + 1$ to n /* subtract a multiple of
                                                row i from row j ... */
        for $k = i$ to n /* ... in columns i through n ... */
            $a_{jk} = a_{jk} - \frac{a_{ji}}{a_{ii}} a_{ik}$ /* ... to zero out column i
                                                below the diagonal */
        end for
    end for
end for

We will now make some improvements to this algorithm, modifying it until it becomes identical to Algorithm 2.3 (except for pivoting, which we omit). First, we recognize that we need not compute the zero entries below the diagonal, because we know they are zero. This shortens the k loop to yield

for $i = 1$ to $n - 1$
    for $j = i + 1$ to n
        for $k = i + 1$ to n
            $a_{jk} = a_{jk} - \frac{a_{ji}}{a_{ii}} a_{ik}$
        end for
    end for
end for

The next performance improvement is to compute $\frac{a_{ji}}{a_{ii}}$ outside the inner loop, since it is constant within the inner loop.

for $i = 1$ to $n - 1$
    for $j = i + 1$ to n
        $l_{ji} = \frac{a_{ji}}{a_{ii}}$
    end for
    for $j = i + 1$ to n
        for $k = i + 1$ to n
            $a_{jk} = a_{jk} - l_{ji} a_{ik}$
        end for
    end for
end for

Finally, we store the multipliers l_{ji} in the subdiagonal entries a_{ji} that we originally zeroed out; they are not needed for anything else. This yields Algorithm 2.3 (except for pivoting).

The operation count of LU is done by replacing loops by summations over the same range, and inner loops by their operation counts:

$$\sum_{i=1}^{n-1} \left(\sum_{j=i+1}^{n} 1 + \sum_{j=i+1}^{n} \sum_{k=i+1}^{n} 2 \right)$$

$$= \sum_{i=1}^{n-1} ((n-i) + 2(n-i)^2) = \frac{2}{3}n^3 + O(n^2).$$

The forward and back substitutions with L and U to complete the solution of $Ax = b$ cost $O(n^2)$, so overall solving $Ax = b$ with Gaussian elimination costs $\frac{2}{3}n^3 + O(n^2)$ operations. Here we have used the fact that $\sum_{i=1}^{m} i^k = m^{k+1}/(k+1) + O(m^k)$. This formula is enough to get the high-order term in the operation count.

There is more to implementing Gaussian elimination than writing the nested loops of Algorithm 2.2. Indeed, depending on the computer, programming language, and matrix size, merely interchanging the last two loops on j and k can change the execution time by orders of magnitude. We discuss this at length in section 2.6.

2.4. Error Analysis

Recall our two-step paradigm for obtaining error bounds for the solution of $Ax = b$:

1. Analyze roundoff errors to show that the result of solving $Ax = b$ is the exact solution $\hat{x}$ of the perturbed linear system $(A + \delta A)\hat{x} = b + \delta b$, where δA and δb are small. This is an example of *backward error analysis*, and δA and δb are called the *backward errors*.

2. Apply the perturbation theory of section 2.2 to bound the error, for example by using bound (2.3) or (2.5).

We have two goals in this section. The first is to show how to implement Gaussian elimination in order to keep the backward errors δA and δb small. In particular, we would like to keep $\frac{\|\delta A\|}{\|A\|}$ and $\frac{\|\delta b\|}{\|b\|}$ as small as $O(\varepsilon)$. This is as small as we can expect to make them, since merely rounding the largest entries of A (or b) to fit into the floating point format can make $\frac{\|\delta A\|}{\|A\|} \geq \varepsilon$ (or $\frac{\|\delta b\|}{\|b\|} \geq \varepsilon$). It turns out that unless we are careful about pivoting, δA and δb need not be small. We discuss this in the next section.

The second goal is to derive practical error bounds which are simultaneously cheap to compute and "tight," i.e., close to the true errors. It turns out that the best bounds for $\|\delta A\|$ that we can formally prove are generally much larger than the errors encountered in practice. Therefore, our practical error bounds (in section 2.4.4) will rely on the computed residual $r = A\hat{x} - b$ and bound (2.5), instead of bound (2.3). We also need to be able to estimate $\kappa(A)$ inexpensively; this is discussed in section 2.4.3.

Unfortunately, we do not have error bounds that *always* satisfy our twin goals of cheapness and tightness, i.e., that simultaneously

1. cost a negligible amount compared to solving $Ax = b$ in the first place (for example, that cost $O(n^2)$ flops versus Gaussian elimination's $O(n^3)$ flops),

2. provide an error bound that is always at least as large as the true error and never more than a constant factor larger (100 times larger, say).

The practical bounds in section 2.4.4 will cost $O(n^2)$ but will on very rare occasions provide error bounds that are much too small or much too large. The probability of getting a bad error bound is so small that these bounds are widely used in practice. The only truly guaranteed bounds use either interval arithmetic, very high precision arithmetic, or both, and are several times more expensive than just solving $Ax = b$ (see section 1.5).

It has in fact been conjectured that no bound satisfying our twin goals of cheapness and tightness exist, but this remains an open problem.

2.4.1. The Need for Pivoting

Let us apply LU factorization without pivoting to $A = \begin{bmatrix} .0001 & 1 \\ 1 & 1 \end{bmatrix}$ in three-decimal-digit floating point arithmetic and see why we get the wrong answer. Note that $\kappa(A) = \|A\|_\infty \cdot \|A^{-1}\|_\infty \approx 4$, so A is well conditioned and thus we should expect to be able to solve $Ax = b$ accurately.

$$L = \begin{bmatrix} 1 & 0 \\ \text{fl}(1/10^{-4}) & 1 \end{bmatrix}, \quad \text{fl}(1/10^{-4}) \text{ rounds to } 10^4,$$

$$U = \begin{bmatrix} 10^{-4} & 1 \\ & \text{fl}(1 - 10^4 \cdot 1) \end{bmatrix}, \quad \text{fl}(1 - 10^4 \cdot 1) \text{ rounds to } -10^4,$$

$$\text{so} \quad LU = \begin{bmatrix} 1 & 0 \\ 10^4 & 1 \end{bmatrix} \begin{bmatrix} 10^{-4} & 1 \\ & -10^4 \end{bmatrix} = \begin{bmatrix} 10^{-4} & 1 \\ 1 & 0 \end{bmatrix}$$

$$\text{but} \quad A = \begin{bmatrix} 10^{-4} & 1 \\ 1 & 1 \end{bmatrix}.$$

Note that the original a_{22} has been entirely "lost" from the computation by subtracting 10^4 from it. We would have gotten the same LU factors whether a_{22} had been 1, 0, -2, or any number such that $\text{fl}(a_{22} - 10^4) = -10^4$. Since the algorithm proceeds to work only with L and U, it will get the same answer for all these different a_{22}, which correspond to completely different A and so completely different $x = A^{-1}b$; there is no way to guarantee an accurate answer. This is called *numerical instability*, since L and U are *not* the exact factors of a matrix close to A. (Another way to say this is that $\|A - LU\|$ is about as large as $\|A\|$, rather than $\varepsilon\|A\|$.)

Let us see what happens when we go on to solve $Ax = [1, 2]^T$ for x using this LU factorization. The correct answer is $x \approx [1, 1]^T$. Instead we get the following. Solving $Ly = [1, 2]^T$ yields $y_1 = \text{fl}(1/1) = 1$ and $y_2 = \text{fl}(2 - 10^4 \cdot 1) = -10^4$; note that the value 2 has been "lost" by subtracting 10^4 from it. Solving $U\hat{x} = y$ yields $\hat{x}_2 = \text{fl}((-10^4)/(-10^4)) = 1$ and $\hat{x}_1 = \text{fl}((1 - 1)/10^{-4}) = 0$, a completely erroneous solution.

Another warning of the loss of accuracy comes from comparing the condition number of A to the condition numbers of L and U. Recall that we transform the problem of solving $Ax = b$ into solving two other systems with L and U, so we do not want the condition numbers of L or U to be much larger than that of A. But here, the condition number of A is about 4, whereas the condition numbers of L and U are about 10^8.

In the next section we will show that doing GEPP nearly always eliminates the instability just illustrated. In the above example, GEPP would have reversed the order of the two equations before proceeding. The reader is invited to confirm that in this case we would get

$$L = \begin{bmatrix} 1 & 0 \\ \text{fl}(.0001/1) & 1 \end{bmatrix} = \begin{bmatrix} 1 & 0 \\ .0001 & 1 \end{bmatrix}$$

and

$$U = \begin{bmatrix} 1 & 1 \\ 0 & \text{fl}(1 - .0001 \cdot 1) \end{bmatrix} = \begin{bmatrix} 1 & 1 \\ 0 & 1 \end{bmatrix}$$

so that LU approximates A quite accurately. Both L and U are quite well-conditioned, as is A. The computed solution vector is also quite accurate.

2.4.2. Formal Error Analysis of Gaussian Elimination

Here is the intuition behind our error analysis of LU decomposition. If intermediate quantities arising in the product $L \cdot U$ are very large compared to $\|A\|$, the information in entries of A will get "lost" when these large values are subtracted from them. This is what happened to a_{22} in the example in section 2.4.1. If the intermediate quantities in the product $L \cdot U$ were instead comparable to those of A, we would expect a tiny backward error $A - LU$ in the factorization. Therefore, we want to bound the largest intermediate quantities in the product $L \cdot U$. We will do this by bounding the entries of the matrix $|L| \cdot |U|$ (see section 1.1 for notation).

Our analysis is analogous to the one we used for polynomial evaluation in section 1.6. There we considered $p = \sum_i a_i x^i$ and showed that if $|p|$ were comparable to the sum of absolute values $\sum_i |a_i x^i|$, then p would be computed accurately.

After presenting a general analysis of Gaussian elimination, we will use it to show that GEPP (or, more expensively, GECP) will keep the entries of $|L| \cdot |U|$ comparable to $\|A\|$ in almost all practical circumstances.

Unfortunately, the best bounds on $\|\delta A\|$ that we can prove in general are still much larger than the errors encountered in practice. Therefore, the error bounds that we use in practice will be based on the computed residual r and bound (2.5) (or bound (2.9)) instead of the rigorous but pessimistic bound in this section.

Now suppose that matrix A has already been pivoted, so the notation is simpler. We simplify Algorithm 2.2 to two equations, one for a_{jk} with $j \le k$ and one for $j > k$. Let us first trace what Algorithm 2.2 does to a_{jk} when $j \le k$: this element is repeatedly updated by subtracting $l_{ji}u_{ik}$ for $i = 1$ to $j - 1$ and is finally assigned to u_{jk} so that

$$u_{jk} = a_{jk} - \sum_{i=1}^{j-1} l_{ji}u_{ik}.$$

When $j > k$, a_{jk} again has $l_{ji}u_{ik}$ subtracted for $i = 1$ to $k - 1$, and then the resulting sum is divided by u_{kk} and assigned to l_{jk}:

$$l_{jk} = \frac{a_{jk} - \sum_{i=1}^{k-1} l_{ji}u_{ik}}{u_{kk}}.$$

To do the roundoff error analysis of these two formulas, we use the result from Question 1.10 that a dot product computed in floating point arithmetic satisfies

$$\text{fl}\left(\sum_{i=1}^{d} x_i y_i\right) = \sum_{i=1}^{d} x_i y_i (1 + \delta_i) \quad \text{with } |\delta_i| \le d\varepsilon.$$

We apply this to the formula for u_{jk}, yielding[9]

$$u_{jk} = \left(a_{jk} - \sum_{i=1}^{j-1} l_{ji}u_{ik}(1 + \delta_i)\right)(1 + \delta')$$

with $|\delta_i| \le (j-1)\varepsilon$ and $|\delta'| \le \varepsilon$. Solving for a_{jk} we get

$$
\begin{aligned}
a_{jk} &= \frac{1}{1+\delta'} u_{jk} \cdot l_{jj} + \sum_{i=1}^{j-1} l_{ji}u_{ik}(1 + \delta_i) \quad \text{since } l_{jj} = 1 \\
&= \sum_{i=1}^{j} l_{ji}u_{ik} + \sum_{i=1}^{j} l_{ji}u_{ik}\delta_i \\
&\qquad \text{with } |\delta_i| \le (j-1)\varepsilon \text{ and } 1 + \delta_j \equiv \frac{1}{1+\delta'} \\
&\equiv \sum_{i=1}^{j} l_{ji}u_{ik} + E_{jk},
\end{aligned}
$$

---

[9]Strictly speaking, the next formula assumes that we compute the sum first and then subtract from a_{jk}. But the final bound does not depend on the order of summation.

where we can bound E_{jk} by

$$|E_{jk}| = \left| \sum_{i=1}^{j} l_{ji} \cdot u_{ik} \cdot \delta_i \right| \leq \sum_{i=1}^{j} |l_{ji}| \cdot |u_{ik}| \cdot n\varepsilon = n\varepsilon(|L| \cdot |U|)_{jk}.$$

Doing the same analysis for the formula for l_{jk} yields

$$l_{jk} = (1 + \delta'') \left(\frac{(1 + \delta')(a_{jk} - \sum_{i=1}^{k-1} l_{ji} u_{ik}(1 + \delta_i))}{u_{kk}} \right)$$

with $|\delta_i| \leq (k-1)\varepsilon$, $|\delta'| \leq \varepsilon$, and $|\delta''| \leq \varepsilon$. We solve for a_{jk} to get

$$
\begin{aligned}
a_{jk} &= \frac{1}{(1 + \delta')(1 + \delta'')} u_{kk} l_{jk} + \sum_{i=1}^{k-1} l_{ji} u_{ik}(1 + \delta_i) \\
&= \sum_{i=1}^{k} l_{ji} u_{ik} + \sum_{i=1}^{k} l_{ji} u_{ik} \delta_i \quad \text{with } 1 + \delta_k \equiv \frac{1}{(1 + \delta')(1 + \delta'')} \\
&\equiv \sum_{i=1}^{k} l_{ji} u_{ik} + E_{jk}
\end{aligned}
$$

with $|\delta_i| \leq n\varepsilon$, and so $|E_{jk}| \leq n\varepsilon(|L| \cdot |U|)_{jk}$ as before.

Altogether, we can summarize this error analysis with the simple formula $A = LU + E$ where $|E| \leq n\varepsilon|L| \cdot |U|$. Taking norms we get $\|E\| \leq n\varepsilon\| |L| \| \cdot \| |U| \|$. If the norm does not depend on the signs of the matrix entries (true for the Frobenius, infinity-, and one-norms but not the two-norm), we can simplify this to $\|E\| \leq n\varepsilon\|L\| \cdot \|U\|$.

Now we consider the rest of the problem: solving $LUx = b$ via $Ly = b$ and $Ux = y$. The result of Question 1.11 shows that solving $Ly = b$ by forward substitution yields a computed solution $\hat{y}$ satisfying $(L + \delta L)\hat{y} = b$ with $|\delta L| \leq n\varepsilon|L|$. Similarly when solving $Ux = \hat{y}$ we get $\hat{x}$ satisfying $(U + \delta U)\hat{x} = \hat{y}$ with $|\delta U| \leq n\varepsilon|U|$.

Combining these yields

$$
\begin{aligned}
b &= (L + \delta L)\hat{y} \\
&= (L + \delta L)(U + \delta U)\hat{x} \\
&= (LU + L\delta U + \delta L U + \delta L \delta U)\hat{x} \\
&= (A - E + L\delta U + \delta L U + \delta L \delta U)\hat{x} \\
&\equiv (A + \delta A)\hat{x}, \quad \text{where} \quad \delta A = -E + L\delta U + \delta L U + \delta L \delta U.
\end{aligned}
$$

Now we combine our bounds on E, δL, and δU and use the triangle inequality to bound δA:

$$|\delta A| = |-E + L\delta U + \delta L U + \delta L \delta U|$$

$$\begin{aligned}
&\leq &&|E| + |L\delta U| + |\delta LU| + |\delta L\delta U| \\
&\leq &&|E| + |L| \cdot |\delta U| + |\delta L| \cdot |U| + |\delta L| \cdot |\delta U| \\
&\leq &&n\varepsilon|L| \cdot |U| + n\varepsilon|L| \cdot |U| + n\varepsilon|L| \cdot |U| + n^2\varepsilon^2|L| \cdot |U| \\
&\approx &&3n\varepsilon|L| \cdot |U|.
\end{aligned}$$

Taking norms and assuming $\|\,|X|\,\| = \|X\|$ (true as before for the Frobenius, infinity-, and one-norms but not the two-norm) we get $\|\delta A\| \leq 3n\varepsilon\|L\| \cdot \|U\|$.

Thus, to see when Gaussian elimination is backward stable, we must ask when $3n\varepsilon\|L\| \cdot \|U\| = O(\varepsilon)\|A\|$; then the $\frac{\|\delta A\|}{\|A\|}$ in the perturbation theory bounds will be $O(\varepsilon)$ as we desire (note that $\delta b = 0$).

The main empirical observation, justified by decades of experience, is that GEPP *almost* always keeps $\|L\| \cdot \|U\| \approx \|A\|$. GEPP guarantees that each entry of L is bounded by 1 in absolute value, so we need consider only $\|U\|$. We define the *pivot growth factor for GEPP*[10] as $g_{\text{PP}} = \|U\|_{\max}/\|A\|_{\max}$, where $\|A\|_{\max} = \max_{ij}|a_{ij}|$, so stability is equivalent to g_{PP} being small or growing slowly as a function of n. In practice, g_{PP} is almost always n or less. The average behavior seems to be $n^{2/3}$ or perhaps even just $n^{1/2}$ [242]. (See Figure 2.1.) This makes GEPP the algorithm of choice for many problems. Unfortunately, there are rare examples in which g_{PP} can be as large as 2^{n-1}.

PROPOSITION 2.1. *GEPP guarantees that $g_{\text{PP}} \leq 2^{n-1}$. This bound is attainable.*

Proof. The first step of GEPP updates $\tilde{a}_{jk} = a_{jk} - l_{ji} \cdot u_{ik}$, where $|l_{ji}| \leq 1$ and $|u_{ik}| = |a_{ik}| \leq \max_{rs}|a_{rs}|$, so $|\tilde{a}_{jk}| \leq 2 \cdot \max_{rs}|a_{rs}|$. So each of the $n-1$ major steps of GEPP can double the size of the remaining matrix entries, and we get 2^{n-1} as the overall bound. See the example in Question 2.14 to see that this is attainable. $\square$

Putting all these bounds together, we get

$$\|\delta A\|_\infty \leq 3g_{\text{PP}}n^3\varepsilon\|A\|_\infty, \tag{2.11}$$

since $\|L\|_\infty \leq n$ and $\|U\|_\infty \leq ng_{\text{PP}}\|A\|_\infty$. The factor $3g_{\text{PP}}n^3$ in the bound causes it to almost always greatly overestimate the true $\|\delta A\|$, even if $g_{\text{PP}} = 1$. For example, if $\varepsilon = 10^{-7}$ and $n = 150$, a very modest-sized matrix, then $3n^3\varepsilon > 1$, meaning that all precision is potentially lost. Example 2.3 graphs $3g_{\text{PP}}n^3\varepsilon$ along with the true backward error to show how it can be pessimistic; $\|\delta A\|$ is usually $O(\varepsilon)\|A\|$, so we can say that GEPP is *backward stable in practice*, even though we can construct examples where it fails. Section 2.4.4 presents practical error bounds for the computed solution of $Ax = b$ that are much smaller than what we get from using $\|\delta A\|_\infty \leq 3g_{\text{PP}}n^3\varepsilon\|A\|_\infty$.

---

[10]This definition is slightly different from the usual one in the literature but essentially equivalent [121, p. 115].

It can be shown that GECP is even more stable than GEPP, with its pivot growth g_{CP} satisfying the worst-case bound [262, p. 213]

$$g_{CP} = \frac{\max_{ij} |u_{ij}|}{\max_{ij} |a_{ij}|} \leq \sqrt{n \cdot 2 \cdot 3^{1/2} \cdot 4^{1/3} \cdots n^{1/(n-1)}} \approx n^{1/2 + \log_e n/4}.$$

This upper bound is also much too large in practice. The average behavior of g_{CP} is $n^{1/2}$. It was an old open conjecture that $g_{CP} \leq n$, but this was recently disproved [99, 122]. It remains an open problem to find a good upper bound for g_{CP} (which is still widely suspected to be $O(n)$.)

The extra $O(n^3)$ comparisons that GECP uses to find the pivots ($O(n^2)$ comparisons per step, versus $O(n)$ for GEPP) makes GECP significantly slower than GEPP, especially on high-performance machines that perform floating point operations about as fast as comparisons. Therefore, using GECP is seldom warranted (but see sections 2.4.4, 2.5.1, and 5.4.3).

EXAMPLE 2.3. Figures 2.1 and 2.2 illustrate these backward error bounds. For both figures, five random matrices A of each dimension were generated, with independent normally distributed entries, of mean 0 and standard deviation 1. (Testing such random matrices can sometimes be misleading about the behavior on some real problems, but it is still informative.) For each matrix, a similarly random vector b was generated. Both GEPP and GECP were used to solve $Ax = b$. Figure 2.1 plots the pivot growth factors g_{PP} and g_{CP}. In both cases they grow slowly with dimension, as expected. Figure 2.2 shows our two upper bounds for the backward error, $3n^3 \varepsilon g_{PP}$ (or $3n^3 \varepsilon g_{CP}$) and $3n\varepsilon \frac{\||L|\cdot|U|\|_\infty}{\|A\|_\infty}$. It also shows the true backward error, computed as described in Theorem 2.2. Machine epsilon is indicated by a solid horizontal line at $\varepsilon = 2^{-53} \approx 1.1 \cdot 10^{-16}$. Both bounds are indeed bounds on the true backward error but are too large by several order of magnitude. For the Matlab program that produced these plots, see HOMEPAGE/Matlab/pivot.m. ◇

2.4.3. Estimating Condition Numbers

To compute a practical error bound based on a bound like (2.5), we need to estimate $\|A^{-1}\|$. This is also enough to estimate the condition number $\kappa(A) = \|A^{-1}\| \cdot \|A\|$, since $\|A\|$ is easy to compute. One approach is to compute A^{-1} explicitly and compute its norm. However, this would cost $2n^3$, more than the original $\frac{2}{3}n^3$ for Gaussian elimination. (Note that this implies that it is not cheaper to solve $Ax = b$ by computing A^{-1} and then multiplying it by b. This is true even if one has many different b vectors. See Question 2.2.) It is a fact that most users will not bother to compute error bounds if they are expensive.

So instead of computing A^{-1} we will devise a much cheaper algorithm to *estimate* $\|A^{-1}\|$. Such an algorithm is called a *condition estimator* and should have the following properties:

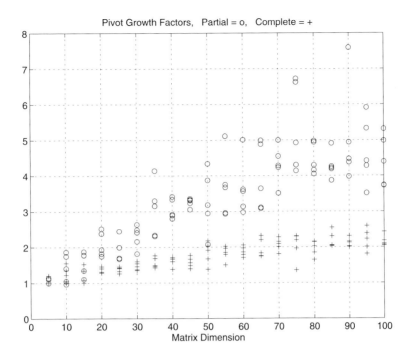

Fig. 2.1. *Pivot growth for random matrices,* $\circ = g_{\text{PP}}$, $+ = g_{\text{CP}}$.

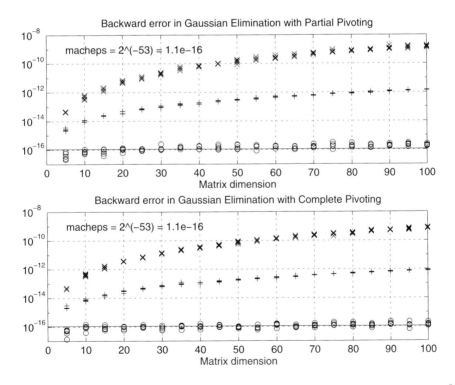

Fig. 2.2. *Backward error in Gaussian elimination on random matrices,* $\times = 3n^3 \varepsilon g$, $+ = 3n \| |L| \cdot |U| \|_\infty / \|A\|_\infty$, $\circ = \|Ax - b\|_\infty / (\|A\|_\infty \|x\|_\infty)$.

1. Given the L and U factors of A, it should cost $O(n^2)$, which for large enough n is negligible compared to the $\frac{2}{3}n^3$ cost of GEPP.

2. It should provide an estimate which is almost always within a factor of 10 of $\|A^{-1}\|$. This is all one needs for an error bound which tells you about how many decimal digits of accuracy that you have. (A factor-of-10 error is one decimal digit.[11])

There are a variety of such estimators available (see [146] for a survey). We choose to present one that is widely applicable to problems besides solving $Ax = b$, at the cost of being slightly slower than algorithms specialized for $Ax = b$ (but it is still reasonably fast). Our estimator, like most others, is guaranteed to produce only a *lower* bound on $\|A^{-1}\|$, not an upper bound. Empirically, it is almost always within a factor of 10, and usually 2 to 3, of $\|A^{-1}\|$. For the matrices in Figures 2.1 and 2.2, where the condition numbers varied from 10 to 10^5, the estimator equaled the condition number to several decimal places 83% of the time and was .43 times too small at worst. This is more than accurate enough to estimate the number of correct decimal digits in the final answer.

The algorithm estimates the one-norm $\|B\|_1$ of a matrix B, provided that we can compute Bx and $B^T y$ for arbitrary x and y. We will apply the algorithm to $B = A^{-1}$, so we need to compute $A^{-1}x$ and $A^{-T}y$, i.e., solve linear systems. This costs just $O(n^2)$ given the LU factorization of A. The algorithm was developed in [138, 146, 148], with the latest version in [147]. Recall that $\|B\|_1$ is defined by

$$\|B\|_1 \neq \max_{x \neq 0} \frac{\|Bx\|_1}{\|x\|_1} = \max_j \sum_{i=1}^{n} |b_{ij}|.$$

It is easy for us to show that the maximum over $x \neq 0$ is attained at $x = e_{j_0} = [0, \ldots, 0, 1, 0, \ldots, 0]^T$. (The single nonzero entry is component j_0, where $\max_j \sum_i |b_{ij}|$ occurs at $j = j_0$.)

Searching over all $e_j, j = 1, \ldots, n$, means computing all columns of $B = A^{-1}$; this is too expensive. Instead, since $\|B\|_1 = \max_{\|x\|_1 \leq 1} \|Bx\|_1$, we can use *hill climbing* or *gradient ascent* on $f(x) \equiv \|Bx\|_1$ inside the set $\|x\|_1 \leq 1$. $\|x\|_1 \leq 1$ is clearly a convex set of vectors, and $f(x)$ is a convex function, since $0 \leq \alpha \leq 1$ implies $f(\alpha x + (1-\alpha)y) = \|\alpha Bx + (1-\alpha)By\|_1 \leq \alpha\|Bx\|_1 + (1 - \alpha)\|By\|_1 = \alpha f(x) + (1-\alpha)f(y)$.

Doing gradient ascent to maximize $f(x)$ means moving x in the direction of the gradient $\nabla f(x)$ (if it exists) as long as $f(x)$ increases. The convexity of $f(x)$ means $f(y) \geq f(x) + \nabla f(x) \cdot (y - x)$ (if $\nabla f(x)$ exists). To compute ∇f we *assume* all $\sum_j b_{ij}x_j \neq 0$ in $f(x) = \sum_i |\sum_j b_{ij}x_j|$ (this is almost always

---

[11]As stated earlier, no one has ever found an estimator that approximates $\|A^{-1}\|$ with some guaranteed accuracy and is simultaneously significantly cheaper than explicitly computing A^{-1}. It has been been conjectured that no such estimator exists, but this has not been proven.

true). Let $\zeta_i = \text{sign}(\sum_j b_{ij}x_j)$, so $\zeta_i = \pm 1$ and $f(x) = \sum_i \sum_j \zeta_i b_{ij}x_j$. Then $\frac{\partial f}{\partial x_k} = \sum_i \zeta_i b_{ik}$ and $\bigtriangledown f = \zeta^T B = (B^T \zeta)^T$.

In summary, to compute $\bigtriangledown f(x)$ takes three steps: $w = Bx$, $\zeta = \text{sign}(w)$, and $\bigtriangledown f = \zeta^T B$.

ALGORITHM 2.5. *Hager's condition estimator returns a lower bound $\|w\|_1$ on $\|B\|_1$:*

> *choose any x such that $\|x\|_1 = 1$*   /* *e.g.* $x_i = \frac{1}{n}$ */
> *repeat*
> $w = Bx$, $\zeta = \text{sign}(w)$, $z = B^T \zeta$ /* $z^T = \bigtriangledown f$ */
>   *if* $\|z\|_\infty \le z^T x$ *then*
>   *return* $\|w\|_1$
>   *else*
> $x = e_j$ *where* $|z_j| = \|z\|_\infty$
>   *endif*
> *end repeat*

THEOREM 2.6.   1. *When $\|w\|_1$ is returned, $\|w\|_1 = \|Bx\|_1$ is a local maximum of $\|Bx\|_1$.*

   2. *Otherwise, $\|Be_j\|$ (at end of loop) $> \|Bx\|$ (at start), so the algorithm has made progress in maximizing $f(x)$.*

Proof.

1. In this case, $\|z\|_\infty \le z^T x$. Near x, $f(x) = \|Bx\|_1 = \sum_i \sum_j \zeta_i b_{ij}x_j$ is linear in x so $f(y) = f(x) + \bigtriangledown f(x) \cdot (y - x) = f(x) + z^T(y - x)$, where $z^T = \bigtriangledown f(x)$. To show x is a local maximum we want $z^T(y - x) \le 0$ when $\|y\|_1 = 1$. We compute

$$z^T(y - x) = z^T y - z^T x = \sum_i z_i \cdot y_i - z^T x \le \sum_i |z_i| \cdot |y_i| - z^T x$$
$$\le \|z\|_\infty \cdot \|y\|_1 - z^T x = \|z\|_\infty - z^T x \le 0 \quad \text{as desired.}$$

2. In this case $\|z\|_\infty > z^T x$. Choose $\tilde{x} = e_j \cdot \text{sign}(z_j)$, where j is chosen so that $|z_j| = \|z\|_\infty$. Then

$$f(\tilde{x}) \ge f(x) + \bigtriangledown f \cdot (\tilde{x} - x) = f(x) + z^T(\tilde{x} - x)$$
$$= f(x) + z^T \tilde{x} - z^T x = f(x) + |z_j| - z^T x > f(x),$$

where the last inequality is true by construction. $\square$

Higham [147, 148] tested a slightly improved version of this algorithm by trying many random matrices of sizes $10, 25, 50$ and condition numbers $\kappa = 10, 10^3, 10^6, 10^9$; in the worst case the computed κ underestimated the true κ by a factor .44. The algorithm is available in LAPACK as subroutine `slacon`. LAPACK routines like `sgesvx` call `slacon` internally and return the estimated condition number. (They actually return the reciprocal of the estimated condition number, to avoid overflow on exactly singular matrices.) A different condition estimator is available in Matlab as `rcond`. The Matlab routine `cond` computes the exact condition number $\|A^{-1}\|_2\|A\|_2$, using algorithms discussed in section 5.4; it is much more expensive than `rcond`.

Estimating the Relative Condition Number

We can also use the algorithm from the last section to estimate the relative condition number $\kappa_{CR}(A) = \| |A^{-1}| \cdot |A| \|_\infty$ from bound (2.8) or to evaluate the bound $\| |A^{-1}| \cdot |r| \|_\infty$ from (2.9). We can reduce both to the same problem, that of estimating $\| |A^{-1}| \cdot g \|_\infty$, where g is a vector of nonnegative entries. To see why, let e be the vector of all ones. From part 5 of Lemma 1.7, we see that $\|X\|_\infty = \|Xe\|_\infty$ if the matrix X has nonnegative entries. Then

$$\| |A^{-1}| \cdot |A| \|_\infty = \| |A^{-1}| \cdot |A|e \|_\infty = \| |A^{-1}| \cdot g \|_\infty, \quad \text{where} \quad g = |A|e.$$

Here is how we estimate $\| |A^{-1}| \cdot g \|_\infty$. Let $G = \text{diag}(g_1, \ldots, g_n)$; then $g = Ge$. Thus

$$
\begin{aligned}
\| |A^{-1}| \cdot g \|_\infty &= \| |A^{-1}| \cdot Ge \|_\infty = \| |A^{-1}| \cdot G \|_\infty = \| |A^{-1}G| \|_\infty \\
&= \|A^{-1}G\|_\infty. \tag{2.12}
\end{aligned}
$$

The last equality is true because $\|Y\|_\infty = \| |Y| \|_\infty$ for any matrix Y. Thus, it suffices to estimate the infinity norm of the matrix $A^{-1}G$. We can do this by applying Hager's algorithm, Algorithm 2.5, to the matrix $(A^{-1}G)^T = GA^{-T}$, to estimate $\|(A^{-1}G)^T\|_1 = \|A^{-1}G\|_\infty$ (see part 6 of Lemma 1.7). This requires us to multiply by the matrix GA^{-T} and its transpose $A^{-1}G$. Multiplying by G is easy since it is diagonal, and we multiply by A^{-1} and A^{-T} using the LU factorization of A, as we did in the last section.

2.4.4. Practical Error Bounds

We present two practical error bounds for our approximate solution $\hat{x}$ of $Ax = b$. For the first bound we use inequality (2.5) to get

$$\text{error} = \frac{\|\hat{x} - x\|_\infty}{\|\hat{x}\|_\infty} \leq \|A^{-1}\|_\infty \cdot \frac{\|r\|_\infty}{\|\hat{x}\|_\infty}, \tag{2.13}$$

where $r = A\hat{x} - b$ is the residual. We estimate $\|A^{-1}\|_\infty$ by applying Algorithm 2.5 to $B = A^{-T}$, estimating $\|B\|_1 = \|A^{-T}\|_1 = \|A^{-1}\|_\infty$ (see parts 5 and 6 of Lemma 1.7).

Our second error bound comes from the tighter inequality (2.9):

$$\text{error} = \frac{\|\hat{x} - x\|_\infty}{\|\hat{x}\|_\infty} \leq \frac{\| |A^{-1}| \cdot |r| \|_\infty}{\|\hat{x}\|_\infty}. \tag{2.14}$$

We estimate $\| |A^{-1}| \cdot |r| \|_{\infty}$ using the algorithm based on equation (2.12). Error bound (2.14) (modified as described below in the subsection "What can go wrong") is computed by LAPACK routines like sgesvx. The LAPACK variable name for the error bound is FERR, for Forward ERRor.

EXAMPLE 2.4. We have computed the first error bound (2.13) and the true error for the same set of examples as in Figures 2.1 and 2.2, plotting the result in Figure 2.3. For each problem $Ax = b$ solved with GEPP we plot a ∘ at the point (true error, error bound), and for each problem $Ax = b$ solved with GECP we plot a + at the point (true error, error bound). If the error bound were equal to the true error, the ∘ or + would lie on the solid diagonal line. Since the error bound always exceeds the true error, the ∘s and +s lie above this diagonal. When the error bound is less than 10 times larger than the true error, the ∘ or + appears between the solid diagonal line and the first superdiagonal dashed line. When the error bound is between 10 and 100 times larger than the true error, the ∘ or + appears between the first two superdiagonal dashed lines. Most error bounds are in this range, with a few error bounds as large as 1000 times the true error. Thus, our computed error bound underestimates the number of correct decimal digits in the answer by one or two and in rare cases by as much as three. The Matlab code for producing these graphs is the same as before, HOMEPAGE/Matlab/pivot.m. ◇

EXAMPLE 2.5. We present an example chosen to illustrate the difference between the two error bounds (2.13) and (2.14). This example will also show that GECP can sometimes be more accurate than GEPP. We choose a set of badly scaled examples constructed as follows. Each test matrix is of the form $A = DB$, with the dimension running from 5 to 100. B is equal to an identity matrix plus very small random offdiagonal entries, around 10^{-7}, so it is very well-conditioned. D is a diagonal matrix with entries scaled geometrically from 1 up to 10^{14}. (In other words, $d_{i+1,i+1}/d_{i,i}$ is the same for all i.) The A matrices have condition numbers $\kappa(A) = \|A^{-1}\|_{\infty} \cdot \|A\|_{\infty}$ nearly equal to 10^{14}, which is very ill-conditioned, although their relative condition numbers $\kappa_{CR}(A) = \| |A^{-1}| \cdot |A| \|_{\infty} = \| |B^{-1}| \cdot |B| \|_{\infty}$ are all nearly 1. As before, machine precision is $\varepsilon = 2^{-53} \approx 10^{-16}$. The examples were computed using the same Matlab code HOMEPAGE/Matlab/pivot.m.

The pivot growth factors g_{PP} and g_{CP} were never larger than about 1.33 for any example, and the backward error from Theorem 2.2 never exceeded 10^{-15} in any case. Hager's estimator was very accurate in all cases, returning the true condition number 10^{14} to many decimal places.

Figure 2.4 plots the error bounds (2.13) and (2.14) for these examples, along with the componentwise relative backward error, as given by the formula in Theorem 2.3. The cluster of plus signs in the upper left corner of Figure 2.4(a) shows that while GECP computes the answer with a tiny error near 10^{-15}, the error bound (2.13) is usually closer to 10^{-2}, which is very pessimistic. This

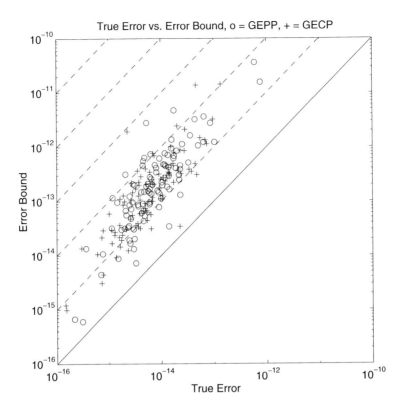

Fig. 2.3. *Error bound (2.13) plotted versus true error, o = GEPP, + = GECP.*

is because the condition number is 10^{14}, and so unless the backward error is much smaller than $\varepsilon \approx 10^{-16}$, which is unlikely, the error bound will be close to $10^{-16}10^{14} = 10^{-2}$. The cluster of circles in the middle top of the same figure shows that GEPP gets a larger error of about 10^{-8}, while the error bound (2.13) is again usually near 10^{-2}.

In contrast, the error bound (2.14) is nearly perfectly accurate, as illustrated by the pluses and circles on the diagonal in Figure 2.4(b). This graph again illustrates that GECP is nearly perfectly accurate, whereas GEPP loses about half the accuracy. This difference in accuracy is explained by Figure 2.4(c), which shows the componentwise relative backward error from Theorem 2.3 for GEPP and GECP. This graph makes it clear that GECP has nearly perfect backward error in the componentwise relative sense, so since the corresponding componentwise relative condition number is 1, the accuracy is perfect. GEPP on the other hand is not completely stable in this sense, losing from 5 to 10 decimal digits.

In section 2.5 we show how to iteratively improve the computed solution $\hat{x}$. One step of this method will make the solution computed by GEPP as accurate as the solution from GECP. Since GECP is significantly more expensive than GEPP in practice, it is very rarely used.   ⋄

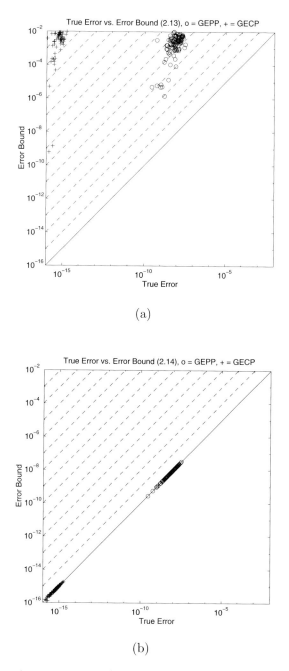

Fig. 2.4. (a) *plots the error bound* (2.13) *versus the true error.* (b) *plots the error bound* (2.14) *versus the true error.*

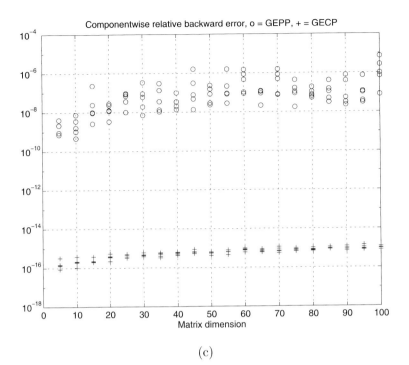

(c)

Fig. 2.4. *Continued.* (c) *plots the componentwise relative backward error from The-orem* 2.3.

What Can Go Wrong

Unfortunately, as mentioned in the beginning of section 2.4, error bounds (2.13) and (2.14) are *not* guaranteed to provide tight bounds in all cases when implemented in practice. In this section we describe the (rare!) ways they can fail, and the partial remedies used in practice.

First, as described in section 2.4.3, the estimate of $\|A^{-1}\|$ from Algorithm 2.5 (or similar algorithms) provides only a lower bound, although the probability is very low that it is more than 10 times too small.

Second, there is a small but nonnegligible probability that roundoff in the evaluation of $r = A\hat{x} - b$ might make $\|r\|$ artificially small, in fact zero, and so also make our computed error bound too small. To take this possibility into account, one can add a small quantity to $|r|$ to account for it: From Question 1.10 we know that the roundoff in evaluating r is bounded by

$$|(A\hat{x} - b) - \mathrm{fl}(A\hat{x} - b)| \leq (n+1)\varepsilon(|A| \cdot |\hat{x}| + |b|), \qquad (2.15)$$

so we can replace $|r|$ with $|r| + (n+1)\varepsilon(|A| \cdot |\hat{x}| + |b|)$ in bound (2.14) (this is done in the LAPACK code `sgesvx`) or $\|r\|$ with $\|r\| + (n+1)\varepsilon(\|A\| \cdot \|\hat{x}\| + \|b\|)$ in bound (2.13). The factor $n+1$ is usually much too large and can be omitted if desired.

Third, roundoff in performing Gaussian elimination on very ill-conditioned matrices can yield such inaccurate L and U that bound (2.14) is much too low.

EXAMPLE 2.6. We present an example, discovered by W. Kahan, that illustrates the difficulties in getting truly guaranteed error bounds. In this example the matrix A will be *exactly* singular. Therefore any error bound on $\frac{\|x-\hat{x}\|}{\|x\|}$ should be one or larger to indicate that no digits in the computed solution are correct, since the true solution does not exist.

Roundoff error during Gaussian elimination will yield nonsingular but very ill-conditioned factors L and U. With this example, computing using Matlab with IEEE double precision arithmetic, the computed residual r turns out to be *exactly* zero because of roundoff, so both error bounds (2.13) and (2.14) return zero. If we repair bound (2.13) by adding $4\varepsilon(\|A\| \cdot \|\hat{x}\| + \|b\|)$, it will be larger than 1 as desired.

Unfortunately our second, "tighter" error bound (2.14) is about 10^{-7}, erroneously indicating that seven digits of the computed solution are correct.

Here is how the example is constructed. Let $\chi = 3/2^{29}$, $\zeta = 2^{14}$,

$$
\begin{aligned}
A &= \begin{bmatrix} \chi \cdot \zeta & -\zeta & \zeta \\ \zeta^{-1} & \zeta^{-1} & 0 \\ \zeta^{-1} & -\chi \cdot \zeta^{-1} & \zeta^{-1} \end{bmatrix} \\
&\approx \begin{bmatrix} 9.1553 \cdot 10^{-5} & -1.6384 \cdot 10^4 & 1.6384 \cdot 10^4 \\ 6.1035 \cdot 10^{-5} & 6.1035 \cdot 10^{-5} & 0 \\ 6.1035 \cdot 10^{-5} & -3.4106 \cdot 10^{-13} & 6.1035 \cdot 10^{-5} \end{bmatrix},
\end{aligned}
$$

and $b = A \cdot [1, 1 + \varepsilon, 1]^T$. A can be computed without any roundoff error, but b has a bit of roundoff, which means that it is not exactly in the space spanned by the columns of A, so $Ax = b$ has no solution. Performing Gaussian elimination, we get

$$
L \approx \begin{bmatrix} 1 & 0 & 0 \\ .66666 & 1 & 0 \\ .66666 & 1.0000 & 1 \end{bmatrix}
$$

and

$$
U \approx \begin{bmatrix} 9.1553 \cdot 10^{-5} & -1.6384 \cdot 10^4 & 1.6384 \cdot 10^4 \\ 0 & 1.0923 \cdot 10^4 & -1.0923 \cdot 10^4 \\ 0 & 0 & 1.8190 \cdot 10^{-12} \end{bmatrix},
$$

yielding a computed value of

$$
A^{-1} \approx \begin{bmatrix} 2.0480 \cdot 10^3 & 5.4976 \cdot 10^{11} & -5.4976 \cdot 10^{11} \\ -2.0480 \cdot 10^3 & -5.4976 \cdot 10^{11} & 5.4976 \cdot 10^{11} \\ -2.0480 \cdot 10^3 & -5.4976 \cdot 10^{11} & 5.4976 \cdot 10^{11} \end{bmatrix}.
$$

This means the computed value of $|A^{-1}| \cdot |A|$ has all entries approximately equal to $6.7109 \cdot 10^7$, so $\kappa_{CR}(A)$ is computed to be $O(10^7)$. In other words, the error bound indicates that about $16 - 7 = 9$ digits of the computed solution are accurate, whereas none are.

Barring large pivot growth, one can prove that bound (2.13) (with $\|r\|$ appropriately increased) cannot be made artificially small by the phenomenon illustrated here.

Similarly, Kahan has found a family of n-by-n singular matrices, where changing one tiny entry (about 2^{-n}) to zero lowers $\kappa_{CR}(A)$ to $O(n^3)$. One could similarly construct examples where A was not exactly singular, so that bounds (2.13) and (2.14) were correct in exact arithmetic, but where roundoff made them much too small. ◇

2.5. Improving the Accuracy of a Solution

We have just seen that the error in solving $Ax = b$ may be as large as $\kappa(A)\varepsilon$. If this error is too large, what can we do? One possibility is to rerun the entire computation in higher precision, but this may be quite expensive in time and space. Fortunately, as long as $\kappa(A)$ is not too large, there are much cheaper methods available for getting a more accurate solution.

To solve any equation $f(x) = 0$, we can try to use Newton's method to improve an approximate solution x_i to get $x_{i+1} = x_i - \frac{f(x_i)}{f'(x_i)}$. Applying this to $f(x) = Ax - b$ yields one step of *iterative refinement*:

$$r = Ax_i - b$$
$$\text{solve } Ad = r \text{ for } d$$
$$x_{i+1} = x_i - d$$

If we could compute $r = Ax_i - b$ exactly and solve $Ad = r$ exactly, we would be done in one step, which is what we expect from Newton applied to a linear problem. Roundoff error prevents this immediate convergence. The algorithm is interesting and of use precisely when A is so ill-conditioned that solving $Ad = r$ (and $Ax_0 = b$) is rather inaccurate.

THEOREM 2.7. *Suppose that r is computed in* double *precision and $\kappa(A) \cdot \varepsilon < c \equiv \frac{1}{3n^3 g + 1} < 1$, where n is the dimension of A and g is the pivot growth factor. Then repeated iterative refinement converges with*

$$\frac{\|x_i - A^{-1}b\|_\infty}{\|A^{-1}b\|_\infty} = O(\varepsilon).$$

Note that the condition number does not appear in the final error bound. This means that we compute the answer accurately independent of the condition number, provided that $\kappa(A)\varepsilon$ is sufficiently less than 1. (In practice, c is too conservative an upper bound, and the algorithm often succeeds even when $\kappa(A)\varepsilon$ is greater than c.)

Sketch of Proof. In order to keep the proof transparent, we will take only the most important rounding errors into account. For brevity, we abbreviate $\| \cdot \|_\infty$ by $\| \cdot \|$. Our goal is to show that

$$\|x_{i+1} - x\| \leq \frac{\kappa(A)\varepsilon}{c}\|x_i - x\| \equiv \zeta\|x_i - x\|.$$

By assumption, $\zeta < 1$, so this inequality implies that the error $\|x_{i+1} - x\|$ decreases monotonically to zero. (In practice it will not decrease all the way to zero because of rounding error in the assignment $x_{i+1} = x_i - d$, which we are ignoring.)

We begin by estimating the error in the computed residual r. We get $r = \text{fl}(Ax_i - b) = Ax_i - b + f$, where by the result of Question 1.10 $|f| \leq n\varepsilon^2(|A| \cdot |x_i| + |b|) + \varepsilon|Ax_i - b| \approx \varepsilon|Ax_i - b|$. The ε^2 term comes from the double precision computation of r, and the ε term comes from rounding the double precision result back to single precision. Since $\varepsilon^2 \ll \varepsilon$, we will neglect the ε^2 term in the bound on $|f|$.

Next we get $(A + \delta A)d = r$, where from bound (2.11) we know that $\|\delta A\| \leq \gamma \cdot \varepsilon \cdot \|A\|$, where $\gamma = 3n^3 g$, although this is usually much too large. As mentioned earlier, we simplify matters by assuming $x_{i+1} = x_i - d$ exactly.

Continuing to ignore all ε^2 terms, we get

$$
\begin{aligned}
d &= (A + \delta A)^{-1}r = (I + A^{-1}\delta A)^{-1}A^{-1}r \\
&= (I + A^{-1}\delta A)^{-1}A^{-1}(Ax_i - b + f) \\
&= (I + A^{-1}\delta A)^{-1}(x_i - x + A^{-1}f) \\
&\approx (I - A^{-1}\delta A)(x_i - x + A^{-1}f) \\
&\approx x_i - x - A^{-1}\delta A(x_i - x) + A^{-1}f.
\end{aligned}
$$

Therefore $x_{i+1} - x = x_i - d - x = A^{-1}\delta A(x_i - x) - A^{-1}f$ and so

$$
\begin{aligned}
\|x_{i+1} - x\| &\leq \|A^{-1}\delta A(x_i - x)\| + \|A^{-1}f\| \\
&\leq \|A^{-1}\| \cdot \|\delta A\| \cdot \|x_i - x\| + \|A^{-1}\| \cdot \varepsilon \cdot \|Ax_i - b\| \\
&\leq \|A^{-1}\| \cdot \|\delta A\| \cdot \|x_i - x\| + \|A^{-1}\| \cdot \varepsilon \cdot \|A(x_i - x)\| \\
&\leq \|A^{-1}\| \cdot \gamma\varepsilon \cdot \|A\| \cdot \|x_i - x\| \\
&\quad + \|A^{-1}\| \cdot \|A\| \cdot \varepsilon \cdot \|x_i - x\| \\
&= \|A^{-1}\| \cdot \|A\| \cdot \varepsilon \cdot (\gamma + 1) \cdot \|x_i - x\|,
\end{aligned}
$$

so if

$$\zeta = \|A^{-1}\| \cdot \|A\| \cdot \varepsilon(\gamma + 1) = \kappa(A)\varepsilon/c < 1,$$

then we have convergence. $\square$

Iterative refinement (or other variations of Newton's method) can be used to improve accuracy for many other problems of linear algebra as well.

2.5.1.   Single Precision Iterative Refinement

This section may be skipped on a first reading.

Sometimes double precision is not available to run iterative refinement. For example, if the input data is already in double precision, we would need to compute the residual r in *quadruple* precision, which may not be available. On some machines, such as the Intel Pentium, double-extended precision is available, which provides 11 more bits of fraction than double precision (see section 1.5). This is not as accurate as quadruple precision (which would need at least $2 \cdot 53 = 106$ fraction bits) but still improves the accuracy noticeably.

But if none of these options are available, one could still run iterative refinement while computing the residual r in single precision (i.e., the same precision as the input data). In this case, Theorem 2.7 does not hold any more. On the other hand, the following theorem shows that under certain technical assumptions, one step of iterative refinement in single precision is still worth doing because it reduces the componentwise relative backward error as defined in Theorem 2.3 to $O(\varepsilon)$. If the corresponding relative condition number $\kappa_{CR}(A) = \| \, |A^{-1}| \cdot |A| \, \|_\infty$ from section 2.2.1 is significantly smaller than the usual condition number $\kappa(A) = \|A^{-1}\|_\infty \cdot \|A\|_\infty$, then the answer will also be more accurate.

THEOREM 2.8. *Suppose that r is computed in single precision and*

$$\|A^{-1}\|_\infty \ \cdot \|A\|_\infty \cdot \frac{\max_i(|A| \cdot |x|)_i}{\min_i(|A| \cdot |x|)_i} \cdot \varepsilon \ < 1.$$

Then one step of iterative refinement yields x_1 such that $(A + \delta A)x_1 = b + \delta b$ with $|\delta a_{ij}| = O(\varepsilon)|a_{ij}|$ and $|\delta b_i| = O(\varepsilon)|b_i|$. In other words, the componentwise relative backward error is as small as possible. For example, this means that if A and b are sparse, then δA and δb have the same sparsity structures as A and b, respectively.

For a proof, see [149] as well as [14, 225, 226, 227] for more details.

Single precision iterative refinement and the error bound (2.14) are implemented in LAPACK routines like `sgesvx`.

EXAMPLE 2.7. We consider the same matrices as in Example 2.5 and perform one step of iterative refinement in the same precision as the rest of the computation ($\varepsilon \approx 10^{-16}$). For these examples, the usual condition number is $\kappa(A) \approx 10^{14}$, whereas $\kappa_{CR}(A) \approx 1$, so we expect a large accuracy improvement. Indeed, the componentwise relative error for GEPP is driven below 10^{-15}, and the corresponding error from (2.14) is driven below 10^{-15} as well. The Matlab code for this example is HOMEPAGE/Matlab/pivot.m. $\diamond$

2.5.2.   Equilibration

There is one more common technique for improving the error in solving a linear system: *equilibration*. This refers to choosing an appropriate diagonal matrix

D and solving $DAx = Db$ instead of $Ax = b$. D is chosen to try to make the condition number of DA smaller than that of A. In Example 2.7 for instance, choosing d_{ii} to be the reciprocal of the two-norm of row i of A would make DA nearly equal to the identity matrix, reducing its condition number from 10^{14} to 1. It is possible to show that choosing D this way reduces the condition number of DA to within a factor of $\sqrt{n}$ of its smallest possible value for any diagonal D [244]. In practice we may also choose two diagonal matrices D_{row} and D_{col} and solve $(D_{row}AD_{col})\bar{x} = D_{row}b$, $x = D_{col}\bar{x}$.

The techniques of iterative refinement and equilibration are implemented in the LAPACK subroutines like `sgerfs` and `sgeequ`, respectively. These are in turn used by driver routines like `sgesvx`.

2.6.   Blocking Algorithms for Higher Performance

At the end of section 2.3, we said that changing the order of the three nested loops in the implementation of Gaussian elimination in Algorithm 2.2 could change the execution speed by orders of magnitude, depending on the computer and the problem being solved. In this section we will explore why this is the case and describe some carefully written linear algebra software which takes these matters into account. These implementations use so-called *block algorithms*, because they operate on square or rectangular subblocks of matrices in their innermost loops rather than on entire rows or columns. These codes are available in public-domain software libraries such as LAPACK (in Fortran, at NETLIB/lapack)[12] and ScaLAPACK (at NETLIB/scalapack). LAPACK (and its versions in other languages) are suitable for PCs, workstations, vector computers, and shared-memory parallel computers. These include the Sun SPARC-center 2000 [238], SGI Power Challenge [223], DEC AlphaServer 8400 [103], and Cray C90/J90 [253, 254]. ScaLAPACK is suitable for distributed-memory parallel computers, such as the IBM SP-2 [256], Intel Paragon [257], Cray T3 series [255], and networks of workstations [9]. These libraries are available on NETLIB, including comprehensive manuals [10, 34].

A more comprehensive discussion of algorithms for high performance (especially parallel) machines may be found on the World Wide Web at PARALLEL_HOMEPAGE.

LAPACK was originally motivated by the poor performance of its predecessors LINPACK and EISPACK (also available on NETLIB) on some high-performance machines. For example, consider the table below, which presents the speed in Mflops of LINPACK's Cholesky routine `spofa` on a Cray YMP, a supercomputer of the late 1980s. Cholesky is a variant of Gaussian elimination suitable for symmetric positive definite matrices. It is discussed in depth in

---

[12]A C translation of LAPACK, called CLAPACK (at NETLIB/clapack), is also available. LAPACK++ (at NETLIB/c++/lapack++)) and LAPACK90 (at NETLIB/lapack90)) are C++ and Fortran 90 interfaces to LAPACK, respectively.

section 2.7; here it suffices to know that it is very similar to Algorithm 2.2. The table also includes the speed of several other linear algebra operations. The Cray YMP is a parallel computer with up to 8 processors that can be used simultaneously, so we include one column of data for 1 processor and another column where all 8 processors are used.

	1 Proc.	8 Procs.
Maximum speed	330	2640
Matrix-matrix multiply ($n = 500$)	312	2425
Matrix-vector multiply ($n = 500$)	311	2285
Solve $TX = B$ ($n = 500$)	309	2398
Solve $Tx = b$ ($n = 500$)	272	584
LINPACK (Cholesky, $n = 500$)	72	72
LAPACK (Cholesky, $n = 500$)	290	1414
LAPACK (Cholesky, $n = 1000$)	301	2115

The top line, the maximum speed of the machine, is an upper bound on the numbers that follow. The basic linear algebra operations on the next four lines have been measured using subroutines especially designed for high speed on the Cray YMP. They all get reasonably close to the maximum possible speed, except for solving $Tx = b$, a single triangular system of linear equations, which does not use 8 processors effectively. Solving $TX = B$ refers to solving triangular systems with many right-hand sides (B is a square matrix). These numbers are for large matrices and vectors ($n = 500$).

The Cholesky routine from LINPACK in the sixth line of the table executes significantly more slowly than these other operations, even though it is working on as large a matrix as the previous operations and doing mathematically similar operations. This poor performance leads us to try to reorganize Cholesky and other linear algebra routines to go as fast as their simpler counterparts like matrix-matrix multiplication. The speeds of these reorganized codes from LAPACK are given in the last two lines of the table. It is apparent that the LAPACK routines come much closer to the maximum speed of the machine. We emphasize that the LAPACK and LINPACK Cholesky routines perform the same floating operations, but in a different order.

To understand how these speedups were attained, we must understand how the time is spent by the computer while executing. This in turn requires us to understand how computer memories operate. It turns out that all computer memories, from the cheapest personal computer to the biggest supercomputer, are built as *hierarchies*, with a series of different kinds of memories ranging from very fast, expensive, and therefore small memory at the top of the hierarchy down to slow, cheap, and very large memory at the bottom.

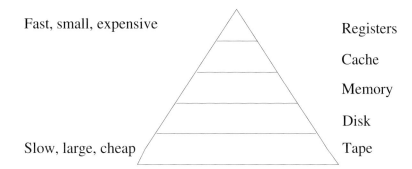

Fast, small, expensive — Registers, Cache, Memory, Disk

Slow, large, cheap — Tape

For example, registers form the fastest memory, then cache, main memory, disks, and finally tape as the slowest, largest, and cheapest. Useful arithmetic and logical operations can be done *only* on data at the top of the hierarchy, in the registers. Data at one level of the memory hierarchy can move to adjacent levels—for example, moving between main memory and disk. The speed at which data move is high near the top of the hierarchy (between registers and cache) and slow near the bottom (between and disk and main memory). In particular, the speed at which arithmetic is done is much faster than the speed at which data is transferred between lower levels in the memory hierarchy, by factors of 10s or even 10000s, depending on the level. This means that an ill-designed algorithm may spend most of its time moving data from the bottom of the memory hierarchy to the registers in order to perform useful work rather than actually doing the work.

Here is an example of a simple algorithm which unfortunately cannot avoid spending most of its time moving data rather than doing useful arithmetic. Suppose that we want to add two large n-by-n matrices, large enough so that they fit only in a large, slow level of the memory hierarchy. To add them, they must be transferred a piece at a time up to the registers to do the additions, and the sums must be transferred back down. Thus, there are exactly 3 memory transfers between fast and slow memory (reading 2 summands into fast memory and writing 1 sum back to slow memory) for every addition performed. If the time to do a floating point operation is t_{arith} seconds and the time to move a word of data between memory levels is t_{mem} seconds, where $t_{\text{mem}} \gg t_{\text{arith}}$, then the execution time of this algorithm is $n^2(t_{\text{arith}} + 3t_{\text{mem}})$, which is much larger than than the time $n^2 t_{\text{arith}}$ required for the arithmetic alone. This means that matrix addition is doomed to run at the speed of the slowest level of memory in which the matrices reside, rather than the much higher speed of addition. In contrast, we will see later that other operations, such as matrix-matrix multiplication, can be made to run at the speed of the fastest level of the memory, even if the data are originally stored in the slowest.

LINPACK's Cholesky routine runs so slowly because it was *not* designed to minimize memory movement on machines such as the Cray YMP.[13] In contrast, matrix-matrix multiplication and the three other basic linear algebra

---

[13]It was designed to reduce another kind of memory movement, *page faults* between main memory and disk.

algorithms measured in the table were specialized to minimize data movement on a Cray YMP.

2.6.1. Basic Linear Algebra Subroutines (BLAS)

Since it is not cost-effective to write a special version of every routine like Cholesky for every new computer, we need a more systematic approach. Since operations like matrix-matrix multiplication are so common, computer manufacturers have standardized them as the *Basic Linear Algebra Subroutines*, or *BLAS* [169, 89, 87], and optimized them for their machines. In other words, a library of subroutines for matrix-matrix multiplication, matrix-vector multiplication, and other similar operations is available with a standard Fortran or C interface on high performance machines (and many others), but underneath they have been optimized for each machine. Our goal is to take advantage of these optimized BLAS by reorganizing algorithms like Cholesky so that they call the BLAS to perform most of their work.

In this section we will discuss the BLAS in general. In section 2.6.2, we will describe how to optimize matrix multiplication in particular. Finally, in section 2.6.3, we show how to reorganize Gaussian elimination so that most of its work is performed using matrix multiplication.

Let us examine the BLAS more carefully. Table 2.1 counts the number of memory references and floating points operations performed by three related BLAS. For example, the number of memory references needed to implement the `saxpy` operation in line 1 of the table is $3n + 1$, because we need to read n values of x_i, n values of y_i, and 1 value of α from slow memory to registers, and then write n values of y_i back to slow memory. The last column gives the ratio q of flops to memory references (its highest-order term in n only).

The significance of q is that it tells us roughly how many flops that we can perform per memory reference or how much useful work we can do compared to the time moving data. This tells us how fast the algorithm can *potentially* run. For example, suppose that an algorithm performs f floating points operations, each of which takes t_{arith} seconds, and m memory references, each of which takes t_{mem} seconds. Then the total running time is as large as

$$f \cdot t_{\text{arith}} + m \cdot t_{\text{mem}} = f \cdot t_{\text{arith}} \cdot \left(1 + \frac{m}{f} \frac{t_{\text{mem}}}{t_{\text{arith}}} \right) = f \cdot t_{\text{arith}} \cdot \left(1 + \frac{1}{q} \frac{t_{\text{mem}}}{t_{\text{arith}}} \right),$$

assuming that the arithmetic and memory references are not performed in parallel. Therefore, the larger the value of q, the closer the running time is to the best possible running time $f \cdot t_{\text{arith}}$, which is how long the algorithm would take if all data were in registers. This means that algorithms with the larger q values are better building blocks for other algorithms.

Table 2.1 reflects a hierarchy of operations: Operations such as `saxpy` perform $O(n^1)$ flops on vectors and offer the worst q values; these are called Level 1 BLAS, or BLAS1 [169], and include inner products, multiplying a

Operation	Definition	f	m	$q = f/m$
saxpy (BLAS1)	$y = \alpha \cdot x + y$ or $y_i = \alpha x_i + y_i$ $i = 1, \ldots, n$	$2n$	$3n + 1$	$2/3$
Matrix-vector mult (BLAS2)	$y = A \cdot x + y$ or $y_i = \sum_{j=1}^{n} a_{ij} x_j + y_i$ $i = 1, \ldots, n$	$2n^2$	$n^2 + 3n$	2
Matrix-matrix mult (BLAS3)	$C = A \cdot B + C$ or $c_{ij} = \sum_{k=1}^{n} a_{ik} b_{jk} + c_{ij}$ $i, j = 1, \ldots, n$	$2n^3$	$4n^2$	$n/2$

Table 2.1. *Counting floating point operations and memory references for the BLAS. f is the number of floating point operations, and m is the number of memory references.*

scalar times a vector and other simple operations. Operations such as matrix-vector multiplication perform $O(n^2)$ flops on matrices and vectors and offer slightly better q values; these are called Level 2 BLAS, or BLAS2 [89, 88], and include solving triangular systems of equations and rank-1 updates of matrices $(A + xy^T$, x and y column vectors). Operations such as matrix-matrix multiplication perform $O(n^3)$ flops on pairs of matrices and offer the best q values; these are called Level 3 BLAS, or BLAS3 [87, 86], and include solving triangular systems of equations with many right-hand sides.

The directory NETLIB/blas includes documentation and (unoptimized) implementations of all the BLAS. For a quick summary of all the BLAS, see NETLIB/blas/blasqr.ps. This summary also appears in [10, App. C] (or NETLIB/lapack/lug/lapack_lug.html).

Since the Level 3 BLAS have the highest q values, we endeavor to reorganize our algorithms in terms of operations such as matrix-matrix multiplication rather than saxpy or matrix-vector multiplication. (LINPACK's Cholesky is constructed in terms of calls to saxpy.) We emphasize that such reorganized algorithms will only be faster when using BLAS that have been optimized.

2.6.2. How to Optimize Matrix Multiplication

Let us examine in detail how to implement matrix multiplication $C = A \cdot B + C$ to minimize the number of memory moves and so optimize its performance. We will see that the performance is sensitive to the implementation details. To simplify our discussion, we will use the following machine model. We assume that matrices are stored columnwise, as in Fortran. (It is easy to modify the examples below if matrices are stored rowwise as in C.) We assume that there are two levels of memory hierarchy, fast and slow, where the slow memory is large enough to contain the three $n \times n$ matrices A, B, and C, but the fast memory contains only M words where $2n < M \ll n^2$; this means that the fast memory is large enough to hold two matrix columns or rows but not a whole matrix. We further assume that the data movement is under programmer control. (In practice, data movement may be done automatically by hardware, such as the cache controller. Nonetheless, the basic optimization scheme remains the same.)

The simplest matrix-multiplication algorithm that one might try consists of three nested loops, which we have annotated to indicate the data movements.

ALGORITHM 2.6. *Unblocked matrix multiplication (annotated to indicate memory activity):*

> *for i = 1 to n*
>   { *Read row i of A into fast memory* }
>   *for j = 1 to n*
>   { *Read C_{ij} into fast memory* }
>   { *Read column j of B into fast memory* }
>   *for k = 1 to n*
> $C_{ij} = C_{ij} + A_{ik} \cdot B_{kj}$
>   *end for*
>   { *Write C_{ij} back to slow memory* }
>   *end for*
> *end for*

The innermost loop is doing a dot product of row i of A and column j of B to compute C_{ij}, as shown in the following figure:

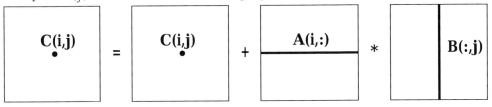

One can also describe the two innermost loops (on j and k) as doing a vector-matrix multiplication of the ith row of A times the matrix B to get the ith row of C. This is a hint that we will not perform any better than these BLAS1 and BLAS2 operations, since they are within the innermost loops.

Here is the detailed count of memory references: n^3 for reading B n times (once for each value of i); n^2 for reading A one row at a time and keeping it in fast memory until it is no longer needed; and $2n^2$ for reading one entry of C at a time, keeping it in fast memory until it is completely computed, and then moving it back to slow memory. This comes to $n^3 + 3n^2$ memory moves, or $q = 2n^3/(n^2+3n^2) \approx 2$, which is no better than the Level 2 BLAS and far from the maximum possible $n/2$ (see Table 2.1). If $M \ll n$, so that we cannot keep a full row of A in fast memory, q further decreases to 1, since the algorithm reduces to a sequence of inner products, which are Level 1 BLAS. For every permutation of the three loops on i, j, and k, one gets another algorithm with q about the same.

Our preferred algorithm uses *blocking*, where C is broken into an $N \times N$ block matrix with $n/N \times n/N$ blocks C^{ij}, and A and B are similarly partitioned, as shown below for $N = 4$. The algorithm becomes

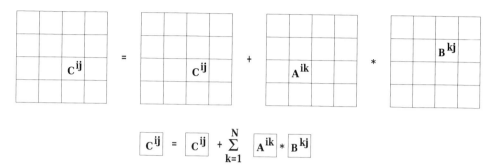

ALGORITHM 2.7. *Blocked matrix multiplication (annotated to indicate memory activity):*

for $i = 1$ to N
    for $j = 1$ to N
    { Read C^{ij} into fast memory }
        for $k = 1$ to N
            { Read A^{ik} into fast memory }
            { Read B^{kj} into fast memory }
            $C^{ij} = C^{ij} + A^{ik} \cdot B^{kj}$
        end for
    { Write C^{ij} back to slow memory }
    end for
end for

Our memory reference count is as follows: $2n^2$ for reading and writing each block of C once, Nn^2 for reading A N times (reading each n/N-by-n/N submatrix A^{ik} N^3 times), and Nn^2 for reading B N times (reading each n/N-by-n/N submatrix B^{kj} N^3 times), for a total of $(2N + 2)n^2 \approx 2Nn^2$ memory references. So we want to choose N as small as possible to minimize the number of memory references. But N is subject to the constraint $M \geq 3(n/N)^2$, which means that one block each from A, B, and C must fit in fast memory simultaneously. This yields $N \approx n\sqrt{3/M}$, and so $q \approx (2n^3)/(2Nn^2) \approx \sqrt{M/3}$, which is much better than the previous algorithm. In particular q grows independently of n as M grows, which means that we expect the algorithm to be fast for any matrix size n and to go faster if the fast memory size M is increased. These are both attractive properties.

In fact, it can be shown that Algorithm 2.7 is asymptotically optimal [151]. In other words, no reorganization of matrix-matrix multiplication (that performs the same $2n^3$ arithmetic operations) can have a q larger than $O(\sqrt{M})$.

On the other hand, this brief analysis ignores a number of practical issues:

1. A real code will have to deal with nonsquare matrices, for which the optimal block sizes may not be square.

2. The cache and register structure of a machine will strongly affect the best shapes of submatrices.

3. There may be special hardware instructions that perform both a multiplication and an addition in one cycle. It may also be possible to execute several multiply-add operations simultaneously if they do not interfere.

For a detailed discussion of these issues for one high-performance workstation, the IBM RS6000/590, see [1], PARALLEL_HOMEPAGE, or http://www.rs6000.ibm.com/resource/technology/essl.html. Figure 2.5 shows the speeds of the three basic BLAS for this machine. The horizontal axis is matrix size, and the vertical axis is speed in Mflops. The peak machine speed is 266 Mflops. The top curve (peaking near 250 Mflops) is square matrix-matrix multiplication. The middle curve (peaking near 100 Mflops) is square matrix-vector multiplication, and the bottom curve (peaking near 75 Mflops) is saxpy. Note that the speed increases for larger matrices. This is a common phenomenon and means that we will try to develop algorithms whose internal matrix-multiplications use as large matrices as reasonable.

Both the above matrix-matrix multiplication algorithms perform $2n^3$ arithmetic operations. It turns out that there are other implementations of matrix-matrix multiplication that use far fewer operations. Strassen's method [3] was the first of these algorithms to be discovered and is the simplest to explain. This algorithm multiplies matrices recursively by dividing them into 2×2 block matrices and multiplying the subblocks using seven matrix multiplications (recursively) and 18 matrix additions of half the size; this leads to an asymptotic complexity of $n^{\log_2 7} \approx n^{2.81}$ instead of n^3.

ALGORITHM 2.8. *Strassen's matrix multiplication algorithm:*

$C = Strassen(A,B,n)$
    /* Return $C = A * B$, where A and B are n-by-n;
        Assume n is a power of 2 */
    if $n = 1$
        return $C = A * B$ */* scalar multiplication */*
    else

$$\text{Partition } A = \begin{bmatrix} A_{11} & A_{12} \\ A_{21} & A_{22} \end{bmatrix} \text{ and } B = \begin{bmatrix} B_{11} & B_{12} \\ B_{21} & B_{22} \end{bmatrix}$$

where the subblocks A_{ij} and B_{ij} are $n/2$-by-$n/2$

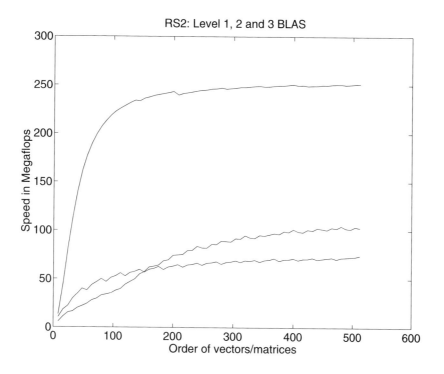

Fig. 2.5. *BLAS speed on the IBM RS* 6000/590.

$$P_1 = Strassen(\ A_{12} - A_{22},\ B_{21} + B_{22},\ n/2\ )$$
$$P_2 = Strassen(\ A_{11} + A_{22},\ B_{11} + B_{22},\ n/2\ )$$
$$P_3 = Strassen(\ A_{11} - A_{21},\ B_{11} + B_{12},\ n/2\ )$$
$$P_4 = Strassen(\ A_{11} + A_{12},\ B_{22},\ n/2\ )$$
$$P_5 = Strassen(\ A_{11},\ B_{12} - B_{22},\ n/2\ )$$
$$P_6 = Strassen(\ A_{22},\ B_{21} - B_{11},\ n/2\ )$$
$$P_7 = Strassen(\ A_{21} + A_{22},\ B_{11},\ n/2\ )$$
$$C_{11} = P_1 + P_2 - P_4 + P_6$$
$$C_{12} = P_4 + P_5$$
$$C_{21} = P_6 + P_7$$
$$C_{22} = P_2 - P_3 + P_5 - P_7$$
$$return\ C = \begin{bmatrix} C_{11} & C_{12} \\ C_{21} & C_{22} \end{bmatrix}$$
$$end\ if$$

It is tedious but straightforward to confirm by induction that this algorithm multiplies matrices correctly (see Question 2.21). To show that its complexity is $O(n^{\log_2 7})$, we let $T(n)$ be the number of additions, subtractions, and multiplications performed by the algorithm. Since the algorithm performs seven recursive calls on matrices of size $n/2$, and 18 additions of $n/2$-by-$n/2$ matrices, we can write down the recurrence $T(n) = 7T(n/2) + 18(n/2)^2$. Changing variables from n to $m = \log_2 n$, we get a new recurrence $\bar{T}(m) = 7\bar{T}(m-1) + 18(2^{m-1})^2$, where $\bar{T}(m) = T(2^m)$. We can confirm that this linear recurrence for $\bar{T}$ has a solution $\bar{T}(m) = O(7^m) = O(n^{\log_2 7})$.

The value of Strassen's algorithm is not just this asymptotic complexity but its reduction of the problem to smaller subproblems which eventually fit in fast memory; once the subproblems fit in fast memory, standard matrix multiplication may be used. This approach has led to speedups on relatively large matrices on some machines [22]. A drawback is the need for significant workspace and somewhat lower numerical stability, although it is adequate for many purposes [77]. There are a number of other even faster matrix multiplication algorithms; the current record is about $O(n^{2.376})$, due to Winograd and Coppersmith [263]. But these algorithms only perform fewer operations than Strassen for impractically large values of n. For a survey see [195].

2.6.3. Reorganizing Gaussian Elimination to Use Level 3 BLAS

We will reorganize Gaussian elimination to use, first, the Level 2 BLAS and, then, the Level 3 BLAS. For simplicity, we assume that no pivoting is necessary.

Indeed, Algorithm 2.4 is already a Level 2 BLAS algorithm, because most of the work is done in the second line, $A(i + 1 : n, i + 1 : n) = A(i + 1 : n, i + 1 : n) - A(i + 1 : n, i) * A(i, i + 1 : n)$, which is a *rank-1 update* of the submatrix $A(i + 1 : n, i + 1 : n)$. The other arithmetic in the algorithm, $A(i + 1 : n, i) = A(i + 1 : n, i)/A(i, i)$, is actually done by multiplying the vector $A(i + 1 : n, i)$ by the scalar $1/A(i, i)$, since multiplication is much faster than division; this is also a Level 1 BLAS operation. We need to modify Algorithm 2.4 slightly because we will use it within the Level 3 version.

ALGORITHM 2.9. *Level 2 BLAS implementation of LU factorization without pivoting for an m-by-n matrix A, where $m \geq n$: Overwrite A by the m-by-n matrix L and m-by-m matrix U. We have numbered the important lines for later reference.*

$\quad$ *for $i = 1$ to $\min(m - 1, n)$*
(1)$\quad$ $A(i + 1 : m, i) = A(i + 1 : m, i)/A(i, i)$
$\quad$ *if $i < n$*
(2)$\quad\quad$ $A(i + 1 : m, i + 1 : n) = A(i + 1 : m, i + 1 : n) -$
$\quad\quad\quad$ $A(i + 1 : m, i) \cdot A(i, i + 1 : n)$
$\quad$ *end for*

The left side of Figure 2.6 illustrates Algorithm 2.9 applied to a square matrix. At step i of the algorithm, columns 1 to $i - 1$ of L and rows 1 to $i - 1$ of U are already done, column i of L and row i of U are to be computed, and the trailing submatrix of A is to be updated by a rank-1 update. On the left side of Figure 2.6, the submatrices are labeled by the lines of the algorithm ((1) or (2)) that update them. The rank-1 update in line (2) is to subtract the

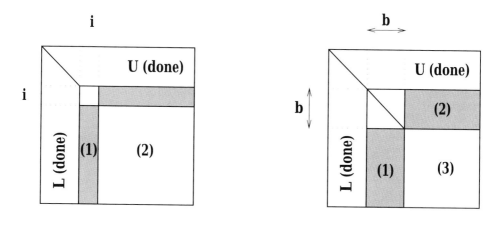

Step i of Level 2 BLAS
Implementation of LU

Step i of Level 3 BLAS
Implementation of LU

Fig. 2.6. *Level 2 and Level 3 BLAS implementations of LU factorization.*

product of the shaded column and the shaded row from the submatrix labeled (2).

The Level 3 BLAS algorithm will reorganize this computation by *delaying the update* of submatrix (2) for b steps, where b is a small integer called the *block size*, and later applying b rank-1 updates all at once in a single matrix-matrix multiplication. To see how to do this, suppose that we have already computed the first $i - 1$ columns of L and rows of U, yielding

$$
A = \begin{matrix} i-1 \\ b \\ n-b-i+1 \end{matrix} \overset{\begin{matrix} i-1 & b & n-b-i+1 \end{matrix}}{\begin{pmatrix} A_{11} & A_{12} & A_{13} \\ A_{21} & A_{22} & A_{23} \\ A_{31} & A_{32} & A_{33} \end{pmatrix}}
$$

$$
= \begin{bmatrix} L_{11} & 0 & 0 \\ L_{21} & I & 0 \\ L_{31} & 0 & I \end{bmatrix} \cdot \begin{bmatrix} U_{11} & U_{21} & U_{31} \\ 0 & \tilde{A}_{22} & \tilde{A}_{23} \\ 0 & \tilde{A}_{32} & \tilde{A}_{33} \end{bmatrix},
$$

where all the matrices are partitioned the same way. This is shown on the right side of Figure 2.6. Now apply Algorithm 2.9 to the submatrix $\left[\begin{smallmatrix} \tilde{A}_{22} \\ \tilde{A}_{32} \end{smallmatrix} \right]$ to get

$$
\begin{bmatrix} \tilde{A}_{22} \\ \tilde{A}_{32} \end{bmatrix} = \begin{bmatrix} L_{22} \\ L_{32} \end{bmatrix} \cdot U_{22} = \begin{bmatrix} L_{22}U_{22} \\ L_{32}U_{22} \end{bmatrix}.
$$

This lets us write

$$
\begin{bmatrix} \tilde{A}_{22} & \tilde{A}_{23} \\ \tilde{A}_{32} & \tilde{A}_{33} \end{bmatrix} = \begin{bmatrix} L_{22}U_{22} & \tilde{A}_{23} \\ L_{32}U_{22} & \tilde{A}_{33} \end{bmatrix}
$$

$$= \begin{bmatrix} L_{22} & 0 \\ L_{32} & I \end{bmatrix} \cdot \begin{bmatrix} U_{22} & L_{22}^{-1}\tilde{A}_{23} \\ 0 & \tilde{A}_{33} - L_{32} \cdot (L_{22}^{-1}\tilde{A}_{23}) \end{bmatrix}$$

$$\equiv \begin{bmatrix} L_{22} & 0 \\ L_{32} & I \end{bmatrix} \cdot \begin{bmatrix} U_{22} & U_{23} \\ 0 & \tilde{A}_{33} - L_{32} \cdot U_{23} \end{bmatrix}$$

$$\equiv \begin{bmatrix} L_{22} & 0 \\ L_{32} & I \end{bmatrix} \cdot \begin{bmatrix} U_{22} & U_{23} \\ 0 & \tilde{\tilde{A}}_{33} \end{bmatrix}.$$

Altogether, we get an updated factorization with b more columns of L and rows of U completed:

$$\begin{bmatrix} A_{11} & A_{12} & A_{13} \\ A_{21} & A_{22} & A_{23} \\ A_{31} & A_{32} & A_{33} \end{bmatrix} = \begin{bmatrix} L_{11} & 0 & 0 \\ L_{21} & L_{22} & 0 \\ L_{31} & L_{23} & I \end{bmatrix} \cdot \begin{bmatrix} U_{11} & U_{21} & U_{31} \\ 0 & U_{22} & U_{23} \\ 0 & 0 & \tilde{\tilde{A}}_{33} \end{bmatrix}.$$

This defines an algorithm with the following three steps, which are illustrated on the right of Figure 2.6:

(1) Use Algorithm 2.9 to factorize $\begin{bmatrix} \tilde{A}_{22} \\ \tilde{A}_{32} \end{bmatrix} = \begin{bmatrix} L_{22} \\ L_{32} \end{bmatrix} \cdot U_{22}$.

(2) Form $U_{23} = L_{22}^{-1}\tilde{A}_{23}$. This means solving a triangular linear system with many right-hand sides ($\tilde{A}_{23}$), a single Level 3 BLAS operation.

(3) Form $\tilde{\tilde{A}}_{33} = \tilde{A}_{33} - L_{32} \cdot U_{23}$, a matrix-matrix multiplication.

More formally, we have the following algorithm.

ALGORITHM 2.10. *Level 3 BLAS implementation of LU factorization without pivoting for an n-by-n matrix A. Overwrite L and U on A. The lines of the algorithm are numbered as above and to correspond to the right part of Figure 2.6.*

> *for $i = 1$ to $n - 1$ step b*
>
> (1)   *Use Algorithm 2.9 to factorize $A(i:n, i:i+b-1) = \begin{bmatrix} L_{22} \\ L_{32} \end{bmatrix} U_{22}$*
>
> (2) $A(i:i+b-1, i+b:n) = L_{22}^{-1} \cdot A(i:i+b-1, i+b:n)$
>   /* form U_{23} */
>
> (3) $A(i+b:n, i+b:n) = A(i+b:n, i+b:n)$
> $- A(i+b:n, i:i+b-1) \cdot A(i:i+b-1, i+b:n)$
>   /* form $\tilde{\tilde{A}}_{33}$ */
>
> *end for*

We still need to choose the block size b in order to maximize the speed of the algorithm. On the one hand, we would like to make b large because we have seen that speed increases when multiplying larger matrices. On the other hand, we can verify that the number of floating point operations performed by the slower Level 2 and Level 1 BLAS in line (1) of the algorithm is about $n^2b/2$ for small b, which grows as b grows, so we do not want to pick b too large. The optimal value of b is machine dependent and can be tuned for each machine. Values of $b = 32$ or $b = 64$ are commonly used.

To see detailed implementations of Algorithms 2.9 and 2.10, see subroutines `sgetf2` and `sgetrf`, respectively, in LAPACK (NETLIB/lapack). For more information on block algorithms, including detailed performance number on a variety of machines, see also [10] or the course notes at PARAL-LEL_HOMEPAGE.

2.6.4. More About Parallelism and Other Performance Issues

In this section we briefly survey other issues involved in implementing Gaussian elimination (and other linear algebra routines) as efficiently as possible.

A *parallel computer* contains $p > 1$ processors capable of simultaneously working on the same problem. One may hope to solve any given problem p times faster on such a machine than on a conventional uniprocessor. But such "perfect efficiency" is rarely achieved, even if there are always at least p independent tasks available to do, because of the overhead of coordinating p processors and the cost of sending data from the processor that may store it to the processor that needs it. This last problem is another example of a *memory hierarchy*: from the point of view of processor i, its own memory is fast, but getting data from the memory owned by processor j is slower, sometimes thousands of times slower.

Gaussian elimination offers many opportunities for parallelism, since each entry of the trailing submatrix may be updated independently and in parallel at each step. But some care is needed to be as efficient as possible. Two standard pieces of software are available. The LAPACK routine `sgetrf` described in the last section [10] runs on *shared-memory parallel machines*, provided that one has available implementations of the BLAS that run in parallel. A related library called ScaLAPACK, for *Scalable LAPACK* [34, 53], is designed for *distributed-memory parallel machines*, i.e., those that require special operations to move data between different processors. All software is available on NETLIB in the LAPACK and ScaLAPACK subdirectories. ScaLAPACK is described in more detail in the notes at PARALLEL_HOMEPAGE. Extensive performance data for linear equation solvers are available as the LINPACK Benchmark [85], with an up-to-date version available at NETLIB/benchmark/performance.ps, or in the Performance Database Server.[14] As of May 1997, the fastest that any linear system had been solved using Gaussian elimination was one with $n = 215000$ on an Intel ASCI Option Red with $p = 7264$ processors; the problem ran at just over 1068 Gflops (gigaflops), out of a maximum 1453 Gflops.

---

[14]http://performance.netlib.org/performance/html/PDStop.html

There are some matrices too large to fit in the main memory of any available machine. These matrices are stored on disk and must be read into main memory piece by piece in order to perform Gaussian elimination. The organization of such routines is largely similar to the technique described above, and they are included in ScaLAPACK.

Finally, one might hope that compilers would become sufficiently clever to take the simplest implementation of Gaussian elimination using three nested loops and automatically "optimize" the code to look like the blocked algorithm discussed in the last subsection. While there is much current research on this topic (see the bibliography in the recent compiler textbook [264]), there is still no reliably fast alternative to optimized libraries such as LAPACK and ScaLAPACK.

2.7. Special Linear Systems

As mentioned in section 1.2, it is important to exploit any special structure of the matrix to increase speed of solution and decrease storage. In practice, of course, the cost of the extra programming effort required to exploit this structure must be taken into account. For example, if our only goal is to minimize the time to get the desired solution, and it takes an extra week of programming effort to decrease the solution time from 10 seconds to 1 second, it is worth doing only if we are going to use the routine more than (1 week * 7 days/week * 24 hours/day * 3600 seconds/hour) / (10 seconds − 1 second) = 67200 times. Fortunately, there are some special structures that turn up frequently enough that standard solutions exist, and we should certainly use them. The ones we consider here are

1. s.p.d. matrices,

2. symmetric indefinite matrices,

3. band matrices,

4. general sparse matrices,

5. dense matrices depending on fewer than n^2 independent parameters.

We will consider only real matrices; extensions to complex matrices are straightforward.

2.7.1. Real Symmetric Positive Definite Matrices

Recall that a real matrix A is s.p.d. if and only if $A = A^T$ and $x^T A x > 0$ for all $x \neq 0$. In this section we will show how to solve $Ax = b$ in half the time and half the space of Gaussian elimination when A is s.p.d.

PROPOSITION 2.2.   1. *If X is nonsingular, then A is s.p.d. if and only if $X^T A X$ is s.p.d.*

2. *If A is s.p.d. and H is any principal submatrix of A ($H = A(j : k, j : k)$ for some $j \leq k$), then H is s.p.d.*

3. *A is s.p.d. if and only if $A = A^T$ and all its eigenvalues are positive.*

4. *If A is s.p.d., then all $a_{ii} > 0$, and $\max_{ij} |a_{ij}| = \max_i a_{ii} > 0$.*

5. *A is s.p.d. if and only if there is a unique lower triangular nonsingular matrix L, with positive diagonal entries, such that $A = LL^T$. $A = LL^T$ is called the* Cholesky factorization *of A, and L is called the* Cholesky factor *of A.*

Proof.

1. X nonsingular implies $Xx \neq 0$ for all $x \neq 0$, so $x^T X^T A X x > 0$ for all $x \neq 0$. So A s.p.d. implies $X^T A X$ is s.p.d. Use X^{-1} to deduce the other implication.

2. Suppose first that $H = A(1 : m, 1 : m)$. Then given any m-vector y, the n-vector $x = [y^T, 0]^T$ satisfies $y^T H y = x^T A x$. So if $x^T A x > 0$ for all nonzero x, then $y^T H y > 0$ for all nonzero y, and so H is s.p.d. If H does not lie in the upper left corner of A, let P be a permutation so that H does lie in the upper left corner of $P^T A P$ and apply Part 1.

3. Let X be the real, orthogonal eigenvector matrix of A so that $X^T A X = \Lambda$ is the diagonal matrix of real eigenvalues λ_i. Since $x^T \Lambda x = \sum_i \lambda_i x_i^2$, Λ is s.p.d if and only if each $\lambda_i > 0$. Now apply Part 1.

4. Let e_i be the ith column of the identity matrix. Then $e_i^T A e_i = a_{ii} > 0$ for all i. If $|a_{kl}| = \max_{ij} |a_{ij}|$ but $k \neq l$, choose $x = e_k - \text{sign}(a_{kl}) e_l$. Then $x^T A x = a_{kk} + a_{ll} - 2|a_{kl}| \leq 0$, contradicting positive-definiteness.

5. Suppose $A = LL^T$ with L nonsingular. Then $x^T A x = (x^T L)(L^T x) = \|L^T x\|_2^2 > 0$ for all $x \neq 0$, so A is s.p.d. If A is s.p.d., we show that L exists by induction on the dimension n. If we choose each $l_{ii} > 0$, our construction will determine L uniquely. If $n = 1$, choose $l_{11} = \sqrt{a_{11}}$, which exists since $a_{11} > 0$. As with Gaussian elimination, it suffices to understand the block 2-by-2 case. Write

$$
\begin{aligned}
A &= \begin{bmatrix} a_{11} & A_{12} \\ A_{12}^T & A_{22} \end{bmatrix} \\
&= \begin{bmatrix} \sqrt{a_{11}} & 0 \\ \frac{A_{12}^T}{\sqrt{a_{11}}} & I \end{bmatrix} \begin{bmatrix} 1 & 0 \\ 0 & \tilde{A}_{22} \end{bmatrix} \begin{bmatrix} \sqrt{a_{11}} & \frac{A_{12}}{\sqrt{a_{11}}} \\ 0 & I \end{bmatrix}
\end{aligned}
$$

$$= \begin{bmatrix} a_{11} & A_{12} \\ A_{12}^T & \tilde{A}_{22} + \frac{A_{12}^T A_{12}}{a_{11}} \end{bmatrix},$$

so the $(n-1)$-by-$(n-1)$ matrix $\tilde{A}_{22} = A_{22} - \frac{A_{12}^T A_{12}}{a_{11}}$ is symmetric.

By Part 1 above, $\begin{bmatrix} 1 & 0 \\ 0 & \tilde{A}_{22} \end{bmatrix}$ is s.p.d, so by Part 2 $\tilde{A}_{22}$ is s.p.d.

Thus by induction there exists an $\tilde{L}$ such that $\tilde{A}_{22} = \tilde{L}\tilde{L}^T$ and

$$\begin{aligned} A &= \begin{bmatrix} \sqrt{a_{11}} & 0 \\ \frac{A_{12}^T}{\sqrt{a_{11}}} & I \end{bmatrix} \begin{bmatrix} 1 & 0 \\ 0 & \tilde{L}\tilde{L}^T \end{bmatrix} \begin{bmatrix} \sqrt{a_{11}} & \frac{A_{12}}{\sqrt{a_{11}}} \\ 0 & I \end{bmatrix} \\ &= \begin{bmatrix} \sqrt{a_{11}} & 0 \\ \frac{A_{12}^T}{\sqrt{a_{11}}} & \tilde{L} \end{bmatrix} \begin{bmatrix} \sqrt{a_{11}} & \frac{A_{12}}{\sqrt{a_{11}}} \\ 0 & \tilde{L}^T \end{bmatrix} \equiv LL^T. \quad \square \end{aligned}$$

We may rewrite this induction as the following algorithm.

ALGORITHM 2.11. *Cholesky algorithm:*

for $j = 1$ to n
  $l_{jj} = (a_{jj} - \sum_{k=1}^{j-1} l_{jk}^2)^{1/2}$
  for $i = j+1$ to n
    $l_{ij} = (a_{ij} - \sum_{k=1}^{j-1} l_{ik}l_{jk})/l_{jj}$
  end for
end for

If A is not positive definite, then (in exact arithmetic) this algorithm will fail by attempting to compute the square root of a negative number or by dividing by zero; this is the cheapest way to test if a symmetric matrix is positive definite.

As with Gaussian elimination, L can overwrite the lower half of A. Only the lower half of A is referred to by the algorithm, so in fact only $n(n+1)/2$ storage is needed instead of n^2. The number of flops is

$$\sum_{j=1}^{n} \left(2j + \sum_{i=j+1}^{n} 2j \right) = \frac{1}{3}n^3 + O(n^2),$$

or just half the flops of Gaussian elimination. Just as with Gaussian elimination, Cholesky may be reorganized to perform most of its floating point operations using Level 3 BLAS; see LAPACK routine `spotrf`.

Pivoting is not necessary for Cholesky to be numerically stable (equivalently, we could also say any diagonal pivot order is numerically stable). We show this as follows. The same analysis as for Gaussian elimination in section 2.4.2 shows that the computed solution $\hat{x}$ satisfies $(A + \delta A)\hat{x} = b$ with

$|\delta A| \le 3n\varepsilon|L| \cdot |L^T|$. But by the Cauchy–Schwartz inequality and Part 4 of Proposition 2.2

$$(|L| \cdot |L^T|)_{ij} = \sum_k |l_{ik}| \cdot |l_{jk}|$$

$$\le \sqrt{\sum l_{ik}^2}\sqrt{\sum l_{jk}^2}$$

$$= \sqrt{a_{ii}} \cdot \sqrt{a_{jj}}$$

$$\le \max_{ij}|a_{ij}|, \tag{2.16}$$

so $\| \, |L| \cdot |L^T| \, \|_\infty \le n\|A\|_\infty$ and $\|\delta A\|_\infty \le 3n^2\varepsilon\|A\|_\infty$.

2.7.2. Symmetric Indefinite Matrices

The question of whether we can still save half the time and half the space when solving a symmetric but indefinite (neither positive definite nor negative definite) linear system naturally arises. It turns out to be possible, but a more complicated pivoting scheme and factorization is required. If A is nonsingular, one can show that there exists a permutation P, a unit lower triangular matrix L, and a block diagonal matrix D with 1-by-1 and 2-by-2 blocks such that $PAP^T = LDL^T$. To see why 2-by-2 blocks are needed in D, consider the matrix $\begin{bmatrix} 0 & 1 \\ 1 & 0 \end{bmatrix}$. This factorization can be computed stably, saving about half the work and space compared to standard Gaussian elimination. The name of the LAPACK subroutine which does this operation is `ssysv`. The algorithm is described in [44].

2.7.3. Band Matrices

A matrix A is called a *band matrix* with *lower bandwidth* b_L and *upper bandwidth* b_U if $a_{ij} = 0$ whenever $i > j + b_L$ or $i < j - b_U$:

$$A = \begin{bmatrix} a_{11} & \cdots & a_{1,b_U+1} & & & 0 \\ \vdots & & & a_{2,b_U+2} & & \\ a_{b_L+1,1} & & & & \ddots & \\ & a_{b_L+2,2} & & & & a_{n-b_U,n} \\ & & \ddots & & & \vdots \\ 0 & & & a_{n,n-b_L} & \cdots & a_{n,n} \end{bmatrix}.$$

Band matrices arise often in practice (we give an example later) and are useful to recognize because their L and U factors are also "essentially banded," making them cheaper to compute and store. We explain what we mean by "essentially banded" below. But first, we consider LU factorization without pivoting and show that L and U are banded in the usual sense, with the same bandwidths as A.

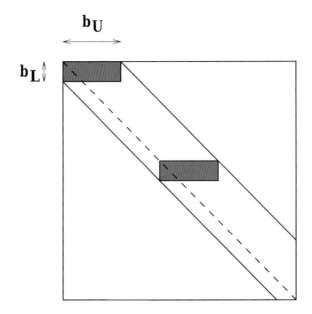

Fig. 2.7. *Band LU factorization without pivoting.*

PROPOSITION 2.3. *Let A be banded with lower bandwidth b_L and upper bandwidth b_U. Let $A = LU$ be computed without pivoting. Then L has lower bandwidth b_L and U has upper bandwidth b_U. L and U can be computed in about $2n \cdot b_U \cdot b_L$ arithmetic operations when b_U and b_L are small compared to n. The space needed is $(b_L + b_U + 1)$. The full cost of solving $Ax = b$ is $2nb_U \cdot b_L + 2nb_U + 2nb_L$.*

Sketch of Proof.   It suffices to look at one step; see Figure 2.7. At step j of Gaussian elimination, the shaded region is modified by subtracting the product of the first column and first row of the shaded region; note that this does not enlarge the bandwidth. □

PROPOSITION 2.4. *Let A be banded with lower bandwidth b_L and upper bandwidth b_U. Then after Gaussian elimination with partial pivoting, U is banded with upper bandwidth at most $b_L + b_U$, and L is "essentially banded" with lower bandwidth b_L. This means that L has at most $b_L + 1$ nonzeros in each column and so can be stored in the same space as a band matrix with lower bandwidth b_L.*

Sketch of Proof.   Again a picture of the region changed by one step of the algorithm illustrates the proof. As illustrated in Figure 2.8, pivoting can increase the upper bandwidth by at most b_L. Later permutations can reorder the entries of earlier columns so that entries of L may lie below subdiagonal b_L but no new nonzeros can be introduced, so the storage needed for L remains b_L per column. □

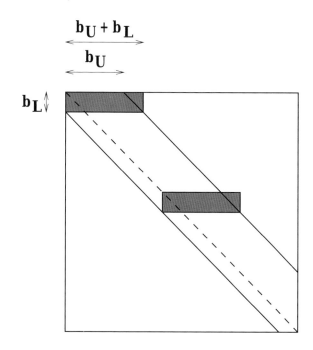

Fig. 2.8. *Band LU factorization with partial pivoting.*

Gaussian elimination and Cholesky for band matrices are available in LA-PACK routines like `ssbsv` and `sspsv`.

Band matrices often arise from discretizing physical problems with nearest neighbor interactions on a mesh (provided the unknowns are ordered rowwise or columnwise; see also Example 2.9 and section 6.3).

EXAMPLE 2.8. Consider the ordinary differential equation (ODE) $y''(x) - p(x)y'(x) - q(x)y(x) = r(x)$ on the interval $[a, b]$ with boundary conditions $y(a) = \alpha$, $y(b) = \beta$. We also assume $q(x) \geq q > 0$. This equation may be used to model the heat flow in a long, thin rod, for example. To solve the differential equation numerically, we *discretize* it by seeking its solution only at the evenly spaced mesh points $x_i = a + ih$, $i = 0, \ldots, N+1$, where $h = (b-a)/(N+1)$ is the mesh spacing. Define $p_i = p(x_i)$, $r_i = r(x_i)$, and $q_i = q(x_i)$. We need to derive equations to solve for our desired approximations $y_i \approx y(x_i)$, where $y_0 = \alpha$ and $y_{N+1} = \beta$. To derive these equations, we approximate the derivative $y'(x_i)$ by the following *finite difference approximation*:

$$y'(x_i) \approx \frac{y_{i+1} - y_{i-1}}{2h}.$$

(Note that as h gets smaller, the right-hand side approximates $y'(x_i)$ more and more accurately.) We can similarly approximate the second derivative by

$$y''(x_i) \approx \frac{y_{i+1} - 2y_i + y_{i-1}}{h^2}.$$

(See section 6.3.1 in Chapter 6 for a more detailed derivation.)

Inserting these approximations into the differential equation yields

$$\frac{y_{i+1} - 2y_i + y_{i-1}}{h^2} - p_i \frac{y_{i+1} - y_{i-1}}{2h} - q_i y_i = r_i, \quad 1 \le i \le N.$$

Rewriting this as a linear system we get $Ay = b$, where

$$y = \begin{bmatrix} y_1 \\ \vdots \\ y_N \end{bmatrix}, \quad b = \frac{-h^2}{2} \begin{bmatrix} r_1 \\ \vdots \\ r_N \end{bmatrix} + \begin{bmatrix} (\frac{1}{2} + \frac{h}{4}p_1)\alpha \\ 0 \\ \vdots \\ 0 \\ (\frac{1}{2} - \frac{h}{4}p_N)\beta \end{bmatrix},$$

and

$$A = \begin{bmatrix} a_1 & -c_1 & & & \\ -b_2 & \ddots & \ddots & & \\ & \ddots & \ddots & c_{N-1} \\ & & -b_N & a_N \end{bmatrix}, \quad \begin{aligned} a_i &= 1 + \frac{h^2}{2}q_i, \\ b_i &= \frac{1}{2}[1 + \frac{h}{2}p_i], \\ c_i &= \frac{1}{2}[1 - \frac{h}{2}p_i]. \end{aligned}$$

Note that $a_i > 0$, and also $b_i > 0$ and $c_i > 0$ if h is small enough.

This is a nonsymmetric *tridiagonal* system to solve for y. We will show how to change it to a symmetric positive definite tridiagonal system, so that we may use *band Cholesky* to solve it.

Choose $D = \mathrm{diag}(1, \sqrt{\frac{c_1}{b_2}}, \sqrt{\frac{c_1 c_2}{b_2 b_3}}, \ldots, \sqrt{\frac{c_1 c_2 \cdots c_{N-1}}{b_2 b_3 \cdots b_N}})$. Then we may change $Ay = b$ to $(DAD^{-1})(Dy) = Db$ or $\tilde{A}\tilde{y} = \tilde{b}$, where

$$\tilde{A} = \begin{bmatrix} a_1 & -\sqrt{c_1 b_2} & & & \\ -\sqrt{c_1 b_2} & a_2 & -\sqrt{c_2 b_3} & & \\ & -\sqrt{c_2 b_3} & \ddots & & \\ & & \ddots & \ddots & -\sqrt{c_{N-1} b_N} \\ & & & -\sqrt{c_{N-1} b_N} & a_N \end{bmatrix}.$$

It is easy to see that $\tilde{A}$ is symmetric, and it has the same eigenvalues as A because A and $\tilde{A} = DAD^{-1}$ are *similar*. (See section 4.2 in Chapter 4 for details.) We will use the next theorem to show it is also positive definite.

THEOREM 2.9. *Gershgorin. Let B be an arbitrary matrix. Then the eigenvalues λ of B are located in the union of the n disks*

$$|\lambda - b_{kk}| \le \sum_{j \ne k} |b_{kj}|.$$

Proof. Given λ and $x \neq 0$ such that $Bx = \lambda x$, let $1 = \|x\|_\infty = x_k$ by scaling x if necessary. Then $\sum_{j=1}^{N} b_{kj} x_j = \lambda x_k = \lambda$, so $\lambda - b_{kk} = \sum_{\substack{j=1 \\ j \neq k}}^{N} b_{kj} x_j$, implying

$$|\lambda - b_{kk}| \leq \sum_{j \neq k} |b_{kj} x_j| \leq \sum_{j \neq k} |b_{kj}|. \quad \square$$

Now if h is so small that for all i, $|\frac{h}{2} p_i| < 1$, then

$$|b_i| + |c_i| = \frac{1}{2}\left(1 + \frac{h}{2}p_i\right) + \frac{1}{2}\left(1 - \frac{h}{2}p_i\right) = 1 < 1 + \frac{h^2}{2}\underline{q} \leq 1 + \frac{h^2}{2}q_i = a_i.$$

Therefore all eigenvalues of A lie inside the disks centered at $1 + h^2 q_i/2 \geq 1 + h^2 \underline{q}/2$ with radius 1; in particular, they must all have positive real parts. Since $\tilde{A}$ is symmetric, its eigenvalues are real and hence positive, so $\tilde{A}$ is positive definite. Its smallest eigenvalue is bounded below by $\underline{q}h^2/2$. Thus, it can be solved by Cholesky. The LAPACK subroutine for solving a symmetric positive definite tridiagonal system is `sptsv`.

In section 4.3 we will again use Gershgorin's theorem to compute perturbation bounds for eigenvalues of matrices. $\diamond$

2.7.4. General Sparse Matrices

A sparse matrix is defined to be a matrix with a large number of zero entries. In practice, this means a matrix with enough zero entries that it is worth using an algorithm that avoids storing or operating on the zero entries. Chapter 6 is devoted to methods for solving sparse linear systems other than Gaussian elimination and its variants. There are a large number of sparse methods, and choosing the best one often requires substantial knowledge about the matrix [24]. In this section we will only sketch the basic issues in sparse Gaussian elimination and give pointers to the literature and available software.

To give a very simple example, consider the following matrix, which is ordered so that GEPP does not permute any rows:

$$A = \begin{bmatrix} 1 & & & & .1 \\ & 1 & & & .1 \\ & & 1 & & .1 \\ & & & 1 & .1 \\ .1 & .1 & .1 & .1 & 1 \end{bmatrix} = LU$$

$$= \begin{bmatrix} 1 & & & & \\ & 1 & & & \\ & & 1 & & \\ & & & 1 & \\ .1 & .1 & .1 & .1 & 1 \end{bmatrix} \cdot \begin{bmatrix} 1 & & & & .1 \\ & 1 & & & .1 \\ & & 1 & & .1 \\ & & & 1 & .1 \\ & & & & .96 \end{bmatrix}.$$

A is called an *arrow matrix* because of the pattern of its nonzero entries. Note that none of the zero entries of A were *filled in* by GEPP so that L and U together can be stored in the same space as the nonzero entries of A. Also, if we count the number of essential arithmetic operations, i.e., not multiplication by zero or adding zero, there are only 12 of them (4 divisions to compute the last row of L and 8 multiplications and additions to update the (5,5) entry), instead of $\frac{2}{3}n^3 \approx 83$. More generally, if A were an n-by-n arrow matrix, it would take only $3n - 2$ locations to store it instead of n^2, and $3n - 3$ floating point operations to perform Gaussian elimination instead of $\frac{2}{3}n^3$. When n is large, both the space and operation count become tiny compared to a dense matrix.

Suppose that instead of A we were given A', which is A with the order of its rows and columns reversed. This amounts to reversing the order of the equations and of the unknowns in the linear system $Ax = b$. GEPP applied to A' again permutes no rows, and to two decimal places we get

$$A' = \begin{bmatrix} 1 & .1 & .1 & .1 & .1 \\ .1 & 1 & & & \\ .1 & & 1 & & \\ .1 & & & 1 & \\ .1 & & & & 1 \end{bmatrix} = L'U'$$

$$= \begin{bmatrix} 1 & & & & \\ .1 & 1 & & & \\ .1 & -.01 & 1 & & \\ .1 & -.01 & -.01 & 1 & \\ .1 & -.01 & -.01 & -.01 & 1 \end{bmatrix} \cdot \begin{bmatrix} 1 & .1 & .1 & .1 & .1 \\ & .99 & -.01 & -.01 & -.01 \\ & & .99 & -.01 & -.01 \\ & & & .99 & -.01 \\ & & & & .99 \end{bmatrix}.$$

Now we see that L' and U' have filled in completely and require n^2 storage. Indeed, after the first step of the algorithm all the nonzeros of A' have filled in, so we must do the same work as dense Gaussian elimination, $\frac{2}{3}n^3$.

This illustrates that the order of the rows and columns is extremely important for saving storage and work. Even if we do not have to worry about pivoting for numerical stability (such as in Cholesky), choosing the optimal permutations of rows and columns to minimize storage or work is an extremely hard problem. In fact, it is NP-complete [111], which means that all known algorithms for finding the optimal permutation run in time which grows *exponentially* with n and so are vastly more expensive than even dense Gaussian elimination for large n. Thus we must settle for using heuristics, of which there are several successful candidates. We illustrate some of these below.

In addition to the complication of choosing a good row and column permutation, there are other reasons sparse Gaussian elimination or Cholesky are much more complicated than their dense counterparts. First, we need to design a data structure that holds only the nonzero entries of A; there are several in common use [93]. Next, we need a data structure to accommodate new entries of L and U that fill in during elimination. This means that either the data structure must grow dynamically during the algorithm or we must cheaply precompute it without actually performing the elimination. Finally, we must use the data structure to perform only the minimum number of floating point operations and at most proportionately many integer and logical operations. In other words, we cannot afford to do $O(n^3)$ integer and logical operations to discover the few floating point operations that we want to do. A more complete discussion of these algorithms is beyond the scope of this book [114, 93], but we will indicate available software.

EXAMPLE 2.9. We illustrate sparse Cholesky on a more realistic example that arises from modeling the displacement of a mechanical structure subject to external forces. Figure 2.9 shows a simple mesh of a two-dimensional slice of a mechanical structure with two internal cavities. The mathematical problem is to compute the displacements of all the grid points of the mesh (which are internal to the structure) subject to some forces applied to the boundary of the structure. The mesh points are numbered from 1 to $n = 483$; more realistic problems would have much larger values of n. The equations relating displacements to forces leads to a system of linear equations $Ax = b$, with one row and column for each of the 483 mesh points and with $a_{ij} \neq 0$ if and only if mesh point i is connected by a line segment to mesh point j. This means that A is a symmetric matrix; it also turns out to be positive definite, so that we can use Cholesky to solve $Ax = b$. Note that A has only $nz = 3971$ nonzeros of a possible $483^2 = 233289$, so A is just $3971/233289 = 1.7\%$ filled. (See Examples 4.1 and 5.1 for similar mechanical modeling problems, where the matrix A is derived in detail.)

Figure 2.10 shows the same mesh (above) along with the nonzero pattern of the matrix A (below), where the 483 nodes are ordered in the "natural" way, with the logically rectangular substructures numbered rowwise, one substructure after the other. The edges in each such substructure have a common color, and these colors match the colors of the nonzeros in the matrix. Each substructure has a label "$(i : j)$" to indicate that it corresponds to rows and columns i through j of A. The corresponding submatrix $A(i : j, i : j)$ is a narrow band matrix. (Example 2.8 and section 6.3 describe other situations in which a mesh leads to a band matrix.) The edges connecting different substructures are red and correspond to the red entries of A, which are farthest from the diagonal of A.

The top pair of plots in Figure 2.11 again shows the sparsity structure of A in the natural order, along with the sparsity structure of its Cholesky factor L. Nonzero entries of L corresponding to nonzero entries of A are black; new nonzeros of L, called *fill-in*, are red. L has 11533 nonzero entries, over five times as many as the lower triangle of A. Computing L by Cholesky costs just 296923 flops, just .8% of the $\frac{1}{3}n^3 = 3.76 \cdot 10^7$ flops that dense Cholesky would have required.

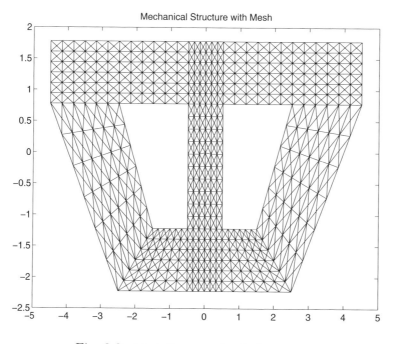

Fig. 2.9. *Mesh for a mechanical structure.*

The number of nonzeros in L and the number of flops required to compute L can be changed significantly by reordering the rows and columns of A. The middle pair of plots in Figure 2.11 shows the results of one such popular re-ordering, called *reverse Cuthill–McKee* [114, 93], which is designed to make A a narrow band matrix. As can be seen, it is quite successful at this, reducing the fill-in of L 21% (from 11533 to 9073) and reducing the flop count almost 39% (from 296923 to 181525).

Another popular ordering algorithm is called *minimum degree ordering* [114, 93], which is designed to create as little fill-in at each step of Cholesky as possible. The results are shown in the bottom pair of plots in Figure 2.11: the fill-in of L is reduced a further 7% (from 9073 to 8440) but the flop count is increased 9% (from 181525 to 198236). $\diamond$

Many sparse matrix examples are available as built-in demos in Matlab, which also has many sparse matrix operations built into it (type "help sparfun" in Matlab for a list). To see the examples, type demo in Matlab, then click on "continue," then on "Matlab/Visit," and then on either "Matrices/Select a demo/Sparse" or "Matrices/Select a demo/Cmd line demos." For example, Figure 2.12 shows a Matlab example of a mesh around a wing, where the goal is to compute the airflow around the wing at the mesh points. The corresponding partial differential equations of airflow lead to a nonsymmetric linear system whose sparsity pattern is also shown.

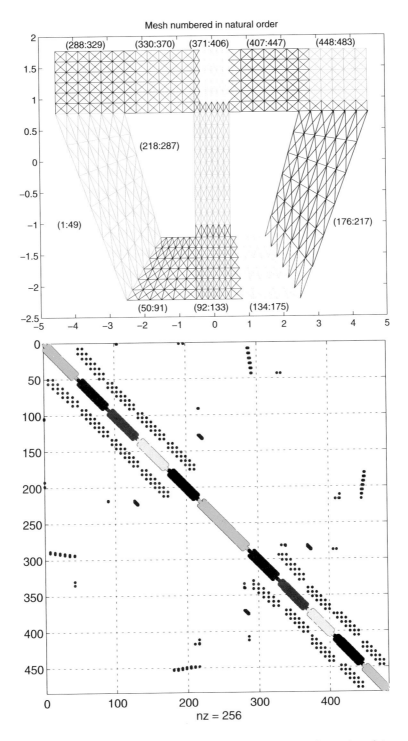

Fig. 2.10. *The edges in the mesh at the top are colored and numbered to match the sparse matrix A at the bottom. For example the first 49 nodes of the mesh (the leftmost green nodes) correspond to rows and columns 1 through 49 of A.*

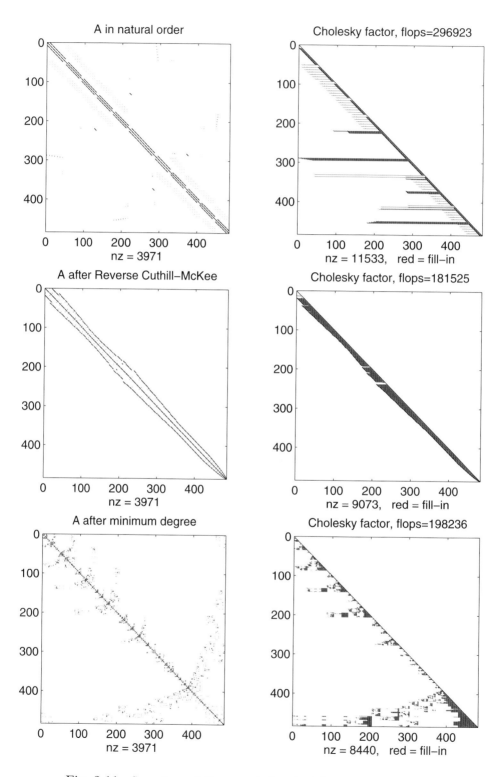

Fig. 2.11. *Sparsity and flop counts for A with various orderings.*

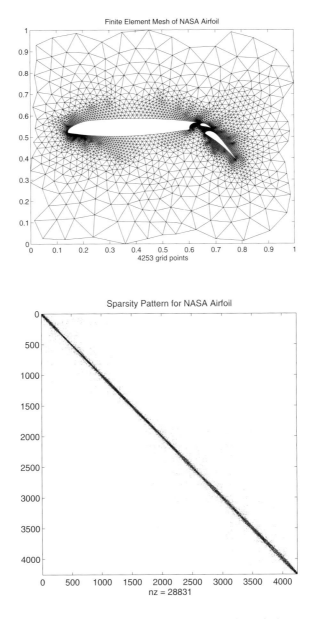

Fig. 2.12. *Mesh around the NASA airfoil.*

Sparse Matrix Software

Besides Matlab, there is a variety of public domain and commercial sparse matrix software available in Fortran or C. Since this is still an active research area (especially with regard to high-performance machines), it is impossible to recommend a single best algorithm. Table 2.2 [177] gives a list of available software, categorized in several ways. We restrict ourselves to supported codes (either public or commercial) or else research codes when no other software is available for that type of problem or machine. We refer to [177, 94] for more complete lists and explanations of the algorithms below.

Table 2.2 is organized as follows. The top group of routines, labeled *serial algorithms*, are designed for single-processor workstations and PCs. The *shared-memory algorithms* are for symmetric multiprocessors, such as the Sun SPARCcenter 2000 [238], SGI Power Challenge [223], DEC AlphaServer 8400 [103], and Cray C90/J90 [253, 254]. The *distributed-memory algorithms* are for machines such as the IBM SP-2 [256], Intel Paragon [257], Cray T3 series [255], and networks of workstations [9]. As you can see, most software has been written for serial machines, some for shared-memory machines, and very little (besides research software) for distributed memory.

The first column gives the *matrix type*. The possibilities include nonsymmetric, symmetric pattern (i.e., either $a_{ij} = a_{ji} \neq 0$, or both can be nonzero and unequal), symmetric (and possibly indefinite), and symmetric positive definite (s.p.d.). The second column gives the name of the routine or of the authors.

The third column gives some detail on the algorithm, indeed more than we have explained in detail in the text: LL (left looking), RL (right looking), frontal, MF (multifrontal), and LDL^T refer to different ways to organize the three nested loops defining Gaussian elimination. Partial, Markowitz, and threshold refer to different pivoting strategies. 2D-blocking refers to which parallel processors are responsible for which parts of the matrix. CAPSS assumes that the linear system is defined by a grid and requires the x, y, and z coordinates of the grid points in order to distribute the matrix among the processors.

The third column also describes the organization of the innermost loop, which could be BLAS1, BLAS2, BLAS3, or scalar. SD refers to the algorithm switching to dense Gaussian elimination after step k when the trailing ($n - k$-by-($n - k$) submatrix is dense enough.

The fifth column describes the status and availability of the software, including whether it is public or commercial and how to get it.

2.7.5. Dense Matrices Depending on Fewer Than $O(n^2)$ Parameters

This is a catch-all heading, which includes a large variety of matrices that arise in practice. We mention just a few cases.

Matrix type	Name	Algorithm	Status/ source
Serial algorithms			
nonsym.	SuperLU	LL, partial, BLAS-2.5	Pub/NETLIB
nonsym.	UMFPACK [62, 63]	MF, Markowitz, BLAS-3	Pub/NETLIB
	MA38 (same as UMFPACK)		Com/HSL
nonsym.	MA48 [96]	Anal: RL, Markowitz	Com/HSL
		Fact: LL, partial, BLAS-1, SD	
nonsym.	SPARSE [167]	RL, Markowitz, scalar	Pub/NETLIB
sym-pattern	⎱⎰ MUPS [5]	MF, threshold, BLAS-3	Com/HSL
	⎰⎱ MA42 [98]	Frontal, BLAS-3	Com/HSL
sym.	MA27 [97]/MA47 [95]	MF, LDL^T, BLAS-1/BLAS-3	Com/HSL
s.p.d.	Ng & Peyton [191]	LL, BLAS-3	Pub/Author
Shared-memory algorithms			
nonsym.	SuperLU	LL, partial, BLAS-2.5	Pub/UCB
nonsym.	PARASPAR [270, 271]	RL, Markowitz, BLAS-1, SD	Res/Author
sym-pattern	MUPS [6]	MF, threshold, BLAS-3	Res/Author
nonsym.	George & Ng [115]	RL, partial, BLAS-1	Res/Author
s.p.d.	Gupta et al. [133]	LL, BLAS-3	Com/SGI
			Pub/Author
s.p.d.	SPLASH [155]	RL, 2-D block, BLAS-3	Pub/Stanford
Distributed-memory algorithms			
sym.	van der Stappen [245]	RL, Markowitz, scalar	Res/Author
sym-pattern	Lucas et al. [180]	MF, no pivoting, BLAS-1	Res/Author
s.p.d.	Rothberg & Schreiber [207]	RL, 2-D block, BLAS-3	Res/Author
s.p.d.	Gupta & Kumar [132]	MF, 2-D block, BLAS-3	Res/Author
s.p.d.	CAPSS [143]	MF, full parallel, BLAS-1 (require coordinates)	Pub/NETLIB

Table 2.2. *Software to solve sparse linear systems using direct methods.*
Abbreviations used in the table:
nonsym. = nonsymmetric.
sym-pattern = symmetric nonzero structure, nonsymmetric values.
sym. = symmetric and may be indefinite.
s.p.d. = symmetric and positive definite.
MF, LL, and RL = multifrontal, left-looking, and right-looking.
SD = switches to a dense code on a sufficiently dense trailing submatrix.
Pub = publicly available; authors may help use the code.
Res = published in literature but may not be available from the authors.
Com = commercial.
HSL = Harwell Subroutine Library:
            http://www.rl.ac.uk/departments/ccd/numerical/hsl/hsl.html.
UCB = http://www.cs.berkeley.edu/~xiaoye/superlu.html.
Stanford = http://www-flash.stanford.edu/apps/SPLASH/.

Vandermonde matrices are of the form

$$
V = \begin{bmatrix}
1 & 1 & \cdots & 1 \\
x_0 & x_1 & & x_n \\
x_0^2 & x_1^2 & & x_n^2 \\
\vdots & \vdots & & \vdots \\
x_0^{n-1} & x_1^{n-1} & & x_n^{n-1}
\end{bmatrix}.
$$

Note that the matrix-vector multiplication

$$
V^T \cdot [a_0, \ldots, a_n]^T = \left[\sum a_i x_0^i, \ldots, \sum a_i x_n^i \right]^T
$$

is equivalent to polynomial evaluation; therefore, solving $V^T a = y$ is polynomial interpolation. Using Newton interpolation we can solve $V^T a = y$ in $\frac{5}{2}n^2$ instead of $\frac{2}{3}n^3$ flops. There is a similar trick to solve $V a = y$ in $\frac{5}{2}n^2$ flops too. See [121, p. 178].

Cauchy matrices C have entries

$$
c_{ij} = \frac{\alpha_i \beta_j}{\xi_i - \eta_j},
$$

where $\alpha = [\alpha_1, \ldots, \alpha_n]$, $\beta = [\beta_1, \ldots, \beta_n]$, $\xi = [\xi_1, \ldots, \xi_n]$, and $\eta = [\eta_1, \ldots, \eta_n]$ are given vectors. The best-known example is the notoriously ill-conditioned *Hilbert matrix* H, with $h_{ij} = 1/(i+j-1)$. These matrices arise in interpolating data by rational functions: Suppose that we want to find the coefficients x_j of the rational function with fixed poles η_j

$$
f(z) = \sum_{j=1}^{n} \frac{x_j}{z - \eta_j}
$$

such that $f(\xi_i) = y_i$ for $i = 1$ to n. Taken together these n equations $f(\xi_i) = y_i$ form an n-by-n linear system with a coefficient matrix that is Cauchy. The inverse of a Cauchy matrix turns out to be a Cauchy matrix, and there is a closed form expression for C^{-1}, based on its connection with interpolation:

$$
(C^{-1})_{ij} = \beta_i^{-1} \alpha_j^{-1} (\xi_j - \eta_i) P_j(\eta_i) Q_i(-\xi_j),
$$

where $P_j(\cdot)$ and $Q_i(\cdot)$ are the Lagrange interpolation polynomials

$$
P_j(z) = \prod_{k \neq j} \frac{\xi_k - z}{\xi_k - \xi_j} \quad \text{and} \quad Q_i(z) = \prod_{k \neq i} \frac{-\eta_k - z}{-\eta_k + \eta_i}.
$$

Toeplitz matrices look like

$$
\begin{bmatrix}
a_0 & a_1 & a_2 & \cdots & a_n \\
a_{-1} & \ddots & \ddots & \ddots & \vdots \\
a_{-2} & \ddots & \ddots & \ddots & a_2 \\
\vdots & \ddots & \ddots & \ddots & a_1 \\
a_{-n} & \cdots & a_{-2} & a_{-1} & a_0
\end{bmatrix};
$$

i.e., they are constant along diagonals. They arise in problems of signal processing. There are algorithms for solving such systems that take only $O(n^2)$ operations.

All these methods generalize to many other similar matrices depending on only $O(n)$ parameters. See [121, p. 183] or [160] for a recent survey.

2.8.   References and Other Topics for Chapter 2

Further details about linear equation solving in general may be found in chapters 3 and 4 of [121]. The reciprocal relationship between condition numbers and distance to the nearest ill-posed problem is further explored in [71]. An average case analysis of pivot growth is described in [242], and an example of bad pivot growth with complete pivoting is given in [122]. Condition estimators are described in [138, 146, 148]. Single precision iterative refinement is analyzed in [14, 225, 226]. A comprehensive discussion of error analysis for linear equation solvers, which covers most of these topics, can be found in [149].

For symmetric indefinite factorization, see [44]. Sparse matrix algorithms are described in [114, 93] as well as the numerous references in Table 2.2. Implementations of many of the algorithms for dense and band matrices described in this chapter are available in LAPACK and CLAPACK [10], which includes a discussion of block algorithms suitable for high-performance computers. Parallel implementations are available in ScaLAPACK [34]. The BLAS are described in [87, 89, 169]. These and other routines are available electronically in NETLIB. An analysis of blocking strategies for matrix multiplication is given in [151]. Strassen's matrix multiplication algorithm is presented in [3], its performance in practice is described in [22], and its numerical stability is described in [77, 149]. A survey of parallel and other block algorithms is given in [76]. For a recent survey of algorithms for structured dense matrices depending only on $O(n)$ parameters, see [160]. For more material on sparse direct methods, see [93, 94, 114, 177].

2.9.   Questions for Chapter 2

QUESTION 2.1. *(Easy)* Using your favorite World Wide Web browser, go to NETLIB (http://www.netlib.org), and answer the following questions.

1. You need a Fortran subroutine to compute the eigenvalues and eigenvectors of real symmetric matrices in double precision. Find one using the search facility in the NETLIB repository. Report the name and URL of the subroutine as well as how you found it.

2. Using the Performance Database Server, find out the current world speed record for solving 100-by-100 dense linear systems using Gaussian elimi-

nation. What is the speed in Mflops, and which machine attained it? Do the same for 1000-by-1000 dense linear systems and "big as you want" dense linear systems. Using the same database, find out how fast your workstation can solve 100-by-100 dense linear systems. Hint: Look at the LINPACK benchmark.

QUESTION 2.2. *(Easy)* Consider solving $AX = B$ for X, where A is n-by-n, and X and B are n-by-m. There are two obvious algorithms. The first algorithm factorizes $A = PLU$ using Gaussian elimination and then solves for each column of X by forward and back substitution. The second algorithm computes A^{-1} using Gaussian elimination and then multiplies $X = A^{-1}B$. Count the number of flops required by each algorithm, and show that the first one requires fewer flops.

QUESTION 2.3. *(Medium)* Let $\| \cdot \|$ be the two-norm. Given a nonsingular matrix A and a vector b, show that for sufficiently small $\|\delta A\|$, there are nonzero δA and δb such that inequality (2.2) is an equality. This justifies calling $\kappa(A) = \|A^{-1}\| \cdot \|A\|$ the condition number of A. Hint: Use the ideas in the proof of Theorem 2.1.

QUESTION 2.4. *(Hard)* Show that bounds (2.7) and (2.8) are attainable.

QUESTION 2.5. *(Medium)* Prove Theorem 2.3. Given the residual $r = A\hat{x} - b$, use Theorem 2.3 to show that bound (2.9) is no larger than bound (2.7). This explains why LAPACK computes a bound based on (2.9), as described in section 2.4.4.

QUESTION 2.6. *(Easy)* Prove Lemma 2.2.

QUESTION 2.7. *(Easy; Z. Bai)* If A is a nonsingular symmetric matrix and has the factorization $A = LDM^T$, where L and M are unit lower triangular matrices and D is a diagonal matrix, show that $L = M$.

QUESTION 2.8. *(Hard)* Consider the following two ways of solving a 2-by-2 linear system of equations:

$$Ax = \begin{bmatrix} a_{11} & a_{12} \\ a_{21} & a_{22} \end{bmatrix} \cdot \begin{bmatrix} x_1 \\ x_2 \end{bmatrix} = \begin{bmatrix} b_1 \\ b_2 \end{bmatrix} = b.$$

Algorithm 1. Gaussian elimination with partial pivoting (GEPP).

Algorithm 2. Cramer's rule:

$$\begin{aligned} \det &= a_{11} * a_{22} - a_{12} * a_{21}, \\ x_1 &= (a_{22} * b_1 - a_{12} * b_2)/\det, \\ x_2 &= (-a_{21} * b_1 + a_{11} * b_2)/\det. \end{aligned}$$

Show by means of a numerical example that Cramer's rule is not backward stable. Hint: Choose the matrix nearly singular and $[b_1 \ b_2]^T \approx [a_{12} \ a_{22}]^T$. What does backward stability imply about the size of the residual? Your numerical example can be done by hand on paper (for example, with four-decimal-digit floating point), on a computer, or a hand calculator.

QUESTION 2.9. *(Medium)* Let B be an n-by-n upper bidiagonal matrix, i.e., nonzero only on the main diagonal and first superdiagonal. Derive an algorithm for computing $\kappa_\infty(B) \equiv \|B\|_\infty \|B^{-1}\|_\infty$ *exactly* (ignoring roundoff). In other words, you should not use an iterative algorithm such as Hager's estimator. Your algorithm should be as cheap as possible; it should be possible to do using no more than $2n - 2$ additions, n multiplications, n divisions, $4n - 2$ absolute values, and $2n - 2$ comparisons. (Anything close to this is acceptable.)

QUESTION 2.10. *(Easy; Z. Bai)* Let A be n-by-m with $n \geq m$. Show that $\|A^T A\|_2 = \|A\|_2^2$ and $\kappa_2(A^T A) = \kappa_2(A)^2$.

Let M be n-by-n and positive definite and L be its Cholesky factor so that $M = LL^T$. Show that $\|M\|_2 = \|L\|_2^2$ and $\kappa_2(M) = \kappa_2(L)^2$.

QUESTION 2.11. *(Easy; Z. Bai)* Let A be symmetric and positive definite. Show that $|a_{ij}| < (a_{ii}a_{jj})^{1/2}$.

QUESTION 2.12. *(Easy; Z. Bai)* Show that if

$$Y = \begin{pmatrix} I & Z \\ 0 & I \end{pmatrix},$$

where I is an n-by-n identity matrix, then $\kappa_F(Y) = \|Y\|_F \|Y^{-1}\|_F = 2n + \|Z\|_F^2$.

QUESTION 2.13. *(Medium)* In this question we will ask how to solve $By = c$ given a fast way to solve $Ax = b$, where $A - B$ is "small" in some sense.

1. Prove the *Sherman–Morrison formula*: Let A be nonsingular, u and v be column vectors, and $A + uv^T$ be nonsingular. Then $(A + uv^T)^{-1} = A^{-1} - (A^{-1}uv^T A^{-1})/(1 + v^T A^{-1}u)$.

   More generally, prove the *Sherman–Morrison–Woodbury formula*: Let U and V be n-by-k rectangular matrices, where $k \leq n$ and A is n-by-n. Then $T = I + V^T A^{-1}U$ is nonsingular if and only if $A + UV^T$ is nonsingular, in which case $(A + UV^T)^{-1} = A^{-1} - A^{-1}UT^{-1}V^T A^{-1}$.

2. If you have a fast algorithm to solve $Ax = b$, show how to build a fast solver for $By = c$, where $B = A + uv^T$.

3. Suppose that $\|A - B\|$ is "small" and you have a fast algorithm for solving $Ax = b$. Describe an iterative scheme for solving $By = c$. How fast do you expect your algorithm to converge? Hint: Use iterative refinement.

QUESTION 2.14. *(Medium; Programming)* Use Netlib to obtain a subroutine to solve $Ax = b$ using Gaussian elimination with partial pivoting. You should get it from either LAPACK (in Fortran, NETLIB/lapack) or CLAPACK (in C, NETLIB/clapack); sgesvx is the main routine in both cases. (There is also a simpler routine sgesv that you might want to look at.) Modify sgesvx (and possibly other subroutines that it calls) to perform complete pivoting instead of partial pivoting; call this new routine gecp. It is probably simplest to modify sgetf2 and use it in place of sgetrf. See HOMEPAGE/Matlab/gecp.m for a Matlab implementation. Test sgesvx and gecp on a number of randomly generated matrices of various sizes up to 30 or so. By choosing x and forming $b = Ax$, you can use examples for which you know the right answer. Check the accuracy of the computed answer $\hat{x}$ as follows. First, examine the error bounds FERR ("Forward ERRor") and BERR ("Backward ERRor") returned by the software; in your own words, say what these bounds mean. Using your knowledge of the exact answer, verify that FERR is correct. Second, compute the exact condition number by inverting the matrix explicitly, and compare this to the estimate RCOND returned by the software. (Actually, RCOND is an estimate of the reciprocal of the condition number.) Third, confirm that $\frac{\|\hat{x}-x\|}{\|\hat{x}\|}$ is bounded by a modest multiple of $macheps$/RCOND. Fourth, you should verify that the (scaled) backward error $R \equiv \|A\hat{x} - b\|/((\|A\| \cdot \|x\| + \|b\|) \cdot macheps)$ is of order unity in each case.

More specifically, your solution should consist of a well-documented program listing of gecp, an explanation of which random matrices you generated (see below), and a table with the following columns (or preferably graphs of each column of data, plotted against the first column):

- test matrix number (to identify it in your explanation of how it was generated);
- its dimension;
- from sgesvx:
    - the pivot growth factor returned by the code
      (this should ideally not be much larger than 1),
    - its estimated condition number (1/RCOND),
    - the ratio of 1/RCOND to your explicitly computed condition number (this should ideally be close to 1),
    - the error bound FERR,
    - the ratio of FERR to the true error
      (this should ideally be at least 1 but not much larger unless you are "lucky" and the true error is zero),
    - the ratio of the true error to ε/RCOND
      (this should ideally be at most 1 or a little less, unless you are "lucky" and the true error is zero),
    - the scaled backward error R/ε
      (this should ideally be O(1) or perhaps O(n)),

    —the backward error BERR/ε
      (this should ideally be O(1) or perhaps O(n)),
    —the run time in seconds;
  • the same data for gecp as for sgesvx.

You need to print the data to only one decimal place, since we care only about approximate magnitudes. Do the error bounds really bound the errors? How do the speeds of sgesvx and gecp compare?

    It is difficult to obtain accurate timings on many systems, since many timers have low resolution, so you should compute the run time as follows:

t_1 = time-so-far
for i = 1 to m
    set up problem
    solve the problem
endfor
t_2 = time-so-far
for i = 1 to m
    set up problem
endfor
t_3 = time-so-far
$t = ((t_2 - t_1) - (t_3 - t_2))/m$

m should be chosen large enough so that $t_2 - t_1$ is at least a few seconds. Then t should be a reliable estimate of the time to solve the problem.

    You should test some well-conditioned problems as well as some that are ill-conditioned. To generate a well-conditioned matrix, let P be a permutation matrix, and add a small random number to each entry. To generate an ill-conditioned matrix, let L be a random lower triangular matrix with tiny diagonal entries and moderate subdiagonal entries. Let U be a similar upper triangular matrix, and let $A = LU$. (There is also an LAPACK subroutine slatms for generating random matrices with a given condition number, which you may use if you like.)

    Also try both solvers on the following class of n-by-n matrices for $n = 1$ up to 30. (If you run in double precision, you may need to run up to $n = 60$.) Shown here is just the case $n = 5$; the others are similar:

$$\begin{bmatrix} 1 & 0 & 0 & 0 & 1 \\ -1 & 1 & 0 & 0 & 1 \\ -1 & -1 & 1 & 0 & 1 \\ -1 & -1 & -1 & 1 & 1 \\ -1 & -1 & -1 & -1 & 1 \end{bmatrix}.$$

Explain the accuracy of the results in terms of the error analysis in section 2.4.

    Your solution should *not* contain any tables of matrix entries or solution components.

In addition to teaching about error bounds, one purpose of this question is to show you what well-engineered numerical software looks like. In practice, one will often use or modify existing software instead of writing one's own from scratch.

QUESTION 2.15. *(Medium; Programming)* This problem depends on Question 2.14. Write another version of sgesvx called sgesvxdouble that computes the residual in double precision during iterative refinement. Modify the error bound FERR in sgesvx to reflect this improved accuracy. Explain your modification. (This may require you to explain how sgesvx computes its error bound in the first place.) On the same set of examples as in the last question, produce a similar table of data. When is sgesvxdouble more accurate than sgesvx?

QUESTION 2.16. *(Hard)* Show how to reorganize the Cholesky algorithm (Algorithm 2.11) to do most of its operations using Level 3 BLAS. Mimic Algorithm 2.10.

QUESTION 2.17. *(Easy)* Suppose that, in Matlab, you have an n-by-n matrix A and an n-by-1 matrix b. What do $A\backslash b$, b'/A, and A/b mean in Matlab? How does $A\backslash b$ differ from $\text{inv}(A) * b$?

QUESTION 2.18. *(Medium)* Let

$$A = \begin{bmatrix} A_{11} & A_{12} \\ A_{21} & A_{22} \end{bmatrix},$$

where A_{11} is k-by-k and nonsingular. Then $S = A_{22} - A_{21}A_{11}^{-1}A_{12}$ is called the *Schur complement of A_{11} in A*, or just Schur complement for short.

1. Show that after k steps of Gaussian elimination without pivoting, A_{22} has been overwritten by S.

2. Suppose $A = A^T$, A_{11} is positive definite, and A_{22} is negative definite $(-A_{22}$ is positive definite). Show that A is nonsingular, that Gaussian elimination without pivoting will work in exact arithmetic, but (by means of a 2-by-2 example) that Gaussian elimination without pivoting may be numerically unstable.

QUESTION 2.19. *(Medium)* Matrix A is called *strictly column diagonally dominant*, or diagonally dominant for short, if

$$|a_{ii}| > \sum_{j=1,\ j\neq i}^{n} |a_{ji}|.$$

• Show that A is nonsingular. Hint: Use Gershgorin's theorem.

- Show that Gaussian elimination with partial pivoting does not actually permute any rows, i.e., that it is identical to Gaussian elimination without pivoting. Hint: Show that after one step of Gaussian elimination, the trailing $(n-1)$-by-$(n-1)$ submatrix, the *Schur complement of a_{11} in A*, is still diagonally dominant. (See Question 2.18 for more discussion of the Schur complement.)

QUESTION 2.20. *(Easy; Z. Bai)* Given an n-by-n nonsingular matrix A, how do you efficiently solve the following problems, using Gaussian elimination with partial pivoting?

(a) Solve the linear system $A^k x = b$, where k is a positive integer.

(b) Compute $\alpha = c^T A^{-1} b$.

(c) Solve the matrix equation $AX = B$, where B is n-by-m.

You should (1) describe your algorithms, (2) present them in pseudocode (using a Matlab-like language; you should not write down the algorithm for GEPP), and (3) give the required flops.

QUESTION 2.21. *(Medium)* Prove that Strassen's algorithm (Algorithm 2.8) correctly multiplies n-by-n matrices, where n is a power of 2.

3

Linear Least Squares Problems

3.1. Introduction

Given an m-by-n matrix A and an m-by-1 vector b, the *linear least squares problem* is to find an n-by-1 vector x minimizing $\|Ax - b\|_2$. If $m = n$ and A is nonsingular, the answer is simply $x = A^{-1}b$. But if $m > n$ so that we have more equations than unknowns, the problem is called *overdetermined*, and generally no x satisfies $Ax = b$ exactly. One occasionally encounters the *underdetermined* problem, where $m < n$, but we will concentrate on the more common overdetermined case.

This chapter is organized as follows. The rest of this introduction describes three applications of least squares problems, to *curve fitting*, to *statistical modeling* of noisy data, and to *geodetic modeling*. Section 3.2 discusses three standard ways to solve the least squares problem: the *normal equations*, the *QR decomposition*, and the *singular value decomposition (SVD)*. We will frequently use the SVD as a tool in later chapters, so we derive several of its properties (although algorithms for the SVD are left to Chapter 5). Section 3.3 discusses perturbation theory for least squares problems, and section 3.4 discusses the implementation details and roundoff error analysis of our main method, QR decomposition. The roundoff analysis applies to many algorithms using orthogonal matrices, including many algorithms for eigenvalues and the SVD in Chapters 4 and 5. Section 3.5 discusses the particularly ill-conditioned situation of rank-deficient least squares problem and how to solve them accurately. Section 3.7 and the questions at the end of the chapter give pointers to other kinds of least squares problems and to software for sparse problems.

EXAMPLE 3.1. A typical application of least squares is *curve fitting*. Suppose that we have m pairs of numbers $(y_1, b_1), \ldots, (y_m, b_m)$ and that we want to find the "best" cubic polynomial fit to b_i as a function of y_i. This means finding polynomial coefficients $x_1, \ldots, x_4$ so that the polynomial $p(y) = \sum_{j=1}^{4} x_j y^{j-1}$ minimizes the residual $r_i \equiv p(y_i) - b_i$ for $i = 1$ to m. We can also write this as minimizing

$$
r \equiv \begin{bmatrix} r_1 \\ r_2 \\ \vdots \\ r_m \end{bmatrix} = \begin{bmatrix} p(y_1) \\ p(y_2) \\ \vdots \\ p(y_m) \end{bmatrix} - \begin{bmatrix} b_1 \\ b_2 \\ \vdots \\ b_m \end{bmatrix}
$$

$$
= \begin{bmatrix} 1 & y_1 & y_1^2 & y_1^3 \\ 1 & y_2 & y_2^2 & y_2^3 \\ \vdots & \vdots & \vdots & \vdots \\ 1 & y_m & y_m^2 & y_m^3 \end{bmatrix} \cdot \begin{bmatrix} x_1 \\ x_2 \\ x_3 \\ x_4 \end{bmatrix} - \begin{bmatrix} b_1 \\ b_2 \\ \vdots \\ b_m \end{bmatrix}
$$

$$
\equiv A \cdot x - b,
$$

where r and b are m-by-1, A is m-by-4, and x is 4-by-1. To minimize r, we could choose any norm, such as $\|r\|_\infty$, $\|r\|_1$, or $\|r\|_2$. The last one, which corresponds to minimizing the sum of the squared residuals $\sum_{i=1}^m r_i^2$, is a *linear least squares problem*.

Figure 3.1 shows an example, where we fit polynomials of increasing degree to the smooth function $b = \sin(\pi y/5) + y/5$ at the 23 points $y = -5, -4.5, -4, \ldots, 5.5, 6$. The left side of Figure 3.1 plots the data points as circles, and four different approximating polynomials of degrees 1, 3, 6, and 19. The right side of Figure 3.1 plots the residual norm $\|r\|_2$ versus degree for degrees from 1 to 20. Note that as the degree increases from 1 to 17, the residual norm decreases. We expect this behavior, since increasing the polynomial degree should let us fit the data better.

But when we reach degree 18, the residual norm suddenly increases dramatically. We can see how erratic the plot of the degree 19 polynomial is on the left (the blue line). This is due to ill-conditioning, as we will later see. Typically, one does polynomial fitting only with relatively low degree polynomials, avoiding ill-conditioning [61]. Polynomial fitting is available as the function `polyfit` in Matlab.

Here is an alternative to polynomial fitting. More generally, one has a set of independent functions $f_1(y), \ldots, f_n(y)$ from $\mathbb{R}^k$ to $\mathbb{R}$ and a set of points $(y_1, b_1), \ldots, (y_m, b_m)$ with $y_i \in \mathbb{R}^k$ and $b_i \in \mathbb{R}$, and one wishes to find a best fit to these points of the form $b = \sum_{j=1}^n x_j f_j(y)$. In other words one wants to choose $x = [x_1, \ldots, x_n]^T$ to minimize the residuals $r_i \equiv \sum_{j=1}^n x_j f_j(y_i) - b_i$ for $1 \le i \le m$. Letting $a_{ij} = f_j(y_i)$, we can write this as $r = Ax - b$, where A is m-by-n, x is n-by-1, and b and r are m-by-1. A good choice of basis functions $f_i(y)$ can lead to better fits and less ill-conditioned systems than using polynomials [33, 84, 168]. $\diamond$

EXAMPLE 3.2. In statistical modeling, one often wishes to estimate certain parameters x_j based on some observations, where the observations are contaminated by noise. For example, suppose that one wishes to predict the college grade point average (GPA) (b) of freshman applicants based on their

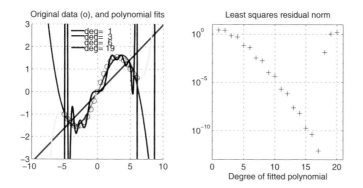

Fig. 3.1. *Polynomial fit to curve* $b = \sin(\pi y/5) + y/5$ *and residual norms.*

high school GPA (a_1) and two Scholastic Aptitude Test scores, verbal (a_2) and quantitative (a_3), as part of the college admissions process. Based on past data from admitted freshmen one can construct a *linear model* of the form $b = \sum_{j=1}^{3} a_j x_j$. The observations are a_{i1}, a_{i2}, a_{i3}, and b_i, one set for each of the m students in the database. Thus, one wants to minimize

$$
r \equiv \begin{bmatrix} r_1 \\ r_2 \\ \vdots \\ r_m \end{bmatrix} = \begin{bmatrix} a_{11} & a_{12} & a_{13} \\ a_{21} & a_{22} & a_{23} \\ \vdots & \vdots & \vdots \\ a_{m1} & a_{m2} & a_{m3} \end{bmatrix} \cdot \begin{bmatrix} x_1 \\ x_2 \\ x_3 \end{bmatrix} - \begin{bmatrix} b_1 \\ b_2 \\ \vdots \\ b_m \end{bmatrix} \equiv A \cdot x - b,
$$

which we can do as a least squares problem.

Here is a statistical justification for least squares, which is called *linear regression* by statisticians: assume that the a_i are known exactly so that only b has noise in it, and that the noise in each b_i is independent and normally distributed with 0 mean and the same standard deviation σ. Let x be the solution of the least squares problem and x_T be the true value of the parameters. Then x is called a *maximum-likelihood estimate* of x_T, and the error $x - x_T$ is normally distributed, with zero mean in each component and *covariance matrix* $\sigma^2 (A^T A)^{-1}$. We will see the matrix $(A^T A)^{-1}$ again below when we solve the least squares problem using the normal equations. For more details on the connection to statistics,[15] see, for example, [33, 259]. ◇

EXAMPLE 3.3. The least squares problem was first posed and formulated by Gauss to solve a practical problem for the German government. There are important economic and legal reasons to know exactly where the boundaries lie between plots of land owned by different people. Surveyors would go out and try to establish these boundaries, measuring certain angles and distances

---

[15]The standard notation in statistics differs from linear algebra: statisticians write $X\beta = y$ instead of $Ax = b$.

and then triangulating from known landmarks. As time passed, it became necessary to improve the accuracy to which the locations of the landmarks were known. So the surveyors of the day went out and remeasured many angles and distances between landmarks, and it fell to Gauss to figure out how to take these more accurate measurements and update the government database of locations. For this he invented least squares, as we will explain shortly [33].

The problem that Gauss solved did not go away and must be periodically revisited. In 1974 the US National Geodetic Survey undertook to update the US geodetic database, which consisted of about 700,000 points. The motivations had grown to include supplying accurate enough data for civil engineers and regional planners to plan construction projects and for geophysicists to study the motion of tectonic plates in the earth's crust (which can move up to 5 cm per year). The corresponding least squares problem was the largest ever solved at the time: about 2.5 million equations in 400,000 unknowns. It was also very sparse, which made it tractable on the computers available in 1978, when the computation was done [164].

Now we briefly discuss the formulation of this problem. It is actually nonlinear and is solved by approximating it by a sequence of linear ones, each of which is a linear least squares problem. The data base consists of a list of points (landmarks), each labeled by location: latitude, longitude, and possibly elevation. For simplicity of exposition, we assume that the earth is flat and suppose that each point i is labeled by linear coordinates $z_i = (x_i, y_i)^T$. For each point we wish to compute a correction $\delta z_i = (\delta x_i, \delta y_i)^T$ so that the corrected location $z_i' = (x_i', y_i')^T = z_i + \delta z_i$ more nearly matches the new, more accurate measurements. These measurements include both distances between selected pairs of points and angles between the line segment from point i to j and i to k (see Figure 3.2). To see how to turn these new measurements into constraints, consider the triangle in Figure 3.2. The corners are labeled by their (corrected) locations, and the angles θ and edge lengths L are also shown. From this data, it is easy to write down constraints based on simple trigonometric identities. For example, an accurate measurement of θ_i leads to the constraint

$$\cos^2 \theta_i = \frac{[(z_j' - z_i')^T (z_k' - z_i')]^2}{(z_j' - z_i')^T (z_j' - z_i') \cdot (z_k' - z_i')^T (z_k' - z_i')},$$

where we have expressed $\cos \theta_i$ in terms of dot products of certain sides of the triangle. If we assume that δz_i is small compared to z_i, then we can linearize this constraint as follows: multiply through by the denominator of the fraction, multiply out all the terms to get a quartic polynomial in all the "δ-variables" (like δx_i), and throw away all terms containing more than one δ-variable as a factor. This yields an equation in which all δ-variables appear linearly. If we collect all these linear constraints from all the new angle and distance measurements together, we get an overdetermined linear system of

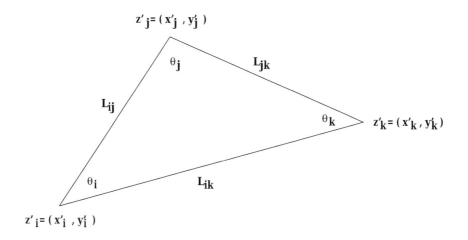

Fig. 3.2. *Constraints in updating a geodetic database.*

equations for all the δ-variables. We wish to find the smallest corrections, i.e., the smallest values of δx_i, etc., that most nearly satisfy these constraints. This is a least squares problem. $\diamond$

Later, after we introduce more machinery, we will also show how *image compression* can be interpreted as a least squares problem (see Example 3.4).

3.2. Matrix Factorizations That Solve the Linear Least Squares Problem

The linear least squares problem has several explicit solutions that we now discuss:

1. normal equations,

2. QR decomposition,

3. SVD,

4. transformation to a linear system (see Question 3.3).

The first method is the fastest but least accurate; it is adequate when the condition number is small. The second method is the standard one and costs up to twice as much as the first method. The third method is of most use on an ill-conditioned problem, i.e., when A is not of full rank; it is several times more expensive again. The last method lets us do iterative refinement to improve the solution when the problem is ill-conditioned. All methods but the third can be adapted to deal efficiently with sparse matrices [33]. We will discuss each solution in turn. We assume initially for methods 1 and 2 that A has full column rank n.

3.2.1. Normal Equations

To derive the *normal equations*, we look for the x where the gradient of $\|Ax - b\|_2^2 = (Ax - b)^T(Ax - b)$ vanishes. So we want

$$
\begin{aligned}
0 &= \lim_{e \to 0} \frac{(A(x + e) - b)^T(A(x + e) - b) - (Ax - b)^T(Ax - b)}{\|e\|_2} \\
&= \lim_{e \to 0} \frac{2e^T(A^T Ax - A^T b) + e^T A^T A e}{\|e\|_2}.
\end{aligned}
$$

The second term $\frac{|e^T A^T A e|}{\|e\|_2} \leq \frac{\|A\|_2^2 \|e\|_2^2}{\|e\|_2} = \|A\|_2^2 \|e\|_2$ approaches 0 as e goes to 0, so the factor $A^T Ax - A^T b$ in the first term must also be zero, or $A^T Ax = A^T b$. This is a system of n linear equations in n unknowns, the normal equations.

Why is $x = (A^T A)^{-1} A^T b$ the minimizer of $\|Ax - b\|_2^2$? We can note that the Hessian $A^T A$ is positive definite, which means that the function is strictly convex and any critical point is a global minimum. Or we can complete the square by writing $x' = x + e$ and simplifying

$$
\begin{aligned}
(Ax' - b)^T(Ax' - b) &= (Ae + Ax - b)^T(Ae + Ax - b) \\
&= (Ae)^T(Ae) + (Ax - b)^T(Ax - b) \\
&\quad + 2(Ae)^T(Ax - b) \\
&= \|Ae\|_2^2 + \|Ax - b\|_2^2 + 2e^T(A^T Ax - A^T b) \\
&= \|Ae\|_2^2 + \|Ax - b\|_2^2.
\end{aligned}
$$

This is clearly minimized by $e = 0$. This is just the Pythagorean theorem, since the residual $r = Ax - b$ is orthogonal to the space spanned by the columns of A, i.e., $0 = A^T r = A^T Ax - A^T b$ as illustrated below (the plane shown is the span of the column vectors of A so that Ax, Ae, and $Ax' = A(x + e)$ all lie in the plane):

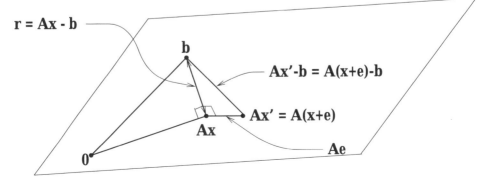

Since $A^T A$ is symmetric and positive definite, we can use the Cholesky decomposition to solve the normal equations. The total cost of computing $A^T A$, $A^T b$, and the Cholesky decomposition is $n^2 m + \frac{1}{3}n^3 + O(n^2)$ flops. Since $m \geq n$, the $n^2 m$ cost of forming $A^T A$ dominates the cost.

3.2.2. QR Decomposition

THEOREM 3.1. QR decomposition. *Let A be m-by-n with $m \geq n$. Suppose that A has full column rank. Then there exist a unique m-by-n orthogonal matrix Q ($Q^T Q = I_n$) and a unique n-by-n upper triangular matrix R with positive diagonals $r_{ii} > 0$ such that $A = QR$.*

Proof. We give two proofs of this theorem. First, this theorem is a restatement of the Gram–Schmidt orthogonalization process [139]. If we apply Gram–Schmidt to the columns a_i of $A = [a_1, a_2, \ldots, a_n]$ from left to right, we get a sequence of orthonormal vectors q_1 through q_n spanning the same space: these orthogonal vectors are the columns of Q. Gram–Schmidt also computes coefficients $r_{ji} = q_j^T a_i$ expressing each column a_i as a linear combination of q_1 through q_i: $a_i = \sum_{j=1}^{i} r_{ji} q_j$. The r_{ji} are just the entries of R.

ALGORITHM 3.1. *The classical Gram–Schmidt (CGS) and modified Gram–Schmidt (MGS) Algorithms for factoring $A = QR$:*

for $i = 1$ to n /* compute ith columns of Q and R */
$\quad$ $q_i = a_i$
$\quad$ *for $j = 1$ to $i - 1$ /* subtract component in q_j direction from a_i */*
$\qquad$ $\begin{cases} r_{ji} = q_j^T a_i & \text{CGS} \\ r_{ji} = q_j^T q_i & \text{MGS} \end{cases}$
$\qquad$ $q_i = q_i - r_{ji} q_j$
$\quad$ *end for*
$\quad$ $r_{ii} = \|q_i\|_2$
$\quad$ *if $r_{ii} = 0$ /* a_i is linearly dependent on $a_1, \ldots, a_{i-1}$ */*
$\qquad$ *quit*
$\quad$ *end if*
$\quad$ $q_i = q_i / r_{ii}$
end for

We leave it as an exercise to show that the two formulas for r_{ji} in the algorithm are mathematically equivalent (see Question 3.1). If A has full column rank, r_{ii} will not be zero. The following figure illustrates Gram–Schmidt when A is 2-by-2:

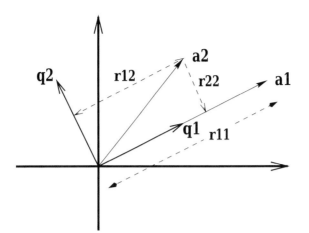

The second proof of this theorem will use Algorithm 3.2, which we present in section 3.4.1. □

Unfortunately, CGS is numerically unstable in floating point arithmetic when the columns of A are nearly linearly dependent. MGS is more stable and will be used in algorithms later in this book but may still result in Q being far from orthogonal ($\|Q^T Q - I\|$ being far larger than ε) when A is ill-conditioned [31, 32, 33, 149]. Algorithm 3.2 in section 3.4.1 is a stable alternative algorithm for factoring $A = QR$. See Question 3.2.

We will derive the formula for the x that minimizes $\|Ax - b\|_2$ using the decomposition $A = QR$ in three slightly different ways. First, we can always choose $m - n$ more orthonormal vectors $\tilde{Q}$ so that $[Q, \tilde{Q}]$ is a square orthogonal matrix (for example, we can choose any $m - n$ more independent vectors $\tilde{X}$ that we want and then apply Algorithm 3.1 to the n-by-n nonsingular matrix $[Q, \tilde{X}]$). Then

$$
\begin{aligned}
\|Ax - b\|_2^2 &= \|[Q, \tilde{Q}]^T (Ax - b)\|_2^2 \quad \text{by part 4 of Lemma 1.7} \\
&= \left\| \begin{bmatrix} Q^T \\ \tilde{Q}^T \end{bmatrix} (QRx - b) \right\|_2^2 \\
&= \left\| \begin{bmatrix} I^{n \times n} \\ O^{(m-n) \times n} \end{bmatrix} Rx - \begin{bmatrix} Q^T b \\ \tilde{Q}^T b \end{bmatrix} \right\|_2^2 \\
&= \left\| \begin{bmatrix} Rx - Q^T b \\ -\tilde{Q}^T b \end{bmatrix} \right\|_2^2 \\
&= \|Rx - Q^T b\|_2^2 + \|\tilde{Q}^T b\|_2^2 \\
&\geq \|\tilde{Q}^T b\|_2^2.
\end{aligned}
$$

We can solve $Rx - Q^T b = 0$ for x, since A and R have the same rank, n, and so R is nonsingular. Then $x = R^{-1} Q^T b$, and the minimum value of $\|Ax - b\|_2$ is $\|\tilde{Q}^T b\|_2$.

Here is a second, slightly different derivation that does not use the matrix

$\tilde{Q}$. Rewrite $Ax - b$ as

$$
\begin{aligned}
Ax - b &= QRx - b = QRx - (QQ^T + I - QQ^T)b \\
&= Q(Rx - Q^T b) - (I - QQ^T)b.
\end{aligned}
$$

Note that the vectors $Q(Rx - Q^T b)$ and $(I - QQ^T)b$ are orthogonal, because $(Q(Rx - Q^T b))^T((I - QQ^T)b) = (Rx - Q^T b)^T[Q^T(I - QQ^T)]b = (Rx - Q^T b)^T[0]b = 0$. Therefore, by the Pythagorean theorem,

$$
\begin{aligned}
\|Ax - b\|_2^2 &= \|Q(Rx - Q^T b)\|_2^2 + \|(I - QQ^T)b\|_2^2 \\
&= \|Rx - Q^T b\|_2^2 + \|(I - QQ^T)b\|_2^2,
\end{aligned}
$$

where we have used part 4 of Lemma 1.7 in the form $\|Qy\|_2^2 = \|y\|_2^2$. This sum of squares is minimized when the first term is zero, i.e., $x = R^{-1}Q^T b$.

Finally, here is a third derivation that starts from the normal equations solution:

$$
\begin{aligned}
x &= (A^T A)^{-1} A^T b \\
&= (R^T Q^T Q R)^{-1} R^T Q^T b = (R^T R)^{-1} R^T Q^T b \\
&= R^{-1} R^{-T} R^T Q^T b = R^{-1} Q^T b.
\end{aligned}
$$

Later we will show that the cost of this decomposition and subsequent least squares solution is $2n^2 m - \frac{2}{3}n^3$, about twice the cost of the normal equations if $m \gg n$ and about the same if $m = n$.

3.2.3. Singular Value Decomposition

The SVD is a very important decomposition which is used for many purposes other than solving least squares problems.

THEOREM 3.2. SVD. *Let A be an arbitrary m-by-n matrix with $m \geq n$. Then we can write $A = U\Sigma V^T$, where U is m-by-n and satisfies $U^T U = I$, V is n-by-n and satisfies $V^T V = I$, and $\Sigma = \mathrm{diag}(\sigma_1, \ldots, \sigma_n)$, where $\sigma_1 \geq \cdots \geq \sigma_n \geq 0$. The columns $u_1, \ldots, u_n$ of U are called* left singular vectors. *The columns $v_1, \ldots, v_n$ of V are called* right singular vectors. *The σ_i are called* singular values. *(If $m < n$, the SVD is defined by considering A^T.)*

A geometric restatement of this theorem is as follows. Given any m-by-n matrix A, think of it as mapping a vector $x \in \mathbb{R}^n$ to a vector $y = Ax \in \mathbb{R}^m$. Then we can choose one orthogonal coordinate system for $\mathbb{R}^n$ (where the unit axes are the columns of V) and another orthogonal coordinate system for $\mathbb{R}^m$ (where the units axes are the columns of U) such that A is diagonal (Σ), i.e., maps a vector $x = \sum_{i=1}^n \beta_i v_i$ to $y = Ax = \sum_{i=1}^n \sigma_i \beta_i u_i$. In other words, any matrix is diagonal, provided that we pick appropriate orthogonal coordinate systems for its domain and range.

Proof of Theorem 3.2. We use induction on m and n: we assume that the SVD exists for $(m-1)$-by-$(n-1)$ matrices and prove it for m-by-n. We assume $A \neq 0$; otherwise we can take $\Sigma = 0$ and let U and V be arbitrary orthogonal matrices.

The basic step occurs when $n = 1$ (since $m \geq n$). We write $A = U\Sigma V^T$ with $U = A/\|A\|_2$, $\Sigma = \|A\|_2$, and $V = 1$.

For the induction step, choose v so $\|v\|_2 = 1$ and $\|A\|_2 = \|Av\|_2 > 0$. Such a v exists by the definition of $\|A\|_2 = \max_{\|v\|_2=1} \|Av\|_2$. Let $u = \frac{Av}{\|Av\|_2}$, which is a unit vector. Choose $\tilde{U}$ and $\tilde{V}$ so that $U = [u, \tilde{U}]$ is an m-by-m orthogonal matrix, and $V = [v, \tilde{V}]$ is an n-by-n orthogonal matrix. Now write

$$U^T A V = \begin{bmatrix} u^T \\ \tilde{U}^T \end{bmatrix} \cdot A \cdot [\, v \quad \tilde{V} \,] = \begin{bmatrix} u^T A v & u^T A \tilde{V} \\ \tilde{U}^T A v & \tilde{U}^T A \tilde{V} \end{bmatrix}.$$

Then

$$u^T A v = \frac{(Av)^T (Av)}{\|Av\|_2} = \frac{\|Av\|_2^2}{\|Av\|_2} = \|Av\|_2 = \|A\|_2 \equiv \sigma$$

and $\tilde{U}^T A v = \tilde{U}^T u \|Av\|_2 = 0$. We claim $u^T A \tilde{V} = 0$ too because otherwise $\sigma = \|A\|_2 = \|U^T A V\|_2 \geq \|[1,0,\dots,0]U^T A V\|_2 = \|[\sigma | u^T A \tilde{V}]\|_2 > \sigma$, a contradiction. (We have used part 7 of Lemma 1.7.)

So $U^T A V = \begin{bmatrix} \sigma & 0 \\ 0 & \tilde{U}^T A \tilde{V} \end{bmatrix} = \begin{bmatrix} \sigma & 0 \\ 0 & \tilde{A} \end{bmatrix}$. We may now apply the induction hypothesis to $\tilde{A}$ to get $\tilde{A} = U_1 \Sigma_1 V_1^T$, where U_1 is $(m-1)$-by-$(n-1)$, Σ_1 is $(n-1)$-by-$(n-1)$, and V_1 is $(n-1)$-by-$(n-1)$. So

$$U^T A V = \begin{bmatrix} \sigma & 0 \\ 0 & U_1 \Sigma_1 V_1^T \end{bmatrix} = \begin{bmatrix} 1 & 0 \\ 0 & U_1 \end{bmatrix} \begin{bmatrix} \sigma & 0 \\ 0 & \Sigma_1 \end{bmatrix} \begin{bmatrix} 1 & 0 \\ 0 & V_1 \end{bmatrix}^T$$

or

$$A = \left(U \begin{bmatrix} 1 & 0 \\ 0 & U_1 \end{bmatrix} \right) \begin{bmatrix} \sigma & 0 \\ 0 & \Sigma_1 \end{bmatrix} \left(V \begin{bmatrix} 1 & 0 \\ 0 & V_1 \end{bmatrix} \right)^T,$$

which is our desired decomposition. $\square$

The SVD has a large number of important algebraic and geometric properties, the most important of which we state here.

THEOREM 3.3. *Let $A = U\Sigma V^T$ be the SVD of the m-by-n matrix A, where $m \geq n$. (There are analogous results for $m < n$.)*

1. *Suppose that A is symmetric, with eigenvalues λ_i and orthonormal eigenvectors u_i. In other words $A = U\Lambda U^T$ is an eigendecomposition of A, with $\Lambda = \mathrm{diag}(\lambda_1, \dots, \lambda_n)$, $U = [u_1, \dots, u_n]$, and $UU^T = I$. Then an SVD of A is $A = U\Sigma V^T$, where $\sigma_i = |\lambda_i|$ and $v_i = \mathrm{sign}(\lambda_i)u_i$, where $\mathrm{sign}(0) = 1$.*

2. *The eigenvalues of the symmetric matrix $A^T A$ are σ_i^2. The right singular vectors v_i are corresponding orthonormal eigenvectors.*

3. *The eigenvalues of the symmetric matrix $A A^T$ are σ_i^2 and $m - n$ zeroes. The left singular vectors u_i are corresponding orthonormal eigenvectors for the eigenvalues σ_i^2. One can take any $m - n$ other orthogonal vectors as eigenvectors for the eigenvalue 0.*

4. *Let $H = \begin{bmatrix} 0 & A^T \\ A & 0 \end{bmatrix}$, where A is square and $A = U \Sigma V^T$ is the SVD of A. Let $\Sigma = \operatorname{diag}(\sigma_1, \ldots, \sigma_n)$, $U = [u_1, \ldots, u_n]$, and $V = [v_1, \ldots, v_n]$. Then the $2n$ eigenvalues of H are $\pm \sigma_i$, with corresponding unit eigenvectors $\frac{1}{\sqrt{2}} \begin{bmatrix} v_i \\ \pm u_i \end{bmatrix}$.*

5. *If A has full rank, the solution of $\min_x \|Ax - b\|_2$ is $x = V \Sigma^{-1} U^T b$.*

6. *$\|A\|_2 = \sigma_1$. If A is square and nonsingular, then $\|A^{-1}\|_2^{-1} = \sigma_n$ and $\|A\|_2 \cdot \|A^{-1}\|_2 = \frac{\sigma_1}{\sigma_n}$.*

7. *Suppose $\sigma_1 \geq \cdots \geq \sigma_r > \sigma_{r+1} = \cdots = \sigma_n = 0$. Then the rank of A is r. The null space of A, i.e., the subspace of vectors v such that $Av = 0$, is the space spanned by columns $r + 1$ through n of V: $\operatorname{span}(v_{r+1}, \ldots, v_n)$. The range space of A, the subspace of vectors of the form Aw for all w, is the space spanned by columns 1 through r of U: $\operatorname{span}(u_1, \ldots, u_r)$.*

8. *Let S^{n-1} be the unit sphere in $\mathbb{R}^n$: $S^{n-1} = \{ x \in \mathbb{R}^n : \|x\|_2 = 1 \}$. Let $A \cdot S^{n-1}$ be the image of S^{n-1} under A: $A \cdot S^{n-1} = \{ Ax : x \in \mathbb{R}^n \text{ and } \|x\|_2 = 1 \}$. Then $A \cdot S^{n-1}$ is an ellipsoid centered at the origin of $\mathbb{R}^m$, with principal axes $\sigma_i u_i$.*

9. *Write $V = [v_1, v_2, \ldots, v_n]$ and $U = [u_1, u_2, \ldots, u_n]$, so $A = U \Sigma V^T = \sum_{i=1}^n \sigma_i u_i v_i^T$ (a sum of rank-1 matrices). Then a matrix of rank $k < n$ closest to A (measured with $\|\cdot\|_2$) is $A_k = \sum_{i=1}^k \sigma_i u_i v_i^T$, and $\|A - A_k\|_2 = \sigma_{k+1}$. We may also write $A_k = U \Sigma_k V^T$, where $\Sigma_k = \operatorname{diag}(\sigma_1, \ldots, \sigma_k, 0, \ldots, 0)$.*

Proof.

1. This is true by the definition of the SVD.

2. $A^T A = V \Sigma U^T U \Sigma V^T = V \Sigma^2 V^T$. This is an eigendecomposition of $A^T A$, with the columns of V the eigenvectors and the diagonal entries of Σ^2 the eigenvalues.

3. Choose an m-by-$(m-n)$ matrix $\tilde{U}$ so that $[U, \tilde{U}]$ is square and orthogonal. Then write

$$AA^T = U \Sigma V^T V \Sigma U^T = U \Sigma^2 U^T = \begin{bmatrix} U, \tilde{U} \end{bmatrix} \begin{bmatrix} \Sigma^2 & 0 \\ 0 & 0 \end{bmatrix} \begin{bmatrix} U, \tilde{U} \end{bmatrix}^T.$$

This is an eigendecomposition of AA^T.

4. See Question 3.14.

5. $\|Ax - b\|_2^2 = \|U\Sigma V^T x - b\|_2^2$. Since A has full rank, so does Σ, and thus Σ is invertible. Now let $[U, \tilde{U}]$ be square and orthogonal as above so

$$
\|U\Sigma V^T x - b\|_2^2 = \left\| \begin{bmatrix} U^T \\ \tilde{U}^T \end{bmatrix} (U\Sigma V^T x - b) \right\|_2^2
$$

$$
= \left\| \begin{bmatrix} \Sigma V^T x - U^T b \\ -\tilde{U}^T b \end{bmatrix} \right\|_2^2
$$

$$
= \|\Sigma V^T x - U^T b\|_2^2 + \|\tilde{U}^T b\|_2^2.
$$

This is minimized by making the first term zero, i.e., $x = V\Sigma^{-1} U^T b$.

6. It is clear from its definition that the two-norm of a diagonal matrix is the largest absolute entry on its diagonal. Thus, by part 3 of Lemma 1.7, $\|A\|_2 = \|U^T AV\|_2 = \|\Sigma\|_2 = \sigma_1$ and $\|A^{-1}\|_2 = \|V^T A^{-1} U\|_2 = \|\Sigma^{-1}\|_2 = \sigma_n^{-1}$.

7. Again choose an m-by-$(m - n)$ matrix $\tilde{U}$ so that the m-by-m matrix $\hat{U} = [U, \tilde{U}]$ is orthogonal. Since $\hat{U}$ and V are nonsingular, A and $\hat{U}^T AV = \begin{bmatrix} \Sigma^{n \times n} \\ 0^{(m-n) \times n} \end{bmatrix} \equiv \hat{\Sigma}$ have the same rank—namely, r—by our assumption about Σ. Also, v is in the null space of A if and only if $V^T v$ is in the null space of $\hat{U}^T AV = \hat{\Sigma}$, since $Av = 0$ if and only if $\hat{U}^T AV(V^T v) = 0$. But the null space of $\hat{\Sigma}$ is clearly spanned by columns $r + 1$ through n of the n-by-n identity matrix I_n, so the null space of A is spanned by V times these columns, i.e., v_{r+1} through v_n. A similar argument shows that the range space of A is the same as $\hat{U}$ times the range space of $\hat{U}^T AV = \hat{\Sigma}$, i.e., $\hat{U}$ times the first r columns of I_m, or u_1 through u_r.

8. We "build" the set $A \cdot S^{n-1}$ by multiplying by one factor of $A = U\Sigma V^T$ at a time. The figure below illustrates what happens when

$$
A = \begin{bmatrix} 3 & 1 \\ 1 & 3 \end{bmatrix}
$$

$$
= \begin{bmatrix} 2^{-1/2} & -2^{-1/2} \\ 2^{-1/2} & 2^{-1/2} \end{bmatrix} \cdot \begin{bmatrix} 4 & 0 \\ 0 & 2 \end{bmatrix} \cdot \begin{bmatrix} 2^{-1/2} & -2^{-1/2} \\ 2^{-1/2} & 2^{-1/2} \end{bmatrix}^T
$$

$$
\equiv U\Sigma V^T.
$$

Assume for simplicity that A is square and nonsingular. Since V is orthogonal and so maps unit vectors to other unit vectors, $V^T \cdot S^{n-1} = S^{n-1}$. Next, since $v \in S^{n-1}$ if and only if $\|v\|_2 = 1$, $w \in \Sigma S^{n-1}$ if and only if $\|\Sigma^{-1} w\|_2 = 1$ or $\sum_{i=1}^n (w_i/\sigma_i)^2 = 1$. This defines an ellipsoid with principal axes $\sigma_i e_i$, where e_i is the ith column of the identity matrix. Finally, multiplying each $w = \Sigma v$ by U just rotates the ellipse so that each e_i becomes u_i, the ith column of U.

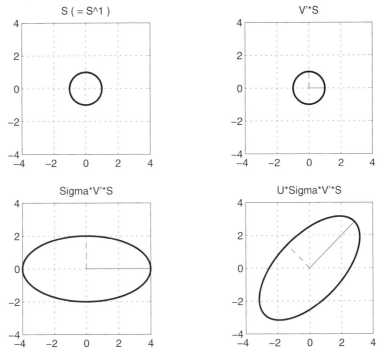

9. A_k has rank k by construction and

$$\|A - A_k\|_2 = \left\| \sum_{i=k+1}^{n} \sigma_i u_i v_i^T \right\| = \left\| U \begin{bmatrix} 0 & & & \\ & \sigma_{k+1} & & \\ & & \ddots & \\ & & & \sigma_n \end{bmatrix} V^T \right\|_2 = \sigma_{k+1}.$$

It remains to show that there is no closer rank k matrix to A. Let B be any rank k matrix, so its null space has dimension $n - k$. The space spanned by $\{v_1, \ldots, v_{k+1}\}$ has dimension $k + 1$. Since the sum of their dimensions is $(n - k) + (k + 1) > n$, these two spaces must overlap. Let h be a unit vector in their intersection. Then

$$\begin{aligned}
\|A - B\|_2^2 &\geq \|(A - B)h\|_2^2 = \|Ah\|_2^2 = \|U\Sigma V^T h\|_2^2 \\
&= \|\Sigma(V^T h)\|_2^2 \\
&\geq \sigma_{k+1}^2 \|V^T h\|_2^2 \\
&= \sigma_{k+1}^2. \quad \square
\end{aligned}$$

EXAMPLE 3.4. We illustrate the last part of Theorem 3.3 by using it for *image compression*. In particular, we will illustrate it with low-rank approximations of a clown. An m-by-n image is just an m-by-n matrix, where entry (i, j) is interpreted as the brightness of pixel (i, j). In other words, matrix entries ranging from 0 to 1 (say) are interpreted as pixels ranging from black ($=0$) through various shades of gray to white ($=1$). (Colors also are possible.) Rather than storing or transmitting all $m \cdot n$ matrix entries to represent the image, we often prefer to *compress* the image by storing many fewer numbers, from which we can still approximately reconstruct the original image. We may use Part 9 of Theorem 3.3 to do this, as we now illustrate.

Consider the image in Figure 3.3(a). This 320-by-200 pixel image corresponds to a 320-by-200 matrix A. Let $A = U \Sigma V^T$ be the SVD of A. Part 9 of Theorem 3.3 tells us that $A_k = \sum_{i=1}^{k} \sigma_i u_i v_i^T$ is the best rank-k approximation of A, in the sense of minimizing $\|A - A_k\|_2 = \sigma_{k+1}$. Note that it only takes $m \cdot k + n \cdot k = (m + n) \cdot k$ words to store u_1 through u_k and $\sigma_1 v_1$ through $\sigma_k v_k$, from which we can reconstruct A_k. In contrast, it takes $m \cdot n$ words to store A (or A_k explicitly), which is much larger when k is small. So we will use A_k as our compressed image, stored using $(m + n) \cdot k$ words. The other images in Figure 3.3 show these approximations for various values of k, along with the relative errors σ_{k+1}/σ_1 and compression ratios $(m + n) \cdot k/(m \cdot n) = 520 \cdot k/64000 \approx k/123$.

k	Relative error $= \sigma_{k+1}/\sigma_k$	Compression ratio $= 520k/64000$
3	.155	.024
10	.077	.081
20	.040	.163

These images were produced by the following commands (the clown and other images are available in Matlab among the visualization demonstration files; check your local installation for location):

```
load clown.mat; [U,S,V]=svd(X); colormap('gray');
image(U(:,1:k)*S(1:k,1:k)*V(:,1:k)')
```

There are also many other, cheaper image-compression techniques available than the SVD [189, 152]. ◇

Later we will see that the cost of solving a least squares problem with the SVD is about the same as with QR when $m \gg n$, and about $4n^2 m - \frac{4}{3}n^3 + O(n^2)$ for smaller m. A precise comparison of the costs of QR and the SVD also depends on the machine being used. See section 3.6 for details.

DEFINITION 3.1. *Suppose that A is m-by-n with $m \geq n$ and has full rank, with $A = QR = U \Sigma V^T$ being A's QR decomposition and SVD, respectively. Then*

$$A^+ \equiv (A^T A)^{-1} A^T = R^{-1} Q^T = V \Sigma^{-1} U^T$$

is called the (Moore–Penrose) pseudoinverse *of A. If $m < n$, then $A^+ \equiv A^T (AA^T)^{-1}$.*

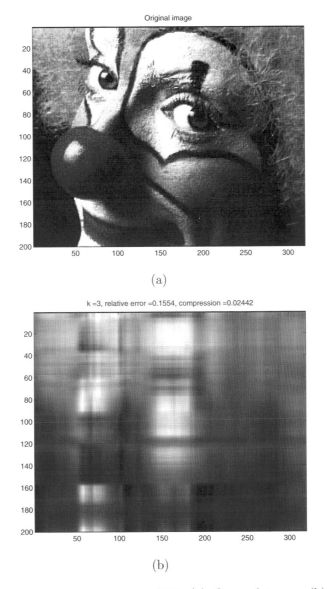

Fig. 3.3. *Image compression using the SVD.* (a) *Original image.* (b) *Rank* $k = 3$ *approximation.*

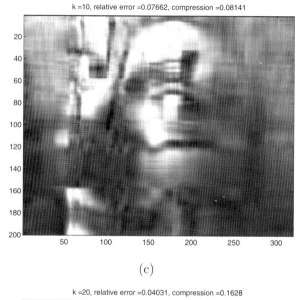

(c)

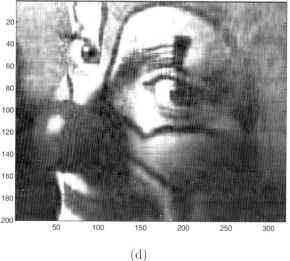

(d)

Fig. 3.3. *Continued.* (c) *Rank $k = 10$ approximation.* (d) *Rank $k = 20$ approximation.*

The pseudoinverse lets us write the solution of the full-rank, overdetermined least squares problem as simply $x = A^+b$. If A is square and full rank, this formula reduces to $x = A^{-1}b$ as expected. The pseudoinverse of A is computed as `pinv(A)` in Matlab. When A is not full rank, the Moore–Penrose pseudoinverse is given by Definition 3.2 in section 3.5.

3.3. Perturbation Theory for the Least Squares Problem

When A is not square, we define its condition number with respect to the 2-norm to be $\kappa_2(A) \equiv \sigma_{\max}(A)/\sigma_{\min}(A)$. This reduces to the usual condition number when A is square. The next theorem justifies this definition.

THEOREM 3.4. *Suppose that A is m-by-n with $m \geq n$ and has full rank. Suppose that x minimizes $\|Ax-b\|_2$. Let $r = Ax-b$ be the residual. Let $\tilde{x}$ minimize $\|(A + \delta A)\tilde{x} - (b + \delta b)\|_2$. Assume $\epsilon \equiv \max(\frac{\|\delta A\|_2}{\|A\|_2}, \frac{\|\delta b\|_2}{\|b\|_2}) < \frac{1}{\kappa_2(A)} = \frac{\sigma_{\min}(A)}{\sigma_{\max}(A)}$. Then*

$$\frac{\|\tilde{x} - x\|_2}{\|x\|_2} \leq \epsilon \cdot \left\{ \frac{2 \cdot \kappa_2(A)}{\cos \theta} + \tan \theta \cdot \kappa_2^2(A) \right\} + O(\epsilon^2) \equiv \epsilon \cdot \kappa_{LS} + O(\epsilon^2),$$

where $\sin \theta = \frac{\|r\|_2}{\|b\|_2}$. In other words, θ is the angle between the vectors b and Ax and measures whether the residual norm $\|r\|_2$ is large (near $\|b\|$) or small (near 0). κ_{LS} is the condition number for the least squares problem.

Sketch of Proof. Expand $\tilde{x} = \left((A + \delta A)^T(A + \delta A)\right)^{-1}(A + \delta A)^T(b + \delta b)$ in powers of δA and δb, and throw away all but the linear terms in δA and δb. □

We have assumed that $\epsilon \cdot \kappa_2(A) < 1$ for the same reason as in the derivation of bound (2.4) for the perturbed solution of the square linear system $Ax = b$: it guarantees that $A + \delta A$ has full rank so that $\tilde{x}$ is uniquely determined.

We may interpret this bound as follows. If θ is 0 or very small, then the residual is small and the effective condition number is about $2\kappa_2(A)$, much like ordinary linear equation solving. If θ is not small but not close to $\pi/2$, the residual is moderately large, and then the effective condition number can be much larger: $\kappa_2^2(A)$. If θ is close to $\pi/2$, so the true solution is nearly zero, then the effective condition number becomes unbounded even if $\kappa_2(A)$ is small. These three cases are illustrated below. The right-hand picture makes it easy to see why the condition number is infinite when $\theta = \pi/2$: in this case the solution $x = 0$, and almost any arbitrarily small change in A or b will yield a nonzero solution x, an "infinitely" large relative change.

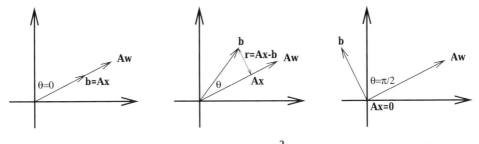

An alternative form for the bound in Theorem 3.4 that eliminates the $O(\epsilon^2)$ term is as follows [258, 149] (here $\tilde{r}$ is the perturbed residual $\tilde{r} = (A + \delta A)\tilde{x} - (b + \delta b)$):

$$\frac{\|\tilde{x} - x\|_2}{\|x\|_2} \leq \frac{\epsilon \kappa_2(A)}{1 - \epsilon \kappa_2(A)} \left(2 + (\kappa_2(A) + 1) \frac{\|r\|_2}{\|A\|_2 \|x\|_2} \right),$$

$$\frac{\|\tilde{r} - r\|_2}{\|r\|_2} \leq (1 + 2\epsilon\kappa_2(A)).$$

We will see that, properly implemented, both the QR decomposition and SVD are numerically stable; i.e., they yield a solution $\tilde{x}$ minimizing $\|(A + \delta A)\tilde{x} - (b + \delta b)\|_2$ with

$$\max \left(\frac{\|\delta A\|}{\|A\|}, \frac{\|\delta b\|}{\|b\|} \right) = O(\varepsilon).$$

We may combine this with the above perturbation bounds to get error bounds for the solution of the least squares problem, much as we did for linear equation solving.

The normal equations are not as accurate. Since they involve solving $(A^T A)x = A^T b$, the accuracy depends on the condition number $\kappa_2(A^T A) = \kappa_2^2(A)$. Thus the error is always bounded by $\kappa_2^2(A)\varepsilon$, never just $\kappa_2(A)\varepsilon$. Therefore we expect that the normal equations can lose twice as many digits of accuracy as methods based on the QR decomposition and SVD.

Furthermore, solving the normal equations is *not* necessarily stable; i.e., the computed solution $\tilde{x}$ does not generally minimize $\|(A + \delta A)\tilde{x} - (b + \delta b)\|_2$ for small δA and δb. Still, when the condition number is small, we expect the normal equations to be about as accurate as the QR decomposition or SVD. Since the normal equations are the fastest way to solve the least squares problem, they are the method of choice when the matrix is well-conditioned.

We return to the problem of solving very ill-conditioned least squares problems in section 3.5.

3.4. Orthogonal Matrices

As we said in section 3.2.2, Gram–Schmidt orthogonalization (Algorithm 3.1) may not compute an orthogonal matrix Q when the vectors being orthogonalized are nearly linearly dependent, so we cannot use it to compute the QR decomposition stably.

Instead, we base our algorithms on certain easily computable orthogonal matrices called *Householder reflections* and *Givens rotations*, which we can choose to introduce zeros into vectors that they multiply. Later we will show that any algorithm that uses these orthogonal matrices to introduce zeros is automatically stable. This error analysis will apply to our algorithms for the QR decomposition as well as many SVD and eigenvalue algorithms in Chapters 4 and 5.

Despite the possibility of nonorthogonal Q, the MGS algorithm has important uses in numerical linear algebra. (There is little use for its less stable version, CGS.) These uses include finding eigenvectors of symmetric tridiagonal matrices using bisection and inverse iteration (section 5.3.4) and the Arnoldi and Lanczos algorithms for reducing a matrix to certain "condensed" forms (sections 6.6.1, 6.6.6, and 7.4). Arnoldi and Lanczos algorithms are used as the basis of algorithms for solving sparse linear systems and finding eigenvalues of sparse matrices. MGS can also be modified to solve the least squares problem stably, but Q may still be far from orthogonal [33].

3.4.1. Householder Transformations

A Householder transformation (or reflection) is a matrix of the form $P = I - 2uu^T$ where $\|u\|_2 = 1$. It is easy to see that $P = P^T$ and $PP^T = (I - 2uu^T)(I - 2uu^T) = I - 4uu^T + 4uu^T uu^T = I$, so P is a symmetric, orthogonal matrix. It is called a reflection because Px is reflection of x in the plane through 0 perpendicular to u.

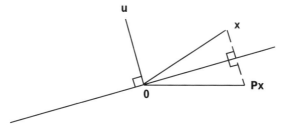

Given a vector x, it is easy to find a Householder reflection $P = I - 2uu^T$ to zero out all but the first entry of x: $Px = [c, 0, \ldots, 0]^T = c \cdot e_1$. We do this as follows. Write $Px = x - 2u(u^T x) = c \cdot e_1$ so that $u = \frac{1}{2(u^T x)}(x - ce_1)$; i.e., u is a linear combination of x and e_1. Since $\|x\|_2 = \|Px\|_2 = |c|$, u must be parallel to the vector $\tilde{u} = x \pm \|x\|_2 e_1$, and so $u = \tilde{u}/\|\tilde{u}\|_2$. One can verify that either choice of sign yields a u satisfying $Px = ce_1$, as long as $\tilde{u} \neq 0$. We will use $\tilde{u} = x + \text{sign}(x_1)e_1$, since this means that there is no cancellation in computing the first component of $\tilde{u}$. In summary, we get

$$\tilde{u} = \begin{bmatrix} x_1 + \text{sign}(x_1) \cdot \|x\|_2 \\ x_2 \\ \vdots \\ x_n \end{bmatrix} \qquad \text{with } u = \frac{\tilde{u}}{\|\tilde{u}\|_2}.$$

We write this as $u = \text{House}(x)$. (In practice, we can store $\tilde{u}$ instead of u to save the work of computing u, and use the formula $P = I - (2/\|\tilde{u}\|_2^2)\tilde{u}\tilde{u}^T$ instead of $P = I - 2uu^T$.)

EXAMPLE 3.5. We show how to compute the QR decomposition of a 5-by-4 matrix A using Householder transformations. This example will make the pattern for general m-by-n matrices evident. In the matrices below, P_i is a 5-by-5 orthogonal matrix, x denotes a generic nonzero entry, and o denotes a zero entry.

1. Choose P_1 so
$$A_1 \equiv P_1 A = \begin{bmatrix} x & x & x & x \\ o & x & x & x \\ o & x & x & x \\ o & x & x & x \\ o & x & x & x \end{bmatrix}.$$

2. Choose $P_2 = \left[\begin{array}{c|c} 1 & 0 \\ \hline 0 & P_2' \end{array} \right]$ so
$$A_2 \equiv P_2 A_1 = \begin{bmatrix} x & x & x & x \\ o & x & x & x \\ o & o & x & x \\ o & o & x & x \\ o & o & x & x \end{bmatrix}.$$

3. Choose $P_3 = \left[\begin{array}{cc|c} 1 & & \\ & 1 & 0 \\ \hline & 0 & P_3' \end{array} \right]$ so
$$A_3 \equiv P_3 A_2 = \begin{bmatrix} x & x & x & x \\ o & x & x & x \\ o & o & x & x \\ o & o & o & x \\ o & o & o & x \end{bmatrix}.$$

4. Choose $P_4 = \left[\begin{array}{ccc|c} 1 & & & \\ & 1 & & 0 \\ & & 1 & \\ \hline & 0 & & P_4' \end{array} \right]$ so $A_4 \equiv P_4 A_3 = \begin{bmatrix} x & x & x & x \\ o & x & x & x \\ o & o & x & x \\ o & o & o & x \\ o & o & o & o \end{bmatrix}.$

Here, we have chosen a Householder matrix P_i' to zero out the subdiagonal entries in column i; this does not disturb the zeros already introduced in previous columns.

Let us call the final 5-by-4 upper triangular matrix $\tilde{R} \equiv A_4$. Then $A = P_1^T P_2^T P_3^T P_4^T \tilde{R} = QR$, where Q is the first four columns of $P_1^T P_2^T P_3^T P_4^T = P_1 P_2 P_3 P_4$ (since all P_i are symmetric) and R is the first four rows of $\tilde{R}$. $\diamond$

Here is the general algorithm for QR decomposition using Householder transformations.

ALGORITHM 3.2. *QR factorization using Householder reflections:*

> *for* $i = 1$ *to* $\min(m - 1, n)$
> $\quad u_i = \text{House}(A(i : m, i))$
> $\quad P_i' = I - 2u_i u_i^T$
> $\quad A(i : m, i : n) = P_i' A(i : m, i : n)$
> *end for*

Here are some more implementation details. We never need to form P_i explicitly but just multiply

$$(I - 2u_i u_i^T)A(i : m, i : n) = A(i : m, i : n) - 2u_i(u_i^T A(i : m, i : n)),$$

which costs less. To store P_i, we need only u_i, or $\tilde{u}_i$ and $\|\tilde{u}_i\|$. These can be stored in column i of A; in fact it need not be changed! Thus QR can be "overwritten" on A, where Q is stored in factored form $P_1 \cdots P_{n-1}$, and P_i is stored as $\tilde{u}_i$ below the diagonal in column i of A. (We need an extra array of length n for the top entry of $\tilde{u}_i$, since the diagonal entry is occupied by R_{ii}.)

Recall that to solve the least squares problem $\min \|Ax - b\|_2$ using $A = QR$, we need to compute $Q^T b$. This is done as follows: $Q^T b = P_n P_{n-1} \cdots P_1 b$, so we need only keep multiplying b by $P_1, P_2, \ldots, P_n$:

> *for* $i = 1$ *to* n
> $\quad \gamma = -2 \cdot u_i^T b(i : m)$
> $\quad b(i : m) = b(i : m) + \gamma u_i$
> *end for*

The cost is n dot products $\gamma = -2 \cdot u_i^T b$ and n "saxpys" $b + \gamma u_i$. The cost of computing $A = QR$ this way is $2n^2 m - \frac{2}{3}n^3$, and the subsequent cost of solving the least squares problem given QR is just an additional $O(mn)$.

The LAPACK routine for solving the least squares problem using QR is `sgels`. Just as Gaussian elimination can be reorganized to use matrix-matrix multiplication and other Level 3 BLAS (see section 2.6), the same can be done for the QR decomposition; see Question 3.17. In Matlab, if the m-by-n matrix A has more rows than columns and b is m by 1, `A\b` solves the least squares problem. The QR decomposition itself is also available via [Q,R]=qr(A).

3.4.2. Givens Rotations

A Givens rotation $R(\theta) \equiv \begin{bmatrix} \cos\theta & -\sin\theta \\ \sin\theta & \cos\theta \end{bmatrix}$ rotates any vector $x \in \mathbb{R}^2$ counterclockwise by θ:

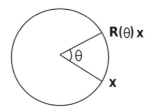

We also need to define the Givens rotation by θ in coordinates i and j:

$$R(i,j,\theta) \equiv \begin{array}{c} \\ \\ \\ i \\ \\ j \\ \\ \\ \end{array} \begin{bmatrix} 1 & & & & & & & & \\ & 1 & & & & & & & \\ & & \ddots & & & & & & \\ & & & \cos\theta & & -\sin\theta & & & \\ & & & & \ddots & & & & \\ & & & \sin\theta & & \cos\theta & & & \\ & & & & & & \ddots & & \\ & & & & & & & 1 & \\ & & & & & & & & 1 \end{bmatrix}$$

Given x, i, and j, we can zero out x_j by choosing $\cos\theta$ and $\sin\theta$ so that

$$\begin{bmatrix} \cos\theta & -\sin\theta \\ \sin\theta & \cos\theta \end{bmatrix} \begin{bmatrix} x_i \\ x_j \end{bmatrix} = \begin{bmatrix} \sqrt{x_i^2 + x_j^2} \\ 0 \end{bmatrix}$$

or $\cos\theta = \dfrac{x_i}{\sqrt{x_i^2+x_j^2}}$ and $\sin\theta = \dfrac{-x_j}{\sqrt{x_i^2+x_j^2}}$.

The QR algorithm using Givens rotations is analogous to using Householder reflections, but when zeroing out column i, we zero it out one entry at a time (bottom to top, say).

EXAMPLE 3.6. We illustrate two intermediate steps in computing the QR decomposition of a 5-by-4 matrix using Givens rotations. To progress from

$$\begin{bmatrix} x & x & x & x \\ o & x & x & x \\ o & o & x & x \\ o & o & x & x \\ o & o & x & x \end{bmatrix} \quad \text{to} \quad \begin{bmatrix} x & x & x & x \\ o & x & x & x \\ o & o & x & x \\ o & o & o & x \\ o & o & o & x \end{bmatrix}$$

we multiply

$$\begin{bmatrix} 1 & & & & \\ & 1 & & & \\ & & 1 & & \\ & & & c & -s \\ & & & s & c \end{bmatrix} \begin{bmatrix} x & x & x & x \\ o & x & x & x \\ o & o & x & x \\ o & o & x & x \\ o & o & x & x \end{bmatrix} = \begin{bmatrix} x & x & x & x \\ o & x & x & x \\ o & o & x & x \\ o & o & x & x \\ o & o & o & x \end{bmatrix}$$

and

$$
\begin{bmatrix} 1 & & & & \\ & 1 & & & \\ & & c' & -s' & \\ & & s' & c' & \\ & & & & 1 \end{bmatrix}
\begin{bmatrix} x & x & x & x \\ o & x & x & x \\ o & o & x & x \\ o & o & x & x \\ o & o & o & x \end{bmatrix}
=
\begin{bmatrix} x & x & x & x \\ o & x & x & x \\ o & o & x & x \\ o & o & o & x \\ o & o & o & x \end{bmatrix}. \quad \diamond
$$

The cost of the QR decomposition using Givens rotations is twice the cost of using Householder reflections. We will need Givens rotations for other applications later.

Here are some implementation details. Just as we overwrote A with Q and R when using Householder reflections, we can do the same with Givens rotations. We use the same trick, storing the information describing the transformation in the entries zeroed out. Since a Givens rotation zeros out just one entry, we must store the information about the rotation there. We do this as follows. Let $s = \sin \theta$ and $c = \cos \theta$. If $|s| < |c|$, store $s \cdot \text{sign}(c)$ and otherwise store $\frac{\text{sign}(s)}{c}$. To recover s and c from the stored value (call it p) we do the following: if $|p| < 1$, then $s = p$ and $c = \sqrt{1 - s^2}$; otherwise $c = \frac{1}{p}$ and $s = \sqrt{1 - c^2}$. The reason we do not just store s and compute $c = \sqrt{1 - s^2}$ is that when s is close to 1, c would be inaccurately reconstructed. Note also that we may recover either s and c or $-s$ and $-c$; this is adequate in practice.

There is also a way to apply a sequence of Givens rotations while performing fewer floating point operations than described above. These are called *fast Givens rotations* [7, 8, 33]. Since they are still slower than Householder reflections for the purposes of computing the QR factorization, we will not consider them further.

3.4.3. Roundoff Error Analysis for Orthogonal Matrices

This analysis proves backward stability for the QR decomposition and for many of the algorithms for eigenvalues and singular values that we will discuss.

LEMMA 3.1. *Let P be an exact Householder (or Givens) transformation, and $\tilde{P}$ be its floating point approximation. Then*

$$
\text{fl}(\tilde{P}A) = P(A + E) \quad \|E\|_2 = O(\varepsilon) \cdot \|A\|_2
$$

and

$$
\text{fl}(A\tilde{P}) = (A + F)P \quad \|F\|_2 = O(\varepsilon) \cdot \|A\|_2.
$$

Sketch of Proof. Apply the usual formula $\text{fl}(a \odot b) = (a \odot b)(1 + \varepsilon)$ to the formulas for computing and applying $\tilde{P}$. See Question 3.16. □

In words, this says that applying a single orthogonal matrix is backward stable.

THEOREM 3.5. *Consider applying a sequence of orthogonal transformations to A_0. Then the computed product is an exact orthogonal transformation of $A_0 + \delta A$, where $\|\delta A\|_2 = O(\varepsilon)\|A\|_2$. In other words, the entire computation is backward stable:*

$$\mathrm{fl}(\tilde{P}_j \tilde{P}_{j-1} \cdots \tilde{P}_1 A_0 \tilde{Q}_1 \tilde{Q}_2 \cdots \tilde{Q}_j) = P_j \cdots P_1 (A_0 + E) Q_1 \cdots Q_j$$

with $\|E\|_2 = j \cdot O(\varepsilon) \cdot \|A\|_2$. Here, as in Lemma 3.1, $\tilde{P}_i$ and $\tilde{Q}_i$ are floating point orthogonal matrices and P_i and Q_i are exact orthogonal matrices.

Proof. Let $\bar{P}_j \equiv P_j \cdots P_1$ and $\bar{Q}_j \equiv Q_1 \cdots Q_j$. We wish to show that $A_j \equiv \mathrm{fl}(\tilde{P}_j A_{j-1} \tilde{Q}_j) = \bar{P}_j (A + E_j) \bar{Q}_j$ for some $\|E_j\|_2 = j O(\varepsilon)\|A\|_2$. We use Lemma 3.1 recursively. The result is vacuously true for $j = 0$. Now assume that the result is true for $j - 1$. Then we compute

$$
\begin{aligned}
B &= \mathrm{fl}(\tilde{P}_j A_{j-1}) \\
&= P_j(A_{j-1} + E') \text{ by Lemma 3.1} \\
&= P_j(\bar{P}_{j-1}(A + E_{j-1})\bar{Q}_{j-1} + E') \text{ by induction} \\
&= \bar{P}_j(A + E_{j-1} + \bar{P}_{j-1}^T E' \bar{Q}_{j-1}^T)\bar{Q}_{j-1} \\
&\equiv \bar{P}_j(A + E'')\bar{Q}_{j-1},
\end{aligned}
$$

where

$$
\begin{aligned}
\|E''\|_2 &= \|E_{j-1} + \bar{P}_{j-1}^T E' \bar{Q}_{j-1}^T\|_2 \leq \|E_{j-1}\|_2 + \|\bar{P}_{j-1}^T E' \bar{Q}_{j-1}^T\|_2 \\
&= \|E_{j-1}\|_2 + \|E'\|_2 \\
&= j O(\varepsilon)\|A\|_2
\end{aligned}
$$

since $\|E_{j-1}\|_2 = (j-1)O(\varepsilon)\|A\|_2$ and $\|E'\|_2 = O(\varepsilon)\|A\|_2$. Postmultiplication by $\tilde{Q}_j$ is handled in the same way. $\square$

3.4.4. Why Orthogonal Matrices?

Let us consider how the error would grow if we were to multiply by a sequence of *nonorthogonal* matrices in Theorem 3.5 instead of orthogonal matrices. Let X be the exact nonorthogonal transformation and $\tilde{X}$ be its floating point approximation. Then the usual floating point error analysis of matrix multiplication tells us that

$$\mathrm{fl}(\tilde{X} A) = XA + E = X(A + X^{-1}E) \equiv X(A + F),$$

where $\|E\|_2 \leq O(\varepsilon)\|X\|_2 \cdot \|A\|_2$ and so $\|F\|_2 \leq \|X^{-1}\|_2 \cdot \|E\|_2 \leq O(\varepsilon) \cdot \kappa_2(X) \cdot \|A\|_2$.

So the error $\|E\|_2$ is magnified by the condition number $\kappa_2(X) \geq 1$. In a larger product $\tilde{X}_k \cdots \tilde{X}_1 A \tilde{Y}_1 \cdots \tilde{Y}_k$ the error would be magnified by $\prod_i \kappa_2(X_i) \cdot \kappa_2(Y_i)$. This factor is minimized if and only if all X_i and Y_i are orthogonal (or scalar multiples of orthogonal matrices), in which case the factor is one.

3.5.   Rank-Deficient Least Squares Problems

So far we have assumed that A has full rank when minimizing $\|Ax - b\|_2$. What happens when A is rank deficient or "close" to rank deficient? Such problems arise in practice in many ways, such as extracting signals from noisy data, solution of some integral equations, digital image restoration, computing inverse Laplace transforms, and so on [141, 142]. These problems are very ill-conditioned, so we will need to impose extra conditions on their solutions to make them well-conditioned. Making an ill-conditioned problem well-conditioned by imposing extra conditions on the solution is called *regularization* and is also done in other fields of numerical analysis when ill-conditioned problems arise.

For example, the next proposition shows that if A is exactly rank deficient, then the least squares solution is not even unique.

PROPOSITION 3.1. *Let A be m-by-n with $m \geq n$ and rank $A = r < n$. Then there is an $n - r$ dimensional set of vectors x that minimize $\|Ax - b\|_2$.*

Proof. Let $Az = 0$. Then if x minimizes $\|Ax - b\|_2$, so does $x + z$. $\square$

Because of roundoff in the entries of A, or roundoff during the computation, it is most often the case that A will have one or more very small computed singular values, rather than some exactly zero singular values. The next proposition shows that in this case, the unique solution is likely to be very large and is certainly very sensitive to error in the right-hand side b (see also Theorem 3.4).

PROPOSITION 3.2. *Let $\sigma_{\min} = \sigma_{\min}(A)$, the smallest singular value of A. Assume $\sigma_{\min} > 0$. Then*

1. *if x minimizes $\|Ax - b\|_2$, then $\|x\|_2 \geq |u_n^T b|/\sigma_{\min}$, where u_n is the last column of U in $A = U\Sigma V^T$.*

2. *changing b to $b + \delta b$ can change x to $x + \delta x$, where $\|\delta x\|_2$ is as large as $\|\delta b\|_2/\sigma_{\min}$.*

In other words, if A is nearly rank deficient ($\sigma_{\min}$ is small), then the solution x is ill-conditioned and possibly very large.

Proof.   For part 1, $x = A^+ b = V\Sigma^{-1}U^T b$, so $\|x\|_2 = \|\Sigma^{-1}U^T b\|_2 \geq |(\Sigma^{-1}U^T b)_n| = |u_n^T b|/\sigma_{\min}$. For part 2, choose δb parallel to u_n. $\square$

We begin our discussion of regularization by showing how to regularize an *exactly* rank-deficient least squares problem: Suppose A is m-by-n with rank $r < n$. Within the $(n - r)$-dimensional solution space, we will look for the unique solution of smallest norm. This solution is characterized by the following proposition.

PROPOSITION 3.3. *When A is exactly singular, the x that minimize $\|Ax - b\|_2$ can be characterized as follows. Let $A = U\Sigma V^T$ have rank $r < n$, and write the SVD of A as*

$$A = [U_1, U_2] \begin{bmatrix} \Sigma_1 & 0 \\ 0 & 0 \end{bmatrix} [V_1, V_2]^T = U_1 \Sigma_1 V_1^T, \tag{3.1}$$

where Σ_1 is $r \times r$ and nonsingular and U_1 and V_1 have r columns. Let $\sigma = \sigma_{\min}(\Sigma_1)$, the smallest nonzero singular value of A. Then

1. *all solutions x can be written $x = V_1 \Sigma_1^{-1} U_1^T b + V_2 z$, z an arbitrary vector.*

2. *the solution x has minimal norm $\|x\|_2$ precisely when $z = 0$, in which case $x = V_1 \Sigma_1^{-1} U_1^T b$ and $\|x\|_2 \le \|b\|_2/\sigma$.*

3. *changing b to $b + \delta b$ can change the minimal norm solution x by at most $\|\delta b\|_2/\sigma$.*

In other words, the norm and condition number of the unique minimal norm solution x depend on the smallest nonzero singular value of A.

Proof. Choose $\tilde{U}$ so $[U, \tilde{U}] = [U_1, U_2, \tilde{U}]$ is an $m \times m$ orthogonal matrix. Then

$$
\begin{aligned}
\|Ax - b\|_2^2 &= \|[U, \tilde{U}]^T(Ax - b)\|_2^2 \\
&= \left\| \begin{bmatrix} U_1^T \\ U_2^T \\ \tilde{U}^T \end{bmatrix} (U_1 \Sigma_1 V_1^T x - b) \right\|_2^2 \\
&= \left\| \begin{bmatrix} \Sigma_1 V_1^T x - U_1^T b \\ U_2^T b \\ \tilde{U}^T b \end{bmatrix} \right\|_2^2 \\
&= \|\Sigma_1 V_1^T x - U_1^T b\|_2^2 + \|U_2^T b\|_2^2 + \|\tilde{U}^T b\|_2^2.
\end{aligned}
$$

1. $\|Ax - b\|_2$ is minimized when $\Sigma_1 V_1^T x = U_1^T b$, or $x = V_1 \Sigma_1^{-1} U_1^T b + V_2 z$ since $V_1^T V_2 z = 0$ for all z.

2. Since the columns of V_1 and V_2 are mutually orthogonal, the Pythagorean theorem implies that $\|x\|_2^2 = \|V_1 \Sigma_1^{-1} U_1^T b\|_2^2 + \|V_2 z\|_2^2$, and this is minimized by $z = 0$.

3. Changing b by δb changes x by at most $\|V_1 \Sigma_1^{-1} U_1^T \delta b\|_2 \le \|\Sigma_1^{-1}\|_2 \|\delta b\|_2 = \|\delta b\|_2/\sigma$. $\square$

Proposition 3.3 tells us that the minimum norm solution x is unique and may be well-conditioned if the smallest *nonzero* singular value is not too small. This is key to a practical algorithm, discussed in the next section.

EXAMPLE 3.7. Suppose that we are doing medical research on the effect of a certain drug on blood sugar level. We collect data from each patient (numbered from $i = 1$ to m) by recording his or her initial blood sugar level $(a_{i,1})$, final blood sugar level (b_i), the amount of drug administered $(a_{i,2})$, and other medical quantities, including body weights on each day of a week-long treatment $(a_{i,3}$ through $a_{i,9})$. In total, there are $n < m$ medical quantities measured for each patient. Our goal is to predict b_i given $a_{i,1}$ through $a_{i,n}$, and we formulate this as the least squares problem $\min_x \|Ax - b\|_2$. We plan to use x to predict the final blood sugar level b_j of future patient j by computing the dot product $\sum_{k=1}^{n} a_{jk} x_k$.

Since people's weight generally does not change significantly from day to day, it is likely that columns 3 through 9 of matrix A, which contain the weights, are very similar. For the sake of argument, suppose that columns 3 and 4 are *identical* (which may be the case if the weights are rounded to the nearest pound). This means that matrix A is rank deficient and that $x_0 = [0, 0, 1, -1, 0, \ldots, 0]^T$ is a right null vector of A. So if x is a (minimum norm) solution of the least squares problem $\min_x \|Ax - b\|_2$, then $x + \beta x_0$ is also a (nonminimum norm) solution for *any* scalar β, including, say, $\beta = 0$ and $\beta = 10^6$. Is there any reason to prefer one value of β over another? The value 10^6 is clearly not a good one, since future patient j, who gains one pound between days 1 and 2, will have that difference of one pound multiplied by 10^6 in the predictor $\sum_{k=1}^{n} a_{jk} x_k$ of final blood sugar level. It is much more reasonable to choose $\beta = 0$, corresponding to the minimum norm solution x.
$\diamond$

For further justification of using the minimum norm solution for rank-deficient problems, see [141, 142].

When A is square and nonsingular, the unique solution of $Ax = b$ is of course $b = A^{-1}x$. If A has more rows than columns and is possibly rank-deficient, the unique minimum-norm least squares solution may be similarly written $b = A^+b$, where the *Moore–Penrose pseudoinverse* A^+ is defined as follows.

DEFINITION 3.2. *(Moore–Penrose pseudoinverse A^+ for possibly rank-deficient A)*

Let $A = U\Sigma V^T = U_1 \Sigma_1 V_1^T$ as in equation (3.1). Then $A^+ \equiv V_1 \Sigma_1^{-1} U_1^T$. This is also written $A^+ = V^T \Sigma^+ U$, where $\Sigma^+ = \begin{bmatrix} \Sigma_1 & 0 \\ 0 & 0 \end{bmatrix}^+ = \begin{bmatrix} \Sigma_1^{-1} & 0 \\ 0 & 0 \end{bmatrix}$.

So the solution of the least squares problem is always $x = A^+b$, and when A is rank deficient, x has minimum norm.

3.5.1. Solving Rank-Deficient Least Squares Problems Using the SVD

Our goal is to compute the minimum norm solution x, despite roundoff. In the last section, we saw that the minimal norm solution was unique and had a condition number depending on the smallest nonzero singular value. Therefore, computing the minimum norm solution requires knowing the smallest nonzero singular value and hence also the rank of A. The main difficulty is that the rank of a matrix changes discontinuously as a function of the matrix.

For example, the 2-by-2 matrix $A = \text{diag}(1, 0)$ is exactly singular, and its smallest nonzero singular value is $\sigma = 1$. As described in Proposition 3.3, the minimum norm least squares solution to $\min_x \|Ax - b\|_2$ with $b = [1, 1]^T$ is $x = [1, 0]^T$, with condition number $1/\sigma = 1$. But if we make an arbitrarily tiny perturbation to get $\hat{A} = \text{diag}(1, \epsilon)$, then σ drops to ϵ and $x = [1, 1/\epsilon]^T$ becomes enormous, as does its condition number $1/\epsilon$. In general, roundoff will make such tiny perturbations, of magnitude $O(\varepsilon)\|A\|_2$. As we just saw, this can increase the condition number from $1/\sigma$ to $1/\varepsilon$.

We deal with this discontinuity algorithmically as follows. In general each computed singular value $\hat{\sigma}_i$ satisfies $|\hat{\sigma}_i - \sigma_i| \leq O(\varepsilon)\|A\|_2$. This is a consequence of backward stability: the computed SVD will be the exact SVD of a slightly different matrix: $\hat{A} = \hat{U}\hat{\Sigma}\hat{V}^T = A + \delta A$, with $\|\delta A\| = O(\varepsilon) \cdot \|A\|$. (This is discussed in detail in Chapter 5.) This means that any $\hat{\sigma}_i \leq O(\varepsilon)\|A\|_2$ can be treated as zero, because roundoff makes it indistinguishable from 0. In the above 2-by-2 example, this means we would set the ϵ in $\hat{A}$ to zero before solving the least squares problem. This would raise the smallest nonzero singular value from ϵ to 1 and correspondingly decrease the condition number from $1/\epsilon$ to $1/\sigma = 1$.

More generally, let tol be a user-supplied measure of uncertainty in the data A. Roundoff implies that $\text{tol} \geq \varepsilon \cdot \|A\|$, but it may be larger, depending on the source of the data in A. Now set $\tilde{\sigma}_i = \hat{\sigma}_i$ if $\hat{\sigma}_i > \text{tol}$, and $\tilde{\sigma}_i = 0$ otherwise. Let $\tilde{\Sigma} = \text{diag}(\tilde{\sigma}_i)$. We call $\hat{U}\tilde{\Sigma}\hat{V}^T$ the *truncated SVD* of A, because we have set singular values smaller than tol to zero. Now we solve the least squares problem using the truncated SVD instead of the original SVD. This is justified since $\|\hat{U}\tilde{\Sigma}\hat{V}^T - \hat{U}\hat{\Sigma}\hat{V}^T\|_2 = \|\hat{U}(\tilde{\Sigma} - \hat{\Sigma})\hat{V}^T\|_2 < \text{tol}$; i.e., the change in A caused by changing each $\hat{\sigma}_i$ to $\tilde{\sigma}_i$ is less than the user's inherent uncertainty in the data. The motivation for using $\tilde{\Sigma}$ instead of $\hat{\Sigma}$ is that of all matrices within distance tol of $\hat{\Sigma}$, $\tilde{\Sigma}$ maximizes the smallest nonzero singular value σ. In other words, it minimizes both the norm of the minimum norm least squares solution x and its condition number.   The picture below illustrates the geometric relationships among the input matrix A, $\hat{A} = \hat{U}\hat{\Sigma}\hat{V}^T$, and $\tilde{A} = \hat{U}\tilde{\Sigma}\hat{V}^T$, where we we think of each matrix as a point in Euclidean space $\mathbb{R}^{m \cdot n}$. In this space, the rank-deficient matrices form a surface, as shown below:

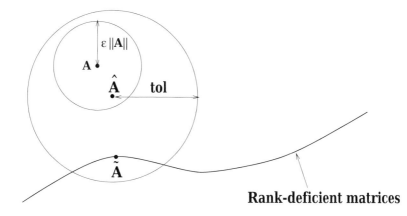

Rank-deficient matrices

EXAMPLE 3.8. We illustrate the above procedure on two 20-by-10 rank-deficient matrices A_1 (of rank $r_1 = 5$) and A_2 (of rank $r_2 = 7$). We write the SVDs of either A_1 or A_2 as $A_i = U_i \Sigma_i V_i^T$, where the common dimension of U_i, Σ_i, and V_i is the rank r_i of A_i; this is the same notation as in Proposition 3.3. The r_i nonzero singular values of A_i (singular values of Σ_i) are shown as red x's in Figure 3.4 (for A_1) and Figure 3.5 (for A_2). Note that A_1 in Figure 3.4 has five large nonzero singular values (all slightly exceeding 1 and so plotted on top of one another, on the right edge the graph), whereas the seven nonzero singular values of A_2 in Figure 3.5 range down to $1.2 \cdot 10^{-9} \approx$ tol.

We then choose an r_i-dimensional vector x_i', and let $x_i = V_i x_i'$ and $b_i = A_i x_i = U_i \Sigma_i x_i'$, so x_i is the exact minimum norm solution minimizing $\|A_i x_i - b_i\|_2$. Then we consider a sequence of perturbed problems $A_i + \delta A$, where the perturbation δA is chosen randomly to have a range of norms, and solve the least squares problems $\|(A_i + \delta A)y_i - b_i\|_2$ using the truncated least squares procedure with tol $= 10^{-9}$. The blue lines in Figures 3.4 and 3.5 plot the computed rank of $A_i + \delta A$ (number of computed singular values exceeding tol $= 10^{-9}$) versus $\|\delta A\|_2$ (in the top graphs), and the error $\|y_i - x_i\|_2 / \|x_i\|_2$ (in the bottom graphs). The Matlab code for producing these figures is in HOMEPAGE/Matlab/RankDeficient.m.

The simplest case is in Figure 3.4, so we consider it first. $A_1 + \delta A$ will have five singular values near or slightly exceeding 1 and the other five equal to $\|\delta A\|_2$ or less. For $\|\delta A\|_2 <$ tol, the computed rank of $A_1 + \delta A$ stays the same as that of A_1, namely, 5. The error also increases slowly from near machine epsilon ($\approx 10^{-16}$) to about 10^{-10} near $\|\delta A\|_2 =$ tol, and then both the rank and the error jump, to 10 and 1, respectively, for larger $\|\delta A\|_2$. This is consistent with our analysis in Proposition 3.3, which says that the condition number is the reciprocal of the smallest nonzero singular value, i.e., the smallest singular value exceeding tol. For $\|\delta A\|_2 <$ tol, this smallest nonzero singular value is near to, or slightly exceeds, 1. Therefore Proposition 3.3 predicts an error of $\|\delta A\|_2 / O(1) = \|\delta A\|_2$. This well-conditioned situation is confirmed by the small error plotted to the left of $\|\delta A\|_2 =$ tol in the bottom graph of Figure 3.4. On the other hand, when $\|\delta A\|_2 >$ tol, then the smallest nonzero

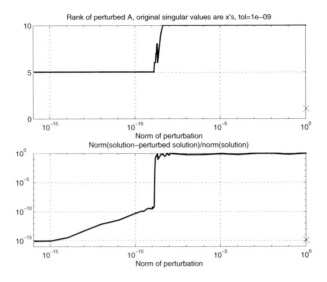

Fig. 3.4. *Graph of truncated least squares solution of* $\min_{y_1} \|(A_1 + \delta A)y_1 - b_1\|_2$, *using* $\mathrm{tol} = 10^{-9}$. *The singular values of* A_1 *are shown as red x's. The norm* $\|\delta A\|_2$ *is the horizontal axis. The top graph plots the rank of* $A_1 + \delta A$, *i.e., the numbers of singular values exceeding* tol. *The bottom graph plots* $\|y_1 - x_1\|_2/\|x_1\|_2$, *where* x_1 *is the solution with* $\delta A = 0$.

singular value is $O(\|\delta A\|_2)$, which is quite small, causing the error to jump to $\|\delta A\|_2/O(\|\delta A\|_2) = O(1)$, as shown to the right of $\|\delta A\|_2 = \mathrm{tol}$ in the bottom graph of Figure 3.4.

In Figure 3.5, the nonzero singular values of A_2 are also shown as red x's; the smallest one, $1.2 \cdot 10^{-9}$, is just larger than tol. So the predicted error when $\|\delta A\|_2 < \mathrm{tol}$ is $\|\delta A\|_2/10^{-9}$, which grows to $O(1)$ when $\|\delta A\|_2 = \mathrm{tol}$. This is confirmed by the bottom graph in Figure 3.5. $\diamond$

3.5.2. Solving Rank-Deficient Least Squares Problems Using QR with Pivoting

A cheaper but sometimes less accurate alternative to the SVD is *QR with pivoting*. In exact arithmetic, if A had rank $r < n$ and its first r columns were independent, then its QR decomposition would look like

$$A = QR = Q \begin{bmatrix} R_{11} & R_{12} \\ 0 & 0 \\ 0 & 0 \end{bmatrix},$$

where R_{11} is r-by-r and nonsingular and R_{12} is r-by-$(n - r)$. With roundoff, we might hope to compute

$$R = \begin{bmatrix} R_{11} & R_{12} \\ 0 & R_{22} \\ 0 & 0 \end{bmatrix}$$

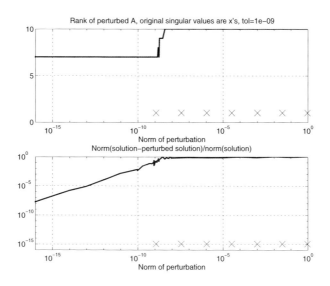

Fig. 3.5. *Graph of truncated least squares solution of* $\min_{y_2} \|(A_2 + \delta A)y_2 - b_2\|_2$, *using* tol $= 10^{-9}$. *The singular values of A_2 are shown as red x's. The norm $\|\delta A\|_2$ is the horizontal axis. The top graph plots the rank of $A_2 + \delta A$, i.e., the numbers of singular values exceeding* tol. *The bottom graph plots $\|y_2 - x_2\|_2/\|x_2\|_2$, where x_2 is the solution with $\delta A = 0$.*

with $\|R_{22}\|_2$ very small, on the order of $\varepsilon\|A\|_2$. In this case we could just set $R_{22} = 0$ and minimize $\|Ax - b\|_2$ as follows: let $[Q, \tilde{Q}]$ be square and orthogonal so that

$$
\|Ax - b\|_2^2 = \left\| \begin{bmatrix} Q^T \\ \tilde{Q}^T \end{bmatrix} (Ax - b) \right\|_2^2 = \left\| \begin{bmatrix} Rx - Q^T b \\ -\tilde{Q}^T b \end{bmatrix} \right\|_2^2
$$
$$
= \|Rx - Q^T b\|_2^2 + \|\tilde{Q}^T b\|_2^2.
$$

Write $Q = [Q_1, Q_2]$ and $x = \begin{bmatrix} x_1 \\ x_2 \end{bmatrix}$ conformally with $R = \begin{bmatrix} R_{11} & R_{12} \\ 0 & 0 \end{bmatrix}$ so that

$$
\|Ax - b\|_2^2 = \|R_{11}x_1 + R_{12}x_2 - Q_1^T b\|_2^2 + \|Q_2^T b\|_2^2 + \|\tilde{Q}^T b\|_2^2
$$

is minimized by choosing $x = \begin{bmatrix} R_{11}^{-1}(Q_1^T b - R_{12}x_2) \\ x_2 \end{bmatrix}$ for any x_2. Note that the choice $x_2 = 0$ does not necessarily minimize $\|x\|_2$, but it is a reasonable choice, especially if R_{11} is well-conditioned and $R_{11}^{-1}R_{12}$ is small.

Unfortunately, this method is not reliable since R may be nearly rank deficient even if no R_{22} is small. For example, the n-by-n bidiagonal matrix

$$
A = \begin{bmatrix} \frac{1}{2} & 1 & & & \\ & \ddots & \ddots & & \\ & & \ddots & & 1 \\ & & & & \frac{1}{2} \end{bmatrix}
$$

has $\sigma_{\min}(A) \approx 2^{-n}$, but $A = Q \cdot R$ with $Q = I$ and $R = A$, and no R_{22} is small.

To deal with this failure to recognize rank deficiency, we may do *QR with column pivoting*. This means that we factorize $AP = QR$, P being a permutation matrix. This idea is that at step i (which ranges from 1 to n, the number of columns) we select from the unfinished part of A (columns i to n and rows i to m) the column of largest norm and exchange it with the ith column. We then proceed to compute the usual Householder transformation to zero out column i in entries $i+1$ to m. This pivoting strategy attempts to keep R_{11} as well-conditioned as possible and R_{22} as small as possible.

EXAMPLE 3.9. If we compute the QR decomposition with column pivoting to the last example (.5 on the diagonal and 1 on the superdiagonal) with $n = 11$, we get $R_{11,11} = 4.23 \cdot 10^{-4}$, a reasonable approximation to $\sigma_{\min}(A) = 3.66 \cdot 10^{-4}$. Note that $R_{nn} \geq \sigma_{\min}(A)$ since $\sigma_{\min}(A)$ is the norm of the smallest perturbation that can lower the rank, and setting R_{nn} to 0 lowers rank. ◇

One can show only $\frac{R_{nn}}{\sigma_{\min}(A)} \lesssim 2^n$, but usually R_{nn} is a reasonable approximation to $\sigma_{\min}(A)$. The worst case, however, is as bad as worst-case pivot growth in GEPP.

More sophisticated pivoting schemes than QR with column pivoting, called *rank-revealing* QR algorithms, have been a subject of much recent study. Rank-revealing QR algorithms that detect rank more reliably and sometimes also faster than QR with column pivoting have been developed [28, 30, 48, 50, 109, 126, 128, 150, 196, 236]. We discuss them further in the next section.

QR decomposition with column pivoting is available as subroutine `sgeqpf` in LAPACK. LAPACK also has several similar factorizations available: RQ (`sgerqf`), LQ (`sgelqf`), and QL (`sgeqlf`). Future LAPACK releases will contain improved versions of QR.

3.6. Performance Comparison of Methods for Solving Least Squares Problems

What is the fastest algorithm for solving dense least squares problems? As discussed in section 3.2, solving the normal equations is fastest, followed by QR and the SVD. If A is quite well-conditioned, then the normal equations are about as accurate as the other methods, so even though the normal equations are not numerically stable, they may be used as well. When A is not well-conditioned but far from rank deficient, we should use QR.

Since the design of fast algorithms for rank-deficient least squares problems is a current research area, it is difficult to recommend a single algorithm to use. We summarize a recent study [206] that compared the performance of several algorithms, comparing them to the fastest stable algorithm for the non–rank-deficient case: QR without pivoting, implemented using Householder transformations as described in section 3.4.1, with memory hierarchy optimizations described in Question 3.17. These comparisons were made in double precision arithmetic on an IBM RS6000/590. Included in the comparison were the rank-revealing QR algorithms mentioned in section 3.5.2 and various implementations of the SVD (see section 5.4). Matrices of various sizes and with various singular value distributions were tested. We present results for two singular value distributions:

Type 1: random matrices, where each entry is uniformly distributed from -1 to 1;

Type 2: matrices with singular values distributed geometrically from 1 to ε (in other words, the ith singular value is γ^i, where γ is chosen so that $\gamma^n = \varepsilon$).

Type 1 matrices are generally well-conditioned, and Type 2 matrices are rank-deficient. We tested small square matrices ($n = m = 20$) and large square matrices ($m = n = 1600$). We tested square matrices because if m is sufficiently greater than n in the m-by-n matrix A, it is cheaper to do a QR decomposition as a "preprocessing step" and then perform rank-revealing QR or the SVD on R. (This is done in LAPACK.) If $m \gg n$, then the initial QR decomposition dominates the the cost of the subsequent operations on the n-by-n matrix R, and all the algorithms cost about the same.

The fastest version of rank-revealing QR was that of [30, 196]. On Type 1 matrices, this algorithm ranged from 3.2 times slower than QR without pivoting for $n = m = 20$ to just 1.1 times slower for $n = m = 1600$. On Type 2 matrices, it ranged from 2.3 times slower (for $n = m = 20$) to 1.2 times slower (for $n = m = 1600$). In contrast, the current LAPACK algorithm, `dgeqpf`, was 2 times to 2.5 times slower for both matrix types.

The fastest version of the SVD was the one in [58], although one based on divide-and-conquer (see section 5.3.3) was about equally fast for $n = m = 1600$. (The one based on divide-and-conquer also used much less memory.) For Type 1 matrices, the SVD algorithm was 7.8 times slower (for $n = m = 20$) to 3.3 times slower (for $n = m = 1600$). For Type 2 matrices, the SVD algorithm was 3.5 times slower (for $n = m = 20$) to 3.0 times slower (for $n = m = 1600$). In contrast, the current LAPACK algorithm, `dgelss`, ranged from 4 times slower (for Type 2 matrices with $n = m = 20$) to 97 times slower (for Type 1 matrices with $n = m = 1600$). This enormous slowdown is apparently due to memory hierarchy effects.

Thus, we see that there is a tradeoff between reliability and speed in solving rank-deficient least squares problems: QR without pivoting is fastest but least reliable, the SVD is slowest but most reliable, and rank-revealing QR is in-between. If $m \gg n$, all algorithms cost about the same. The choice of algorithm depends on the relative importance of speed and reliability to the user.

Future LAPACK releases will contain improved versions of both rank-revealing QR and SVD algorithms for the least squares problem.

3.7. References and Other Topics for Chapter 3

The best recent reference on least squares problems is [33], which also discusses variations on the basic problem discussed here (such as constrained, weighted, and updating least squares), different ways to regularize rank-deficient problems, and software for sparse least squares problems. See also chapter 5 of [121] and [168]. Perturbation theory and error bounds for the least squares solution are discussed in detail in [149]. Rank-revealing QR decompositions are discussed in [28, 30, 48, 50, 126, 150, 196, 206, 236]. In particular, these papers examine the tradeoff between cost and accuracy in rank determination, and in [206] there is a comprehensive performance comparison of the available methods for rank-deficient least squares problems.

3.8. Questions for Chapter 3

QUESTION 3.1. *(Easy)* Show that the two variations of Algorithm 3.1, CGS and MGS, are mathematically equivalent by showing that the two formulas for r_{ji} yield the same results in exact arithmetic.

QUESTION 3.2. *(Easy)* This question will illustrate the difference in numerical stability among three algorithms for computing the QR factorization of a matrix: Householder QR (Algorithm 3.2), CGS (Algorithm 3.1), and MGS (Algorithm 3.1). Obtain the Matlab program QRStability.m from HOMEPAGE/Matlab/QRStability.m. This program generates random matrices with user-specified dimensions m and n and condition number cnd, computes their QR decomposition using the three algorithms, and measures the accuracy of the results. It does this with the *residual* $\|A - Q \cdot R\| / \|A\|$, which should be around machine epsilon ε for a stable algorithm, and the *orthogonality of Q* $\|Q^T \cdot Q - I\|$, which should also be around ε. Run this program for small matrix dimensions (such as m= 6 and n= 4), modest numbers of random matrices (samples= 20), and condition numbers ranging from cnd= 1 up to cnd= 10^{15}. Describe what you see. Which algorithms are more stable than others? See if you can describe how large $\|Q^T \cdot Q - I\|$ can be as a function of choice of algorithm, cnd and ε.

QUESTION 3.3. *(Medium; Hard)* Let A be m-by-n, $m \geq n$, and have full rank.

1. *(Medium)* Show that $\begin{bmatrix} I & A \\ A^T & 0 \end{bmatrix} \cdot \begin{bmatrix} r \\ x \end{bmatrix} = \begin{bmatrix} b \\ 0 \end{bmatrix}$ has a solution where x minimizes $\|Ax - b\|_2$. One reason for this formulation is that we can apply iterative refinement to this linear system if we want a more accurate answer (see section 2.5).

2. *(Medium)* What is the condition number of the coefficient matrix in terms of the singular values of A? Hint: Use the SVD of A.

3. *(Medium)* Give an explicit expression for the inverse of the coefficient matrix, as a block 2-by-2 matrix. Hint: Use 2-by-2 block Gaussian elimination. Where have we previously seen the (2,1) block entry?

4. *(Hard)* Show how to use the QR decomposition of A to implement an iterative refinement algorithm to improve the accuracy of x.

QUESTION 3.4. *(Medium) Weighted least squares:* If some components of $Ax - b$ are more important than others, we can weight them with a scale factor d_i and solve the weighted least squares problem min $\|D(Ax - b)\|_2$ instead, where D has diagonal entries d_i. More generally, recall that if C is symmetric positive definite, then $\|x\|_C \equiv (x^T C x)^{1/2}$ is a norm, and we can consider minimizing $\|Ax - b\|_C$. Derive the normal equations for this problem, as well as the formulation corresponding to the previous question.

QUESTION 3.5. *(Medium; Z. Bai)* Let $A \in \mathbb{R}^{n \times n}$ be positive definite. Two vectors u_1 and u_2 are called A-orthogonal if $u_1^T A u_2 = 0$. If $U \in \mathbb{R}^{n \times r}$ and $U^T A U = I$, then the columns of U are said to be A-orthonormal. Show that every subspace has an A-orthonormal basis.

QUESTION 3.6. *(Easy; Z. Bai)* Let A have the form

$$A = \begin{bmatrix} R \\ S \end{bmatrix},$$

where R is n-by-n and upper triangular, and S is m-by-n and dense. Describe an algorithm using Householder transformations for reducing A to upper triangular form. Your algorithm should not "fill in" the zeros in R and thus require fewer operations than would Algorithm 3.2 applied to A.

QUESTION 3.7. *(Medium; Z. Bai)* If $A = R + uv^T$, where R is an upper triangular matrix, and u and v are column vectors, describe an efficient algorithm to compute the QR decomposition of A. Hint: Using Givens rotations, your algorithm should take $O(n^2)$ operations. In contrast, Algorithm 3.2 would take $O(n^3)$ operations.

QUESTION 3.8. *(Medium; Z. Bai)* Let $x \in \mathbb{R}^n$ and let P be a Householder matrix such that $Px = \pm\|x\|_2 e_1$. Let $G_{1,2}, \ldots, G_{n-1,n}$ be Givens rotations, and let $Q = G_{1,2} \cdots G_{n-1,n}$. Suppose $Qx = \pm\|x\|_2 e_1$. Must P equal Q? (You need to give a proof or a counterexample.)

QUESTION 3.9. *(Easy; Z. Bai)* Let A be m-by-n, with SVD $A = U\Sigma V^T$. Compute the SVDs of the following matrices in terms of U, Σ, and V:

1. $(A^T A)^{-1}$,

2. $(A^T A)^{-1} A^T$,

3. $A(A^T A)^{-1}$,

4. $A(A^T A)^{-1} A^T$.

QUESTION 3.10. *(Medium; R. Schreiber)* Let A_k be a best rank-k approximation of the matrix A, as defined in Part 9 of Theorem 3.3. Let σ_i be the ith singular value of A. Show that A_k is unique if $\sigma_k > \sigma_{k+1}$.

QUESTION 3.11. *(Easy; Z. Bai)* Let A be m-by-n. Show that $X = A^+$ (the Moore–Penrose pseudoinverse) minimizes $\|AX - I\|_F$ over all n-by-m matrices X. What is the value of this minimum?

QUESTION 3.12. *(Medium; Z. Bai)* Let A, B, and C be matrices with dimensions such that the product $A^T C B^T$ is well defined. Let $\mathcal{X}$ be the set of matrices X minimizing $\|AXB - C\|_F$, and let X_0 be the unique member of $\mathcal{X}$ minimizing $\|X\|_F$. Show that $X_0 = A^+ C B^+$. Hint: Use the SVDs of A and B.

QUESTION 3.13. *(Medium; Z. Bai)* Show that the Moore–Penrose pseudoinverse of A satisfies the following identities:

$$
\begin{aligned}
AA^+ A &= A, \\
A^+ AA^+ &= A^+, \\
A^+ A &= (A^+ A)^T, \\
AA^+ &= (AA^+)^T.
\end{aligned}
$$

QUESTION 3.14. *(Medium)* Prove part 4 of Theorem 3.3: Let $H = \begin{bmatrix} 0 & A^T \\ A & 0 \end{bmatrix}$, where A is square and $A = U\Sigma V^T$ is its SVD. Let $\Sigma = \operatorname{diag}(\sigma_1, \ldots, \sigma_n)$, $U = [u_1, \ldots, u_n]$, and $V = [v_1, \ldots, v_n]$. Prove that the $2n$ eigenvalues of H are $\pm\sigma_i$, with corresponding unit eigenvectors $\frac{1}{\sqrt{2}}\begin{bmatrix} v_i \\ \pm u_i \end{bmatrix}$. Extend to the case of rectangular A.

QUESTION 3.15. *(Medium)* Let A be m-by-n, $m < n$, and of full rank. Then $\min \|Ax - b\|_2$ is called an *underdetermined least squares problem*. Show that the solution is an $(n - m)$-dimensional set. Show how to compute the unique minimum norm solution using appropriately modified normal equations, QR decomposition, and SVD.

QUESTION 3.16. *(Medium)* Prove Lemma 3.1.

QUESTION 3.17. *(Hard)* In section 2.6.3, we showed how to reorganize Gaussian elimination to perform Level 2 BLAS and Level 3 BLAS at each step in order to exploit the higher speed of these operations. In this problem, we will show how to apply a sequence of Householder transformations using Level 2 and Level 3 BLAS.

1. Let $u_1, \ldots, u_b$ be a sequence of vectors of dimension n, where $\|u_i\|_2 = 1$ and the first $i - 1$ components of u_i are zero. Let $P = P_b \cdot P_{b-1} \cdots P_1$, where $P_i = I - 2u_i u_i^T$ is a Householder transformation. Show that there is a b-by-b lower triangular matrix T such that $P = I - UTU^T$, where $U = [u_1, \ldots, u_b]$. In particular, provide an algorithm for computing the entries of T. This identity shows that we can replace multiplication by b Householder transformations P_1 through P_b by three matrix multiplications by U, T, and U^T (plus the cost of computing T).

2. Let House(x) be a function of the vector x which returns a unit vector u such that $(I - 2uu^T)x = \|x\|_2 e_1$; we showed how to implement House(x) in section 3.4. Then Algorithm 3.2 for computing the QR decomposition of the m-by-n matrix A may be written as

   for $i = 1 : m$
       $u_i = \text{House}(A(i : m, i))$
       $P_i = I - 2u_i u_i^T$
       $A(i : m, i : n) = P_i A(i : m, i : n)$
   endfor

   Show how to implement this in terms of the Level 2 BLAS in an efficient way (in particular, matrix-vector multiplications and rank-1 updates). What is the floating point operation count? (Just the high-order terms in n and m are enough.) It is sufficient to write a short program in the same notation as above (although trying it in Matlab and comparing with Matlab's own QR factorization are a good way to make sure that you are right!).

3. Using the results of step (1), show how to implement QR decomposition in terms of Level 3 BLAS. What is the operation count? This technique is used to accelerate the QR decomposition, just as we accelerated Gaussian elimination in section 2.6. It is used in the LAPACK routine `sgeqrf`.

QUESTION 3.18. *(Medium)* It is often of interest to solve *constrained least squares problems*, where the solution x must satisfy a linear or nonlinear constraint in addition to minimizing $\|Ax - b\|_2$. We consider one such problem here. Suppose that we want to choose x to minimize $\|Ax - b\|_2$ subject to the linear constraint $Cx = d$. Suppose also that A is m-by-n, C is p-by-n, and C has full rank. We also assume that $p \leq n$ (so $Cx = d$ is guaranteed to be consistent) and $n \leq m + p$ (so the system is not underdetermined). Show that there is a unique solution under the assumption that $\left[\begin{smallmatrix} A \\ C \end{smallmatrix} \right]$ has full column rank. Show how to compute x using two QR decompositions and some matrix-vector multiplications and solving some triangular systems of equations. Hint: Look at LAPACK routine `sgglse` and its description in the LAPACK manual [10] (NETLIB/lapack/lug/lapack_lug.html).

QUESTION 3.19. *(Hard; Programming)* Write a program (in Matlab or any other language) to update a geodetic database using least squares, as described in Example 3.3. Take as input a set of "landmarks," their approximate coordinates (x_i, y_i), and a set of new angle measurements θ_j and distance measurements L_{ij}. The output should be corrections $(\delta x_i, \delta y_i)$ for each landmark, an error bound for the corrections, and a picture (triangulation) of the old and new landmarks.

QUESTION 3.20. *(Hard)* Prove Theorem 3.4.

QUESTION 3.21. *(Medium)* Redo Example 3.1, using a rank-deficient least squares technique from section 3.5.1. Does this improve the accuracy of the high-degree approximating polynomials?

4

Nonsymmetric Eigenvalue Problems

4.1. Introduction

We discuss canonical forms (in section 4.2), perturbation theory (in section 4.3), and algorithms for the eigenvalue problem for a single nonsymmetric matrix A (in section 4.4). Chapter 5 is devoted to the special case of real symmetric matrices $A = A^T$ (and the SVD). Section 4.5 discusses generalizations to eigenvalue problems involving more than one matrix, including motivating applications from the analysis of vibrating systems, the solution of linear differential equations, and computational geometry. Finally, section 4.6 summarizes all the canonical forms, algorithms, costs, applications, and available software in a list.

One can roughly divide the algorithms for the eigenproblem into two groups: *direct methods* and *iterative methods*. This chapter considers only direct methods, which are intended to compute all of the eigenvalues, and (optionally) eigenvectors. Direct methods are typically used on dense matrices and cost $O(n^3)$ operations to compute all eigenvalues and eigenvectors; this cost is relatively insensitive to the actual matrix entries.

The main direct method used in practice is *QR iteration with implicit shifts* (see section 4.4.8). It is interesting that after more than 30 years of dependable service, convergence failures of this algorithm have quite recently been observed, analyzed, and patched [25, 65]. But there is still no global convergence proof, even though the current algorithm is considered quite reliable. So the problem of devising an algorithm that is numerically stable and globally (and quickly!) convergent remains open. (Note that "direct" methods must still iterate, since finding eigenvalues is mathematically equivalent to finding zeros of polynomials, for which no noniterative methods can exist. We call a method *direct* if experience shows that it (nearly) never fails to converge in a fixed number of iterations.)

Iterative methods, which are discussed in Chapter 7, are usually applied to sparse matrices or matrices for which matrix-vector multiplication is the only convenient operation to perform. Iterative methods typically provide approximations only to a subset of the eigenvalues and eigenvectors and are usually run only long enough to get a few adequately accurate eigenvalues rather than a large number. Their convergence properties depend strongly on the matrix entries.

4.2. Canonical Forms

DEFINITION 4.1. *The polynomial $p(\lambda) = \det(A - \lambda I)$ is called the* character-istic polynomial *of A. The roots of $p(\lambda) = 0$ are the* eigenvalues *of A.*

Since the degree of the characteristic polynomial $p(\lambda)$ equals n, the dimension of A, it has n roots, so A has n eigenvalues.

DEFINITION 4.2. *A nonzero vector x satisfying $Ax = \lambda x$ is a* (right) *eigen-vector for the eigenvalue λ. A nonzero vector y such that $y^* A = \lambda y^*$ is a* left *eigenvector. (Recall that $y^* = (\overline{y})^T$ is the* conjugate transpose *of y.)*

Most of our algorithms will involve transforming the matrix A into simpler, or *canonical* forms, from which it is easy to compute its eigenvalues and eigenvectors. These transformations are called *similarity transformations* (see below). The two most common canonical forms are called the *Jordan form* and *Schur form*.   The Jordan form is useful theoretically but is very hard to compute in a numerically stable fashion, which is why our algorithms will aim to compute the Schur form instead.

To motivate Jordan and Schur forms, let us ask which matrices have the property that their eigenvalues are easy to compute. The easiest case would be a *diagonal matrix*, whose eigenvalues are simply its diagonal entries. Equally easy would be a *triangular matrix*, whose eigenvalues are also its diagonal entries. Below we will see that a matrix in Jordan or Schur form is triangular. But recall that a real matrix can have complex eigenvalues, since the roots of its characteristic polynomial may be real or complex. Therefore, there is not always a *real* triangular matrix with the same eigenvalues as a *real* general matrix, since a real triangular matrix can only have real eigenvalues. Therefore, we must either use complex numbers or look beyond real triangular matrices for our canonical forms for real matrices. It will turn out to be sufficient to consider *block triangular matrices*, i.e., matrices of the form

$$
A = \begin{bmatrix}
A_{11} & A_{12} & \cdots & A_{1b} \\
 & A_{22} & \cdots & A_{2b} \\
 & & \ddots & \vdots \\
 & & & A_{bb}
\end{bmatrix},
\tag{4.1}
$$

where each A_{ii} is square and all entries below the A_{ii} blocks are zero. One can easily show that the characteristic polynomial $\det(A - \lambda I)$ of A is the product

$\prod_{i=1}^{b} \det(A_{ii} - \lambda I)$ of the characteristic polynomials of the A_{ii} and therefore that the set $\lambda(A)$ of eigenvalues of A is the union $\cup_{i=1}^{b}\lambda(A_{ii})$ of the sets of eigenvalues of the diagonal blocks A_{ii} (see Question 4.1). The canonical forms that we compute will be block triangular and will proceed computationally by breaking up large diagonal blocks into smaller ones. If we start with a complex matrix A, the final diagonal blocks will be 1-by-1, so the ultimate canonical form will be triangular. If we start with a real matrix A, the ultimate canonical form will have 1-by-1 diagonal blocks (corresponding to real eigenvalues) and 2-by-2 diagonal blocks (corresponding to complex conjugate pairs of eigenvalues); such a block triangular matrix is called *quasi-triangular*.

It is also easy to find the eigenvectors of a (block) triangular matrix; see section 4.2.1.

DEFINITION 4.3. *Let S be any nonsingular matrix. Then A and $B = S^{-1}AS$ are called* similar *matrices, and S is a* similarity transformation.

PROPOSITION 4.1. *Let $B = S^{-1}AS$, so A and B are similar. Then A and B have the same eigenvalues, and x (or y) is a right (or left) eigenvector of A if and only if $S^{-1}x$ (or S^*y) is a right (or left) eigenvector of B.*

Proof. Using the fact that $\det(X \cdot Y) = \det(X) \cdot \det(Y)$ for any square matrices X and Y, we can write $\det(A - \lambda I) = \det(S^{-1}(A - \lambda I)S) = \det(B - \lambda I)$, so A and B have the same characteristic polynomials. $Ax = \lambda x$ holds if and only if $S^{-1}ASS^{-1}x = \lambda S^{-1}x$ or $B(S^{-1}x) = \lambda(S^{-1}x)$. Similarly, $y^*A = \lambda y^*$ if and only if $y^*SS^{-1}AS = \lambda y^*S$ or $(S^*y)^*B = \lambda(S^*y)^*$. $\square$

THEOREM 4.1. *Jordan canonical form. Given A, there exists a nonsingular S such that $S^{-1}AS = J$, where J is in* Jordan canonical form. *This means that J is block diagonal, with $J = \mathrm{diag}(J_{n_1}(\lambda_1), J_{n_2}(\lambda_2), \ldots, J_{n_k}(\lambda_k))$ and*

$$
J_{n_i}(\lambda_i) = \begin{bmatrix} \lambda_i & 1 & & 0 \\ & \ddots & \ddots & \\ & & \ddots & 1 \\ 0 & & & \lambda_i \end{bmatrix}^{n_i \times n_i}.
$$

J is unique, up to permutations of its diagonal blocks.

For a proof of this theorem, see a book on linear algebra such as [110] or [139].

Each $J_m(\lambda)$ is called a *Jordan block* with eigenvalue λ of *algebraic multiplicity* m. If some $n_i = 1$, and λ_i is an eigenvalue of only that one Jordan block, then λ_i is called a *simple eigenvalue*. If all $n_i = 1$, so that J is diagonal, A is called *diagonalizable*; otherwise it is called *defective*. An n-by-n defective matrix does *not* have n eigenvectors, as described in more detail in the next

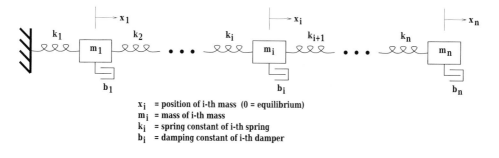

Fig. 4.1. *Damped, vibrating mass-spring system.*

proposition. Although defective matrices are "rare" in a certain well-defined sense, the fact that some matrices do not have n eigenvectors is a fundamental fact confronting anyone designing algorithms to compute eigenvectors and eigenvalues. In section 4.3, we will see some of the difficulties that such matrices cause. Symmetric matrices, discussed in Chapter 5, are never defective.

PROPOSITION 4.2. *A Jordan block has one right eigenvector, $e_1 = [1, 0, \ldots, 0]^T$, and one left eigenvector, $e_n = [0, \ldots, 0, 1]^T$. Therefore, a matrix has n eigenvectors matching its n eigenvalues if and only if it is diagonalizable. In this case, $S^{-1}AS = \mathrm{diag}(\lambda_i)$. This is equivalent to $AS = S\,\mathrm{diag}(\lambda_i)$, so the ith column of S is a right eigenvector for λ_i. It is also equivalent to $S^{-1}A = \mathrm{diag}(\lambda_i)S^{-1}$, so the conjugate transpose of the ith row of S^{-1} is a left eigenvector for λ_i. If all n eigenvalues of a matrix A are distinct, then A is diagonalizable.*

Proof. Let $J = J_m(\lambda)$ for ease of notation. It is easy to see $Je_1 = \lambda e_1$ and $e_n^T J = \lambda e_n^T$, so e_1 and e_n are right and left eigenvectors of J, respectively. To see that J has only one right eigenvector (up to scalar multiples), note that any eigenvector x must satisfy $(J - \lambda I)x = 0$, so x is in the null space of

$$
J - \lambda I = \begin{bmatrix} 0 & 1 & & & \\ & \ddots & \ddots & & \\ & & \ddots & 1 \\ & & & 0 \end{bmatrix}.
$$

But the null space of $J - \lambda I$ is clearly span(e_1), so there is just one eigenvector. If all eigenvalues of A are distinct, then all its Jordan blocks must be 1-by-1, so $J = \mathrm{diag}(\lambda_1, \ldots, \lambda_n)$ is diagonal. $\square$

EXAMPLE 4.1. We illustrate the concepts of eigenvalue and eigenvector with a problem of *mechanical vibrations*. We will see a defective matrix arise in a natural physical context. Consider the damped mass spring system in Figure 4.1, which we will use to illustrate a variety of eigenvalue problems.

Newton's law $F = ma$ applied to this system yields

$$m_i \ddot{x}_i(t) = k_i(x_{i-1}(t) - x_i(t))$$

force on mass i from spring i

$$+k_{i+1}(x_{i+1}(t) - x_i(t)) \qquad (4.2)$$

force on mass i from spring $i + 1$

$$-b_i \dot{x}_i(t)$$

force on mass i from damper i

or

$$M\ddot{x}(t) = -B\dot{x}(t) - Kx(t), \qquad (4.3)$$

where $M = \text{diag}(m_1, \ldots, m_n)$, $B = \text{diag}(b_1, \ldots, b_n)$, and

$$K = \begin{bmatrix} k_1 + k_2 & -k_2 & & & \\ -k_2 & k_2 + k_3 & -k_3 & & \\ & \ddots & \ddots & \ddots & \\ & & -k_{n-1} & k_{n-1} + k_n & -k_n \\ & & & -k_n & k_n \end{bmatrix}.$$

We assume that all the masses m_i are positive. M is called the *mass matrix*, B is the *damping matrix*, and K is the *stiffness matrix*.

Electrical engineers analyzing linear circuits arrive at an analogous equation by applying Kirchoff's and related laws instead of Newton's law. In this case x represents branch currents, M represent inductances, B represents resistances, and K represents admittances (reciprocal capacitances).

We will use a standard trick to change this second-order differential equation to a first-order differential equation, changing variables to

$$y(t) = \begin{bmatrix} \dot{x}(t) \\ x(t) \end{bmatrix}.$$

This yields

$$\begin{aligned} \dot{y}(t) &= \begin{bmatrix} \ddot{x}(t) \\ \dot{x}(t) \end{bmatrix} = \begin{bmatrix} -M^{-1}B\dot{x}(t) - M^{-1}Kx(t) \\ \dot{x}(t) \end{bmatrix} \\ &= \begin{bmatrix} -M^{-1}B & -M^{-1}K \\ I & 0 \end{bmatrix} \cdot \begin{bmatrix} \dot{x}(t) \\ x(t) \end{bmatrix} \\ &= \begin{bmatrix} -M^{-1}B & -M^{-1}K \\ I & 0 \end{bmatrix} \cdot y(t) \equiv Ay(t). \qquad (4.4) \end{aligned}$$

To solve $\dot{y}(t) = Ay(t)$, we assume that $y(0)$ is given (i.e., the initial positions $x(0)$ and velocities $\dot{x}(0)$ are given).

One way to write down the solution of this differential equation is $y(t) = e^{At}y(0)$, where e^{At} is the matrix exponential. We will give another more elementary solution in the special case where A is diagonalizable; this will be

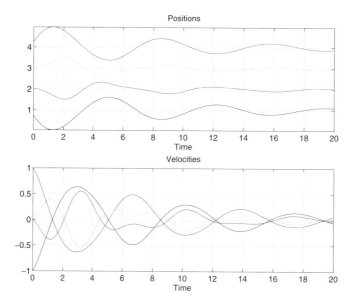

Fig. 4.2. *Positions and velocities of a mass-spring system with four masses $m_1 = m_4 = 2$ and $m_2 = m_3 = 1$. The spring constants are all $k_i = 1$. The damping constants are all $b_i = .4$. The initial displacements are $x_1(0) = -.25$, $x_2(0) = x_3(0) = 0$, and $x_4(0) = .25$. The initial velocities are $v_1(0) = -1$, $v_2(0) = v_3(0) = 0$, and $v_4(0) = 1$. The equilibrium positions are 1, 2, 3, and 4. The software for solving and plotting an arbitrary mass-spring system is HOMEPAGE/Matlab/massspring.m.*

true for almost all choices of m_i, k_i, and b_i. We will return to consider other situations later. (The general problem of computing matrix functions such as e^{At} is discussed further in section 4.5.1 and Question 4.4.)

When A is diagonalizable, we can write $A = S\Lambda S^{-1}$, where $\Lambda = \text{diag}(\lambda_1, \ldots, \lambda_n)$. Then $\dot{y}(t) = Ay(t)$ is equivalent to $\dot{y}(t) = S\Lambda S^{-1}y(t)$ or $S^{-1}\dot{y}(t) = \Lambda S^{-1}y(t)$ or $\dot{z}(t) = \Lambda z(t)$, where $z(t) \equiv S^{-1}y(t)$. This diagonal system of differential equations $\dot{z}_i(t) = \lambda_i z_i(t)$ has solutions $z_i(t) = e^{\lambda_i t}z_i(0)$, so $y(t) = S\text{diag}(e^{\lambda_1 t}, \ldots, e^{\lambda_n t})S^{-1}y(0) = Se^{\Lambda t}S^{-1}y(0)$. A sample numerical solution for four masses and springs is shown in Figure 4.2.

To see the physical significance of the nondiagonalizability of A for a mass-spring system, consider the case of a single mass, spring, and damper, whose differential equation we can simplify to $m\ddot{x}(t) = -b\dot{x}(t) - kx(t)$, and so $A = \begin{bmatrix} -b/m & -k/m \\ 1 & 0 \end{bmatrix}$. The two eigenvalues of A are $\lambda_{\pm} = \frac{b}{2m}(-1 \pm (1 - \frac{4km}{b^2})^{1/2})$. When $\frac{4km}{b^2} < 1$, the system is *overdamped*, and there are two negative real eigenvalues, whose mean value is $-\frac{b}{2m}$. In this case the solution eventually decays monotonically to zero. When $\frac{4km}{b^2} > 1$, the system is *underdamped*, and there are two complex conjugate eigenvalues with real part $-\frac{b}{2m}$. In this case the solution oscillates while decaying to zero. In both cases the system is diagonalizable since the eigenvalues are distinct. When $\frac{4km}{b^2} = 1$, the system is *critically damped*, there are two real eigenvalues equal to $-\frac{b}{2m}$, and A has a single 2-by-2 Jordan block with this eigenvalue. In other words, the non-diagonalizable matrices form the "boundary" between two physical behaviors: oscillation and monotonic decay.

When A is diagonalizable but S is an ill-conditioned matrix, so that S^{-1} is difficult to evaluate accurately, the explicit solution $y(t) = Se^{\Lambda t}S^{-1}y(0)$ will be quite inaccurate and useless numerically. We will use this mechanical system as a running example because it illustrates so many eigenproblems. $\diamond$

To continue our discussion of canonical forms, it is convenient to define the following generalization of an eigenvector.

DEFINITION 4.4. *An* invariant subspace *of A is a subspace $\mathbf{X}$ of $\mathbb{R}^n$, with the property that $x \in \mathbf{X}$ implies that $Ax \in \mathbf{X}$. We also write this as $A\mathbf{X} \subseteq \mathbf{X}$.*

The simplest, one-dimensional invariant subspace is the set span(x) of all scalar multiples of an eigenvector x. Here is the analogous way to build an invariant subspace of larger dimension. Let $X = [x_1, \ldots, x_m]$, where $x_1, \ldots, x_m$ are any set of independent eigenvectors with eigenvalues $\lambda_1, \ldots, \lambda_m$. Then $\mathbf{X} = \text{span}(X)$ is an invariant subspace since $x \in \mathbf{X}$ implies $x = \sum_{i=1}^{m} \alpha_i x_i$ for some scalars α_i, so $Ax = \sum_{i=1}^{m} \alpha_i A x_i = \sum_{i=1}^{m} \alpha_i \lambda_i x_i \in \mathbf{X}$. $A\mathbf{X}$ will equal $\mathbf{X}$ unless some eigenvalue λ_i equals zero. The next proposition generalizes this.

PROPOSITION 4.3. *Let A be n-by-n, let $X = [x_1, \ldots, x_m]$ be any n-by-m matrix with independent columns, and let $\mathbf{X} = \text{span}(X)$, the m-dimensional space spanned by the columns of X. Then $\mathbf{X}$ is an invariant subspace if and only if there is an m-by-m matrix B such that $AX = XB$. In this case the m eigenvalues of B are also eigenvalues of A. (When $m = 1$, $X = [x_1]$ is an eigenvector and B is an eigenvalue.)*

Proof. Assume first that $\mathbf{X}$ is invariant. Then each Ax_i is also in $\mathbf{X}$, so each Ax_i must be a linear combination of a basis of $\mathbf{X}$, say, $Ax_i = \sum_{j=1}^{m} x_j b_{ji}$. This last equation is equivalent to $AX = XB$. Conversely, $AX = XB$ means that each Ax_i is a linear combination of columns of X, so $\mathbf{X}$ is invariant.

Now assume $AX = XB$. Choose any n-by-$(n - m)$ matrix $\tilde{X}$ such that $\hat{X} = [X, \tilde{X}]$ is nonsingular. Then A and $\hat{X}^{-1}A\hat{X}$ are similar and so have the same eigenvalues. Write $\hat{X}^{-1} = [\begin{smallmatrix} Y^{m \times n} \\ \tilde{Y}^{(n-m) \times n} \end{smallmatrix}]$, so $\hat{X}^{-1}\hat{X} = I$ implies

$YX = I$ and $\tilde{Y}X = 0$. Then $\hat{X}^{-1}A\hat{X} = [\begin{smallmatrix} Y \\ \tilde{Y} \end{smallmatrix}] \cdot [AX, A\tilde{X}] = [\begin{smallmatrix} YAX & YA\tilde{X} \\ \tilde{Y}AX & \tilde{Y}A\tilde{X} \end{smallmatrix}] =$

$[\begin{smallmatrix} YXB & YA\tilde{X} \\ \tilde{Y}XB & \tilde{Y}A\tilde{X} \end{smallmatrix}] = [\begin{smallmatrix} B & YA\tilde{X} \\ 0 & \tilde{Y}A\tilde{X} \end{smallmatrix}]$. Thus by Question 4.1 the eigenvalues of A are the union of the eigenvalues of B and the eigenvalues of $\tilde{Y}A\tilde{X}$. $\square$

For example, write the Jordan canonical form $S^{-1}AS = J = \text{diag}(J_{n_i}(\lambda_i))$ as $AS = SJ$, where $S = [S_1, S_2, \ldots, S_k]$ and S_i has n_i columns (the same as $J_{n_i}(\lambda_i)$; see Theorem 4.1 for notation). Then $AS = SJ$ implies $AS_i = S_i J_{n_i}(\lambda_i)$, i.e., span$(S_i)$ is an invariant subspace.

The Jordan form tells everything that we might want to know about a matrix and its eigenvalues, eigenvectors, and invariant subspaces. There are also explicit formulas based on the Jordan form to compute e^A or any other function of a matrix (see section 4.5.1). But it is bad to compute the Jordan form for two numerical reasons:

First reason: It is a discontinuous function of A, so any rounding error can change it completely.

EXAMPLE 4.2. Let

$$
J_n(0) = \begin{bmatrix} 0 & 1 & & \\ & \ddots & \ddots & \\ & & \ddots & 1 \\ & & & 0 \end{bmatrix},
$$

which is in Jordan form. For arbitrarily small ϵ, adding $i \cdot \epsilon$ to the (i, i) entry changes the eigenvalues to the n distinct values $i \cdot \epsilon$, and so the Jordan form changes from $J_n(0)$ to diag$(\epsilon, 2\epsilon, \ldots, n\epsilon)$. $\diamond$

Second reason: It cannot be computed stably in general. In other words, when we have finished computing S and J, we cannot guarantee that $S^{-1}(A + \delta A)S = J$ for some small δA.

EXAMPLE 4.3. Suppose $S^{-1}AS = J$ exactly, where S is very ill-conditioned. ($\kappa(S) = \|S\| \cdot \|S^{-1}\|$ is very large.) Suppose that we are extremely lucky and manage to compute S *exactly* and J with just a tiny error δJ with $\|\delta J\| = O(\varepsilon)\|A\|$. How big is the backward error? In other words, how big must δA be so that $S^{-1}(A + \delta A)S = J + \delta J$? We get $\delta A = S\delta J S^{-1}$, and all that we can conclude is that $\|\delta A\| \leq \|S\| \cdot \|\delta J\| \cdot \|S^{-1}\| = O(\varepsilon)\kappa(S)\|A\|$. Thus $\|\delta A\|$ may be much larger than $\varepsilon\|A\|$, which prevents backward stability. $\diamond$

So instead of computing $S^{-1}AS = J$, where S can be an arbitrarily ill-conditioned matrix, we will restrict S to be orthogonal (so $\kappa_2(S) = 1$) to guarantee stability. We cannot get a canonical form as simple as the Jordan form any more, but we do get something almost as good.

THEOREM 4.2. Schur canonical form. *Given A, there exists a unitary matrix Q and an upper triangular matrix T such that $Q^*AQ = T$. The eigenvalues of A are the diagonal entries of T.*

Proof. We use induction on n. It is obviously true if A is 1 by 1. Now let λ be any eigenvalue and u a corresponding eigenvector normalized so $\|u\|_2 = 1$. Choose $\tilde{U}$ so $U = [u, \tilde{U}]$ is a square unitary matrix. (Note that λ and u may be complex even if A is real.) Then

$$U^* \cdot A \cdot U = \begin{bmatrix} u^* \\ \tilde{U}^* \end{bmatrix} \cdot A \cdot [u, \tilde{U}] = \begin{bmatrix} u^* A u & u^* A \tilde{U} \\ \tilde{U}^* A u & \tilde{U}^* A \tilde{U} \end{bmatrix}.$$

Now as in the proof of Proposition 4.3, we can write $u^* A u = \lambda u^* u = \lambda$, and $\tilde{U}^* A u = \lambda \tilde{U}^* u = 0$ so $U^* A U \equiv [\begin{smallmatrix} \lambda & \tilde{a}_{12} \\ 0 & \tilde{A}_{22} \end{smallmatrix}]$. By induction, there is a unitary P, so $P^* A_{22} P = \tilde{T}$ is upper triangular. Then

$$U^* A U = \begin{bmatrix} \lambda & \tilde{a}_{12} \\ 0 & P\tilde{T}P^* \end{bmatrix} = \begin{bmatrix} 1 & 0 \\ 0 & P \end{bmatrix} \begin{bmatrix} \lambda & \tilde{a}_{12}P \\ 0 & \tilde{T} \end{bmatrix} \begin{bmatrix} 1 & 0 \\ 0 & P^* \end{bmatrix},$$

so

$$\begin{bmatrix} 1 & 0 \\ 0 & P^* \end{bmatrix} U^* A U \begin{bmatrix} 1 & 0 \\ 0 & P \end{bmatrix} = \begin{bmatrix} \lambda & \tilde{a}_{12}P \\ 0 & \tilde{T} \end{bmatrix} = T$$

is upper triangular and $Q = U[\begin{smallmatrix} 1 & 0 \\ 0 & P \end{smallmatrix}]$ is unitary as desired. □

Notice that the Schur form is not unique, because the eigenvalues may appear on the diagonal of T in any order.

This introduces complex numbers even when A is real. When A is real, we prefer a canonical form that uses only real numbers, because it will be cheaper to compute. As mentioned at the beginning of this section, this means that we will have to sacrifice a *triangular* canonical form and settle for a *block-triangular* canonical form.

THEOREM 4.3. *Real Schur canonical form. If A is real, there exists a real orthogonal matrix V such that $V^T A V = T$ is quasi–upper triangular. This means that T is block upper triangular with 1-by-1 and 2-by-2 blocks on the diagonal. Its eigenvalues are the eigenvalues of its diagonal blocks. The 1-by-1 blocks correspond to real eigenvalues, and the 2-by-2 blocks to complex conjugate pairs of eigenvalues.*

Proof. We use induction as before. Let λ be an eigenvalue. If λ is real, it has a real eigenvector u and we proceed as in the last theorem. If λ is complex, let u be a (necessarily) complex eigenvector, so $Au = \lambda u$. Since $\overline{Au} = A\overline{u} = \overline{\lambda}\overline{u}$, $\overline{\lambda}$ and $\overline{u}$ are also an eigenvalue/eigenvector pair. Let $u_R = \frac{1}{2}u + \frac{1}{2}\overline{u}$ be the real part of u and $u_I = \frac{1}{2i}u - \frac{1}{2i}\overline{u}$ be the imaginary part. Then span$\{u_R, u_I\} = $ span$\{u, \overline{u}\}$ is a two-dimensional invariant subspace. Let $\tilde{U} = [u_R, u_I]$ and $\tilde{U} = QR$ be its QR decomposition. Thus span$\{Q\} = $ span$\{u_R, u_I\}$ is invariant. Choose $\tilde{Q}$ so that $U = [Q, \tilde{Q}]$ is real and orthogonal, and compute

$$U^T \cdot A \cdot U = \begin{bmatrix} Q^T \\ \tilde{Q}^T \end{bmatrix} \cdot A \cdot [Q, \tilde{Q}] = \begin{bmatrix} Q^T A Q & Q^T A \tilde{Q} \\ \tilde{Q}^T A Q & \tilde{Q}^T A \tilde{Q} \end{bmatrix}.$$

Since Q spans an invariant subspace, there is a 2-by-2 matrix B such that $AQ = QB$. Now as in the proof of Proposition 4.3, we can write $Q^T A Q = Q^T Q B = B$ and $\tilde{Q}^T A Q = \tilde{Q}^T Q B = 0$, so $U^T A U = [\begin{smallmatrix} B & Q^T A \tilde{Q} \\ 0 & \tilde{Q}^T A \tilde{Q} \end{smallmatrix}]$. Now apply induction to $\tilde{Q}^T A \tilde{Q}$. □

4.2.1. Computing Eigenvectors from the Schur Form

Let $Q^*AQ = T$ be the Schur form. Then if $Tx = \lambda x$, we have $AQx = QTx = \lambda Qx$, so Qx is an eigenvector of A. So to find eigenvectors of A, it suffices to find eigenvectors of T.

Suppose that $\lambda = t_{ii}$ has multiplicity 1 (i.e., it is simple). Write $(T - \lambda I)x = 0$ as

$$0 = \begin{bmatrix} T_{11} - \lambda I & T_{12} & T_{13} \\ 0 & 0 & T_{23} \\ 0 & 0 & T_{33} - \lambda I \end{bmatrix} \begin{bmatrix} x_1 \\ x_2 \\ x_3 \end{bmatrix}$$

$$= \begin{bmatrix} (T_{11} - \lambda I)x_1 + T_{12}x_2 + T_{13}x_3 \\ T_{23}x_3 \\ (T_{33} - \lambda I)x_3 \end{bmatrix},$$

where T_{11} is $(i-1)$-by-$(i-1)$, $T_{22} = \lambda$ is 1-by-1, T_{33} is $(n-i)$-by-$(n-i)$, and x is partitioned conformably. Since λ is simple, both $T_{11} - \lambda I$ and $T_{33} - \lambda I$ are nonsingular, so $(T_{33} - \lambda I)x_3 = 0$ implies $x_3 = 0$. Therefore $(T_{11} - \lambda I)x_1 = -T_{12}x_2$. Choosing (arbitrarily) $x_2 = 1$ means $x_1 = -(T_{11} - \lambda I)^{-1}T_{12}$, so

$$x = \begin{bmatrix} (\lambda I - T_{11})^{-1}T_{12} \\ 1 \\ 0 \end{bmatrix}.$$

In other words, we just need to solve a triangular system for x_1. To find a *real* eigenvector from real Schur form, we get a quasi-triangular system to solve. Computing complex eigenvectors from real Schur form using only real arithmetic also just involves equation solving but is a little trickier. See subroutine `strevc` in LAPACK for details.

4.3. Perturbation Theory

In this section we will concentrate on understanding when eigenvalues are ill-conditioned and thus hard to compute accurately. In addition to providing error bounds for computed eigenvalues, we will also relate eigenvalue condition numbers to related quantities, including the distance to the nearest matrix with an *infinitely* ill-conditioned eigenvalue, and the condition number of the matrix of eigenvectors.

We begin our study by asking when eigenvalues have *infinite* condition numbers. This is the case for multiple eigenvalues, as the following example illustrates.

EXAMPLE 4.4. Let

$$A = \begin{bmatrix} 0 & 1 & & & \\ & \ddots & \ddots & & \\ & & \ddots & 1 \\ \epsilon & & & 0 \end{bmatrix}$$

be an n-by-n matrix. Then A has characteristic polynomial $\lambda^n - \epsilon = 0$ so $\lambda = \sqrt[n]{\epsilon}$ (n possible values). The nth root of ϵ grows much faster than any multiple of ϵ for small ϵ. More formally, the condition number is infinite because $\frac{d\lambda}{d\epsilon} = \frac{\epsilon^{\frac{1}{n}-1}}{n} = \infty$ at $\epsilon = 0$ for $n \geq 2$. For example, take $n = 16$ and $\epsilon = 10^{-16}$. Then for each eigenvalue $|\lambda| = .1$. $\diamond$

So we expect a large condition number if an eigenvalue is "close to multiple"; i.e., there is a small δA such that $A + \delta A$ has exactly a multiple eigenvalue. Having an infinite condition number does not mean that they cannot be computed with any correct digits, however.

PROPOSITION 4.4. *Eigenvalues of A are continuous functions of A, even if they are not differentiable.*

Proof. It suffices to prove the continuity of roots of polynomials, since the coefficients of the characteristic polynomial are continuous (in fact polynomial) functions of the matrix entries. We use the argument principle from complex analysis [2]: the number of roots of a polynomial p inside a simple closed curve γ is $\frac{1}{2\pi i} \oint_\gamma \frac{p'(z)}{p(z)} dz$. If p is changed just a little, $\frac{p'(z)}{p(z)}$ is changed just a little, so $\frac{1}{2\pi i} \oint_\gamma \frac{p'(z)}{p(z)} dz$ is changed just a little. But since it is an integer, it must be constant, so the number of roots inside the curve γ is constant. This means that the roots cannot pass outside the curve γ (no matter how small γ is, provided that we perturb p by little enough), so the roots must be continuous. $\square$

In what follows, we will concentrate on computing the condition number of a simple eigenvalue. If λ is a simple eigenvalue of A and δA is small, then we can identify an eigenvalue $\lambda + \delta\lambda$ of $A + \delta A$ "corresponding to" λ: it is the closest one to λ. We can easily compute the condition number of a simple eigenvalue.

THEOREM 4.4. *Let λ be a simple eigenvalue of A with right eigenvector x and left eigenvector y, normalized so that $\|x\|_2 = \|y\|_2 = 1$. Let $\lambda + \delta\lambda$ be the corresponding eigenvalue of $A + \delta A$. Then*

$$\delta\lambda = \frac{y^* \delta A x}{y^* x} + O(\|\delta A\|^2) \text{ or}$$
$$|\delta\lambda| \leq \frac{\|\delta A\|}{|y^* x|} + O(\|\delta A\|^2) = \sec\Theta(y, x)\|\delta A\| + O(\|\delta A\|^2),$$

where $\Theta(y, x)$ is the acute angle between y and x. In other words, $\sec\Theta(y, x) = 1/|y^* x|$ is the condition number of the eigenvalue λ.

Proof. Subtract $Ax = \lambda x$ from $(A + \delta A)(x + \delta x) = (\lambda + \delta\lambda)(x + \delta x)$ to get

$$A\delta x + \delta A x + \delta A \delta x = \lambda\delta x + \delta\lambda x + \delta\lambda\delta x.$$

Ignore the second-order terms (those with two "δ terms" as factors: $\delta A\delta x$ and $\delta\lambda\delta x$) and multiply by y^* to get $y^* A\delta x + y^*\delta A x = y^*\lambda\delta x + y^*\delta\lambda x$.

Now $y^* A \delta x$ cancels $y^* \lambda \delta x$, so we can solve for $\delta \lambda = (y^* \delta A x)/(y^* x)$ as desired. $\square$

Note that a Jordan block has right and left eigenvectors e_1 and e_n, respectively, so the condition number of its eigenvalue is $1/|e_n^* e_1| = 1/0 = \infty$, which agrees with our earlier analysis.

At the other extreme, in the important special case of symmetric matrices, the condition number is 1, so the eigenvalues are always accurately determined by the data.

COROLLARY 4.1. *Let A be symmetric (or normal: $AA^* = A^* A$). Then $|\delta \lambda| \leq \|\delta A\| + O(\|\delta A\|^2)$.*

Proof. If A is symmetric or normal, then its eigenvectors are all orthogonal, i.e., $Q^* A Q = \Lambda$ with $QQ^* = I$. So the right eigenvectors x (columns of Q) and left eigenvectors y (conjugate transposes of the rows of Q^*) are identical, and $1/|y^* x| = 1$. $\square$

To see a variety of numerical examples, run the Matlab code referred to in Question 4.14.

Later, in Theorem 5.1, we will prove that in fact $|\delta \lambda| \leq \|\delta A\|_2$ if $\delta A = \delta A^T$, no matter how large $\|\delta A\|_2$ is.

Theorem 4.4 is useful only for sufficiently small $\|\delta A\|$. We can remove the $O(\|\delta A\|^2)$ term and so get a simple theorem true for any size perturbation $\|\delta A\|$, at the cost of increasing the condition number by a factor of n.

THEOREM 4.5. *Bauer–Fike. Let A have all simple eigenvalues (i.e., be diagonalizable). Call them λ_i, with right and left eigenvectors x_i and y_i, normalized so $\|x_i\|_2 = \|y_i\|_2 = 1$. Then the eigenvalues of $A + \delta A$ lie in disks B_i, where B_i has center λ_i and radius $n \frac{\|\delta A\|_2}{|y_i^* x_i|}$.*

Our proof will use Gershgorin's theorem (Theorem 2.9), which we repeat here.

GERSHGORIN'S THEOREM. *Let B be an arbitrary matrix. Then the eigenvalues λ of B are located in the union of the n disks defined by $|\lambda - b_{ii}| \leq \sum_{j \neq i} |b_{ij}|$ for $i = 1$ to n.*

We will also need two simple lemmas.

LEMMA 4.1. *Let $S = [x_1, \ldots, x_n]$, the nonsingular matrix of right eigenvectors. Then*

$$S^{-1} = \begin{bmatrix} y_1^*/y_1^* x_1 \\ y_2^*/y_2^* x_2 \\ \vdots \\ y_n^*/y_n^* x_n \end{bmatrix}.$$

Proof of Lemma. We know that $AS = S\Lambda$, where $\Lambda = \text{diag}(\lambda_1, \ldots, \lambda_n)$, since the columns x_i of S are eigenvectors. This is equivalent to $S^{-1}A = \Lambda S^{-1}$, so the rows of S^{-1} are conjugate transposes of the left eigenvectors y_i. So

$$S^{-1} = \begin{bmatrix} y_1^* \cdot c_1 \\ \vdots \\ y_n^* \cdot c_n \end{bmatrix}$$

for some constants c_i. But $I = S^{-1}S$, so $1 = (S^{-1}S)_{ii} = y_i^* x_i \cdot c_i$, and $c_i = \frac{1}{y_i^* x_i}$ as desired. $\square$

LEMMA 4.2. *If each column of (any matrix) S has two-norm equal to 1, $\|S\|_2 \leq \sqrt{n}$. Similarly, if each row of a matrix has two-norm equal to 1, its two-norm is at most $\sqrt{n}$.*

Proof of Lemma. $\|S\|_2 = \|S^T\|_2 = \max_{\|x\|_2 = 1} \|S^T x\|_2$. Each component of $S^T x$ is bounded by 1 by the Cauchy–Schwartz inequality, so $\|S^T x\|_2 \leq \|[1, \ldots, 1]^T\|_2 = \sqrt{n}$. $\square$

Proof of the Bauer–Fike theorem. We will apply Gershgorin's theorem to $S^{-1}(A + \delta A)S = \Lambda + F$, where $\Lambda = S^{-1}AS = \text{diag}(\lambda_1, \ldots, \lambda_n)$ and $F = S^{-1}\delta AS$. The idea is to show that the eigenvalues of $A + \delta A$ lie in balls centered at the λ_i with the given radii. To do this, we take the disks containing the eigenvalues of $\Lambda + F$ that are defined by Gershgorin's theorem,

$$|\lambda - (\lambda_i + f_{ii})| \leq \sum_{j \neq i} |f_{ij}|,$$

and enlarge them slightly to get the disks

$$\begin{aligned} |\lambda - \lambda_i| &\leq \sum_j |f_{ij}| \\ &\leq n^{1/2} \cdot \left(\sum_j |f_{ij}|^2\right)^{1/2} \qquad \text{by Cauchy–Schwarz} \\ &= n^{1/2} \cdot \|F(i,:)\|_2. \end{aligned} \qquad (4.5)$$

Now we need to bound the two-norm of the ith row $F(i,:)$ of $F = S^{-1}\delta AS$:

$$\begin{aligned} \|F(i,:)\|_2 &= \|(S^{-1}\delta AS)(i,:)\|_2 \\ &\leq \|(S^{-1})(i,:)\|_2 \cdot \|\delta A\|_2 \cdot \|S\|_2 \quad \text{by Lemma 1.7} \\ &\leq \frac{n^{1/2}}{|y_i^* x_i|} \cdot \|\delta A\|_2 \quad \text{by Lemmas 4.1 and 4.2.} \end{aligned}$$

Combined with equation (4.5), this proves the theorem. $\square$

We do not want to leave the impression that multiple eigenvalues cannot be computed with any accuracy at all just because they have infinite condition numbers. Indeed, we expect to get a *fraction* of the digits correct rather than lose a fixed number of digits. To illustrate, consider the 2-by-2 matrix with a double eigenvalue at 1: $A = \begin{bmatrix} 1 & 1 \\ 0 & 1 \end{bmatrix}$. If we perturb the (2,1) entry (the most sensitive) from 0 to machine epsilon ε, the eigenvalues change from 1 to $1 \pm \sqrt{\varepsilon}$. In other words the computed eigenvalues agree with the true eigenvalue to half precision. More generally, with a triple root, we expect to get about one third of the digits correct, and so on for higher multiplicities. See also Question 1.20.

We now turn to a geometric property of the condition number shared by other problems. Recall the property of the condition number $\|A\| \cdot \|A^{-1}\|$ for matrix inversion: its reciprocal measured the distance to nearest singular matrix, i.e., matrix with an infinite condition number (see Theorem 2.1). An analogous fact is true about eigenvalues. Since multiple eigenvalues have infinite condition numbers, the set of matrices with multiple eigenvalues plays the same role for computing eigenvalues as the singular matrices did for matrix inversion, where being "close to singular" implied ill-conditioning.

THEOREM 4.6. *Let λ be a simple eigenvalue of A, with unit right and left eigenvectors x and y and condition number $c = 1/|y^*x|$. Then there is a δA such that $A + \delta A$ has a multiple eigenvalue at λ, and*

$$\frac{\|\delta A\|_2}{\|A\|_2} \leq \frac{1}{\sqrt{c^2 - 1}}.$$

When $c \gg 1$, i.e., the eigenvalue is ill-conditioned, then the upper bound on the distance is $1/\sqrt{c^2 - 1} \approx 1/c$, the reciprocal of the condition number.

Proof. First we show that we can assume without loss of generality that A is upper triangular (in Schur form), with $a_{11} = \lambda$. This is because putting A in Schur form is equivalent to replacing A by $T = Q^*AQ$, where Q is unitary. If x and y are eigenvectors of A, then Q^*x and Q^*y are eigenvectors of T. Since $(Q^*y)^*(Q^*x) = y^*QQ^*x = y^*x$, changing to Schur form does not change the condition number of λ. (Another way to say this is that the condition number is the secant of the angle $\Theta(x, y)$ between x and y, and changing x to Q^*x and y to Q^*y just rotates x and y the same way without changing the angle between them.)

So without loss of generality we can assume that $A = \begin{bmatrix} \lambda & A_{12} \\ 0 & A_{22} \end{bmatrix}$. Then $x = e_1$ and y is parallel to $\tilde{y} = [1, A_{12}(\lambda I - A_{22})^{-1}]^*$, or $y = \tilde{y}/\|\tilde{y}\|_2$. Thus

$$c = \frac{1}{|y^*x|} = \frac{\|\tilde{y}\|_2}{|\tilde{y}^*x|} = \|\tilde{y}\|_2 = (1 + \|A_{12}(\lambda I - A_{22})^{-1}\|_2^2)^{1/2}$$

or

$$\begin{aligned} \sqrt{c^2 - 1} &= \|A_{12}(\lambda I - A_{22})^{-1}\|_2 \leq \|A_{12}\|_2 \cdot \|(\lambda I - A_{22})^{-1}\|_2 \\ &\leq \frac{\|A\|_2}{\sigma_{\min}(\lambda I - A_{22})}. \end{aligned}$$

By definition of the smallest singular value, there is a δA_{22} where $\|\delta A_{22}\|_2 = \sigma_{\min}(\lambda I - A_{22})$ such that $A_{22} + \delta A_{22} - \lambda I$ is singular; i.e., λ is an eigenvalue of $A_{22} + \delta A_{22}$. Thus $\begin{bmatrix} \lambda & A_{12} \\ 0 & A_{22} + \delta A_{22} \end{bmatrix}$ has a double eigenvalue at λ, where

$$\|\delta A_{22}\|_2 = \sigma_{\min}(\lambda I - A_{22}) \leq \frac{\|A\|_2}{\sqrt{c^2 - 1}}$$

as desired. $\square$

Finally, we relate the condition numbers of the eigenvalues to the smallest possible condition number $\|S\| \cdot \|S^{-1}\|$ of any similarity S that diagonalizes A: $S^{-1}AS = \Lambda = \mathrm{diag}(\lambda_1, \ldots, \lambda_n)$. The theorem says that if any eigenvalue has a large condition number, then S has to have an approximately equally large condition number. In other words, the condition numbers for finding the (worst) eigenvalue and for reducing the matrix to diagonal form are nearly the same.

THEOREM 4.7. *Let A be diagonalizable with eigenvalues λ_i and right and left eigenvectors x_i and y_i, respectively, normalized so that $\|x_i\|_2 = \|y_i\|_2 = 1$. Let us suppose that S satisfies $S^{-1}AS = \Lambda = \mathrm{diag}(\lambda_1, \ldots, \lambda_n)$. Then $\|S\|_2 \cdot \|S^{-1}\|_2 \geq \max_i 1/|y_i^* x_i|$. If we choose $S = [x_1, \ldots, x_n]$, then $\|S\|_2 \cdot \|S^{-1}\|_2 \leq n \cdot \max_i 1/|y_i^* x_i|$; i.e., the condition number of S is within a factor of n of its smallest value.*

For a proof, see [69].

For an overview of condition numbers for the eigenproblem, including eigenvectors, invariant subspaces, and the eigenvalues corresponding to an invariant subspace, see chapter 4 of the LAPACK manual [10], as well as [161, 237]. Algorithms for computing these condition numbers are available in subroutines `strsna` and `strsen` of LAPACK or by calling the driver routines `sgeevx` and `sgeesx`.

4.4. Algorithms for the Nonsymmetric Eigenproblem

We will build up to our ultimate algorithm, the shifted Hessenberg QR algorithm, by starting with simpler ones. For simplicity of exposition, we assume A is real.

Our first and simplest algorithm is the *power method* (section 4.4.1), which can find only the largest eigenvalue of A in absolute value and the corresponding eigenvector. To find the other eigenvalues and eigenvectors, we apply the power method to $(A - \sigma I)^{-1}$ for some *shift* σ, an algorithm called *inverse iteration* (section 4.4.2); note that the largest eigenvalue of $(A - \sigma I)^{-1}$ is $1/(\lambda_i - \sigma)$, where λ_i is the closest eigenvalue to σ, so we can choose which eigenvalues to find by choosing σ. Our next improvement to the power method lets us compute an entire invariant subspace at a time rather than just a single eigenvector;

we call this *orthogonal iteration* (section 4.4.3). Finally, we reorganize orthogonal iteration to make it convenient to apply to $(A - \sigma I)^{-1}$ instead of A; this is called QR iteration (section 4.4.4).

Mathematically speaking, QR iteration (with a shift σ) is our ultimate algorithm. But several problems remain to be solved to make it sufficiently fast and reliable for practical use (section 4.4.5). Section 4.4.6 discusses the first transformation designed to make QR iteration fast: reducing A from dense to *upper Hessenberg form* (nonzero only on and above the first subdiagonal). Subsequent sections describe how to implement QR iteration efficiently on upper Hessenberg matrices. (Section 4.4.7 shows how upper Hessenberg form simplifies in the cases of the symmetric eigenvalue problem and SVD.)

4.4.1. Power Method

ALGORITHM 4.1. *Power method: Given x_0, we iterate*

$$i = 0$$
$$repeat$$
$$\quad y_{i+1} = Ax_i$$
$$\quad x_{i+1} = y_{i+1}/\|y_{i+1}\|_2 \qquad \text{(approximate eigenvector)}$$
$$\quad \tilde{\lambda}_{i+1} = x_{i+1}^T A x_{i+1} \qquad \text{(approximate eigenvalue)}$$
$$\quad i = i + 1$$
$$until\ convergence$$

Let us first apply this algorithm in the very simple case when $A = \mathrm{diag}(\lambda_1, \ldots, \lambda_n)$, with $|\lambda_1| > |\lambda_2| \geq \cdots \geq |\lambda_n|$. In this case the eigenvectors are just the columns e_i of the identity matrix. Note that x_i can also be written $x_i = A^i x_0 / \|A^i x_0\|_2$, since the factors $1/\|y_{i+1}\|_2$ only scale x_{i+1} to be a unit vector and do not change its direction. Then we get

$$A^i x_0 \equiv A^i \begin{bmatrix} \xi_1 \\ \xi_2 \\ \vdots \\ \xi_n \end{bmatrix} = \begin{bmatrix} \xi_1 \lambda_1^i \\ \xi_2 \lambda_2^i \\ \vdots \\ \xi_n \lambda_n^i \end{bmatrix} = \xi_1 \lambda_1^i \begin{bmatrix} 1 \\ \frac{\xi_2}{\xi_1} \left(\frac{\lambda_2}{\lambda_1} \right)^i \\ \vdots \\ \frac{\xi_n}{\xi_1} \left(\frac{\lambda_n}{\lambda_1} \right)^i \end{bmatrix},$$

where we have assumed $\xi_1 \neq 0$. Since all the fractions λ_j/λ_1 are less than 1 in absolute value, $A^i x_0$ becomes more and more nearly parallel to e_1, so $x_i = A^i x_0 / \|A^i x_0\|_2$ becomes closer and closer to $\pm e_1$, the eigenvector corresponding to the largest eigenvalue λ_1. The rate of convergence depends on how much smaller than 1 the ratios $|\lambda_2/\lambda_1| \geq \cdots \geq |\lambda_n/\lambda_1|$ are, the smaller the faster. Since x_i converges to $\pm e_1$, $\tilde{\lambda}_i = x_i^T A x_i$ converges to λ_1, the largest eigenvalue.

In showing that the power method converges, we have made several assumptions, most notably that A is diagonal. To analyze a more general case, we now assume that $A = S \Lambda S^{-1}$ is diagonalizable, with $\Lambda = \mathrm{diag}(\lambda_1, \ldots, \lambda_n)$

and the eigenvalues sorted so that $|\lambda_1| > |\lambda_2| \geq \cdots \geq |\lambda_n|$. Write $S = [s_1, \ldots, s_n]$, where the columns s_i are the corresponding eigenvectors and also satisfy $\|s_i\|_2 = 1$; in the last paragraph we had $S = I$. This lets us write $x_0 = S(S^{-1}x_0) \equiv S([\xi_1, \ldots, \xi_n]^T)$. Also, since $A = S\Lambda S^{-1}$, we can write

$$A^i = \underbrace{(S\Lambda S^{-1}) \cdots (S\Lambda S^{-1})}_{i \text{ times}} = S\Lambda^i S^{-1}$$

since all the $S^{-1} \cdot S$ pairs cancel. This finally lets us write

$$A^i x_0 = (S\Lambda^i S^{-1})S \begin{bmatrix} \xi_1 \\ \xi_2 \\ \vdots \\ \xi_n \end{bmatrix} = S \begin{bmatrix} \xi_1 \lambda_1^i \\ \xi_2 \lambda_2^i \\ \vdots \\ \xi_n \lambda_n^i \end{bmatrix} = \xi_1 \lambda_1^i S \begin{bmatrix} 1 \\ \frac{\xi_2}{\xi_1}\left(\frac{\lambda_2}{\lambda_1}\right)^i \\ \vdots \\ \frac{\xi_n}{\xi_1}\left(\frac{\lambda_n}{\lambda_1}\right)^i \end{bmatrix}.$$

As before, the vector in brackets converges to e_1, so $A^i x_0$ gets closer and closer to a multiple of $Se_1 = s_1$, the eigenvector corresponding to λ_1. Therefore, $\tilde{\lambda}_i = x_i^T A x_i$ converges to $s_1^T A s_1 = s_1^T \lambda_1 s_1 = \lambda_1$.

A minor drawback of this method is the assumption that $\xi_1 \neq 0$, i.e., that x_0 is not the invariant subspace span$\{s_2, \ldots, s_n\}$; this is true with very high probability if x_0 is chosen at random. A major drawback is that it converges to the eigenvalue/eigenvector pair only for the eigenvalue of largest absolute magnitude, and its convergence rate depends on $|\lambda_2/\lambda_1|$, a quantity which may be close to 1 and thus cause very slow convergence. Indeed, if A is real and the largest eigenvalue is complex, there are two complex conjugate eigenvalues of largest absolute value $|\lambda_1| = |\lambda_2|$, and so the above analysis does not work at all. In the extreme case of an orthogonal matrix, *all* the eigenvalues have the same absolute value, namely, 1.

To plot the convergence of the power method, see HOMEPAGE/Matlab/powerplot.m.

4.4.2. Inverse Iteration

We will overcome the drawbacks of the power method just described by applying the power method to $(A - \sigma I)^{-1}$ instead of A, where σ is called a *shift*. This will let us converge to the eigenvalue closest to σ, rather than just λ_1. This method is called inverse iteration or the inverse power method.

ALGORITHM 4.2. *Inverse iteration: Given x_0, we iterate*

$$i = 0$$
repeat
$$y_{i+1} = (A - \sigma I)^{-1} x_i$$
$$x_{i+1} = y_{i+1}/\|y_{i+1}\|_2 \qquad \textit{(approximate eigenvector)}$$

$$\tilde{\lambda}_{i+1} = x_{i+1}^T A x_{i+1} \qquad \textit{(approximate eigenvalue)}$$
$$i = i + 1$$
until convergence

To analyze the convergence, note that $A = S \Lambda S^{-1}$ implies $A - \sigma I = S(\Lambda - \sigma I)S^{-1}$ and so $(A - \sigma I)^{-1} = S(\Lambda - \sigma I)^{-1}S^{-1}$. Thus $(A - \sigma I)^{-1}$ has the same eigenvectors s_i as A with corresponding eigenvalues $((\Lambda - \sigma I)^{-1})_{jj} = (\lambda_j - \sigma)^{-1}$. The same analysis as before tells us to expect x_i to converge to the eigenvector corresponding to the largest eigenvalue in absolute value. More specifically, assume that $|\lambda_k - \sigma|$ is smaller than all the other $|\lambda_i - \sigma|$ so that $(\lambda_k - \sigma)^{-1}$ is the largest eigenvalue in absolute value. Also, write $x_0 = S[\xi_1, \dots, \xi_n]^T$ as before, and assume $\xi_k \neq 0$. Then

$$
(A - \sigma I)^{-i} x_0 = (S(\Lambda - \sigma I)^{-i} S^{-1}) S \begin{bmatrix} \xi_1 \\ \xi_2 \\ \vdots \\ \xi_n \end{bmatrix} = S \begin{bmatrix} \xi_1 (\lambda_1 - \sigma)^{-i} \\ \vdots \\ \xi_n (\lambda_n - \sigma)^{-i} \end{bmatrix}
$$

$$
= \xi_k (\lambda_k - \sigma)^{-i} S \begin{bmatrix} \frac{\xi_1}{\xi_k} \left(\frac{\lambda_k - \sigma}{\lambda_1 - \sigma} \right)^i \\ \vdots \\ 1 \\ \vdots \\ \frac{\xi_n}{\xi_k} \left(\frac{\lambda_k - \sigma}{\lambda_n - \sigma} \right)^i \end{bmatrix},
$$

where the 1 is in entry k. Since all the fractions $(\lambda_k - \sigma)/(\lambda_i - \sigma)$ are less than one in absolute value, the vector in brackets approaches e_k, so $(A - \sigma I)^{-i} x_0$ gets closer and closer to a multiple of $S e_k = s_k$, the eigenvector corresponding to λ_k. As before, $\tilde{\lambda}_i = x_i^T A x_i$ also converges to λ_k.

The advantage of inverse iteration over the power method is the ability to converge to any desired eigenvalue (the one nearest the shift σ). By choosing σ very close to a desired eigenvalue, we can converge very quickly and thus not be as limited by the proximity of nearby eigenvalues as is the original power method. The method is particularly effective when we have a good approximation to an eigenvalue and want only its corresponding eigenvector (for example, see section 5.3.4). Later we will explain how to choose such a σ without knowing the eigenvalues, which is what we are trying to compute in the first place!

4.4.3. Orthogonal Iteration

Our next improvement will permit us to converge to a $(p > 1)$-dimensional invariant subspace, rather than one eigenvector at a time. It is called *orthogonal iteration* (and sometimes *subspace iteration* or *simultaneous iteration*).

ALGORITHM 4.3. *Orthogonal iteration: Let Z_0 be an $n \times p$ orthogonal matrix. Then we iterate*

> $i = 0$
> *repeat*
> $\quad Y_{i+1} = AZ_i$
> $\quad$ *Factor $Y_{i+1} = Z_{i+1}R_{i+1}$ using Algorithm 3.2 (QR decomposition)*
> $\qquad\qquad\qquad\qquad\quad$ *(Z_{i+1} spans an approximate*
> $\qquad\qquad\qquad\qquad\qquad$ *invariant subspace)*
> $\quad i = i + 1$
> *until convergence*

Here is an informal analysis of this method. Assume $|\lambda_p| > |\lambda_{p+1}|$. If $p = 1$, this method and its analysis are identical to the power method. When $p > 1$, we write $\text{span}\{Z_{i+1}\} = \text{span}\{Y_{i+1}\} = \text{span}\{AZ_i\}$, so $\text{span}\{Z_i\} = \text{span}\{A^i Z_0\} = \text{span}\{S\Lambda^i S^{-1} Z_0\}$. Note that

$$
\begin{aligned}
S\Lambda^i S^{-1} Z_0 &= S \, \text{diag}(\lambda_1^i, \ldots, \lambda_n^i) S^{-1} Z_0 \\
&= \lambda_p^i S
\begin{bmatrix}
(\lambda_1/\lambda_p)^i & & & & \\
& \ddots & & & \\
& & 1 & & \\
& & & \ddots & \\
& & & & (\lambda_n/\lambda_p)^i
\end{bmatrix}
S^{-1} Z_0.
\end{aligned}
$$

Since $|\frac{\lambda_j}{\lambda_p}| \geq 1$ if $j \leq p$, and $|\frac{\lambda_j}{\lambda_p}| < 1$ if $j > p$, we get

$$
\begin{bmatrix}
(\lambda_1/\lambda_p)^i & & \\
& \ddots & \\
& & (\lambda_n/\lambda_p)^i
\end{bmatrix}
S^{-1} Z_0 =
\begin{bmatrix}
V_i^{p \times p} \\
W_i^{(n-p) \times p}
\end{bmatrix},
$$

where W_i approaches zero like $(\lambda_{p+1}/\lambda_p)^i$, and V_i does not approach zero. Indeed, if V_0 has full rank (a generalization of the assumption in section 4.4.1 that $\xi_1 \neq 0$), then V_i will have full rank too. Write the matrix of eigenvectors $S = [s_1, \ldots, s_n] \equiv [S_p^{n \times p}, \hat{S}_p^{n \times (n-p)}]$, i.e., $S_p = [s_1, \ldots, s_p]$. Then $S\Lambda^i S^{-1} Z_0 = \lambda_p^i S[\begin{smallmatrix} V_i \\ W_i \end{smallmatrix}] = \lambda_p^i (S_p V_i + \hat{S}_p W_i)$. Thus

$$
\text{span}(Z_i) = \text{span}(S\Lambda^i S^{-1} Z_0) = \text{span}(S_p V_i + \hat{S}_p W_i) \Rightarrow \text{span}(S_p X_i)
$$

converges to $\text{span}(S_p V_i) = \text{span}(S_p)$, the invariant subspace spanned by the first p eigenvectors, as desired.

The use of the QR decomposition keeps the vectors spanning $\text{span}\{A^i Z_0\}$ of full rank despite roundoff.

Note that if we follow only the first $\tilde{p} < p$ columns of Z_i through the iterations of the algorithm, they are *identical* to the columns that we would compute if we had started with only the first $\tilde{p}$ columns of Z_0 instead of p columns. In other words, orthogonal iteration is effectively running the algorithm for $\tilde{p} = 1, 2, \ldots, p$ all at the same time. So if *all* the eigenvalues have distinct absolute values, the same convergence analysis as before implies that the first $\tilde{p} \leq p$ columns of Z_i converge to span$\{s_1, \ldots, s_{\tilde{p}}\}$ for any $\tilde{p} \leq p$.

Thus, we can let $p = n$ and $Z_0 = I$ in the orthogonal iteration algorithm. The next theorem shows that under certain assumptions, we can use orthogonal iteration to compute the Schur form of A.

THEOREM 4.8. *Consider running orthogonal iteration on matrix A with $p = n$ and $Z_0 = I$. If all the eigenvalues of A have distinct absolute values and if all the principal submatrices $S(1 : j, 1 : j)$ have full rank, then $A_i \equiv Z_i^T A Z_i$ converges to the Schur form of A, i.e., an upper triangular matrix with the eigenvalues on the diagonal. The eigenvalues will appear in decreasing order of absolute value.*

Sketch of Proof. The assumption about nonsingularity of $S(1 : j, 1 : j)$ for all j implies that X_0 is nonsingular, as required by the earlier analysis. Geometrically, this means that no vector in the invariant subspace span$\{s_i, \ldots, s_j\}$ is orthogonal to span$\{e_i, \ldots, e_j\}$, the space spanned by the first j columns of $Z_0 I$. First note that Z_i is a square orthogonal matrix, so A and $A_i = Z_i^T A Z_i$ are similar. Write $Z_i = [Z_{1i}, Z_{2i}]$, where Z_{1i} has p columns, so

$$A_i = Z_i^T A Z_i = \begin{bmatrix} Z_{1i}^T A Z_{1i} & Z_{1i}^T A Z_{2i} \\ Z_{2i}^T A Z_{1i} & Z_{2i}^T A Z_{2i} \end{bmatrix}.$$

Since span$\{Z_{1i}\}$ converges to an invariant subspace of A, span$\{A Z_{1i}\}$ converges to the same subspace, so $Z_{2i}^T A Z_{1i}$ converges to $Z_{2i}^T Z_{1i} = 0$. Since this is true for all $p < n$, every subdiagonal entry of A_i converges to zero, so A_i converges to upper triangular form, i.e., Schur form. $\square$

In fact, this proof shows that the submatrix $Z_{2i}^T A Z_{1i} = A_i(p + 1 : n, 1 : p)$ should converge to zero like $|\lambda_{p+1}/\lambda_p|^i$. Thus, λ_p should appear as the (p, p) entry of A_i and converge like $\max(|\lambda_{p+1}/\lambda_p|^i, |\lambda_p/\lambda_{p-1}|^i)$.

EXAMPLE 4.5. The convergence behavior of orthogonal iteration is illustrated by the following numerical experiment, where we took $\Lambda = \text{diag}(1, 2, 6, 30)$ and a random S (with condition number about 20), formed $A = S \cdot \Lambda \cdot S^{-1}$, and ran orthogonal iteration on A with $p = 4$ for 19 iterations. Figures 4.3 and 4.4 show the convergence of the algorithm. Figure 4.3 plots the actual errors $|A_i(p, p) - \lambda_p|$ in the computed eigenvalues as solid lines and the approximations $\max(|\lambda_{p+1}/\lambda_p|^i, |\lambda_p/\lambda_{p-1}|^i)$ as dotted lines. Since the graphs are (essentially) straight lines with the same slope on a semilog scale, this means that they are both graphs of functions of the form $y = c \cdot r^i$, where c and r are constants and r (the slope) is the same for both, as we predicted above.

Similarly, Figure 4.4 plots the actual values $\|A_i(p+1:n,1:p)\|_2$ as solid lines and the approximations $|\lambda_{p+1}/\lambda_p|^i$ as dotted lines; they also match well. Here are A_0 and A_{19} for comparison:

$$A = A_0 \;=\; \begin{bmatrix} 3.5488 & 15.593 & 8.5775 & -4.0123 \\ 2.3595 & 24.526 & 14.596 & -5.8157 \\ 8.9953 \cdot 10^{-2} & 27.599 & 21.483 & -5.8415 \\ 1.9227 & 55.667 & 39.717 & -10.558 \end{bmatrix},$$

$$A_{19} \;=\; \begin{bmatrix} 30.000 & -32.557 & -70.844 & 14.984 \\ 6.7607 \cdot 10^{-13} & 6.0000 & 1.8143 & -.55754 \\ 1.5452 \cdot 10^{-23} & 1.1086 \cdot 10^{-9} & 2.0000 & -.25894 \\ 7.3360 \cdot 10^{-29} & 3.3769 \cdot 10^{-15} & 4.9533 \cdot 10^{-6} & 1.0000 \end{bmatrix}.$$

See HOMEPAGE/Matlab/qriter.m for Matlab software to run this and similar examples. ◇

EXAMPLE 4.6. To see why the assumption in Theorem 4.8 about nonsingularity of $S(1:j,1:j)$ is necessary, suppose that A is diagonal with the eigenvalues *not* in decreasing order on the diagonal. Then orthogonal iteration yields $Z_i = \mathrm{diag}(\pm 1)$ (a diagonal matrix with diagonal entries ± 1) and $A_i = A$ for all i, so the eigenvalues do not move into decreasing order. To see why the assumption that the eigenvalues have distinct absolute values is necessary, suppose that A is orthogonal, so all its eigenvalues have absolute value 1. Again, the algorithm leaves A_i essentially unchanged. (The rows and columns may be multiplied by -1.)

4.4.4. QR Iteration

Our next goal is to reorganize orthogonal iteration to incorporate shifting and inverting, as in section 4.4.2. This will make it more efficient and eliminate the assumption that eigenvalues differ in magnitude, which was needed in Theorem 4.8 to prove convergence.

ALGORITHM 4.4. *QR iteration: Given A_0, we iterate*

> $i = 0$
> *repeat*
> > *Factor* $A_i = Q_i R_i$ *(the QR decomposition)*
> > $A_{i+1} = R_i Q_i$
> > $i = i + 1$
> *until convergence*

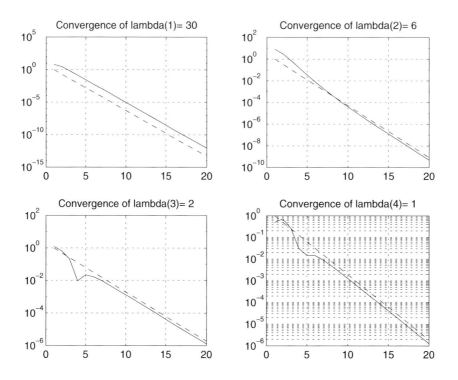

Fig. 4.3. *Convergence of diagonal entries during orthogonal iteration.*

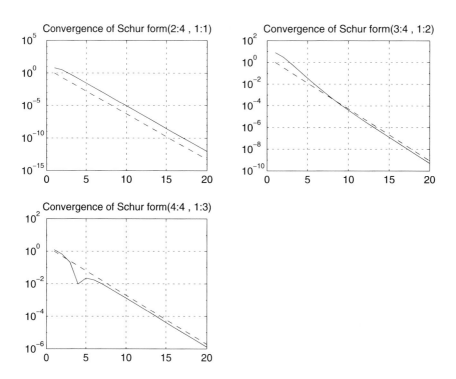

Fig. 4.4. *Convergence to Schur form during orthogonal iteration.*

Since $A_{i+1} = R_i Q_i = Q_i^T (Q_i R_i) Q_i = Q_i^T A_i Q_i$, A_{i+1} and A_i are orthogonally similar.

We claim that the A_i computed by QR iteration is identical to the matrix $Z_i^T A Z_i$ implicitly computed by orthogonal iteration.

LEMMA 4.3. *Let A_i be the matrix computed by Algorithm 4.4. Then $A_i = Z_i^T A Z_i$, where Z_i is the matrix computed from orthogonal iteration (Algorithm 4.3) starting with $Z_0 = I$. Thus A_i converges to Schur form if all the eigenvalues have different absolute values.*

Proof. We use induction. Assume $A_i = Z_i^T A Z_i$. From Algorithm 4.3, we can write $A Z_i = Z_{i+1} R_{i+1}$, where Z_{i+1} is orthogonal and R_{i+1} is upper triangular. Then $Z_i^T A Z_i = Z_i^T (Z_{i+1} R_{i+1})$ is the product of an orthogonal matrix $Q = Z_i^T Z_{i+1}$ and an upper triangular matrix $R = R_{i+1} = Z_{i+1}^T A Z_i$; this must be the QR decomposition $A_i = QR$, since the QR decomposition is unique (except for possibly multiplying each column of Q and row of R by -1). Then

$$Z_{i+1}^T A Z_{i+1} = (Z_{i+1}^T A Z_i)(Z_i^T Z_{i+1}) = R_{i+1}(Z_i^T Z_{i+1}) = RQ.$$

This is precisely how the QR iteration maps A_i to A_{i+1}, so $Z_{i+1}^T A Z_{i+1} = A_{i+1}$ as desired. $\square$

To see a variety of numerical examples illustrating the convergence of QR iteration, run the Matlab code referred to in Question 4.15.

From earlier analysis, we know that the convergence rate depends on the ratios of eigenvalues. To speed convergence, we use shifting and inverting.

ALGORITHM 4.5. *QR iteration with a shift: Given A_0, we iterate*

> $i = 0$
> *repeat*
> > *Choose a shift σ_i near an eigenvalue of A*
> > *Factor $A_i - \sigma_i I = Q_i R_i$ (QR decomposition)*
> > $A_{i+1} = R_i Q_i + \sigma_i I$
> > $i = i + 1$
> *until convergence*

LEMMA 4.4. *A_i and A_{i+1} are orthogonally similar.*

Proof. $A_{i+1} = R_i Q_i + \sigma_i I = Q_i^T Q_i R_i Q_i + \sigma_i Q_i^T Q_i = Q_i^T (Q_i R_i + \sigma_i I) Q_i = Q_i^T A_i Q_i$. $\square$

If R_i is nonsingular, we may also write

$$\begin{aligned} A_{i+1} &= R_i Q_i + \sigma_i I = R_i Q_i R_i R_i^{-1} + \sigma_i R_i R_i^{-1} = R_i(Q_i R_i + \sigma_i I)R_i^{-1} \\ &= R_i A_i R_i^{-1}. \end{aligned}$$

If σ_i is an *exact* eigenvalue of A_i, then we claim that QR iteration converges in one step: since σ_i is an eigenvalue, $A_i - \sigma_i I$ is singular, so R_i is singular, and so some diagonal entry of R_i must be zero. Suppose $R_i(n,n) = 0$. This implies that the last row of $R_i Q_i$ is 0, so the last row of $A_{i+1} = R_i Q_i + \sigma_i I$ equals $\sigma_i e_n^T$, where e_n is the nth column of the n-by-n identity matrix. In other words, the last row of A_{i+1} is zero except for the eigenvalue σ_i appearing in the (n,n) entry. This means that the algorithm has *converged*, because A_{i+1} is block upper triangular, with a trailing 1-by-1 block σ_i; the leading $(n-1)$-by-$(n-1)$ block A' is a new, smaller eigenproblem to which QR iteration can be solved without ever modifying σ_i again: $A_{i+1} = \begin{bmatrix} A' & a \\ 0 & \sigma_i \end{bmatrix}$.

When σ_i is not an exact eigenvalue, then we will accept $A_{i+1}(n,n)$ as having converged when the lower left block $A_{i+1}(n, 1:n-1)$ is small enough. Recall from our earlier analysis that we expect $A_{i+1}(n, 1:n-1)$ to shrink by a factor $|\lambda_k - \sigma_i| / \min_{j \neq k} |\lambda_j - \sigma_i|$, where $|\lambda_k - \sigma_i| = \min_j |\lambda_j - \sigma_i|$. So if σ_i is a very good approximation to eigenvalue λ_k, we expect fast convergence.

Here is another way to see why convergence should be fast, by recognizing that QR iteration is implicitly doing inverse iteration. When σ_i is an exact eigenvalue, the last column $\underline{q}_i$ of Q_i will be a left eigenvector of A_i for eigenvalue σ_i, since $\underline{q}_i^* A_i = \underline{q}_i^* (Q_i R_i + \sigma_i I) = e_n^T R_i + \sigma_i \underline{q}_i^* = \sigma_i \underline{q}_i^*$. When σ_i is close to an eigenvalue, we expect $\underline{q}_i$ to be close to an eigenvector for the following reason: $\underline{q}_i$ is parallel to $((A_i - \sigma_i I)^*)^{-1} e_n$ (we explain why below). In other words $\underline{q}_i$ is the same as would be obtained from inverse iteration on $(A_i - \sigma_i I)^*$ (and so we expect it to be close to a left eigenvector).

Here is the proof that $\underline{q}_i$ is parallel to $((A_i - \sigma_i I)^*)^{-1} e_n$. $A_i - \sigma_i I = Q_i R_i$ implies $(A_i - \sigma_i I) R_i^{-1} = Q_i$. Inverting and taking the conjugate transpose of both sides leave the right-hand side Q_i unchanged and change the left-hand side to $((A_i - \sigma_i I)^*)^{-1} R_i^*$, whose last column is $((A_i - \sigma_i I)^*)^{-1} \cdot [0, \ldots, 0, R_i(n,n)]^T$, which is proportional to the last column of $((A_i - \sigma_i I)^*)^{-1}$.

How do we choose σ_i to be an accurate approximate eigenvalue, when we are trying to compute eigenvalues in the first place? We will say more about this later, but for now note that near convergence to a *real* eigenvalue $A_i(n,n)$ is close to that eigenvalue, so $\sigma_i = A_i(n,n)$ is a good choice of shift. In fact, it yields local *quadratic convergence*, which means that the number of correct digits *doubles* at every step. We explain why quadratic convergence occurs as follows: Suppose at step i that $\|A_i(n, 1:n-1)\| / \|A\| \equiv \eta \ll 1$. If we were to set $A_i(n, 1:n-1)$ to exactly 0, we would make A_i block upper triangular and so perturb a true eigenvalue λ_k to make it equal to $A_i(n,n)$. If this eigenvalue is far from the other eigenvalues, it will not be ill-conditioned, so this perturbation will be $O(\eta \|A\|)$. In other words, $|\lambda_k - A_i(n,n)| = O(\eta \|A\|)$. On the next iteration, if we choose $\sigma_i = A_i(n,n)$, we expect $A_{i+1}(n, 1:n-1)$ to shrink by a factor $|\lambda_k - \sigma_i| / \min_{j \neq k} |\lambda_j - \sigma_i| = O(\eta)$, implying that $\|A_{i+1}(n, 1: n-1)\| = O(\eta^2 \|A\|)$, or $\|A_{i+1}(n, 1:n-1)\| / \|A\| = O(\eta^2)$. Decreasing the error this way from η to $O(\eta^2)$ is quadratic convergence.

EXAMPLE 4.7. Here are some shifted QR iterations starting with the same 4-by-4 matrix A_0 as in Example 4.5, with shift $\sigma_i = A_i(4,4)$. The convergence is a bit erratic at first but eventually becomes quadratic near the end, with $\|A_i(4, 1 : 3)\| \approx |A_i(4,3)|$ approximately squaring at each of the last three steps. Also, the number of correct digits in $A_i(4,4)$ doubles at the fourth through second-to-last steps.

$A_0(4,:) =$	$+1.9$	$+56.$	$+40.$	-10.558
$A_1(4,:) =$	$-.85$	-4.9	$+2.2 \cdot 10^{-2}$	-6.6068
$A_2(4,:) =$	$+.35$	$+.86$	$+.30$	0.74894
$A_3(4,:) =$	$-1.2 \cdot 10^{-2}$	$-.17$	$-.70$	1.4672
$A_4(4,:) =$	$-1.5 \cdot 10^{-4}$	$-1.8 \cdot 10^{-2}$	$-.38$	1.4045
$A_5(4,:) =$	$-3.0 \cdot 10^{-6}$	$-2.2 \cdot 10^{-3}$	$-.50$	1.1403
$A_6(4,:) =$	$-1.4 \cdot 10^{-8}$	$-6.3 \cdot 10^{-5}$	$-7.8 \cdot 10^{-2}$	1.0272
$A_7(4,:) =$	$-1.4 \cdot 10^{-11}$	$-3.6 \cdot 10^{-7}$	$-2.3 \cdot 10^{-3}$	0.99941
$A_8(4,:) =$	$+2.8 \cdot 10^{-16}$	$+4.2 \cdot 10^{-11}$	$+1.4 \cdot 10^{-6}$	0.9999996468853453
$A_9(4,:) =$	$-3.4 \cdot 10^{-24}$	$-3.0 \cdot 10^{-18}$	$-4.8 \cdot 10^{-13}$	0.9999999999998767
$A_{10}(4,:) =$	$+1.5 \cdot 10^{-38}$	$+7.4 \cdot 10^{-32}$	$+6.0 \cdot 10^{-26}$	1.000000000000001

By the time we reach A_{10}, the rest of the matrix has made a lot of progress toward convergence as well, so later eigenvalues will be computed very quickly, in one or two steps each:

$$A_{10} = \begin{bmatrix} 30.000 & -32.557 & -70.844 & 14.985 \\ 6.1548 \cdot 10^{-6} & 6.0000 & 1.8143 & -.55754 \\ 2.5531 \cdot 10^{-13} & 2.0120 \cdot 10^{-6} & 2.0000 & -.25894 \\ 1.4692 \cdot 10^{-38} & 7.4289 \cdot 10^{-32} & 6.0040 \cdot 10^{-26} & 1.0000 \end{bmatrix}. \quad \diamond$$

4.4.5. Making QR Iteration Practical

Here are some remaining problems we have to solve to make the algorithm more practical:

1. The iteration is too expensive. The QR decomposition costs $O(n^3)$ flops, so if we were lucky enough to do only one iteration per eigenvalue, the cost would be $O(n^4)$. But we seek an algorithm with a total cost of only $O(n^3)$.

2. How shall we choose σ_i to accelerate convergence to a complex eigenvalue? Choosing σ_i complex means all arithmetic has to be complex, increasing the cost by a factor of about 4 when A is real. We seek an algorithm that uses all real arithmetic if A is real and converges to real Schur form.

3. How do we recognize convergence?

The solutions to these problems, which we will describe in more detail later, are as follows:

1. We will initially reduce the matrix to *upper Hessenberg form*; this means that A is zero below the first subdiagonal (i.e., $a_{ij} = 0$ if $i > j + 1$) (see section 4.4.6). Then we will apply a step of QR iteration *implicitly*, i.e., without computing Q or multiplying by it explicitly (see section 4.4.8). This will reduce the cost of one QR iteration from $O(n^3)$ to $O(n^2)$ and the overall cost from $O(n^4)$ to $O(n^3)$ as desired

   When A is symmetric we will reduce it to tridiagonal form instead, reducing the cost of a single QR iteration further to $O(n)$. This is discussed in section 4.4.7 and Chapter 5.

2. Since complex eigenvalues of real matrices occur in complex conjugate pairs, we can shift by σ_i and $\bar{\sigma}_i$ simultaneously; it turns out that this will permit us to maintain real arithmetic (see section 4.4.8). If A is symmetric, all eigenvalues are real, and this is not an issue.

3. Convergence occurs when subdiagonal entries of A_i are "small enough." To help choose a practical threshold, we use the notion of backward stability: Since A_i is related to A by a similarity transformation by an orthogonal matrix, we expect A_i to have roundoff errors of size $O(\varepsilon\|A\|)$ in it anyway. Therefore, any subdiagonal entry of A_i smaller than $O(\varepsilon\|A\|)$ in magnitude may as well be zero, so we set it to zero.[16] When A is upper Hessenberg, setting $a_{p+1,p}$ to zero will make A into a block upper triangular matrix $A = \begin{bmatrix} A_{11} & A_{12} \\ 0 & A_{22} \end{bmatrix}$, where A_{11} is p-by-p and A_{11} and A_{22} are both Hessenberg. Then the eigenvalues of A_{11} and A_{22} may be found independently to get the eigenvalues of A. When all these diagonal blocks are 1-by-1 or 2-by-2, the algorithm has finished.

4.4.6. Hessenberg Reduction

Given a real matrix A, we seek an orthogonal Q so that QAQ^T is upper Hessenberg. The algorithm is a simple variation on the idea used for the QR decomposition.

EXAMPLE 4.8. We illustrate the general pattern of Hessenberg reduction with a 5-by-5 example. Each Q_i below is a 5-by-5 Householder reflection, chosen to zero out entries $i + 2$ through n in column i and leaving entries 1 through i unchanged.

---

[16]In practice, we use a slightly more stringent condition, replacing $\|A\|$ with the norm of a submatrix of A, to take into account matrices which may be "graded" with large entries in one place and small entries elsewhere. We can also set a subdiagonal entry to zero when the product $a_{p+1,p}a_{p+2,p+1}$ of two adjacent subdiagonal entries is small enough. See the LAPACK routine slahqr for details.

1. Choose Q_1 so

$$Q_1A = \begin{bmatrix} x & x & x & x & x \\ x & x & x & x & x \\ o & x & x & x & x \\ o & x & x & x & x \\ o & x & x & x & x \end{bmatrix} \quad \text{and} \quad A_1 \equiv Q_1AQ_1^T = \begin{bmatrix} x & x & x & x & x \\ x & x & x & x & x \\ o & x & x & x & x \\ o & x & x & x & x \\ o & x & x & x & x \end{bmatrix}.$$

Q_1 leaves the first row of Q_1A unchanged, and Q_1^T leaves the first column of $Q_1AQ_1^T$ unchanged, including the zeros.

2. Choose Q_2 so

$$Q_2A_1 = \begin{bmatrix} x & x & x & x & x \\ x & x & x & x & x \\ o & x & x & x & x \\ o & o & x & x & x \\ o & o & x & x & x \end{bmatrix} \quad \text{and} \quad A_2 \equiv Q_2A_1Q_2^T = \begin{bmatrix} x & x & x & x & x \\ x & x & x & x & x \\ o & x & x & x & x \\ o & o & x & x & x \\ o & o & x & x & x \end{bmatrix}.$$

Q_2 changes only the last three rows of A_1, and Q_2^T leaves the first two columns of $Q_2A_1Q_2^T$ unchanged, including the zeros.

3. Choose Q_3 so

$$Q_3A_2 = \begin{bmatrix} x & x & x & x & x \\ x & x & x & x & x \\ o & x & x & x & x \\ o & o & x & x & x \\ o & o & o & x & x \end{bmatrix} \quad \text{and} \quad A_3 = Q_3A_2Q_3^T = \begin{bmatrix} x & x & x & x & x \\ x & x & x & x & x \\ o & x & x & x & x \\ o & o & x & x & x \\ o & o & o & x & x \end{bmatrix},$$

which is upper Hessenberg. Altogether $A_3 = (Q_3Q_2Q_1) \cdot A(Q_3Q_2Q_1)^T \equiv QAQ^T$. ◇

The general algorithm for Hessenberg reduction is as follows.

ALGORITHM 4.6. *Reduction to upper Hessenberg form:*

if Q is desired, set $Q = I$
for $i = 1 : n - 2$
  $u_i = \text{House}(A(i + 1 : n, i))$
  $P_i = I - 2u_iu_i^T$ */* $Q_i = \text{diag}(I^{i \times i}, P_i)$ */*
  $A(i + 1 : n, i : n) = P_i \cdot A(i + 1 : n, i : n)$
  $A(1 : n, i + 1 : n) = A(1 : n, i + 1 : n) \cdot P_i$
  if Q is desired
    $Q(i + 1 : n, i : n) = P_i \cdot Q(i + 1 : n, i : n)$ */* $Q = Q_i \cdot Q$ */*
  end if
end for

As with the QR decomposition, one does not form P_i explicitly but instead multiplies by $I - 2u_i u_i^T$ via matrix-vector operations. The u_i vectors can also be stored below the subdiagonal, similar to the QR decomposition. They can be applied using Level 3 BLAS, as described in Question 3.17. This algorithm is available as the Matlab command hess or the LAPACK routine sgehrd.

The number of floating point operations is easily counted to be $\frac{10}{3}n^3 + O(n^2)$, or $\frac{14}{3}n^3 + O(n^2)$ if the product $Q = Q_{n-1} \cdots Q_1$ is computed as well.

The advantage of Hessenberg form under QR iteration is that it costs only $6n^2 + O(n)$ flops per iteration instead of $O(n^3)$, and its form is preserved so that the matrix remains upper Hessenberg.

PROPOSITION 4.5. *Hessenberg form is preserved by QR iteration.*

Proof. It is easy to confirm that the QR decomposition of an upper Hessenberg matrix like $A_i - \sigma I$ yields an upper Hessenberg Q (since the jth column of Q is a linear combination of the leading j columns of $A_i - \sigma I$). Then it is easy to confirm that RQ remains upper Hessenberg and adding σI does not change this. $\square$

DEFINITION 4.5. *An upper Hessenberg matrix H is* unreduced *if all subdiagonals are nonzero.*

It is easy to see that if H is reduced because $h_{i+1,i} = 0$, then its eigenvalues are those of its leading i-by-i Hessenberg submatrix and its trailing $(n-i)$-by-$(n-i)$ Hessenberg submatrix, so we need consider only unreduced matrices.

4.4.7.   Tridiagonal and Bidiagonal Reduction

If A is symmetric, the Hessenberg reduction process leaves A symmetric at each step, so zeros are created in symmetric positions. This means we need work on only half the matrix, reducing the operation count to $\frac{4}{3}n^3 + O(n^2)$ or $\frac{8}{3}n^3 + O(n^2)$ to form $Q_{n-1} \ldots Q_1$ as well. We call this algorithm *tridiagonal reduction*. We will use this algorithm in Chapter 5. This routine is available as LAPACK routine ssytrd.

Looking ahead a bit to our discussion of computing the SVD in section 5.4, we recall from section 3.2.3 that the eigenvalues of the symmetric matrix $A^T A$ are the squares of the singular values of A. Our eventual SVD algorithm will use this fact, so we would like to find a form for A which implies that $A^T A$ is tridiagonal. We will choose A to be *upper bidiagonal*, or nonzero only on the diagonal and first superdiagonal. Thus, we want to compute orthogonal matrices Q and V such that QAV is bidiagonal. The algorithm, called *bidiagonal reduction*, is very similar to Hessenberg and tridiagonal reduction.

EXAMPLE 4.9. Here is a 4-by-4 example of bidiagonal reduction, which illustrates the general pattern:

1. Choose Q_1 so

$$Q_1 A = \begin{bmatrix} x & x & x & x \\ o & x & x & x \\ o & x & x & x \\ o & x & x & x \end{bmatrix} \quad \text{and } V_1 \text{ so } A_1 \equiv Q_1 A V_1 = \begin{bmatrix} x & x & o & o \\ o & x & x & x \\ o & x & x & x \\ o & x & x & x \end{bmatrix}.$$

Q_1 is a Householder reflection, and V_1 is a Householder reflection that leaves the first column of $Q_1 A$ unchanged.

2. Choose Q_2 so

$$Q_2 A_1 = \begin{bmatrix} x & x & o & o \\ o & x & x & x \\ o & o & x & x \\ o & o & x & x \end{bmatrix} \quad \text{and } V_2 \text{ so } A_2 \equiv Q_2 A_1 V_2 = \begin{bmatrix} x & x & o & o \\ o & x & x & o \\ o & o & x & x \\ o & o & x & x \end{bmatrix}.$$

Q_2 is a Householder reflection that leaves the first row of A_1 unchanged. V_2 is a Householder reflection that leaves the first two columns of $Q_2 A_1$ unchanged.

3. Choose Q_3 so

$$Q_3 A_2 = \begin{bmatrix} x & x & o & o \\ o & x & x & o \\ o & o & x & x \\ o & o & o & x \end{bmatrix} \quad \text{and } V_3 = I \text{ so } A_3 = Q_3 A_2.$$

Q_3 is a Householder reflection that leaves the first two rows of A_2 unchanged. ◇

In general, if A is n-by-n, then we get orthogonal matrices $Q = Q_{n-1} \cdots Q_1$ and $V = V_1 \cdots V_{n-2}$ such that $QAV = A'$ is upper bidiagonal.

Note that $A'^T A' = V^T A^T Q^T Q A V = V^T A^T A V$, so $A'^T A'$ has the same eigenvalues as $A^T A$; i.e., A' has the same singular values as A.

The cost of this bidiagonal reduction is $\frac{8}{3}n^3 + O(n^2)$ flops, plus another $4n^3 + O(n^2)$ flops to compute Q and V. This routine is available as LAPACK routine sgebrd.

4.4.8. QR Iteration with Implicit Shifts

In this section we show how to implement QR iteration cheaply on an upper Hessenberg matrix. The implementation will be *implicit* in the sense that we do not explicitly compute the QR factorization of a matrix H but rather construct Q implicitly as a product of Givens rotations and other simple orthogonal matrices. The *implicit Q theorem* described below shows that this implicitly constructed Q is the Q we want. Then we show how to incorporate a single shift σ, which is necessary to accelerate convergence. To retain real arithmetic in the presence of complex eigenvalues, we then show how to do a *double* shift, i.e., combine two consecutive QR iterations with complex conjugate shifts σ and $\bar{\sigma}$; the result after this double shift is again real. Finally, we discuss strategies for choosing shifts σ and $\bar{\sigma}$ to provide reliable quadratic convergence. However, there have been recent discoveries of rare situation where convergence does not occur [25, 65], so finding a completely reliable and fast implementation of QR iteration remains an open problem.

Implicit Q Theorem

Our eventual implementation of QR iteration will depend on the following theorem.

THEOREM 4.9. *Implicit Q theorem. Suppose that $Q^T A Q = H$ is unreduced upper Hessenberg. Then columns 2 through n of Q are determined uniquely (up to signs) by the first column of Q.*

This theorem implies that to compute $A_{i+1} = Q_i^T A_i Q_i$ from A_i in the QR algorithm, we will need only to

1. compute the first column of Q_i (which is parallel to the first column of $A_i - \sigma_i I$ and so can be gotten just by normalizing this column vector).

2. choose other columns of Q_i so Q_i is orthogonal and A_{i+1} is unreduced Hessenberg.

Then by the implicit Q theorem, we know that we will have computed A_{i+1} correctly because Q_i is unique up to signs, which do not matter. (Signs do not matter because changing the signs of the columns of Q_i is the same as changing $A_i - \sigma_i I = Q_i R_i$ to $(Q_i S_i)(S_i R_i)$, where $S_i = \text{diag}(\pm 1, \ldots, \pm 1)$. Then $A_{i+1} = (S_i R_i)(Q_i S_i) + \sigma_i I = S_i(R_i Q_i + \sigma_i I)S_i$, which is an orthogonal similarity that just changes the signs of the columns and rows of A_{i+1}.)

Proof of the implicit Q theorem. Suppose that $Q^T A Q = H$ and $V^T A V = G$ are unreduced upper Hessenberg, Q and V are orthogonal, and the first columns of Q and V are equal. Let $(X)_i$ denote the ith column of X. We wish to show $(Q)_i = \pm (V)_i$ for all $i > 1$, or equivalently, that $W \equiv V^T Q = \text{diag}(\pm 1, \ldots, \pm 1)$.

Since $W = V^T Q$, we get $GW = GV^T Q = V^T A Q = V^T Q H = W H$. Now $GW = WH$ implies $G(W)_i = (GW)_i = (WH)_i = \sum_{j=1}^{i+1} h_{ji}(W)_j$, so $h_{i+1,i}(W)_{i+1} = G(W)_i - \sum_{j=1}^{i} h_{ji}(W)_j$. Since $(W)_1 = [1, 0, \ldots, 0]^T$ and G is upper Hessenberg, we can use induction on i to show that $(W)_i$ is nonzero in entries 1 to i only; i.e., W is upper triangular. Since W is also orthogonal, W is diagonal = $\text{diag}(\pm 1, \ldots, \pm 1)$. $\square$

Implicit Single Shift QR Algorithm

To see how to use the implicit Q theorem to compute A_1 from $A_0 = A$, we use a 5-by-5 example.

EXAMPLE 4.10. 1. Choose

$$
Q_1^T = \begin{bmatrix} c_1 & s_1 & & & \\ -s_1 & c_1 & & & \\ & & 1 & & \\ & & & 1 & \\ & & & & 1 \end{bmatrix} \quad \text{so } A_1 \equiv Q_1^T A Q_1 = \begin{bmatrix} x & x & x & x & x \\ x & x & x & x & x \\ + & x & x & x & x \\ o & o & x & x & x \\ o & o & o & x & x \end{bmatrix}.
$$

We discuss how to choose c_1 and s_1 below; for now they may be any Givens rotation. The $+$ in position (3,1) is called a *bulge* and needs to be gotten rid of to restore Hessenberg form.

2. Choose

$$
Q_2^T = \begin{bmatrix} 1 & & & & \\ & c_2 & s_2 & & \\ & -s_2 & c_2 & & \\ & & & 1 & \\ & & & & 1 \end{bmatrix} \quad \text{so } Q_2^T A_1 = \begin{bmatrix} x & x & x & x & x \\ x & x & x & x & x \\ o & x & x & x & x \\ o & o & x & x & x \\ o & o & o & x & x \end{bmatrix}
$$

and

$$
A_2 \equiv Q_2^T A_1 Q_2 = \begin{bmatrix} x & x & x & x & x \\ x & x & x & x & x \\ o & x & x & x & x \\ o & + & x & x & x \\ o & o & o & x & x \end{bmatrix}.
$$

Thus the bulge has been "chased" from (3,1) to (4,2).

3. Choose

$$
Q_3^T = \begin{bmatrix} 1 & & & & \\ & 1 & & & \\ & & c_3 & s_3 & \\ & & -s_3 & c_3 & \\ & & & & 1 \end{bmatrix} \quad \text{so } Q_3^T A_2 = \begin{bmatrix} x & x & x & x & x \\ x & x & x & x & x \\ o & x & x & x & x \\ o & o & x & x & x \\ o & o & o & x & x \end{bmatrix}
$$

and

$$
A_3 \equiv Q_3^T A_2 Q_3 = \begin{bmatrix} x & x & x & x & x \\ x & x & x & x & x \\ o & x & x & x & x \\ o & o & x & x & x \\ o & o & + & x & x \end{bmatrix}.
$$

The bulge has been chased from (4,2) to (5,3).

4. Choose

$$Q_4^T = \begin{bmatrix} 1 & & & & \\ & 1 & & & \\ & & 1 & & \\ & & & c_4 & s_4 \\ & & & -s_4 & c_4 \end{bmatrix} \quad \text{so } Q_4^T A_3 = \begin{bmatrix} x & x & x & x & x \\ x & x & x & x & x \\ o & x & x & x & x \\ o & o & x & x & x \\ o & o & o & x & x \end{bmatrix}$$

and

$$A_4 = Q_4^T A_3 Q_4 = \begin{bmatrix} x & x & x & x & x \\ x & x & x & x & x \\ o & x & x & x & x \\ o & o & x & x & x \\ o & o & o & x & x \end{bmatrix},$$

so we are back to upper Hessenberg form.

Altogether $Q^T A Q$ is upper Hessenberg, where

$$Q = Q_1 Q_2 Q_3 Q_4 = \begin{bmatrix} c_1 & x & x & x & x \\ s_1 & x & x & x & x \\ & s_2 & x & x & x \\ & & s_3 & x & x \\ & & & s_4 & x \end{bmatrix},$$

so the first column of Q is $[c_1, s_1, 0, \ldots, 0]^T$, which by the implicit Q theorem has uniquely determined the other columns of Q (up to signs). We now choose the first column of Q to be proportional to the first column of $A - \sigma I$, $[a_{11} - \sigma, a_{21}, 0, \ldots, 0]^T$. This means Q is the same as in the QR decomposition of $A - \sigma I$, as desired. ◇

The cost of one implicit QR iteration for an n-by-n matrix is $6n^2 + O(n)$.

Implicit Double Shift QR Algorithm

This section describes how to maintain real arithmetic by shifting by σ and $\bar{\sigma}$ at the same time. This is essential for an efficient practical implementation but not for a mathematical understanding of the algorithm and may be skipped on a first reading.

The results of shifting by σ and $\bar{\sigma}$ in succession are

$$\begin{aligned} A_0 - \sigma I &= Q_1 R_1, \\ A_1 &= R_1 Q_1 + \sigma I \quad \text{so } A_1 = Q_1^T A_0 Q_1, \\ A_1 - \bar{\sigma} I &= Q_2 R_2, \\ A_2 &= R_2 Q_2 + \bar{\sigma} I \quad \text{so } A_2 = Q_2^T A_1 Q_2 = Q_2^T Q_1^T A_0 Q_1 Q_2. \end{aligned}$$

LEMMA 4.5. *We can choose Q_1 and Q_2 so*

(1) $Q_1 Q_2$ is real,

(2) A_2 is therefore real,

(3) the first column of $Q_1 Q_2$ is easy to compute.

Proof. Since $Q_2 R_2 = A_1 - \bar{\sigma} I = R_1 Q_1 + (\sigma - \bar{\sigma}) I$, we get

$$
\begin{aligned}
Q_1 Q_2 R_2 R_1 &= Q_1 (R_1 Q_1 + (\sigma - \bar{\sigma}) I) R_1 \\
&= Q_1 R_1 Q_1 R_1 + (\sigma - \bar{\sigma}) Q_1 R_1 \\
&= (A_0 - \sigma I)(A_0 - \sigma I) + (\sigma - \bar{\sigma})(A_0 - \sigma I) \\
&= A_0^2 - 2(\Re \sigma) A_0 + |\sigma|^2 I \equiv M.
\end{aligned}
$$

Thus $(Q_1 Q_2)(R_2 R_1)$ is the QR decomposition of the real matrix M, and therefore $Q_1 Q_2$, as well as $R_2 R_1$, can be chosen real. This means that $A_2 = (Q_1 Q_2)^T A (Q_1 Q_2)$ also is real.

The first column of $Q_1 Q_2$ is proportional to the first column of $A_0^2 - 2 \Re \sigma A_0 + |\sigma^2| I$, which is

$$
\begin{bmatrix}
a_{11}^2 + a_{12} a_{21} - 2(\Re \sigma) a_{11} + |\sigma|^2 \\
a_{21}(a_{11} + a_{22} - 2(\Re \sigma)) \\
a_{21} a_{32} \\
0 \\
\vdots \\
0
\end{bmatrix} . \quad \square
$$

The rest of the columns of $Q_1 Q_2$ are computed implicitly using the implicit Q theorem. The process is still called "bulge chasing," but now the bulge is 2-by-2 instead of 1-by-1.

EXAMPLE 4.11. Here is a 6-by-6 example of bulge chasing.

1. Choose $Q_1^T = [\begin{smallmatrix} \tilde{Q}_1^T & 0 \\ 0 & I \end{smallmatrix}]$, where the first column of $\tilde{Q}_1^T$ is given as above, so

$$
Q_1^T A =
\begin{bmatrix}
x & x & x & x & x & x \\
x & x & x & x & x & x \\
+ & x & x & x & x & x \\
o & o & x & x & x & x \\
o & o & o & x & x & x \\
o & o & o & o & x & x
\end{bmatrix}
\quad \text{and} \quad
A_1 \equiv Q_1^T A Q_1 =
\begin{bmatrix}
x & x & x & x & x & x \\
x & x & x & x & x & x \\
+ & x & x & x & x & x \\
+ & + & x & x & x & x \\
o & o & o & x & x & x \\
o & o & o & o & x & x
\end{bmatrix} .
$$

We see that there is a 2-by-2 bulge, indicted by plus signs.

2. Choose a Householder reflection Q_2^T, which affects only rows 2, 3, and 4 of $Q_2^T A_1$, zeroing out entries $(3,1)$ and $(4,1)$ of A_1 (this means that Q_2^T is the identity matrix outside rows and columns 2 through 4):

$$Q_2^T A_1 = \begin{bmatrix} x & x & x & x & x & x \\ x & x & x & x & x & x \\ o & x & x & x & x & x \\ o & + & x & x & x & x \\ o & o & o & x & x & x \\ o & o & o & o & x & x \end{bmatrix} \quad \text{and} \quad A_2 \equiv Q_2^T A_1 Q_2 = \begin{bmatrix} x & x & x & x & x & x \\ x & x & x & x & x & x \\ o & x & x & x & x & x \\ o & + & x & x & x & x \\ o & + & + & x & x & x \\ o & o & o & o & x & x \end{bmatrix},$$

and the 2-by-2 bulge has been "chased" one column.

3. Choose a Householder reflection Q_3^T, which affects only rows 3, 4, and 5 of $Q_3^T A_2$, zeroing out entries $(4,2)$ and $(5,2)$ of A_2 (this means that Q_3^T is the identity outside rows and columns 3 through 5):

$$Q_3^T A_2 = \begin{bmatrix} x & x & x & x & x & x \\ x & x & x & x & x & x \\ o & x & x & x & x & x \\ o & o & x & x & x & x \\ o & o & + & x & x & x \\ o & o & o & o & x & x \end{bmatrix} \quad \text{and} \quad A_3 \equiv Q_3^T A_2 Q_3 = \begin{bmatrix} x & x & x & x & x & x \\ x & x & x & x & x & x \\ o & x & x & x & x & x \\ o & o & x & x & x & x \\ o & o & + & x & x & x \\ o & o & + & + & x & x \end{bmatrix}.$$

4. Choose a Householder reflection Q_4^T, which affects only rows 4, 5, and 6 of $Q_4^T A_3$, zeroing out entries $(5,3)$ and $(6,3)$ of A_3 (this means that Q_4^T is the identity matrix outside rows and columns 4 through 6):

$$A_4 \equiv Q_4^T A_3 Q_4 = \begin{bmatrix} x & x & x & x & x & x \\ x & x & x & x & x & x \\ o & x & x & x & x & x \\ o & o & x & x & x & x \\ o & o & o & x & x & x \\ o & o & o & + & x & x \end{bmatrix}.$$

5. Choose

$$Q_5^T = \begin{bmatrix} 1 & & & & & \\ & 1 & & & & \\ & & 1 & & & \\ & & & 1 & & \\ \hline & & & & c & s \\ & & & & -s & c \end{bmatrix} \quad \text{so} \quad A_5 = Q_5^T A_4 Q_5 = \begin{bmatrix} x & x & x & x & x & x \\ x & x & x & x & x & x \\ o & x & x & x & x & x \\ o & o & x & x & x & x \\ o & o & o & x & x & x \\ o & o & o & o & x & x \end{bmatrix}. \diamond$$

Choosing a Shift for the QR Algorithm

To completely specify one iteration of either single shift or double shift Hessenberg QR iteration, we need to choose the shift σ (and $\bar{\sigma}$). Recall from the end of section 4.4.4 that a reasonable choice of single shift, one that resulted in asymptotic quadratic convergence to a real eigenvalue, was $\sigma = a_{n,n}$, the bottom right entry of A_i. The generalization for double shifting is to use the *Francis shift*, which means that σ and $\bar{\sigma}$ are the eigenvalues of the bottom 2-by-2 corner of A_i: $\begin{bmatrix} a_{n-1,n-1} & a_{n-1,n} \\ a_{n,n-1} & a_{n,n} \end{bmatrix}$. This will let us converge to either two real eigenvalues in the bottom 2-by-2 corner or a single 2-by-2 block with complex conjugate eigenvalues. When we are close to convergence, we expect $a_{n-1,n-2}$ (and possibly $a_{n,n-1}$) to be small so that the eigenvalues of this 2-by-2 matrix are good approximations for eigenvalues of A. Indeed, one can show that this choice leads to quadratic convergence asymptotically. This means that once $a_{n-1,n-2}$ (and possibly $a_{n,n-1}$) is small enough, its magnitude will square at each step and quickly approach zero. In practice, this works so well that on average only two QR iterations per eigenvalue are needed for convergence for almost all matrices. This justifies calling QR iteration a "direct" method.

In practice, the QR iteration with the Francis shift can fail to converge (indeed, it leaves

$$\begin{bmatrix} 0 & 0 & 1 \\ 1 & 0 & 0 \\ 0 & 1 & 0 \end{bmatrix}$$

unchanged). So the practical algorithm in use for decades had an "exceptional shift" every 10 shifts if convergence had not occurred. Still, tiny sets of matrices where that algorithm did not converge were discovered only recently [25, 65]; matrices in a small neighborhood of

$$\begin{bmatrix} 0 & 1 & 0 & 0 \\ 1 & 0 & h & 0 \\ 0 & -h & 0 & 1 \\ 0 & 0 & 1 & 0 \end{bmatrix},$$

where h is a few thousand times machine epsilon, form such a set. So another "exceptional shift" was recently added to the algorithm to patch this case. But it is still an open problem to find a shift strategy that guarantees fast convergence for all matrices.

4.5.   Other Nonsymmetric Eigenvalue Problems

4.5.1.   Regular Matrix Pencils and Weierstrass Canonical Form

The standard eigenvalue problem asks for which scalars z the matrix $A - zI$ is singular; these scalars are the eigenvalues. This notion generalizes in several important ways.

DEFINITION 4.6. $A - \lambda B$, where A and B are m-by-n matrices, is called a matrix pencil, or just a pencil. Here λ is an indeterminate, not a particular, numerical value.

DEFINITION 4.7. If A and B are square and $\det(A - \lambda B)$ is not identically zero, the pencil $A - \lambda B$ is called regular. Otherwise it is called singular. When $A - \lambda B$ is regular, $p(\lambda) \equiv \det(A - \lambda B)$ is called the characteristic polynomial of $A - \lambda B$ and the eigenvalues of $A - \lambda B$ are defined to be

(1) the roots of $p(\lambda) = 0$,
(2) ∞ (with multiplicity $n - \deg(p)$) if $\deg(p) < n$.

EXAMPLE 4.12. Let

$$A - \lambda B = \begin{bmatrix} 1 & & \\ & 1 & \\ & & 0 \end{bmatrix} - \lambda \begin{bmatrix} 2 & & \\ & 0 & \\ & & 1 \end{bmatrix}.$$

Then $p(\lambda) = \det(A - \lambda B) = (1 - 2\lambda) \cdot (1 - 0\lambda) \cdot (0 - \lambda) = (2\lambda - 1)\lambda$, so the eigenvalues are $\lambda = \frac{1}{2}, 0$ and ∞. ◇

Matrix pencils arise naturally in many mathematical models of physical systems; we give examples below. The next proposition relates the eigenvalues of a regular pencil $A - \lambda B$ to the eigenvalues of a single matrix.

PROPOSITION 4.6. Let $A - \lambda B$ be regular. If B is nonsingular, all eigenvalues of $A - \lambda B$ are finite and the same as the eigenvalues of AB^{-1} or $B^{-1}A$. If B is singular, $A - \lambda B$ has eigenvalue ∞ with multiplicity $n - \text{rank}(B)$. If A is nonsingular, the eigenvalues of $A - \lambda B$ are the same as the reciprocals of the eigenvalues of $A^{-1}B$ or BA^{-1}, where a zero eigenvalue of $A^{-1}B$ corresponds to an infinite eigenvalue of $A - \lambda B$.

Proof. If B is nonsingular and λ' is an eigenvalue, then $0 = \det(A - \lambda'B) = \det(AB^{-1} - \lambda'I) = \det(B^{-1}A - \lambda'I)$, so λ' is also an eigenvalue of AB^{-1} and $B^{-1}A$. If B is singular, then take $p(\lambda) = \det(A - \lambda B)$, write the SVD of B as $B = U\Sigma V^T$, and substitute to get

$$p(\lambda) = \det(A - \lambda U\Sigma V^T) = \det(U(U^T AV - \lambda\Sigma)V^T) = \pm\det(U^T AV - \lambda\Sigma).$$

Since $\text{rank}(B) = \text{rank}(\Sigma)$, only $\text{rank}(B)$ λ's appear in $U^T AV - \lambda\Sigma$, so the degree of the polynomial $\det(U^T AV - \lambda\Sigma)$ is $\text{rank}(B)$.

If A is nonsingular, $\det(A - \lambda B) = 0$ if and only if $\det(I - \lambda A^{-1}B) = 0$ or $\det(I - \lambda BA^{-1}) = 0$. This equality can hold only if $\lambda \neq 0$ and $1/\lambda$ is an eigenvalue of $A^{-1}B$ or BA^{-1}. □

DEFINITION 4.8. *Let λ' be a finite eigenvalue of the regular pencil $A - \lambda B$. Then $x \neq 0$ is a right eigenvector if $(A - \lambda' B)x = 0$, or equivalently $Ax = \lambda' Bx$. If $\lambda' = \infty$ is an eigenvalue and $Bx = 0$, then x is a right eigenvector. A left eigenvector of $A - \lambda B$ is a right eigenvector of $(A - \lambda B)^*$.*

EXAMPLE 4.13. Consider the pencil $A - \lambda B$ in Example 4.12. Since A and B are diagonal, the right and left eigenvectors are just the columns of the identity matrix. ◇

EXAMPLE 4.14. Consider the damped mass-spring system from Example 4.1. There are two matrix pencils that arise naturally from this problem. First, we can write the eigenvalue problem

$$Ax = \begin{bmatrix} -M^{-1}B & -M^{-1}K \\ I & 0 \end{bmatrix} x = \lambda x$$

as

$$\begin{bmatrix} -B & -K \\ I & 0 \end{bmatrix} x = \lambda \begin{bmatrix} M & 0 \\ 0 & I \end{bmatrix} x.$$

This may be a superior formulation if M is very ill-conditioned, so that $M^{-1}B$ and $M^{-1}K$ are hard to compute accurately.

Second, it is common to consider the case $B = 0$ (no damping), so the original differential equation is $M\ddot{x}(t) + Kx(t) = 0$. Seeking solutions of the form $x_i(t) = e^{\lambda_i t} x_i(0)$, we get $\lambda_i^2 e^{\lambda_i t} M x_i(0) + e^{\lambda_i t} K x_i(0) = 0$, or $\lambda_i^2 M x_i(0) + K x_i(0) = 0$. In other words, $-\lambda_i^2$ is an eigenvalue and $x_i(0)$ is a right eigenvector of the pencil $K - \lambda M$. Since we are assuming that M is nonsingular, these are also the eigenvalue and right eigenvector of $M^{-1}K$. ◇

Infinite eigenvalues also arise naturally in practice. For example, later in this section we will show how infinite eigenvalues correspond to *impulse response* in a system described by ordinary differential equations with linear constraints, or *differential-algebraic equations* [41]. See also Question 4.16 for an application of matrix pencils to computational geometry and computer graphics.

Recall that all of our theory and algorithms for the eigenvalue problem of a single matrix A depended on finding a similarity transformation $S^{-1}AS$ of A that is in "simpler" form than A. The next definition shows how to generalize the notion of similarity to matrix pencils. Then we show how the Jordan form and Schur form generalize to pencils.

DEFINITION 4.9. *Let P_L and P_R be nonsingular matrices. Then pencils $A - \lambda B$ and $P_L A P_R - \lambda P_L B P_R$ are called* equivalent.

PROPOSITION 4.7. *The equivalent regular pencils $A - \lambda B$ and $P_L A P_R - \lambda P_L B P_R$ have the same eigenvalues. The vector x is a right eigenvector of $A - \lambda B$ if*

*and only if $P_R^{-1}x$ is a right eigenvector of $P_L A P_R - \lambda P_L B P_R$. The vector y
is a left eigenvector of $A - \lambda B$ if and only if $(P_L^*)^{-1}y$ is a left eigenvector of
$P_L A P_R - \lambda P_L B P_R$.*

Proof.

$\det(A - \lambda B) = 0$ if and only if $\det(P_L(A - \lambda B)P_R) = 0$.

$(A - \lambda B)x = 0$ if and only if $P_L(A - \lambda B)P_R P_R^{-1}x = 0$.

$(A - \lambda B)^*y = 0$ if and only if $P_R^*(A - \lambda B)^* P_L^*(P_L^*)^{-1}y = 0$. $\square$

The following theorem generalizes the Jordan canonical form to regular matrix pencils.

THEOREM 4.10. *Weierstrass canonical form. Let $A - \lambda B$ be regular. Then
there are nonsingular P_L and P_R such that*

$$P_L(A - \lambda B)P_R = \mathrm{diag}(J_{n_1}(\lambda_1) - \lambda I_{n_1}, \ldots, J_{n_k}(\lambda_{n_k}) - \lambda I_{n_k}, N_{m_1}, \ldots, N_{m_r}),$$

where $J_{n_i}(\lambda_i)$ is an n_i-by-n_i Jordan block with eigenvalue λ_i,

$$J_{n_i}(\lambda_i) = \begin{bmatrix} \lambda_i & 1 & & \\ & \ddots & \ddots & \\ & & \ddots & 1 \\ & & & \lambda_i \end{bmatrix},$$

and N_{m_i} is a "Jordan block for $\lambda = \infty$ with multiplicity m_i,"

$$N_{m_i} = \begin{bmatrix} 1 & \lambda & & \\ & 1 & \ddots & \\ & & \ddots & \lambda \\ & & & 1 \end{bmatrix} = I_{m_i} - \lambda J_{m_i}(0).$$

For a proof, see [110].

Application of Jordan and Weierstrass Forms to Differential Equations

Consider the linear differential equation $\dot{x}(t) = Ax(t) + f(t)$, $x(0) = x_0$. An
explicit solution is given by $x(t) = e^{At}x_0 + \int_0^t e^{A(t-\tau)}f(\tau)d\tau$. If we know
the Jordan form $A = SJS^{-1}$, we may change variables in the differential
equation to $y(t) = S^{-1}x(t)$ to get $\dot{y}(t) = Jy(t) + S^{-1}f(t)$, with solution $y(t) =
e^{Jt}y_0 + \int_0^t e^{J(t-\tau)}S^{-1}f(\tau)d\tau$. There is an explicit formula to compute e^{Jt} or
any other function $f(J)$ of a matrix in Jordan form J. (We should not use this
formula numerically! For the basis of a better algorithm, see Question 4.4.)
Suppose that f is given by its Taylor series $f(z) = \sum_{i=0}^{\infty} \frac{f^{(i)}(0)z^i}{i!}$ and J is a

single Jordan block $J = \lambda I + N$, where N has ones on the first superdiagonal and zeros elsewhere. Then

$$
\begin{aligned}
f(J) &= \sum_{i=0}^{\infty} \frac{f^{(i)}(0)(\lambda I + N)^i}{i!} \\
&= \sum_{i=0}^{\infty} \frac{f^{(i)}(0)}{i!} \sum_{j=0}^{i} \binom{i}{j} \lambda^{i-j} N^j \quad \text{by the binomial theorem} \\
&= \sum_{j=0}^{\infty} \sum_{i=j}^{\infty} \frac{f^{(i)}(0)}{i!} \binom{i}{j} \lambda^{i-j} N^j \quad \text{reversing the order of summation} \\
&= \sum_{j=0}^{n-1} N^j \sum_{i=j}^{\infty} \frac{f^{(i)}(0)}{i!} \binom{i}{j} \lambda^{i-j},
\end{aligned}
$$

where in the last equality we used the fact that $N^j = 0$ for $j > n - 1$. Note that N^j has ones on the jth superdiagonal and zeros elsewhere. Finally, note that $\sum_{i=j}^{\infty} \frac{f^{(i)}(0)}{i!} \binom{i}{j} \lambda^{i-j}$ is the Taylor expansion for $f^{(j)}(\lambda)/j!$. Thus

$$
f(J) = f\left(\begin{bmatrix} \lambda & 1 & & \\ & \ddots & \ddots & \\ & & \ddots & 1 \\ & & & \lambda \end{bmatrix}^{n \times n}\right) = \sum_{j=0}^{n-1} \frac{N^j f^{(j)}(\lambda)}{j!}
$$

$$
= \begin{bmatrix} f(\lambda) & f'(\lambda) & \frac{f''(\lambda)}{2!} & \cdots & \frac{f^{(n-1)}(\lambda)}{(n-1)!} \\ & \ddots & \ddots & \ddots & \vdots \\ & & \ddots & \ddots & \frac{f''(\lambda)}{2!} \\ & & & \ddots & f'(\lambda) \\ & & & & f(\lambda) \end{bmatrix} \tag{4.6}
$$

so that $f(J)$ is upper triangular with $f^{(j)}(\lambda)/j!$ on the jth superdiagonal.

To solve the more general problem $B\dot{x} = Ax + f(t)$, $A - \lambda B$ regular, we use the Weierstrass form: let $P_L(A - \lambda B)P_R$ be in Weierstrass form, and rewrite the equation as $P_L B P_R P_R^{-1} \dot{x} = P_L A P_R P_R^{-1} x + P_L f(t)$. Let $P_R^{-1} x = y$ and $P_L f(t) = g(t)$. Now the problem has been decomposed into subproblems:

$$
\begin{bmatrix} I_{n_1} & & & & & \\ & \ddots & & & & \\ & & I_{n_k} & & & \\ & & & J_{m_1}(0) & & \\ & & & & \ddots & \\ & & & & & J_{m_r}(0) \end{bmatrix} \dot{y}
$$

$$= \begin{bmatrix} J_{n_1}(\lambda_1) & & & & & \\ & \ddots & & & & \\ & & J_{n_k}(\lambda_k) & & & \\ & & & I_{m_1} & & \\ & & & & \ddots & \\ & & & & & I_{m_r} \end{bmatrix} y + g.$$

Each subproblem $\dot{\tilde{y}} = J_{n_i}(\lambda_i)\tilde{y} + \tilde{g}(t) \equiv J\tilde{y} + \tilde{y}(t)$ is a standard linear ODE as above with solution

$$\tilde{y}(t) = \tilde{y}(0)e^{Jt} + \int_0^t e^{J(t-\tau)}\tilde{g}(\tau)d\tau$$

The solution of $J_m(0)\dot{\tilde{y}} = \tilde{y} + \tilde{g}(t)$ is gotten by back substitution starting from the last equation: write $J_m(0)\dot{\tilde{y}} = \tilde{y} + \tilde{g}(t)$ as

$$\begin{bmatrix} 0 & 1 & & \\ & \ddots & \ddots & \\ & & \ddots & 1 \\ & & & 0 \end{bmatrix} \begin{bmatrix} \dot{\tilde{y}}_1 \\ \vdots \\ \vdots \\ \dot{\tilde{y}}_m \end{bmatrix} = \begin{bmatrix} \tilde{y}_1 \\ \vdots \\ \vdots \\ \tilde{y}_m \end{bmatrix} + \begin{bmatrix} \tilde{g}_1 \\ \vdots \\ \vdots \\ \tilde{g}_m \end{bmatrix}.$$

The mth (last) equation says $0 = \tilde{y}_m + \tilde{g}_m$ or $\tilde{y}_m = -\tilde{g}_m$. The ith equation says $\dot{\tilde{y}}_{i+1} = \tilde{y}_i + \tilde{g}_i$, so $\tilde{y}_i = \dot{\tilde{y}}_{i+1} - \tilde{g}_i$ and thus

$$\tilde{y}_i = \sum_{k=i}^{m} -\frac{d^{k-i}}{dt^{k-i}}\tilde{g}_k(t).$$

Therefore the solution depends on derivatives of $\tilde{g}$, *not* an integral of $\tilde{g}$ as in the usual ODE. Thus a continuous $\tilde{g}$ which is not differentiable can cause a discontinuity in the solution; this is sometimes called an *impulse response* and occurs only if there are infinite eigenvalues. Furthermore, to have a continuous solution $\tilde{y}$ must satisfy certain consistency conditions at $t = 0$:

$$y_i(0) = \sum_{k=m}^{i} -\frac{d^{k-i}}{dt^{k-i}}g_k(0).$$

Numerical methods, based on time-stepping, for solving such *differential algebraic equations*, or ODEs with algebraic constraints, are described in [41].

Generalized Schur Form for Regular Pencils

Just as we cannot compute the Jordan form stably, we cannot compute its generalization by Weierstrass stably. Instead, we compute the generalized Schur form.

THEOREM 4.11. Generalized Schur form. *Let $A - \lambda B$ be regular. Then there exist unitary Q_L and Q_R so that $Q_L A Q_R = T_A$ and $Q_L B Q_R = T_B$ are both upper triangular. The eigenvalues of $A - \lambda B$ are then $T_{A_{ii}}/T_{B_{ii}}$, the ratios of the diagonal entries of T_A and T_B.*

Proof. The proof is very much like that for the usual Schur form. Let λ' be an eigenvalue and x be a unit right eigenvector: $\|x\|_2 = 1$. Since $Ax - \lambda' Bx = 0$, both Ax and Bx are multiples of the same unit vector y (even if one of Ax or Bx is zero). Now let $X = [x, \tilde{X}]$ and $Y = [y, \tilde{Y}]$ be unitary matrices with first columns x and y, respectively. Then $Y^* A X = \begin{bmatrix} \tilde{a}_{11} & \tilde{a}_{12} \\ 0 & \tilde{A}_{22} \end{bmatrix}$ and $Y^* B X = \begin{bmatrix} \tilde{b}_{11} & \tilde{b}_{12} \\ 0 & \tilde{B}_{22} \end{bmatrix}$ by construction. Apply this process inductively to $\tilde{A}_{22} - \lambda \tilde{B}_{22}$. $\square$

If A and B are real, there is a generalized real Schur form too: real orthogonal Q_L and Q_R, where $Q_L A Q_R$ is quasi–upper triangular and $Q_L B Q_R$ is upper triangular.

The QR algorithm and all its refinements generalize to compute the generalized (real) Schur form; it is called the QZ algorithm and available in LAPACK subroutine `sgges`. In Matlab one uses the command `eig(A,B)`.

Definite Pencils

A simpler special case that often arises in practice is the pencil $A - \lambda B$, where $A = A^T$, $B = B^T$, and B is positive definite. Such pencils are called *definite pencils*.

THEOREM 4.12. *Let $A = A^T$, and let $B = B^T$ be positive definite. Then there is a real nonsingular matrix X so that $X^T A X = \mathrm{diag}(\alpha_1, \ldots, \alpha_n)$ and $X^T B X = \mathrm{diag}(\beta_1, \ldots, \beta_n)$. In particular, all the eigenvalues α_i/β_i are real and finite.*

Proof. The proof that we give is actually the algorithm used to solve the problem:

(1) Let $LL^T = B$ be the Cholesky decomposition.
(2) Let $H = L^{-1} A L^{-T}$; note that H is symmetric.
(3) Let $H = Q \Lambda Q^T$, with Q orthogonal, Λ real and diagonal.

Then $X = L^{-T} Q$ satisfies $X^T A X = Q^T L^{-1} A L^{-T} Q = \Lambda$ and $X^T B X = Q^T L^{-1} B L^{-T} Q = I$. $\square$

Note that the theorem is also true if $\alpha A + \beta B$ is positive definite for some scalars α and β.

Software for this problem is available as LAPACK routine `ssygv`.

EXAMPLE 4.15. Consider the pencil $K - \lambda M$ from Example 4.14. This is a definite pencil since the stiffness matrix K is symmetric and the mass matrix

M is symmetric and positive definite. In fact, K is tridiagonal and M is diagonal in this very simple example, so M's Cholesky factor L is also diagonal, and $H = L^{-1}KL^{-T}$ is also symmetric and tridiagonal. In Chapter 5 we will consider a variety of algorithms for the symmetric tridiagonal eigenproblem. $\diamond$

4.5.2. Singular Matrix Pencils and the Kronecker Canonical Form

Now we consider singular pencils $A - \lambda B$. Recall that $A - \lambda B$ is singular if either A and B are nonsquare or they are square and $\det(A - \lambda B) = 0$ for all values of λ. The next example shows that care is needed in extending the definition of eigenvalues to this case.

EXAMPLE 4.16. Let $A = \begin{bmatrix} 1 & 0 \\ 0 & 0 \end{bmatrix}$ and $B = \begin{bmatrix} 1 & 0 \\ 0 & 0 \end{bmatrix}$. Then by making arbitrarily small changes to get $A' = \begin{bmatrix} 1 & \epsilon_1 \\ \epsilon_2 & 0 \end{bmatrix}$ and $B' = \begin{bmatrix} 1 & \epsilon_3 \\ \epsilon_4 & 0 \end{bmatrix}$, the eigenvalues become ϵ_1/ϵ_3 and ϵ_2/ϵ_4, which can be arbitrary complex numbers. So the eigenvalues are *infinitely* sensitive. $\diamond$

Despite this extreme sensitivity, singular pencils are used in modeling certain physical systems, as we describe below.

We continue by showing how to generalize the Jordan and Weierstrass forms to singular pencils. In addition to Jordan and "infinite Jordan" blocks, we get two new "singular blocks" in the canonical form.

THEOREM 4.13. *Kronecker canonical form. Let A and B be arbitrary rectangular m-by-n matrices. Then there are square nonsingular matrices P_L and P_R so that $P_L A P_R - \lambda P_L B P_R$ is block diagonal with four kinds of blocks:*

$$J_m(\lambda') - \lambda I \;=\; \begin{bmatrix} \lambda' - \lambda & 1 & & \\ & \ddots & \ddots & \\ & & \ddots & 1 \\ & & & \lambda' - \lambda \end{bmatrix}, \quad m\text{-by-}m \ Jordan \ block;$$

$$N_m \;=\; \begin{bmatrix} 1 & \lambda & & \\ & \ddots & \ddots & \\ & & \ddots & \lambda \\ & & & 1 \end{bmatrix}, \quad \begin{array}{l} m\text{-by-}m \ Jordan \ block \\ for \ \lambda = \infty; \end{array}$$

$$L_m = \begin{bmatrix} 1 & \lambda & & \\ & \ddots & \ddots & \\ & & 1 & \lambda \end{bmatrix}, \qquad \begin{array}{l} \textit{m-by-}(m+1) \textit{ right} \\ \textit{singular block;} \end{array}$$

$$L_m^T = \begin{bmatrix} 1 & & \\ \lambda & \ddots & \\ & \ddots & 1 \\ & & \lambda \end{bmatrix}, \qquad \begin{array}{l} (m+1)\textit{-by-}m \textit{ left} \\ \textit{singular block.} \end{array}$$

We call L_m a right singular block since it has a right null vector $[\lambda^m, -\lambda^{m-1}, \ldots, \pm 1]$ for all λ. L_m^T has an analogous left null vector.

For a proof, see [110].

Just as Schur form generalized to regular matrix pencils in the last section, it can be generalized to arbitrary singular pencils as well. For the canonical form, perturbation theory and software, see [27, 79, 246].

Singular pencils are used to model systems arising in systems and control. We give two examples.

Application of Kronecker Form to Differential Equations

Suppose that we want to solve $B\dot{x} = Ax + f(t)$, where $A - \lambda B$ is a singular pencil. Write $P_L B P_R P_R^{-1}\dot{x} = P_L A P_R P_R^{-1} x + P_L f(t)$ to decompose the problem into independent blocks. There are four kinds, one for each kind in the Kronecker form. We have already dealt with $J_m(\lambda') - \lambda I$ and N_m blocks when we considered regular pencils and Weierstrass form, so we have to consider only L_m and L_m^T blocks. From the L_m blocks we get

$$\begin{bmatrix} 0 & 1 & \\ & \ddots & \ddots \\ & & 0 & 1 \end{bmatrix} \begin{bmatrix} \dot{y}_1 \\ \vdots \\ \dot{y}_{m+1} \end{bmatrix} = \begin{bmatrix} 1 & 0 & \\ & \ddots & \ddots \\ & & 1 & 0 \end{bmatrix} \begin{bmatrix} y_1 \\ \vdots \\ y_{m+1} \end{bmatrix} + \begin{bmatrix} g_1 \\ \vdots \\ g_m \end{bmatrix}$$

or

$$\begin{array}{llll} \dot{y}_2 & = & y_1 + g_1 & \text{or} \quad y_2(t) = y_2(0) + \int_0^t (y_1(\tau) + g_1(\tau))d\tau, \\ \dot{y}_3 & = & y_2 + g_2 & \text{or} \quad y_3(t) = y_3(0) + \int_0^t (y_2(\tau) + g_2(\tau))d\tau, \\ & \vdots & & \\ \dot{y}_{m+1} & = & y_m + g_m & \text{or} \quad y_{m+1}(t) = y_{m+1}(0) + \int_0^t (y_m(\tau) + g_m(\tau))d\tau. \end{array}$$

This means that we can choose y_1 as an arbitrary integrable function and use the above recurrence relations to get a solution. This is because we have one more unknown than equation, so the the ODE is *underdetermined*. From the L_m^T blocks we get

$$\begin{bmatrix} 0 & & & \\ 1 & \ddots & & \\ & \ddots & 0 & \\ & & 1 \end{bmatrix} \begin{bmatrix} \dot{y}_1 \\ \vdots \\ \dot{y}_m \end{bmatrix} = \begin{bmatrix} 1 & & & \\ 0 & \ddots & & \\ & \ddots & 1 & \\ & & 0 \end{bmatrix} \begin{bmatrix} y_1 \\ \vdots \\ y_m \end{bmatrix} + \begin{bmatrix} g_1 \\ \vdots \\ g_{m+1} \end{bmatrix}$$

or

$$\begin{aligned} 0 &= y_1 + g_1, \\ \dot{y}_1 &= y_2 + g_2, \\ &\vdots \\ \dot{y}_{m-1} &= y_m + g_m, \\ \dot{y}_m &= g_{m+1}. \end{aligned}$$

Starting with the first equation, we solve to get

$$\begin{aligned} y_1 &= -g_1, \\ y_2 &= -g_2 - \dot{g}_1, \\ &\vdots \\ y_m &= -g_m - \dot{g}_{m-1} - \cdots - \frac{d^{m-1}}{dt^{m-1}} g_1 \end{aligned}$$

and the *consistency condition* $g_{m+1} = -\dot{g}_m - \cdots - \frac{d^m}{dt^m} g_1$. So unless the g_i satisfy this equation, there is no solution. Here we have one more equation than unknown, and the subproblem is *overdetermined*.

Application of Kronecker Form to Systems and Control Theory

The *controllable subspace* of $\dot{x}(t) = Ax(t) + Bu(t)$ is the space in which the *system state* $x(t)$ can be "controlled" by choosing the *control input* $u(t)$ starting at $x(0) = 0$. This equation is used to model (feedback) control systems, where the $u(t)$ is chosen by the control system engineer to make $x(t)$ have certain desirable properties, such as boundedness. From

$$\begin{aligned} x(t) &= \int_0^t e^{A(t-\tau)} Bu(\tau) d\tau = \int_0^t \sum_{i=0}^\infty \frac{(t-\tau)^i}{i!} A^i Bu(\tau) d\tau \\ &= \sum_{i=0}^\infty A^i B \int_0^t \frac{(t-\tau)^i}{i!} u(\tau) d\tau \end{aligned}$$

one can prove the controllable space is span$\{[B, AB, A^2B, \ldots, A^{n-1}B]\}$; any components of $x(t)$ outside this space cannot be controlled by varying $u(t)$. To compute this space in practice, in order to determine whether the physical system being modeled can in fact be controlled by input $u(t)$, one applies a QR-like algorithm to the singular pencil $[B, A - \lambda I]$. For details, see [78, 246, 247].

4.5.3.   Nonlinear Eigenvalue Problems

Finally, we consider the *nonlinear eigenvalue problem* or *matrix polynomial*

$$\sum_{i=0}^{d} \lambda^i A_i = \lambda^d A_d + \lambda^{d-1} A_{d-1} + \cdots + \lambda A_1 + A_0. \qquad (4.7)$$

Suppose for simplicity that the A_i are n-by-n matrices and A_d is nonsingular.

DEFINITION 4.10.   *The* characteristic polynomial *of the matrix polynomial* (4.7) *is* $p(\lambda) = \det(\sum_{i=0}^{d} \lambda^i A_i)$. *The roots of* $p(\lambda) = 0$ *are defined to be the* eigenvalues. *One can confirm that* $p(\lambda)$ *has degree* $d \cdot n$, *so there are* $d \cdot n$ *eigenvalues. Suppose that* γ *is an eigenvalue. A nonzero vector* x *satisfying* $\sum_{i=0}^{d} \gamma^i A_i x = 0$ *is a* right eigenvector *for* γ. *A* left eigenvector y *is defined analogously by* $\sum_{i=0}^{d} \gamma^i y^* A_i = 0$.

EXAMPLE 4.17. Consider Example 4.1 once again. The ODE arising there in equation (4.3) is $M\ddot{x}(t) + B\dot{x}(t) + Kx(t) = 0$. If we seek solutions of the form $x(t) = e^{\lambda_i t} x_i(0)$, we get $e^{\lambda_i t}(\lambda_i^2 M x_i(0) + \lambda_i B x_i(0) + K x_i(0)) = 0$, or $\lambda_i^2 M x_i(0) + \lambda_i B x_i(0) + K x_i(0) = 0$. Thus λ_i is an eigenvalue and $x_i(0)$ is an eigenvector of the matrix polynomial $\lambda^2 M + \lambda B + K$. ◇

Since we are assuming that A_d is nonsingular, we can multiply through by A_d^{-1} to get the equivalent problem $\lambda^d I + A_d^{-1} A_{d-1} \lambda^{d-1} + \cdots + A_d^{-1} A_0$. Therefore, to keep the notation simple, we will assume $A_d = I$ (see section 4.6 for the general case). In the very simplest case where each A_i is 1-by-1, i.e., a scalar, the original matrix polynomial is equal to the characteristic polynomial.

We can turn the problem of finding the eigenvalues of a matrix polynomial into a standard eigenvalue problem by using a trick analogous to the one used to change a high-order ODE into a first-order ODE. Consider first the simplest case $n = 1$, where each A_i is a scalar. Suppose that γ is a root. Then the vector $x' = [\gamma^{d-1}, \gamma^{d-2}, \ldots, \gamma, 1]^T$ satisfies

$$Cx' \equiv \begin{bmatrix} -A_{d-1} & -A_{d-2} & \cdots & \cdots & \cdots & -A_0 \\ 1 & 0 & \cdots & \cdots & \cdots & 0 \\ 0 & 1 & 0 & \cdots & \cdots & 0 \\ \vdots & \ddots & \ddots & \ddots & \ddots & \vdots \\ 0 & \cdots & \cdots & 0 & 1 & 0 \end{bmatrix} x' = \begin{bmatrix} -\sum_{i=0}^{d-1} \gamma^i A_i \\ \gamma^{d-1} \\ \vdots \\ \gamma^2 \\ \gamma \end{bmatrix}$$

$$= \begin{bmatrix} \gamma^d \\ \gamma^{d-1} \\ \vdots \\ \gamma^2 \\ \gamma \end{bmatrix} = \gamma x'.$$

Thus x' is an eigenvector and γ is an eigenvalue of the matrix C, which is called the *companion matrix* of the polynomial (4.7).

(The Matlab routine `roots` for finding roots of a polynomial applies the Hessenberg QR iteration of section 4.4.8 to the companion matrix C, since this is currently one of the most reliable, if expensive, methods known [100, 117, 241]. Cheaper alternatives are under development.)

The same idea works when the A_i are matrices. C becomes an $(n \cdot d)$-by-$(n \cdot d)$ *block companion matrix*, where the 1's and 0's below the top row become n-by-n identity and zero matrices, respectively. Also, x' becomes

$$
x' = \begin{bmatrix} \gamma^{d-1}x \\ \gamma^{d-2}x \\ \vdots \\ \gamma x \\ x \end{bmatrix},
$$

where x is a right eigenvector of the matrix polynomial. It again turns out that $Cx' = \gamma x'$.

EXAMPLE 4.18. Returning once again to $\lambda^2 M + \lambda B + K$, we first convert it to $\lambda^2 + \lambda M^{-1}B + M^{-1}K$ and then to the companion matrix

$$
C = \begin{bmatrix} -M^{-1}B & -M^{-1}K \\ I & 0 \end{bmatrix}.
$$

This is the same as the matrix A in equation 4.4 of Example 4.1. ◇

Finally, Question 4.16 shows how to use matrix polynomials to solve a problem in *computational geometry*.

4.6. Summary

The following list summarizes all the canonical forms, algorithms, their costs, and applications to ODEs described in this chapter. It also includes pointers to algorithms exploiting symmetry, although these are discussed in more detail in the next chapter. Algorithms for sparse matrices are discussed in Chapter 7.

- $A - \lambda I$

  - Jordan form: For some nonsingular S,

$$
A - \lambda I = S \cdot \mathrm{diag}\left(\ldots, \begin{bmatrix} \lambda_i - \lambda & 1 & & \\ & \ddots & \ddots & \\ & & \ddots & 1 \\ & & & \lambda_i - \lambda \end{bmatrix}^{n_i \times n_i}, \ldots\right) \cdot S^{-1}.
$$

- Schur form: For some unitary Q, $A - \lambda I = Q(T - \lambda I)Q^*$, where T is triangular.

- Real Schur form of real A: For some real orthogonal Q, $A - \lambda I = Q(T - \lambda I)Q^T$, where T is real quasi-triangular.

- Application to ODEs: Provides solution of $\dot{x}(t) = Ax(t) + f(t)$.

- Algorithm: Do Hessenberg reduction (Algorithm 4.6), followed by QR iteration to get Schur form (Algorithm 4.5, implemented as described in section 4.4.8). Eigenvectors can be computed from the Schur form (as described in section 4.2.1).

- Cost: This costs $10n^3$ flops if eigenvalues only are desired, $25n^3$ if T and Q are also desired, and a little over $27n^3$ if eigenvectors are also desired. Since not all parts of the algorithm can take advantage of the Level 3 BLAS, the cost is actually higher than a comparison with the $2n^3$ cost of matrix multiply would indicate: instead of taking $(10n^3)/(2n^3) = 5$ times longer to compute eigenvalues than to multiply matrices, it takes 23 times longer for $n = 100$ and 19 times longer for $n = 1000$ on an IBM RS6000/590 [10, page 62]. Instead of taking $(27n^3)/(2n^3) = 13.5$ times longer to compute eigenvalues and eigenvectors, it takes 41 times longer for $n = 100$ and 60 times longer for $n = 1000$ on the same machine. Thus computing eigenvalues of nonsymmetric matrices is expensive. (The symmetric case is *much* cheaper; see Chapter 5.)

- LAPACK: `sgees` for Schur form or `sgeev` for eigenvalues and eigenvectors; `sgeesx` or `sgeevx` for error bounds too.

- Matlab: `schur` for Schur form or `eig` for eigenvalues and eigenvectors.

- Exploiting symmetry: When $A = A^*$, better algorithms are discussed in Chapter 5, especially section 5.3.

- Regular $A - \lambda B$ ($\det(A - \lambda B) \not\equiv 0$)

  - Weierstrass form: For some nonsingular P_L and P_R,

$$
A - \lambda B = P_L \cdot \operatorname{diag}\left(\text{Jordan,} \begin{bmatrix} 1 & \lambda & & \\ & \ddots & \ddots & \\ & & \ddots & \lambda \\ & & & 1 \end{bmatrix}^{n_i \times n_i} \right) P_R^{-1}.
$$

  - Generalized Schur form: For some unitary Q_L and Q_R, $A - \lambda B = Q_L(T_A - \lambda T_B)Q_R^*$, where T_A and T_B are triangular.

- Generalized real Schur form of real A and B: For some real orthogonal Q_L and Q_R, $A - \lambda B = Q_L(T_A - \lambda T_B)Q_R^T$, where T_A is real quasi-triangular and T_B is real triangular.

- Application to ODEs: Provides solution of $B\dot{x}(t) = Ax(t) + f(t)$, where the solution is uniquely determined but may depend non-smoothly on the data (*impulse response*).

- Algorithm: Hessenberg/triangular reduction followed by QZ iteration (QR applied implicitly to AB^{-1}).

- Cost: Computing T_A and T_B costs $30n^3$. Computing Q_L and Q_R in addition costs $66n^3$. Computing eigenvectors as well costs a little less than $69n^3$ in total. As before, Level 3 BLAS cannot be used in all parts of the algorithm.

- LAPACK: sgges for Schur form or sggev for eigenvalues; sggesx or sggevx for error bounds too.

- Matlab: eig for eigenvalues and eigenvectors.

- Exploiting symmetry: When $A = A^*$, $B = B^*$, and B is positive definite, one can convert the problem to finding the eigenvalues of a single symmetric matrix using Theorem 4.12. This is done in LAPACK routines ssygv, sspgv (for symmetric matrices in "packed storage"), and ssbgv (for symmetric band matrices).

- Singular $A - \lambda B$

  - Kronecker form: For some nonsingular P_L and P_R,

$$
A - \lambda B = P_L
\cdot \mathrm{diag}\left(\mathrm{Weierstrass}, \begin{bmatrix} 1 & \lambda & & \\ & \ddots & \ddots & \\ & & 1 & \lambda \end{bmatrix}^{n_i \times n_i}, \begin{bmatrix} 1 & & & \\ \lambda & \ddots & & \\ & \ddots & 1 & \\ & & & \lambda \end{bmatrix}^{m_i \times m_i}\right) P_R^{-1}.
$$

  - Generalized upper triangular form: For some unitary Q_L and Q_R, $A - \lambda B = Q_L(T_A - \lambda T_B)Q_R^*$, where T_A and T_B are in generalized upper triangular form, with diagonal blocks corresponding to different parts of the Kronecker form. See [79, 246] for details of the form and algorithms.

  - Cost: The most general and reliable version of the algorithm can cost as much as $O(n^4)$, depending on the details of the Kronecker Structure; this is much more than for regular $A - \lambda B$. There is also a slightly less reliable $O(n^3)$ algorithm [27].

- Application to ODEs: Provides solution of $B\dot{x}(t) = Ax(t) + f(t)$, where the solution may be overdetermined or underdetermined.
- Software: NETLIB/linalg/guptri.

- Matrix polynomials $\sum_{i=0}^{d} \lambda^i A_i$ [118]

  - If $A_d = I$ (or A_d is square and well-conditioned enough to replace each A_i by $A_d^{-1} A_i$), then linearize to get the standard problem

$$
\begin{bmatrix}
-A_{d-1} & -A_{d-2} & \cdots & \cdots & \cdots & -A_0 \\
I & 0 & \cdots & \cdots & \cdots & 0 \\
0 & I & 0 & \cdots & \cdots & 0 \\
\vdots & & \ddots & \ddots & \ddots & \vdots \\
0 & & \cdots & & 0 & I & 0
\end{bmatrix} - \lambda I.
$$

  - If A_d is ill-conditioned or singular, linearize to get the pencil

$$
\begin{bmatrix}
-A_{d-1} & -A_{d-2} & \cdots & \cdots & \cdots & -A_0 \\
I & 0 & \cdots & \cdots & \cdots & 0 \\
0 & I & 0 & \cdots & \cdots & 0 \\
\vdots & \ddots & \ddots & \ddots & \ddots & \vdots \\
0 & & \cdots & 0 & I & 0
\end{bmatrix} - \lambda
\begin{bmatrix}
A_d & & & \\
& I & & \\
& & I & \\
& & & \ddots \\
& & & & I
\end{bmatrix}.
$$

4.7. References and Other Topics for Chapter 4

For a general discussion of properties of eigenvalues and eigenvectors, see [139]. For more details about perturbation theory of eigenvalues and eigenvectors, see [161, 237, 52], and chapter 4 of [10]. For a proof of Theorem 4.7, see [69]. For a discussion of Weierstrass and Kronecker canonical forms, see [110, 118]. For their application to systems and control theory, see [246, 247, 78]. For applications to computational geometry, graphics, and mechanical CAD, see [181, 182, 165]. For a discussion of parallel algorithms for the nonsymmetric eigenproblem, see [76].

4.8. Questions for Chapter 4

QUESTION 4.1. *(Easy)* Let A be defined as in equation (4.1). Show that $\det(A) = \prod_{i=1}^{b} \det(A_{ii})$ and then that $\det(A - \lambda I) = \prod_{i=1}^{b} \det(A_{ii} - \lambda I)$. Conclude that the set of eigenvalues of A is the union of the sets of eigenvalues of A_{11} through A_{bb}.

QUESTION 4.2. *(Medium; Z. Bai)* Suppose that A is *normal;* i.e., $AA^* = A^*A$. Show that if A is also triangular, it must be diagonal. Use this to show that an n-by-n matrix is normal if and only if it has n orthonormal eigenvectors. Hint: Show that A is normal if and only if its Schur form is normal.

QUESTION 4.3. *(Easy; Z. Bai)* Let λ and μ be distinct eigenvalues of A, let x be a right eigenvector for λ, and let y be a left eigenvector for μ. Show that x and y are orthogonal.

QUESTION 4.4. *(Medium)* Suppose A has distinct eigenvalues. Let $f(z) = \sum_{i=-\infty}^{+\infty} a_i z^i$ be a function which is defined at the eigenvalues of A. Let $Q^* A Q = T$ be the Schur form of A (so Q is unitary and T upper triangular).

1. Show that $f(A) = Q f(T) Q^*$. Thus to compute $f(A)$ it suffices to be able to compute $f(T)$. In the rest of the problem you will derive a simple recurrence formula for $f(T)$.

2. Show that $(f(T))_{ii} = f(T_{ii})$ so that the diagonal of $f(T)$ can be computed from the diagonal of T.

3. Show that $T f(T) = f(T) T$.

4. From the last result, show that the ith superdiagonal of $f(T)$ can be computed from the $(i-1)$st and earlier subdiagonals. Thus, starting at the diagonal of $f(T)$, we can compute the first superdiagonal, second superdiagonal, and so on.

QUESTION 4.5. *(Easy)* Let A be a square matrix. Apply either Question 4.4 to the Schur form of A or equation (4.6) to the Jordan form of A to conclude that the eigenvalues of $f(A)$ are $f(\lambda_i)$, where the λ_i are the eigenvalues of A. This result is called the *spectral mapping theorem*.

This question is used in the proof of Theorem 6.5 and section 6.5.6.

QUESTION 4.6. *(Medium)* In this problem we will show how to solve the *Sylvester* or *Lyapunov* equation $AX - XB = C$, where X and C are m-by-n, A is m-by-m, and B is n-by-n. This is a system of mn linear equations for the entries of X.

1. Given the Schur decompositions of A and B, show how $AX - XB = C$ can be transformed into a similar system $A'Y - YB' = C'$, where A' and B' are upper triangular.

2. Show how to solve for the entries of Y one at a time by a process analogous to back substitution. What condition on the eigenvalues of A and B guarantees that the system of equations is nonsingular?

3. Show how to transform Y to get the solution X.

QUESTION 4.7. *(Medium)* Suppose that $T = \begin{bmatrix} A & C \\ 0 & B \end{bmatrix}$ is in Schur form. We want to find a matrix S so that $S^{-1} T S = \begin{bmatrix} A & 0 \\ 0 & B \end{bmatrix}$. It turns out we can choose S of the form $\begin{bmatrix} I & R \\ 0 & I \end{bmatrix}$. Show how to solve for R.

QUESTION 4.8. *(Medium; Z. Bai)* Let A be m-by-n and B be n-by-m. Show that the matrices

$$\begin{pmatrix} AB & 0 \\ B & 0 \end{pmatrix} \quad \text{and} \quad \begin{pmatrix} 0 & 0 \\ B & BA \end{pmatrix}$$

are similar. Conclude that the nonzero eigenvalues of AB are the same as those of BA.

QUESTION 4.9. *(Medium; Z. Bai)* Let A be n-by-n with eigenvalues $\lambda_1, \ldots, \lambda_n$. Show that

$$\sum_{i=1}^{n} |\lambda_i|^2 = \min_{\det(S) \neq 0} \|S^{-1}AS\|_F^2.$$

QUESTION 4.10. *(Medium; Z. Bai)* Let A be an n-by-n matrix with eigenvalues $\lambda_1, \ldots, \lambda_n$.

1. Show that A can be written $A = H + S$, where $H = H^*$ is Hermitian and $S = -S^*$ is skew-Hermitian. Give explicit formulas for H and S in terms of A.

2. Show that $\sum_{i=1}^{n} |\Re\lambda_i|^2 \leq \|H\|_F^2$.

3. Show that $\sum_{i=1}^{n} |\Im\lambda_i|^2 \leq \|S\|_F^2$.

4. Show that A is normal ($AA^* = A^*A$) if and only if $\sum_{i=1}^{n} |\lambda_i|^2 = \|A\|_F^2$.

QUESTION 4.11. *(Easy)* Let λ be a simple eigenvalue, and let x and y be right and left eigenvectors. We define the *spectral projection* P corresponding to λ as $P = xy^*/(y^*x)$. Prove that P has the following properties.

1. P is uniquely defined, even though we could use any nonzero scalar multiples of x and y in its definition.

2. $P^2 = P$. (Any matrix satisfying $P^2 = P$ is called a *projection matrix*.)

3. $AP = PA = \lambda P$. (These properties motivate the name *spectral* projection, since P "contains" the left and right invariant subspaces of λ.)

4. $\|P\|_2$ is the condition number of λ.

QUESTION 4.12. *(Easy; Z. Bai)* Let $A = \begin{bmatrix} a & c \\ 0 & b \end{bmatrix}$. Show that the condition numbers of the eigenvalues of A are both equal to $(1 + (\frac{c}{a-b})^2)^{1/2}$. Thus, the condition number is large if the difference $a - b$ between the eigenvalues is small compared to c, the offdiagonal part of the matrix.

QUESTION 4.13. *(Medium; Z. Bai)* Let A be a matrix, x be a unit vector ($\|x\|_2 = 1$), μ be a scalar, and $r = Ax - \mu x$. Show that there is a matrix E with $\|E\|_F = \|r\|_2$ such that $A + E$ has eigenvalue μ and eigenvector x.

QUESTION 4.14. *(Medium; Programming)* In this question we will use a Matlab program to plot eigenvalues of a perturbed matrix and their condition numbers. (It is available at HOMEPAGE/Matlab/eigscat.m.) The input is

    a = input matrix,
    err = size of perturbation,
    m = number of perturbed matrices to compute.

The output consists of three plots in which each symbol is the location of an eigenvalue of a perturbed matrix:

    "o" marks the location of each unperturbed eigenvalue.
    "x" marks the location of each perturbed eigenvalue, where a real perturbation matrix of norm err is added to a.
    "." marks the location of each perturbed eigenvalue, where a complex perturbation matrix of norm err is added to a.

A table of the eigenvalues of A and their condition numbers is also printed.

Here are some interesting examples to try (for as large an m as you want to wait; the larger the m the better, and m equal to a few hundred is good).

```
(1) a = randn(5) (if a does not have complex eigenvalues,
                      try again)
          err=1e-5, 1e-4, 1e-3, 1e-2, .1, .2

(2) a = diag(ones(4,1),1);  err=1e-12, 1e-10, 1e-8

(3) a=[[1 1e6 0       0]; ...
       [0 2     1e-3  0]; ...
       [0 0     3     10]; ...
       [0 0    -1     4]]
          err=1e-8, 1e-7, 1e-6, 1e-5, 1e-4, 1e-3

(4) [q,r]=qr(randn(4,4));a=q*diag(ones(3,1),1)*q'
          err=1e-16, 1e-14, 1e-12, 1e-10, 1e-8

(5) a = [[1 1e3 1e6];[0 1 1e3];[0 0 1]],
          err=1e-7, 1e-6, 5e-6, 8e-6, 1e-5, 1.5e-5, 2e-5

(6) a = [[1   0   0   0   0   0]; ...
         [0   2   1   0   0   0]; ...
         [0   0   2   0   0   0]; ...
         [0   0   0   3 1e2 1e4]; ...
         [0   0   0   0   3 1e2]; ...
         [0   0   0   0   0   3]]
          err= 1e-10, 1e-8, 1e-6, 1e-4, 1e-3
```

Your assignment is to try these examples and compare the regions occupied by the eigenvalues (the so-called *pseudospectrum*) with the bounds described in section 4.3. What is the difference between real perturbations and complex perturbations? What happens to the regions occupied by the eigenvalues as the perturbation err goes to zero? What is limiting size of the regions as err goes to zero (i.e., how many digits of the computed eigenvalues are correct)?

QUESTION 4.15. *(Medium; Programming)* In this question we use a Matlab program to plot the diagonal entries of a matrix undergoing unshifted QR iteration. The values of each diagonal are plotted after each QR iteration, each diagonal corresponding to one of the plotted curves. (The program is available at HOMEPAGE/Matlab/qrplt.m and also shown below.)   The inputs are

   a = input matrix,
   m = number of QR iterations,

and the output is a plot of the diagonals.

   Examples to try this code on are as follows (choose m large enough so that the curves either converge or go into cycles):

```
a = randn(6);
b = randn(6); a = b*diag([1,2,3,4,5,6])*inv(b);
a = [[1 10];[-1 1]]; m = 300
a = diag((1.5*ones(1,5)).\verb+^+(0:4)) +
    .01*(diag(ones(4,1),1)+diag(ones(4,1),-1)); m=30
```

   What happens if there are complex eigenvalues?

   In what order do the eigenvalues appear in the matrix after many iterations?

   Perform the following experiment: Suppose that a is *n*-by-*n* and symmetric. In Matlab, let perm=(n:-1:1). This produces a list of the integers from n down to 1. Run the iteration for m iterations. Let a=a(perm,perm); we call this "flipping" a, because it reverses the order of the rows and columns of a. Run the iteration again for m iterations, and again form a=a(perm,perm). How does this value of a compare with the original value of a? You should not let *m* be too large (try *m* = 5) or else roundoff will obscure the relationship you should see. (See also Corollary 5.4 and Question 5.25.)

   Change the code to compute the error in each diagonal from its final value (do this just for matrices with all real eigenvalues). Plot the log of this error versus the iteration number. What do you get asymptotically?

```
hold off
e=diag(a);
for i=1:m,
   [q,r]=qr(a);dd=diag(sign(diag(r)));r=dd*r;q=q*dd;a=r*q; ...
   e=[e,diag(a)];
end
clg
plot(e','w'),grid
```

QUESTION 4.16. *(Hard; Programming)* This problem describes an application of the nonlinear eigenproblem to computer graphics, computational geometry, and mechanical CAD; see also [181, 182, 165].

Let $F = [f_{ij}(x_1, x_2, x_3)]$ be a matrix whose entries are polynomials in the three variables x_i. Then $\det(F) = 0$ will (generally) define a two-dimensional surface S in 3-space. Let $x_1 = g_1(t)$, $x_2 = g_2(t)$, and $x_3 = g_3(t)$ define a (one-dimensional) curve C parameterized by t, where the g_i are also polynomials. We want to find the intersection $S \cap C$. Show how to express this as an eigenvalue problem (which can then be solved numerically). More generally, explain how to find the intersection of a surface $\det(F(x_1, \ldots, x_n)) = 0$ and curve $\{x_i = g_i(t), \ 1 \le i \le n\}$. At most how many discrete solutions can there be, as a function of n, the dimension d of F, and the maximum of the degrees of the polynomials f_{ij} and g_k?

Write a Matlab program to solve this problem, for $n = 3$ variables, by converting it to an eigenvalue problem. It should take as input a compact description of the entries of each $f_{ij}(x_k)$ and $g_i(t)$ and produce a list of the intersection points. For instance, it could take the following inputs:

- Array NumTerms(1:d,1:d), where NumTerms(i,j) is the number of terms in the polynomial $f_{ij}(x_1, x_2, x_3)$.

- Array Sterms(1:4, 1:TotalTerms), where TotalTerms is the sum of all the entries in NumTerms(.,.). Each column of Sterms represents one term in one polynomial: The first NumTerms(1,1) columns of Sterms represent the terms in f_{11}, the second Numterm(2,1) columns of Sterms represent the terms in f_{21}, and so on. The term represented by Sterms(1:4,k) is Sterm(4, k) $\cdot x_1^{\text{Sterm}(1,k)} \cdot x_2^{\text{Sterm}(2,k)} \cdot x_3^{\text{Sterm}(3,k)}$.

- Array tC(1:3) contains the degrees of polynomials g_1, g_2, and g_3 in that order.

- Array Curve(1: tC(1)+tC(2)+tC(3)+3) contains the coefficients of the polynomials g_1, g_2, and g_3, one polynomial after the other, from the constant term to the highest order coefficient of each.

Your program should also compute error bounds for the computed answers. This will be possible only when the eigenproblem can be reduced to one for which the error bounds in Theorems 4.4 or 4.5 apply. You do not have to provide error bounds when the eigenproblem is a more general one. (For a description of error bounds for more general eigenproblems, see [10, 237].

Write a second Matlab program that plots S and C for the case $n = 3$ and marks the intersection points.

Are there any limitations on the input data for your codes to work? What happens if S and C do not intersect? What happens if S lies in C?

Run your codes on at least the following examples. You should be able to solve the first five by hand to check your code.

1. $g_1 = t$, $g_2 = 1 + t$, $g_3 = 2 + t$, $F = \begin{bmatrix} x_1 + x_2 + x_3 & 0 \\ 0 & 3x_1 + 5x_2 - 7x_3 + 10 \end{bmatrix}$.

2. $g_1 = t^3$, $g_2 = 1 + t^3$, $g_3 = 2 + t^3$, $F = \begin{bmatrix} x_1 + x_2 + x_3 & 0 \\ 0 & 3x_1 + 5x_2 - 7x_3 + 10 \end{bmatrix}$.

3. $g_1 = t^2$, $g_2 = 1 + t^2$, $g_3 = 2 + t^2$, $F = \begin{bmatrix} x_1 + x_2 + x_3 & 0 \\ 0 & 3x_1 + 5x_2 - 7x_3 + 10 \end{bmatrix}$.

4. $g_1 = t^2$, $g_2 = 1 + t^2$, $g_3 = 2 + t^2$, $F = \begin{bmatrix} 1 & 0 \\ 0 & 3x_1 + 5x_2 - 7x_3 + 9 \end{bmatrix}$.

5. $g_1 = t^2$, $g_2 = 1 + t^2$, $g_3 = 2 + t^2$, $F = \begin{bmatrix} x_1 + x_2 + x_3 & 0 \\ 0 & 3x_1 + 5x_2 - 7x_3 + 8 \end{bmatrix}$.

6. $g_1 = t^2$, $g_2 = 1 + t^2$, $g_3 = 2 + t^2$, $F = \begin{bmatrix} x_1 + x_2 + x_3 & x_1 \\ x_3 & 3x_1 + 5x_2 - 7x_3 + 10 \end{bmatrix}$.

7. $g_1 = 7 - 3t + t^5$, $g_2 = 1 + t^2 + t^5$, $g_3 = 2 + t^2 - t^5$,

$$F = \begin{bmatrix} x_1 x_2 + x_3^5 & 3 - x_2^2 & 5 + x_1 + x_2 + x_3 + x_1 x_2 + x_1 x_3 + x_2 x_3 \\ x_2 - 7x_3^5 & 1 - x_1^2 + x_1 x_2 x_3^3 & 3 + x_1 + 3x_3 - 9x_2 x_3 \\ 2 & 3x_1 + 5x_2 - 7x_3 + 8 & x_1^3 - x_2^4 + 4x_3^5 \end{bmatrix}.$$

You should turn in

• mathematical formulation of the solution in terms of an eigenproblem.

• the algorithm in at most two pages, including a road map to your code (subroutine names for each high level operation). It should be easy to see how the mathematical formulation leads to the algorithm and how the algorithm matches the code.

— At most how many discrete solutions can there be?

— Do all compute eigenvalues represent actual intersections? Which ones do?

— What limits does your code place on the input for it to work correctly?

— What happens if S and C do not intersect?

— What happens if S contains C?

• mathematical formulation of the error bounds.

• the algorithm for computing the error bounds in at most two pages, including a road map to your code (subroutine names for each high-level operation). It should be easy to see how the mathematical formulation leads to the algorithm and how the algorithm matches the code.

• program listing.

For each of the seven examples, you should turn in

• the original statement of the problem.

• the resulting eigenproblem.

• the numerical solutions.

• plots of S and C; do your numerical solutions match the plots?

• the result of substituting the computed answers in the equations defining S and C: are they satisfied (to within roundoff)?

5

The Symmetric Eigenproblem and Singular Value Decomposition

5.1. Introduction

We discuss perturbation theory (in section 5.2), algorithms (in sections 5.3 and 5.4), and applications (in section 5.5 and elsewhere) of the symmetric eigenvalue problem. We also discuss its close relative, the SVD. Since the eigendecomposition of the symmetric matrix $H = \begin{bmatrix} 0 & A^T \\ A & 0 \end{bmatrix}$ and the SVD of A are very simply related (see Theorem 3.3), most of the perturbation theorems and algorithms for the symmetric eigenproblem extend to the SVD.

As discussed at the beginning of Chapter 4, one can roughly divide the algorithms for the symmetric eigenproblem (and SVD) into two groups: *direct methods* and *iterative methods*. This chapter considers only direct methods, which are intended to compute all (or a selected subset) of the eigenvalues and (optionally) eigenvectors, costing $O(n^3)$ operations for dense matrices. Iterative methods are discussed in Chapter 7.

Since there has been a great deal of recent progress in algorithms and applications of symmetric eigenproblems, we will highlight three examples:

- A high-speed algorithm for the symmetric eigenproblem based on divide-and-conquer is discussed in section 5.3.3. This is the fastest available algorithm for finding all eigenvalues and all eigenvectors of a large dense or banded symmetric matrix (or the SVD of a general matrix). It is significantly faster than the previous "workhorse" algorithm, QR iteration.[17]

- High-accuracy algorithms based on the dqds and Jacobi algorithms are discussed in sections 5.2.1, 5.4.2, and 5.4.3. These algorithms can find tiny eigenvalues (or singular values) more accurately than alternative

---

[17] There is yet more recent work [201, 203] on an algorithm based on inverse iteration (Algorithm 4.2), which may provide a still faster and more accurate algorithm. But as of June 1997 the theory and software were still under development.

algorithms like divide-and-conquer, although sometimes more slowly, in the sense of Jacobi.

• Section 5.5 discusses a "nonlinear" vibrating system, described by a differential equation called the *Toda flow*. Its continuous solution is closely related to the intermediate steps of the QR algorithm for the symmetric eigenproblem.

Following Chapter 4, we will continue to use a vibrating mass-spring system as a running example to illustrate features of the symmetric eigenproblem.

EXAMPLE 5.1. Symmetric eigenvalue problems often arise in analyzing *mechanical vibrations*. Example 4.1 presented one such example in detail; we will use notation from that example, so the reader is advised to review it now. To make the problem in Example 4.1 symmetric, we need to assume that there is no damping, so the differential equations of motion of the mass-spring system become $M\ddot{x}(t) = -Kx(t)$, where $M = \text{diag}(m_1, \ldots, m_n)$ and

$$
K = \begin{bmatrix}
k_1 + k_2 & -k_2 & & & \\
-k_2 & k_2 + k_3 & -k_3 & & \\
& \ddots & \ddots & \ddots & \\
& & -k_{n-1} & k_{n-1} + k_n & -k_n \\
& & & -k_n & k_n
\end{bmatrix}.
$$

Since M is nonsingular, we can rewrite this as $\ddot{x}(t) = -M^{-1}Kx(t)$. If we seek solutions of the form $x(t) = e^{\gamma t}x(0)$, then we get $e^{\gamma t}\gamma^2 x(0) = -M^{-1}Ke^{\gamma t}x(0)$, or $M^{-1}Kx(0) = -\gamma^2 x(0)$. In other words, $-\gamma^2$ is an eigenvalue and $x(0)$ is an eigenvector of $M^{-1}K$. Now $M^{-1}K$ is not generally symmetric, but we can make it symmetric as follows. Define $M^{1/2} = \text{diag}(m_1^{1/2}, \ldots, m_n^{1/2})$, and multiply $M^{-1}Kx(0) = -\gamma^2 x(0)$ by $M^{1/2}$ on both sides to get

$$
M^{-1/2}Kx(0) = M^{-1/2}K(M^{-1/2}M^{1/2})x(0) = -\gamma^2 M^{1/2}x(0)
$$

or $\hat{K}\hat{x} = -\gamma^2\hat{x}$, where $\hat{x} = M^{1/2}x(0)$ and $\hat{K} = M^{-1/2}KM^{-1/2}$. It is easy to see that

$$
\hat{K} = \begin{bmatrix}
\frac{k_1+k_2}{m_1} & \frac{-k_2}{\sqrt{m_1 m_2}} & & & \\
\frac{-k_2}{\sqrt{m_1 m_2}} & \frac{k_2+k_3}{m_2} & \frac{-k_3}{\sqrt{m_2 m_3}} & & \\
& \ddots & \ddots & \ddots & \\
& & \frac{-k_{n-1}}{\sqrt{m_{n-2} m_{n-1}}} & \frac{k_{n-1}+k_n}{m_{n-1}} & \frac{-k_n}{\sqrt{m_{n-1} m_n}} \\
& & & \frac{-k_n}{\sqrt{m_{n-1} m_n}} & \frac{k_n}{m_n}
\end{bmatrix}
$$

is symmetric. Thus each eigenvalue $-\gamma^2$ of $\hat{K}$ is real, and each eigenvector $\hat{x} = M^{1/2}x(0)$ of $\hat{K}$ is orthogonal to the others.

In fact, $\hat{K}$ is a *tridiagonal* matrix, a special form to which any symmetric matrix can be reduced, using Algorithm 4.6, specialized to symmetric matrices as described in section 4.4.7. Most of the algorithms in section 5.3 for finding the eigenvalues and eigenvectors of a symmetric matrix assume that the matrix has initially been reduced to tridiagonal form.

There is another way to express the solution to this mechanical vibration problem, using the SVD. Define $K_D = \mathrm{diag}(k_1, \ldots, k_n)$ and $K_D^{1/2} = \mathrm{diag}(k_1^{1/2}, \ldots, k_n^{1/2})$. Then K can be factored as $K = BK_DB^T$, where

$$
B = \begin{bmatrix} 1 & -1 & & \\ & \ddots & \ddots & \\ & & \ddots & -1 \\ & & & 1 \end{bmatrix},
$$

as can be confirmed by a small calculation. Thus

$$
\begin{aligned}
\hat{K} &= M^{-1/2}KM^{-1/2} \\
&= M^{-1/2}BK_DB^TM^{-1/2} \\
&= (M^{-1/2}BK_D^{1/2}) \cdot (K_D^{1/2}B^TM^{-1/2}) \\
&= (M^{-1/2}BK_D^{1/2}) \cdot (M^{-1/2}BK_D^{1/2})^T \\
&\equiv GG^T.
\end{aligned} \tag{5.1}
$$

Therefore the singular values of $G = M^{-1/2}BK_D^{1/2}$ are the square roots of the eigenvalues of $\hat{K}$, and the left singular vectors of G are the eigenvectors of $\hat{K}$, as shown in Theorem 3.3. Note that G is nonzero only on the main diagonal and on the first superdiagonal. Such matrices are called *bidiagonal*, and most algorithms for the SVD begin by reducing the matrix to bidiagonal form, using the algorithm in section 4.4.7.

Note that the factorization $\hat{K} = GG^T$ implies that $\hat{K}$ is positive definite, since G is nonsingular. Therefore the eigenvalues $-\gamma^2$ of $\hat{K}$ are all positive. Thus γ is pure imaginary, and the solutions of the original differential equation $x(t) = e^{\gamma t}x(0)$ are oscillatory with frequency $|\gamma|$.

For a Matlab solution of a vibrating mass-spring system, see HOMEPAGE/Matlab/massspring.m. For a Matlab animation of the vibrations of a similar physical system, type demo and then click on continue/ fun-extras/miscellaneous/bending. ◇

5.2. Perturbation Theory

Suppose that A is symmetric, with eigenvalues $\alpha_1 \geq \cdots \geq \alpha_n$ and corresponding unit eigenvectors $q_1, \ldots, q_n$. Suppose E is also symmetric, and let $\hat{A} = A + E$ have perturbed eigenvalues $\hat{\alpha}_1 \geq \cdots \geq \hat{\alpha}_n$ and corresponding perturbed eigenvectors $\hat{q}_1, \ldots, \hat{q}_n$. The major goal of this section is to bound the differences between the eigenvalues α_i and $\hat{\alpha}_i$ and between the eigenvectors q_i and $\hat{q}_i$ in terms of the "size" of E. Most of our bounds will use $\|E\|_2$ as the size of E, except for section 5.2.1, which discusses "relative" perturbation theory.

We already derived our first perturbation bound for eigenvalues in Chapter 4, where we proved Corollary 4.1: *Let A be symmetric with eigenvalues $\alpha_1 \geq \cdots \geq \alpha_n$. Let $A + E$ be symmetric with eigenvalues $\hat{\alpha}_1 \geq \cdots \geq \hat{\alpha}_n$. If α_i is simple, then $|\alpha_i - \hat{\alpha}_i| \leq \|E\|_2 + O(\|E\|_2^2)$.*

This result is weak because it assumes α_i has multiplicity one, and it is useful only for sufficiently small $\|E\|_2$. The next theorem eliminates both weaknesses.

THEOREM 5.1. *Weyl. Let A and E be n-by-n symmetric matrices. Let $\alpha_1 \geq \cdots \geq \alpha_n$ be the eigenvalues of A and $\hat{\alpha}_1 \geq \cdots \geq \hat{\alpha}_n$ be the eigenvalues of $\hat{A} = A + E$. Then $|\alpha_i - \hat{\alpha}_i| \leq \|E\|_2$.*

COROLLARY 5.1. *Let G and F be arbitrary matrices (of the same size) where $\sigma_1 \geq \cdots \geq \sigma_n$ are the singular values of G and $\sigma_1' \geq \cdots \geq \sigma_n'$ are the singular values of $G + F$. Then $|\sigma_i - \sigma_i'| \leq \|F\|_2$.*

We can use Weyl's theorem to get error bounds for the eigenvalues computed by any backward stable algorithm, such as QR iteration: Such an algorithm computes eigenvalues $\hat{\alpha}_i$ that are the exact eigenvalues of $\hat{A} = A + E$ where $\|E\|_2 = O(\varepsilon)\|A\|_2$. Therefore, their errors can be bounded by $|\alpha_i - \hat{\alpha}_i| \leq \|E\|_2 = O(\varepsilon)\|A\|_2 = O(\varepsilon) \max_j |\alpha_j|$. This is a very satisfactory error bound, especially for large eigenvalues (those α_i near $\|A\|_2$ in magnitude), since they will be computed with most of their digits correct. Small eigenvalues ($|\alpha_i| \ll \|A\|_2$) may have fewer correct digits (but see section 5.2.1).

We will prove Weyl's theorem using another useful classical result: the Courant–Fischer minimax theorem. To state this theorem we need to introduce the *Rayleigh quotient*, which will also play an important role in several algorithms, such as Algorithm 5.1.

DEFINITION 5.1. *The* Rayleigh quotient *of a symmetric matrix A and nonzero vector u is $\rho(u, A) \equiv (u^T A u)/(u^T u)$.*

Here are some simple but important properties of $\rho(u, A)$. First, $\rho(\gamma u, A) = \rho(u, A)$ for any nonzero scalar γ. Second, if $A q_i = \alpha_i q_i$, then $\rho(q_i, A) = \alpha_i$. More generally, suppose $Q^T A Q = \Lambda = \text{diag}(\alpha_i)$ is the eigendecomposition of A, with $Q = [q_1, \ldots, q_n]$. Expand u in the basis of eigenvectors q_i as follows: $u = Q(Q^T u) \equiv Q\xi = \sum_i q_i \xi_i$. Then we can write

$$\rho(u, A) = \frac{\xi^T Q^T A Q \xi}{\xi^T Q^T Q \xi} = \frac{\xi^T \Lambda \xi}{\xi^T \xi} = \frac{\sum_i \alpha_i \xi_i^2}{\sum_i \xi_i^2}.$$

In other words, $\rho(u, A)$ is a weighted average of the eigenvalues of A. Its largest value, $\max_{u \neq 0} \rho(u, A)$, occurs for $u = q_1$ ($\xi = e_1$) and equals $\rho(q_1, A) = \alpha_1$.

Its smallest value, $\min_{u \neq 0} \rho(u, A)$, occurs for $u = q_n$ ($\xi = e_n$) and equals $\rho(q_n, A) = \alpha_n$. Together, these facts imply

$$\max_{u \neq 0} |\rho(u, A)| = \max(|\alpha_1|, |\alpha_n|) = \|A\|_2. \tag{5.2}$$

THEOREM 5.2. *Courant–Fischer minimax theorem. Let $\alpha_1 \geq \cdots \geq \alpha_n$ be eigenvalues of the symmetric matrix A and $q_1, \ldots, q_n$ be the corresponding unit eigenvectors.*

$$\max_{\mathbf{R}^j} \min_{0 \neq r \in \mathbf{R}^j} \rho(r, A) = \alpha_j = \min_{\mathbf{S}^{n-j+1}} \max_{0 \neq s \in \mathbf{S}^{n-j+1}} \rho(s, A).$$

The maximum in the first expression for α_j is over all j dimensional subspaces $\mathbf{R}^j$ of $\mathbb{R}^n$, and the subsequent minimum is over all nonzero vectors r in the subspace. The maximum is attained for $\mathbf{R}^j = \operatorname{span}(q_1, q_2, \ldots, q_j)$, and a minimizing r is $r = q_j$.

The minimum in the second expression for α_j is over all $(n - j + 1)$-dimensional subspaces $\mathbf{S}^{n-j+1}$ of $\mathbb{R}^n$, and the subsequent maximum is over all nonzero vectors s in the subspace. The minimum is attained for $\mathbf{S}^{n-j+1} = \operatorname{span}(q_j, q_{j+1}, \ldots, q_n)$, and a maximizing s is $s = q_j$.

EXAMPLE 5.2. Let $j = 1$, so α_j is the largest eigenvalue. Given $\mathbf{R}^1$, $\rho(r, A)$ is the same for all nonzero $r \in \mathbf{R}^1$, since all such r are scalar multiples of one another. Thus the first expression for α_1 simplifies to $\alpha_1 = \max_{r \neq 0} \rho(r, A)$. Similarly, since $n - j + 1 = n$, the only subspace $\mathbf{S}^{n-j+1}$ is $\mathbb{R}^n$, the whole space. Then the second expression for α_1 also simplifies to $\alpha_1 = \max_{s \neq 0} \rho(s, A)$.

One can similarly show that the theorem simplifies to the following expression for the smallest eigenvalue: $\alpha_n = \min_{r \neq 0} \rho(r, A)$. ◇

Proof of the Courant–Fischer minimax theorem. Choose any subspaces $\mathbf{R}^j$ and $\mathbf{S}^{n-j+1}$ of the indicated dimensions. Since the sum of their dimensions $j + (n - j + 1) = n + 1$ exceeds n, there must be a nonzero vector $x_{\mathbf{RS}} \in \mathbf{R}^j \cap \mathbf{S}^{n-j+1}$. Thus

$$\min_{0 \neq r \in \mathbf{R}^j} \rho(r, A) \leq \rho(x_{\mathbf{RS}}, A) \leq \max_{0 \neq s \in \mathbf{S}^{n-j+1}} \rho(s, A).$$

Now choose $\hat{\mathbf{R}}^j$ to maximize the expression on the left, and choose $\hat{\mathbf{S}}^{n-j+1}$ to minimize the expression on the right. Then

$$\begin{aligned}
\max_{\mathbf{R}^j} \min_{0 \neq r \in \mathbf{R}^j} \rho(r, A) &= \min_{0 \neq r \in \hat{\mathbf{R}}^j} \rho(r, A) &&(5.3) \\
&\leq \rho(x_{\hat{\mathbf{R}}\hat{\mathbf{S}}}, A) \\
&\leq \max_{0 \neq s \in \hat{\mathbf{S}}^{n-j+1}} \rho(s, A) \\
&= \min_{\mathbf{S}^{n-j+1}} \max_{0 \neq s \in \mathbf{S}^{n-j+1}} \rho(s, A).
\end{aligned}$$

To see that all these inequalities are actually equalities, we exhibit particular $\mathbf{R}^j$ and $\mathbf{S}^{n-j+1}$ that make the lower bound equal the upper bound. First
choose $\underline{\mathbf{R}}^j = \text{span}(q_1, \ldots, q_j)$ so that

$$\max_{\mathbf{R}^j} \min_{0 \neq r \in \mathbf{R}^j} \rho(r, A) \geq \min_{0 \neq r \in \underline{\mathbf{R}}^j} \rho(r, A)$$

$$= \min_{0 \neq r = \sum_{i \leq j} \xi_i q_i} \rho(r, A)$$

$$= \min_{\text{some } \xi_i \neq 0} \frac{\sum_{i \leq j} \xi_i^2 \alpha_i}{\sum_{i \leq j} \xi_i^2} = \alpha_j.$$

Next choose $\bar{\mathbf{S}}^{n-j+1} = \text{span}(q_j, \ldots, q_n)$ so that

$$\min_{\mathbf{S}^{n-j+1}} \max_{0 \neq s \in \mathbf{S}^{n-j+1}} \rho(s, A) \leq \max_{0 \neq s \in \bar{\mathbf{S}}^{n-j+1}} \rho(s, A)$$

$$= \max_{0 \neq s = \sum_{i \geq j} \xi_i q_i} \rho(s, A)$$

$$= \max_{\text{some } \xi_i \neq 0} \frac{\sum_{i \geq j} \xi_i^2 \alpha_i}{\sum_{i \geq j} \xi_i^2} = \alpha_j.$$

Thus, the lower and upper bounds are sandwiched between α_j below and
α_j above, so they must all equal α_j as desired. $\quad\square$

EXAMPLE 5.3. Figure 5.1 illustrates this theorem graphically for 3-by-3 ma-
trices. Since $\rho(u/\|u\|_2, A) = \rho(u, A)$, we can think of $\rho(u, A)$ as a function on
the unit sphere $\|u\|_2 = 1$. Figure 5.1 shows a contour plot of this function on
the unit sphere for $A = \text{diag}(1, .25, 0)$. For this simple matrix $q_i = e_i$, the ith
column of the identity matrix. The figure is symmetric about the origin since
$\rho(u, A) = \rho(-u, A)$. The small red circles near $\pm q_1$ surround the global maximum $\rho(\pm q_1, A) = 1$, and the small green circles near $\pm q_3$ surround the global
minimum $\rho(\pm q_3, A) = 0$. The two great circles are contours for $\rho(u, A) = .25$,
the second eigenvalue. Within the two narrow (green) "apple slices" defined
by the great circles, $\rho(u, A) < .25$, and within the wide (red) apple slices,
$\rho(u, A) > .25$.
Let us interpret the minimax theorem in terms of this figure. Choosing
a space $\mathbf{R}^2$ is equivalent to choosing a great circle C; every point on C lies
within $\mathbf{R}^2$, and $\mathbf{R}^2$ consists of all scalar multiplicatons of the vectors in C.
Thus $\min_{0 \neq r \in \mathbf{R}^2} \rho(r, A) = \min_{r \in C} \rho(r, A)$. There are four cases to consider to
compute $\min_{r \in C} \rho(r, A)$:

1. C does *not* go through the intersection points $\pm q_2$ of the two great circles
   in Figure 5.1. Then C clearly must intersect both a narrow green apple
   slice and a wide red apple slice, so $\min_{r \in C} \rho(r, A) < .25$.

2. C does go through the two intersection points $\pm q_2$ and otherwise lies in
   the narrow green apple slices. Then $\min_{r \in C} \rho(r, A) < .25$.

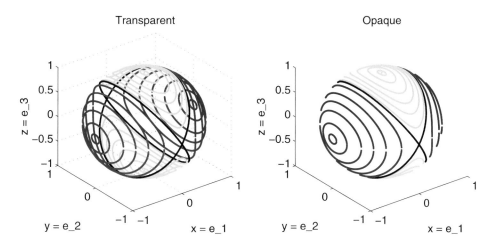

Fig. 5.1. *Contour plot of the Rayleigh quotient on the unit sphere.*

3. C does go through the two intersection points $\pm q_2$ and otherwise lies in the wide red apple slices. Then $\min_{r \in C} \rho(r, A) = .25$, attained for $r = \pm q_2$.

4. C coincides with one of the two great circles. Then $\rho(r, A) = .25$ for all $r \in C$.

The minimax theorem says that $\alpha_2 = .25$ is the maximum of $\min_{r \in C} \rho(r, A)$ over all choices of great circle C. This maximum is attained in cases 3 and 4 above. In particular, for C bisecting the wide red apple slices (case 3), $\mathbf{R}^2 = \mathrm{span}(q_1, q_2)$.

Software to draw contour plots like those in Figure 5.1 for an arbitrary 3-by-3 symmetric matrix may be found at HOMEPAGE/Matlab/RayleighContour.m. $\diamond$

Finally, we can present the *proof of Weyl's theorem.*

$$
\begin{aligned}
\hat{\alpha}_j &= \min_{\mathbf{S}^{n-j+1}} \max_{0 \neq u \in \mathbf{S}^{n-j+1}} \frac{u^T (A + E) u}{u^T u} && \text{by the minimax theorem} \\
&= \min_{\mathbf{S}^{n-j+1}} \max_{0 \neq u \in \mathbf{S}^{n-j+1}} \left(\frac{u^T A u}{u^T u} + \frac{u^T E u}{u^T u} \right) \\
&\leq \min_{\mathbf{S}^{n-j+1}} \max_{0 \neq u \in \mathbf{S}^{n-j+1}} \left(\frac{u^T A u}{u^T u} + \|E\|_2 \right) && \text{by equation (5.2)} \\
&= \alpha_j + \|E\|_2 && \text{by the minimax theorem again.}
\end{aligned}
$$

Reversing the roles of A and $A + E$, we also get $\alpha_j \leq \hat{\alpha}_j + \|E\|_2$. Together, these two inequalities complete the proof of Weyl's theorem. $\square$

A theorem closely related to the Courant–Fischer minimax theorem, one that we will need later to justify the Bisection algorithm in section 5.3.4, is Sylvester's theorem of inertia.

DEFINITION 5.2. *The* inertia *of a symmetric matrix A is the triple of integers* Inertia$(A) \equiv (\nu, \zeta, \pi)$, *where ν is the number of negative eigenvalues of A, ζ is the number of zero eigenvalues of A, and π is the number of positive eigenvalues of A.*

If X is orthogonal, then $X^T A X$ and A are similar and so have the same eigenvalues. When X is only nonsingular, we say $X^T A X$ and A are *congruent*. In this case $X^T A X$ will generally not have the same eigenvalues as A, but the next theorem tells us that the two sets of eigenvalues will at least have the same signs.

THEOREM 5.3. Sylvester's inertia theorem. *Let A be symmetric and X be nonsingular. Then A and $X^T A X$ have the same inertia.*

Proof. Let n be the dimension of A. Now suppose that A has ν negative eigenvalues but that $X^T A X$ has $\nu' < \nu$ negative eigenvalues; we will find a contradiction to prove that this cannot happen. Let $\mathbf{N}$ be the corresponding ν dimensional negative eigenspace of A; i.e., $\mathbf{N}$ is spanned by the eigenvectors of the ν negative eigenvalues of A. This means that for any nonzero $x \in \mathbf{N}$, $x^T A x < 0$. Let $\mathbf{P}$ be the $(n - \nu')$-dimensional nonnegative eigenspace of $X^T A X$; this means that for any nonzero $x \in \mathbf{P}$, $x^T X^T A X x \geq 0$. Since X is nonsingular, the space $X\mathbf{P}$ is also $n - \nu'$ dimensional. Since $\dim(\mathbf{N}) + \dim(X\mathbf{P}) = \nu + n - \nu' > n$, the spaces $\mathbf{N}$ and $X\mathbf{P}$ must contain a nonzero vector x in their intersection. But then $0 > x^T A x$ since $x \in \mathbf{N}$ and $0 \leq x^T A x$ since $x \in X\mathbf{P}$, which is a contradiction. Therefore, $\nu = \nu'$. Reversing the roles of A and $X^T A X$, we also get $\nu' \leq \nu$; i.e., A and $X^T A X$ have the same number of negative eigenvalues. An analogous argument shows they have the same number of positive eigenvalues. Thus, they must also have the same number of zero eigenvalues. $\square$

Now we consider how eigenvectors can change by perturbing A to $A + E$ of A. To state our bound we need to define the *gap* in the spectrum.

DEFINITION 5.3. *Let A have eigenvalues $\alpha_1 \geq \cdots \geq \alpha_n$. Then the* gap *between an eigenvalue α_i and the rest of the spectrum is defined to be* $\mathrm{gap}(i, A) = \min_{j \neq i} |\alpha_j - \alpha_i|$. *We will also write* $\mathrm{gap}(i)$ *if A is understood from the context.*

The basic result is that the sensitivity of an eigenvector depends on the gap of its corresponding eigenvalue: a small gap implies a sensitive eigenvector.

EXAMPLE 5.4. Let $A = \begin{bmatrix} 1+g & \\ & 1 \end{bmatrix}$ and $A + E = \begin{bmatrix} 1+g & \epsilon \\ \epsilon & 1 \end{bmatrix}$, where $0 < \epsilon < g$. Thus $\mathrm{gap}(i, A) = g \approx \mathrm{gap}(i, A + E)$ for $i = 1, 2$. The eigenvectors of A are just $q_1 = e_1$ and $q_2 = e_2$. A small computation reveals that the eigenvectors of $A + E$ are

$$\hat{q}_1 = \beta \cdot \begin{bmatrix} 1 + \sqrt{1 + \left(\frac{2\epsilon}{g}\right)^2} \\ \frac{2\epsilon}{g} \end{bmatrix} \approx \begin{bmatrix} 1 \\ \frac{\epsilon}{g} \end{bmatrix},$$

$$\hat{q}_2 = \beta \cdot \left[\begin{array}{c} -\frac{2\epsilon}{g} \\ 1 + \sqrt{1 + \left(\frac{2\epsilon}{g}\right)^2} \end{array} \right] \approx \left[\begin{array}{c} -\frac{\epsilon}{g} \\ 1 \end{array} \right],$$

where $\beta \approx 1/2$ is a normalization factor. We see that the angle between the perturbed vectors $\hat{q}_i$ and unperturbed vectors q_i equals ϵ/g to first order in ϵ. So the angle is proportional to the reciprocal of the gap g. ◇

The general case is essentially the same as the 2-by-2 case just analyzed.

THEOREM 5.4. *Let $A = Q\Lambda Q^T = Q\mathrm{diag}(\alpha_i)Q^T$ be an eigendecomposition of A. Let $A + E = \hat{A} = \hat{Q}\hat{\Lambda}\hat{Q}^T$ be the perturbed eigendecomposition. Write $Q = [q_1, \ldots, q_n]$ and $\hat{Q} = [\hat{q}_1, \ldots, \hat{q}_n]$, where q_i and $\hat{q}_i$ are the unperturbed and perturbed unit eigenvectors, respectively. Let θ denote the acute angle between q_i and $\hat{q}_i$. Then*

$$\frac{1}{2} \sin 2\theta \le \frac{\|E\|_2}{\mathrm{gap}(i, A)}, \qquad provided \ that \ \mathrm{gap}(i, A) > 0.$$

Similarly

$$\frac{1}{2} \sin 2\theta \le \frac{\|E\|_2}{\mathrm{gap}(i, A + E)}, \qquad provided \ that \ \mathrm{gap}(i, A + E) > 0.$$

Note that when $\theta \ll 1$, then $(1/2)\sin 2\theta \approx \sin \theta \approx \theta$.

The attraction of stating the bound in terms of $\mathrm{gap}(i, A + E)$, as well as $\mathrm{gap}(i, A)$, is that frequently we know only the eigenvalues of $A + E$, since they are typically the output of the eigenvalue algorithm that we have used. In this case it is straightforward to evaluate $\mathrm{gap}(i, A + E)$, whereas we can only estimate $\mathrm{gap}(i, A)$.

When the first upper bound exceeds $1/2$, i.e., $\|E\|_2 \ge \mathrm{gap}(i, A)/2$, the bound reduces to $\sin 2\theta \le 1$, which provides no information about θ. Here is why we cannot bound θ in this situation: If E is this large, then $A + E$'s eigenvalue $\hat{\alpha}_i$ could be sufficiently far from α_i for $A + E$ to have a multiple eigenvalue at $\hat{\alpha}_i$. For example, consider $A = \mathrm{diag}(2, 0)$ and $A + E = I$. But such an $A + E$ does not have a unique eigenvector q_i; indeed, $A + E = I$ has any vector as an eigenvector. Thus, it makes no sense to try to bound θ. The same considerations apply when the second upper bound exceeds $1/2$.

Proof. It suffices to prove the first upper bound, because the second one follows by considering $A + E$ as the unperturbed matrix and $A = (A + E) - E$ as the perturbed matrix.

Let $q_i + d$ be an eigenvector of $A + E$. To make d unique, we impose the restriction that it be orthogonal to q_i (written $d \perp q_i$) as shown below. Note that this means that $q_i + d$ is not a unit vector, so $\hat{q}_i = (q_i + d)/\|q_i + d\|_2$. Then $\tan \theta = \|d\|_2$ and $\sec \theta = \|q_i + d\|_2$.

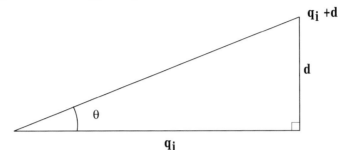

Now write the ith column of $(A + E)\hat{Q} = \hat{Q}\hat{\Lambda}$ as

$$(A + E)(q_i + d) = \hat{\alpha}_i(q_i + d), \tag{5.4}$$

where we have also multiplied each side by $\|q_i + d\|_2$. Define $\eta = \hat{\alpha}_i - \alpha_i$. Subtract $Aq_i = \alpha_i q_i$ from both sides of (5.4) and rearrange to get

$$(A - \alpha_i I)d = (\eta I - E)(q_i + d). \tag{5.5}$$

Since $q_i^T(A - \alpha_i I) = 0$, both sides of (5.5) are orthogonal to q_i. This lets us write $z \equiv (\eta I - E)(q_i + d) = \sum_{j \neq i} \zeta_j q_j$ and $d \equiv \sum_{j \neq i} \delta_j q_j$. Since $(A - \alpha_i I)q_j = (\alpha_j - \alpha_i)q_j$, we can write

$$(A - \alpha_i I)d = \sum_{j \neq i}(\alpha_j - \alpha_i)\delta_j q_j = \sum_{j \neq i} \zeta_j q_j = (\eta I - E)(q_i + d)$$

or

$$d = \sum_{j \neq i} \delta_j q_j = \sum_{j \neq i} \frac{\zeta_j}{\alpha_j - \alpha_i} q_j.$$

Thus

$$
\begin{aligned}
\tan \theta &= \|d\|_2 \\
&= \left\| \sum_{j \neq i} \frac{\zeta_j}{\alpha_j - \alpha_i} q_j \right\|_2 \\
&= \left(\sum_{j \neq i} \left(\frac{\zeta_j}{\alpha_j - \alpha_i} \right)^2 \right)^{1/2} \quad \text{since the } q_j \text{ are orthonormal} \\
&\leq \frac{1}{\text{gap}(i, A)} \left(\sum_{j \neq i} \zeta_j^2 \right)^{1/2} \quad \begin{array}{l} \text{since } \text{gap}(i, A) \text{ is the} \\ \text{smallest denominator} \end{array} \\
&= \frac{\|z\|_2}{\text{gap}(i, A)}.
\end{aligned}
$$

If we were to use Weyl's theorem and the triangle inequality to bound $\|z\|_2 \leq (\|E\|_2 + |\eta|) \cdot \|q_i + d\|_2 \leq 2\|E\|_2 \sec\theta$, then we could conclude that $\sin\theta \leq 2\|E\|_2/\mathrm{gap}(i, A)$.

But we can do a little better than this by bounding $\|z\|_2 = \|(\eta I - E)(q_i + d)\|_2$ more carefully: Multiply (5.4) by q_i^T on both sides, cancel terms, and rearrange to get $\eta = q_i^T E(q_i + d)$. Thus

$$
\begin{aligned}
z &= (q_i + d)\eta - E(q_i + d) = (q_i + d)q_i^T E(q_i + d) - E(q_i + d) \\
&= ((q_i + d)q_i^T - I)E(q_i + d),
\end{aligned}
$$

and so $\|z\|_2 \leq \|(q_i + d)q_i^T - I\| \cdot \|E\|_2 \cdot \|q_i + d\|$. We claim that $\|(q_i + d)q_i^T - I\|_2 = \|q_i + d\|_2$ (see Question 5.7). Thus $\|z\|_2 \leq \|q_i + d\|_2^2 \cdot \|E\|_2$, so

$$
\tan\theta \leq \frac{\|z\|_2}{\mathrm{gap}(i, A)} \leq \frac{\|q_i + d\|_2^2 \|E\|_2}{\mathrm{gap}(i, A)} = \frac{\sec^2\theta \cdot \|E\|_2}{\mathrm{gap}(i, A)}
$$

or

$$
\frac{\|E\|_2}{\mathrm{gap}(i, A)} \geq \frac{\tan\theta}{\sec^2\theta} = \sin\theta \cos\theta = \frac{1}{2}\sin 2\theta
$$

as desired. $\square$

An analogous theorem can be proven for singular vectors (see Question 5.8).

The Rayleigh quotient has other nice properties. The next theorem tells us that the Rayleigh quotient is a "best approximation" to an eigenvalue in a natural sense. This is the basis of the Rayleigh quotient iteration in section 5.3.2 and the iterative algorithms in Chapter 7. It may also be used to evaluate the accuracy of an approximate eigenpair obtained in any way at all, not just by the algorithms discussed here.

THEOREM 5.5. *Let A be symmetric, x be a unit vector, and β be a scalar. Then A has an eigenpair $Aq_i = \alpha_i q_i$ satisfying $|\alpha_i - \beta| \leq \|Ax - \beta x\|_2$. Given x, the choice $\beta = \rho(x, A)$ minimizes $\|Ax - \beta x\|_2$.*

With a little more information about the spectrum of A, we can get tighter bounds. Let $r = Ax - \rho(x, A)x$. Let α_i be the eigenvalue of A closest to $\rho(x, A)$. Let $\mathrm{gap}' \equiv \min_{j \neq i}|\alpha_j - \rho(x, A)|$; this is a variation on the gap defined earlier. Let θ be the acute angle between x and q_i. Then

$$
\sin\theta \leq \frac{\|r\|_2}{\mathrm{gap}'} \tag{5.6}
$$

and

$$
|\alpha_i - \rho(x, A)| \leq \frac{\|r\|_2^2}{\mathrm{gap}'}. \tag{5.7}
$$

See Theorem 7.1 for a generalization of this result to a set of eigenvalues.

Notice that in equation (5.7) the difference between the Rayleigh quotient $\rho(x, A)$ and an eigenvalue α_i is proportional to the *square* of the residual norm

$\|r\|_2$. This high accuracy is the basis of the cubic convergence of the Rayleigh quotient iteration algorithm of section 5.3.2.

Proof. We prove only the first result and leave the others for questions 5.9 and 5.10 at the end of the chapter.

If β is an eigenvalue of A, the result is immediate. So assume instead that $A - \beta I$ is nonsingular. Then $x = (A - \beta I)^{-1}(A - \beta I)x$ and

$$1 = \|x\|_2 \leq \|(A - \beta I)^{-1}\|_2 \cdot \|(A - \beta I)x\|_2.$$

Writing A's eigendecomposition as $A = Q\Lambda Q^T = Q\text{diag}(\alpha_1, \ldots, \alpha_n)Q^T$, we get

$$\|(A - \beta I)^{-1}\|_2 = \|Q(\Lambda - \beta I)^{-1}Q^T\|_2 = \|(\Lambda - \beta I)^{-1}\|_2 = 1/\min_i |\alpha_i - \beta|,$$

so $\min_i |\alpha_i - \beta| \leq \|(A - \beta I)x\|_2$ as desired.

To show that $\beta = \rho(x, A)$ minimizes $\|Ax - \beta x\|_2$ we will show that x is orthogonal to $Ax - \rho(x, A)x$ so that applying the Pythagorean theorem to the sum of orthogonal vectors

$$Ax - \beta x = [Ax - \rho(x, A)x] + [(\rho(x, A) - \beta)x]$$

yields

$$\begin{aligned} \|Ax - \beta x\|_2^2 &= \|Ax - \rho(x, A)x\|_2^2 + \|(\rho(x, A) - \beta)x\|_2^2 \\ &\geq \|Ax - \rho(x, A)x\|_2^2 \end{aligned}$$

with equality only when $\beta = \rho(x, A)$.

To confirm orthogonality of x and $Ax - \rho(x, A)x$ we need to verify that

$$x^T(Ax - \rho(x, A)x) = x^T\left(Ax - \frac{(x^T Ax)}{x^T x}x\right) = x^T Ax - x^T Ax\frac{x^T x}{x^T x} = 0$$

as desired. $\square$

EXAMPLE 5.5. We illustrate Theorem 5.5 using a matrix from Example 5.4. Let $A = \begin{bmatrix} 1+g & \epsilon \\ \epsilon & 1 \end{bmatrix}$, where $0 < \epsilon < g$. Let $x = [1, 0]^T$ and $\beta = \rho(x, A) = 1 + g$. Then $r = Ax - \beta x = [0, \epsilon]^T$ and $\|r\|_2 = \epsilon$. The eigenvalues of A are $\alpha_\pm = 1 + \frac{g}{2} \pm \frac{g}{2}(1 + (\frac{2\epsilon}{g})^2)^{1/2}$, and the eigenvectors are given in Example 5.4 (where the matrix is called $A + E$ instead of A).

Theorem 5.5 predicts that $\|Ax - \beta x\|_2 = \|r\|_2 = \epsilon$ is a bound on the distance from $\beta = 1 + g$ to the nearest eigenvalue α_+ of A; this is also predicted by Weyl's theorem (Theorem 5.1). We will see below that this bound is much looser than bound (5.7).

When ϵ is much smaller than g, there will be one eigenvalue near $1 + g$ with its eigenvector near x and another eigenvalue near 1 with its eigenvector near $[0, 1]^T$. This means gap$' = |\alpha_- - \rho(x, A)| = \frac{g}{2}(1 + (1 + (\frac{2\epsilon}{g})^2)^{1/2})$, and so bound (5.6) implies that the angle θ between x and the true eigenvector is bounded by

$$\sin \theta \le \frac{\|r\|_2}{\text{gap}'} = \frac{2\epsilon/g}{1 + (1 + (\frac{2\epsilon}{g})^2)^{1/2}}.$$

Comparing with the explicit eigenvectors in Example 5.4, we see that the upper bound is actually equal to $\tan \theta$, which is nearly the same as $\sin \theta$ for tiny θ. So bound (5.6) is quite accurate.

Now consider bound (5.7) on the difference $|\beta - \alpha_+|$. It turns out that for this 2-by-2 example both $|\beta - \alpha_+|$ and its bound are *exactly* equal to

$$\frac{\|r\|_2^2}{\text{gap}'} = \epsilon \cdot \frac{2\epsilon/g}{1 + (1 + (\frac{2\epsilon}{g})^2)^{1/2}}.$$

Let us evaluate these bounds in the special case where $g = 10^{-2}$ and $\epsilon = 10^{-5}$. Then the eigenvalues of A are approximately $\alpha_+ = 1.01000001 = 1.01 + 10^{-8}$ and $\alpha_- = .99999999 = 1 - 10^{-8}$. The first bound is $|\beta - \alpha_+| \le \|r\|_2 = 10^{-5}$, which is 10^3 times larger than the actual error 10^{-8}. In contrast, bound (5.7) is $|\beta - \alpha_+| \le \|r\|_2^2/\text{gap}' = (10^{-5})^2/(1.01 - \alpha_-) \approx 10^{-8}$, which is tight. The actual angle θ between x and the true eigenvector for α_+ is about 10^{-3}, as is the bound $\|r\|_2/\text{gap}' = 10^{-5}/(1.01 - \alpha_-) \approx 10^{-3}$. ◇

Finally, we discuss what happens when one has a group of k tightly clustered eigenvalues, and wants to compute their eigenvectors. By "tightly clustered" we mean that the gap between any eigenvalue in the cluster and some other eigenvalue in the cluster is small but that eigenvalues not in the cluster are far away. For example, one could have $k = 20$ eigenvalues in the interval [.9999,1.0001], but all other eigenvalues might be greater than 2. Then Theorems 5.4 and 5.5 indicate that we cannot hope to get the individual eigenvectors accurately. However, it is possible to compute the k-dimensional invariant subspace spanned by these vectors quite accurately. See [197] for details.

5.2.1. Relative Perturbation Theory

This section describes tighter bounds on eigenvalues and eigenvectors than in the last section. These bounds are needed to justify the high-accuracy algorithms for computing singular values and eigenvalues described in sections 5.4.2 and 5.4.3.

To contrast the bounds that we will present here to those in the previous section, let us consider the 1-by-1 case. Given a scalar α, a perturbed scalar $\hat{\alpha} = \alpha + e$ and a bound $|e| \le \epsilon$, we can obviously bound the *absolute error* in $\hat{\alpha}$ by $|\hat{\alpha} - \alpha| \le \epsilon$. This was the approach taken in the last section. Consider instead the perturbed scalar $\hat{\alpha} = x^2 \alpha$ and a bound $|x^2 - 1| \le \epsilon$. This lets us bound the *relative error* in $\hat{\alpha}$ by

$$\frac{|\hat{\alpha} - \alpha|}{|\alpha|} = |x^2 - 1| \le \epsilon.$$

We generalize this simple idea to matrices as follows. In the last section we bounded the *absolute* difference in the eigenvalues α_i of A and $\hat{\alpha}_i$ of $\hat{A} = A + E$ by $|\hat{\alpha}_i - \alpha_i| \leq \|E\|_2$. Here we will bound the *relative* difference between the eigenvalues α_i of A and $\hat{\alpha}_i$ of $\hat{A} = X^T A X$ in terms of $\epsilon \equiv \|X^T X - I\|_2$.

THEOREM 5.6. *"Relative" Weyl. Let A have eigenvalues α_i and $\hat{A} = X^T A X$ have eigenvalues $\hat{\alpha}_i$. Let $\epsilon \equiv \|X^T X - I\|_2$. Then $|\hat{\alpha}_i - \alpha_i| \leq |\alpha_i|\epsilon$. If $\alpha_i \neq 0$, then we can also write*

$$\frac{|\hat{\alpha}_i - \alpha_i|}{|\alpha_i|} \leq \epsilon. \tag{5.8}$$

Proof. Since the ith eigenvalue of $A - \alpha_i I$ is zero, Sylvester's theorem of inertia tells us that the same is true of

$$X^T(A - \alpha_i I)X = (X^T A X - \alpha_i I) + \alpha_i(I - X^T X) \equiv H + F.$$

Weyl's theorem says that $|\lambda_i(H) - 0| \leq \|F\|_2$, or $|\hat{\alpha}_i - \alpha_i| \leq |\alpha_i| \cdot \|X^T X - I\|_2 = |\alpha_i|\epsilon$. $\square$

Note that when X is orthogonal, $\epsilon = \|X^T X - I\|_2 = 0$, so the theorem confirms that $X^T A X$ and A have the same eigenvalues. If X is "nearly" orthogonal, i.e., ϵ is small, the theorem says the eigenvalue are nearly the same, in the sense of relative error.

COROLLARY 5.2. *Let G be an arbitrary matrix with singular values σ_i, and let $\hat{G} = Y^T G X$ have singular values $\hat{\sigma}_i$. Let $\epsilon \equiv \max(\|X^T X - I\|_2, \|Y^T Y - I\|_2)$. Then $|\hat{\sigma}_i - \sigma_i| \leq \epsilon\sigma_i$. If $\sigma_i \neq 0$, then we can write*

$$\frac{|\hat{\sigma}_i - \sigma_i|}{\sigma_i} \leq \epsilon. \tag{5.9}$$

We can similarly extend Theorem 5.4 to bound the difference between eigenvectors q_i of A and eigenvectors $\hat{q}_i$ of $\hat{A} = X^T A X$. To do so, we need to define the *relative gap* in the spectrum.

DEFINITION 5.4. *The* relative gap *between an eigenvalue α_i of A and the rest of the spectrum is defined to be* $\text{rel\_gap}(i, A) = \min_{j \neq i} \frac{|\alpha_j - \alpha_i|}{|\alpha_i|}$.

THEOREM 5.7. *Suppose that A has eigenvalues α_i and corresponding unit eigenvectors q_i. Suppose $\hat{A} = X^T A X$ has eigenvalues $\hat{\alpha}_i$ and corresponding unit eigenvectors $\hat{q}_i$. Let θ be the acute angle between q_i and $\hat{q}_i$. Let $\epsilon_1 = \|I - X^{-T} X^{-1}\|_2$ and $\epsilon_2 = \|X - I\|_2$. Then provided that $\epsilon_1 < 1$ and $\text{rel\_gap}(i, X^T A X) > 0$,*

$$\frac{1}{2}\sin 2\theta \leq \frac{\epsilon_1}{1 - \epsilon_1} \cdot \frac{1}{\text{rel\_gap}(i, X^T A X)} + \epsilon_2.$$

Proof. Let $\eta = \hat{\alpha}_i - \alpha_i$, $H = A - \hat{\alpha}_i I$, and $F = \hat{\alpha}_i(I - X^{-T}X^{-1})$. Note that

$$H + F = A - \hat{\alpha}_i X^{-T}X^{-1} = X^{-T}(X^T A X - \hat{\alpha}_i I)X^{-1}.$$

Thus $Hq_i = -\eta q_i$ and $(H + F)(X\hat{q}_i) = 0$ so that $X\hat{q}_i$ is an eigenvector of $H + F$ with ith eigenvalue 0. Let θ_1 be the acute angle between q_i and $X\hat{q}_i$. By Theorem 5.4, we can bound

$$\frac{1}{2}\sin 2\theta_1 \le \frac{\|F\|_2}{\text{gap}(i, H + F)}. \tag{5.10}$$

We have $\|F\|_2 = |\hat{\alpha}_i|\epsilon_1$. Now $\text{gap}(i, H + F)$ is the magnitude of the smallest nonzero eigenvalue of $H + F$. Since $X^T(H + F)X = X^T A X - \hat{\alpha}_i I$ has eigenvalues $\hat{\alpha}_j - \hat{\alpha}_i$, Theorem 5.6 tells us that the eigenvalues of $H + F$ lie in intervals from $(1 - \epsilon_1)(\hat{\alpha}_j - \hat{\alpha}_i)$ to $(1 + \epsilon_1)(\hat{\alpha}_j - \hat{\alpha}_i)$. Thus $\text{gap}(i, H + F) \ge (1 - \epsilon_1)\text{gap}(i, X^T A X)$, and so substituting into (5.10) yields

$$\frac{1}{2}\sin 2\theta_1 \le \frac{\epsilon_1|\hat{\alpha}_i|}{(1 - \epsilon_1)\text{gap}(i, X^T A X)} = \frac{\epsilon_1}{(1 - \epsilon_1)\text{rel\_gap}(i, X^T A X)}. \tag{5.11}$$

Now let θ_2 be the acute angle between $X\hat{q}_i$ and $\hat{q}_i$ so that $\theta \le \theta_1 + \theta_2$. Using trigonometry we can bound $\sin\theta_2 \le \|(X - I)\hat{q}_i\|_2 \le \|X - I\|_2 = \epsilon_2$, and so by the triangle inequality (see Question 5.11)

$$\begin{aligned}
\frac{1}{2}\sin 2\theta &\le \frac{1}{2}\sin 2\theta_1 + \frac{1}{2}\sin 2\theta_2 \\
&\le \frac{1}{2}\sin 2\theta_1 + \sin\theta_2 \\
&\le \frac{\epsilon_1}{(1 - \epsilon_1)\text{rel\_gap}(i, X^T A X)} + \epsilon_2
\end{aligned}$$

as desired. $\square$

An analogous theorem can be proven for singular vectors [101].

EXAMPLE 5.6. We again consider the mass-spring system of Example 5.1 and use it to show that bounds on eigenvalues provided by Weyl's theorem (Theorem 5.1) can be much worse (looser) than the "relative" version of Weyl's theorem (Theorem 5.6). We will also see that the eigenvector bound of Theorem 5.7 can be much better (tighter) than the bound of Theorem 5.4.

Suppose that $M = \text{diag}(1, 100, 10000)$ and $K_D = \text{diag}(10000, 100, 1)$. Following Example 5.1, we define $K = BK_D B^T$ and $\hat{K} = M^{-1/2}KM^{-1/2}$, where

$$B = \begin{bmatrix} 1 & -1 & & \\ & \ddots & \ddots & \\ & & \ddots & -1 \\ & & & 1 \end{bmatrix}$$

and so

$$\hat{K} = M^{-1/2}KM^{-1/2} = \begin{bmatrix} 10100 & -10 & \\ -10 & 1.01 & -.001 \\ & -.001 & .0001 \end{bmatrix}.$$

To five decimal places, the eigenvalues of $\hat{K}$ are 10100, 1.0001, and .00099. Suppose we now perturb the masses (m_{ii}) and spring constants $(k_{D,ii})$ by at most 1% each. How much can the eigenvalues change? The largest matrix entry is $\hat{K}_{11}$, and changing m_{11} to .99 and $k_{D,11}$ to 10100 will change $\hat{K}_{11}$ to about 10305, a change of 205 in norm. Thus, Weyl's theorem tells us each eigenvalue could change by as much as ± 205, which would change the smaller two eigenvalues utterly. The eigenvector bound from Theorem 5.4 also indicates that the corresponding eigenvectors could change completely.

Now let us apply Theorem 5.6 to $\hat{K}$, or actually Corollary 5.2 to $G = M^{-1/2}BK_D^{1/2}$, where $\hat{K} = GG^T$ as defined in Example 5.1. Changing each mass by at most 1% is equivalent to perturbing G to XG, where X is diagonal with diagonal entries between $1/\sqrt{.99} \approx 1.005$ and $1/\sqrt{1.01} \approx .995$. Then Corollary 5.2 tells us that the singular values of G can change only by factors within the interval $[.995, 1.005]$, so the eigenvalues of M can change only by 1% too. In other words, the smallest eigenvalue can change only in its second decimal place, just like the largest eigenvalue. Similarly, changing the spring constants by at most 1% is equivalent to changing G to GX, and again the eigenvalues cannot change by more than 1%. If we perturb both M and K_D at the same time, the eigenvalues will move by about 2%. Since the eigenvalues differ so much in magnitude, their relative gaps are all quite large, and so their eigenvectors can rotate only by about 3% in angle too.

For a different approach to relative error analysis, more suitable for matrices arising from differential ("unbounded") operators, see [161].

5.3. Algorithms for the Symmetric Eigenproblem

We discuss a variety of algorithms for the symmetric eigenproblem. As mentioned in the introduction, we will discuss only *direct methods*, leaving *iterative methods* for Chapter 7.

In Chapter 4 on the nonsymmetric eigenproblem, the only algorithm that we discussed was QR iteration, which could find all the eigenvalues and optionally all the eigenvectors. We have many more algorithms available for the symmetric eigenproblem, which offer us more flexibility and efficiency. For example, the *Bisection algorithm* described below can be used to find only the eigenvalues in a user-specified interval $[a, b]$ and can do so much faster than it could find all the eigenvalues.

All the algorithms below, except Rayleigh quotient iteration and Jacobi's method, assume that the matrix has first been reduced to tridiagonal form, using the variation of Algorithm 4.6 in section 4.4.7. This is an initial cost of $\frac{4}{3}n^3$ flops, or $\frac{8}{3}n^3$ flops if eigenvectors are also desired.

1. *Tridiagonal QR iteration.* This algorithm finds all the eigenvalues, and optionally all the eigenvectors, of a symmetric tridiagonal matrix. Implemented efficiently, it is currently the fastest practical method to find all the eigenvalues of a symmetric tridiagonal matrix, taking $O(n^2)$ flops. Since reducing a dense matrix to tridiagonal form costs $\frac{4}{3}n^3$ flops, $O(n^2)$ is negligible for large enough n. But for finding all the eigenvectors as well, QR iteration takes a little over $6n^3$ flops on average and is only the fastest algorithm for small matrices, up to about $n = 25$. This is the algorithm underlying the Matlab command $\mathtt{eig}$[18] and the LAPACK routines $\mathtt{ssyev}$ (for dense matrices) and $\mathtt{sstev}$ (for tridiagonal matrices).

2. *Rayleigh quotient iteration.* This algorithm underlies QR iteration, but we present it separately in order to more easily analyze its extremely rapid convergence and because it may be used as an algorithm by itself. In fact, it generally converges cubically (as does QR iteration), which means that the number of correct digits asymptotically *triples* at each step.

3. *Divide-and-conquer.* This is currently the fastest method to find all the eigenvalues and eigenvectors of symmetric tridiagonal matrices larger than $n = 25$. (The implementation in LAPACK, $\mathtt{sstevd}$, defaults to QR iteration for smaller matrices.)

   In the worst case, divide-and-conquer requires $O(n^3)$ flops, but in practice the constant is quite small. Over a large set of random test cases, it appears to take only $O(n^{2.3})$ flops on average, and as low as $O(n^2)$ for some eigenvalue distributions.

   In theory, divide-and-conquer could be implemented to run in $O(n \cdot \log^p n)$ flops, where p is a small integer [131]. This super-fast implementation uses the fast multipole method (FMM) [124], originally invented for the completely different problem of computing the mutual forces on n electrically charged particles. But the complexity of this super-fast implementation means that QR iteration is currently the algorithm of choice for finding all eigenvalues, and divide-and-conquer without the FMM is the method of choice for finding all eigenvalues and all eigenvectors.

4. *Bisection and inverse iteration.* Bisection may be used to find just a subset of the eigenvalues of a symmetric tridiagonal matrix, say, those in an interval $[a, b]$ or $[\alpha_i, \alpha_{i-j}]$. It needs only $O(nk)$ flops, where k is the

---

[18]Matlab checks to see whether the argument of $\mathtt{eig}$ is symmetric or not and uses the symmetric algorithm when appropriate.

number of eigenvalues desired. Thus Bisection can be much faster than QR iteration when $k \ll n$, since QR iteration requires $O(n^2)$ flops. Inverse iteration (Algorithm 4.2) can then be used to find the corresponding eigenvectors. In the best case, when the eigenvalues are "well separated" (we explain this more fully later), inverse iteration also costs only $O(nk)$ flops. This is much less than either QR or divide-and-conquer (without the FMM), even when all eigenvalues and eigenvectors are desired ($k = n$). But in the worst case, when many eigenvalues are clustered close together, inverse iteration takes $O(nk^2)$ flops and does not even guarantee the accuracy of the computed eigenvectors (although in practice it is almost always accurate). So divide-and-conquer and QR are currently the algorithms of choice for finding all (or most) eigenvalues and eigenvectors, especially when eigenvalues may be clustered. Bisection and inverse iteration are available as options in the LAPACK routine ssyevx.

There is current research on inverse iteration addressing the problem of close eigenvalues, which may make it the fastest method to find all the eigenvectors (besides, theoretically, divide-and-conquer with the FMM) [105, 203, 83, 201, 176, 173, 175, 269]. However, software implementing this improved version of inverse iteration is not yet available.

5. *Jacobi's method.* This method is historically the oldest method for the eigenproblem, dating to 1846. It is usually much slower than any of the above methods, taking $O(n^3)$ flops with a large constant. But the method remains interesting, because it is sometimes much more accurate than the above methods. This is because Jacobi's method is sometimes capable of attaining the relative accuracy described in section 5.2.1 and so can sometimes compute tiny eigenvalues much more accurately than the previous methods [82]. We discuss the high-accuracy property of Jacobi's method in section 5.4.3, where we show how to compute the SVD.

Subsequent sections describe these algorithms in more detail. Section 5.3.6 presents comparative performance results.

5.3.1. Tridiagonal QR Iteration

Recall that the QR algorithm for the nonsymmetric eigenproblem had two phases:

1. Given A, use Algorithm 4.6 to find an orthogonal Q so that $QAQ^T = H$ is upper Hessenberg.

2. Apply QR iteration to H (as described in section 4.4.8) to get a sequence $H = H_0, H_1, H_2, \ldots$ of upper Hessenberg matrices converging to real Schur form.

Our first algorithm for the symmetric eigenproblem is completely analogous to this:

1. Given $A = A^T$, use the variation of Algorithm 4.6 in section 4.4.7 to find an orthogonal Q so that $QAQ^T = T$ is tridiagonal.

2. Apply QR iteration to T to get a sequence $T = T_0, T_1, T_2, \ldots$ of tridiagonal matrices converging to diagonal form.

We can see that QR iteration keeps all the T_i tridiagonal by noting that since QAQ^T is symmetric and upper Hessenberg, it must also be lower Hessenberg, i.e., tridiagonal. This keeps each QR iteration very inexpensive. An operation count reveals the following:

1. Reducing A to symmetric tridiagonal form T costs $\frac{4}{3}n^3 + O(n^2)$ flops, or $\frac{8}{3}n^3 + O(n^2)$ flops if eigenvectors are also desired.

2. One tridiagonal QR iteration with a single shift ("bulge chasing") costs $6n$ flops.

3. Finding all eigenvalues of T takes only 2 QR steps per eigenvalue on average, for a total of $6n^2$ flops.

4. Finding all eigenvalues and eigenvectors of T costs $6n^3 + O(n^2)$ flops.

5. The total cost to find just the eigenvalues of A is $\frac{4}{3}n^3 + O(n^2)$ flops.

6. The total cost to find all the eigenvalues and eigenvectors of A is $8\frac{2}{3}n^3 + O(n^2)$ flops.

We must still describe how the shifts are chosen to implement each QR iteration. Denote the ith iterate by

$$
T_i = \begin{bmatrix} a_1 & b_1 & & & \\ b_1 & \ddots & \ddots & & \\ & \ddots & \ddots & b_{n-1} \\ & & b_{n-1} & a_n \end{bmatrix}.
$$

The simplest choice of shift would be $\sigma_i = a_n$; this is the single shift QR iteration discussed in section 4.4.8. It turns out to be cubically convergent for almost all matrices, as shown in the next section. Unfortunately, examples exist where it does not converge [197, p. 76], so to get global convergence a slightly more complicated shift strategy is needed: We let the shift σ_i be the eigenvalue of $\begin{bmatrix} a_{n-1} & b_{n-1} \\ b_{n-1} & a_n \end{bmatrix}$ that is closest to a_n. This is called *Wilkinson's shift*.

THEOREM 5.8. *Wilkinson. QR iteration with Wilkinson's shift is globally, and at least linearly, convergent. It is asymptotically cubically convergent for almost all matrices.*

A proof of this theorem can be found in [197]. In LAPACK this routine is available as ssyev. The inner loop of the algorithm can be organized more efficiently when eigenvalues only are desired (ssterf; see also [104, 200]) than when eigenvectors are also computed (ssteqr).

EXAMPLE 5.7. Here is an illustration of the convergence of tridiagonal QR iteration, starting with the following tridiagonal matrix (diagonals only are shown, in columns):

$$T_0 = \text{tridiag} \begin{bmatrix} & .24929 & \\ 1.263 & & 1.263 \\ & .96880 & \\ -.82812 & & -.82812 \\ & .48539 & \\ -3.1883 & & -3.1883 \\ & -.91563 & \end{bmatrix}.$$

The following table shows the last offdiagonal entry of each T_i, the last diagonal entry of each T_i, and the difference between the last diagonal entry and its ultimate value (the eigenvalue $\alpha \approx -3.54627$). The cubic convergence of the error to zero in the last column is evident.

i	$T_i(4,3)$	$T_i(4,4)$	$T_i(4,4) - \alpha$
0	-3.1883	$-.91563$	2.6306
1	$-5.7 \cdot 10^{-2}$	-3.5457	$5.4 \cdot 10^{-4}$
2	$-2.5 \cdot 10^{-7}$	-3.5463	$1.2 \cdot 10^{-14}$
3	$-6.1 \cdot 10^{-23}$	-3.5463	0

At this point

$$T_3 = \text{tridiag} \begin{bmatrix} & 1.9871 & \\ .77513 & & .77513 \\ & 1.7049 & \\ -1.7207 & & -1.7207 \\ & .64214 & \\ -6.1 \cdot 10^{-23} & & -6.1 \cdot 10^{-23} \\ & -3.5463 & \end{bmatrix},$$

and we set the very tiny (4,3) and (3,4) entries to 0. This is called *deflation* and is stable, perturbing T_3 by only $6.1 \cdot 10^{-23}$ in norm. We now apply QR iteration again to the leading 3-by-3 submatrix of T_3, repeating the process to get the other eigenvalues. See HOMEPAGE/Matlab/tridiQR.m.

5.3.2. Rayleigh Quotient Iteration

Recall from our analysis of QR iteration in section 4.4 that we are implicitly doing inverse iteration at every step. We explore this more carefully when the shift we choose to use in the inverse iteration is the Rayleigh quotient.

ALGORITHM 5.1. *Rayleigh quotient iteration: Given x_0 with $\|x_0\|_2 = 1$, and a user-supplied stopping tolerance* tol, *we iterate*

$\rho_0 = \rho(x_0, A) = \frac{x_0^T A x_0}{x_0^T x_0}$

$i = 0$

repeat

    $y_i = (A - \rho_{i-1}I)^{-1}x_{i-1}$

    $x_i = y_i / \|y_i\|_2$

    $\rho_i = \rho(x_i, A)$

    $i = i + 1$

until convergence ($\|Ax_i - \rho_i x_i\|_2 <$ tol)

When the stopping criterion is satisfied, Theorem 5.5 tells us that ρ_i is within tol of an eigenvalue of A.

If one uses the shift $\sigma_i = a_{nn}$ in QR iteration and starts Rayleigh quotient iteration with $x_0 = [0, \ldots, 0, 1]^T$, then the connection between QR and inverse iteration discussed in section 4.4 can be used to show that the sequence of σ_i and ρ_i from the two algorithms are identical (see Question 5.13). In this case we will prove that convergence is almost always cubic.

THEOREM 5.9. *Rayleigh quotient iteration is locally cubically convergent; i.e., the number of correct digits triples at each step once the error is small enough and the eigenvalue is simple.*

Proof. We claim that it is enough to analyze the case when A is diagonal. To see why, write $Q^T A Q = \Lambda$, where Q is the orthogonal matrix whose columns are eigenvectors, and $\Lambda = \text{diag}(\alpha_1, \ldots, \alpha_n)$ is the diagonal matrix of eigenvalues. Now change variables in Rayleigh quotient iteration to $\hat{x}_i \equiv Q^T x_i$ and $\hat{y}_i \equiv Q^T y_i$. Then

$$\rho_i = \rho(x_i, A) = \frac{x_i^T A x_i}{x_i^T x_i} = \frac{\hat{x}_i^T Q^T A Q \hat{x}_i}{\hat{x}_i^T Q^T Q \hat{x}_i} = \frac{\hat{x}_i^T \Lambda \hat{x}_i}{\hat{x}_i^T \hat{x}_i} = \rho(\hat{x}_i, \Lambda)$$

and $Q\hat{y}_{i+1} = (A - \rho_i I)^{-1} Q \hat{x}_i$, so

$$\hat{y}_{i+1} = Q^T (A - \rho_i I)^{-1} Q \hat{x}_i = (Q^T A Q - \rho_i I)^{-1} \hat{x}_i = (\Lambda - \rho_i I)^{-1} \hat{x}_i.$$

Therefore, running Rayleigh quotient iteration with A and x_0 is equivalent to running Rayleigh quotient iteration with Λ and $\hat{x}_0$. Thus we will assume without loss of generality that $A = \Lambda$ is already diagonal, so the eigenvectors of A are e_i, the columns of the identity matrix.

Suppose without loss of generality that x_i is converging to e_1, so we can write $x_i = e_1 + d_i$, where $\|d_i\|_2 \equiv \epsilon \ll 1$. To prove cubic convergence, we need to show that $x_{i+1} = e_1 + d_{i+1}$ with $\|d_{i+1}\|_2 = O(\epsilon^3)$.

We first note that

$$1 = x_i^T x_i = (e_1 + d_i)^T (e_1 + d_i) = e_1^T e_1 + 2e_1^T d_i + d_i^T d_i = 1 + 2d_{i1} + \epsilon^2$$

so that $d_{i1} = -\epsilon^2/2$. Therefore

$$\rho_i = x_i^T \Lambda x_i = (e_1 + d_i)^T \Lambda (e_1 + d_i) = e_1^T \Lambda e_1 + 2e_1^T \Lambda d_i + d_i^T \Lambda d_i = \alpha_1 - \eta,$$

where $\eta \equiv -2e_1^T \Lambda d_i - d_i^T \Lambda d_i = \alpha_1 \epsilon^2 - d_i^T \Lambda d_i$. We see that

$$|\eta| \leq |\alpha_1|\epsilon^2 + \|\Lambda\|_2 \|d_i\|_2^2 \leq 2\|\Lambda\|_2 \epsilon^2, \tag{5.12}$$

so $\rho_i = \alpha_1 - \eta = \alpha_1 + O(\epsilon^2)$ is a very good approximation to the eigenvalue α_1.

Now we can write

$$\begin{aligned}
y_{i+1} &= (\Lambda - \rho_i I)^{-1} x_i \\
&= \left[\frac{x_{i1}}{\alpha_1 - \rho_i}, \frac{x_{i2}}{\alpha_2 - \rho_i}, \ldots, \frac{x_{in}}{\alpha_n - \rho_i} \right]^T \\
&\qquad \text{because } (\Lambda - \rho_i I)^{-1} = \mathrm{diag}\left(\frac{1}{\alpha_j - \rho_i} \right) \\
&= \left[\frac{1 + d_{i1}}{\alpha_1 - \rho_i}, \frac{d_{i2}}{\alpha_2 - \rho_i}, \ldots, \frac{d_{in}}{\alpha_n - \rho_i} \right]^T \\
&\qquad \text{because } x_i = e_1 + d_i \\
&= \left[\frac{1 - \epsilon^2/2}{\eta}, \frac{d_{i2}}{\alpha_2 - \alpha_1 + \eta}, \ldots, \frac{d_{in}}{\alpha_n - \alpha_1 + \eta} \right]^T \\
&\qquad \text{because } \rho_i = \alpha_1 - \eta \text{ and } d_{i1} = -\epsilon^2/2 \\
&= \frac{1 - \epsilon^2/2}{\eta} \cdot \left[1, \frac{d_{i2}\eta}{(1 - \epsilon^2/2)(\alpha_2 - \alpha_1 + \eta)}, \ldots, \right. \\
&\qquad\qquad \left. \frac{d_{in}\eta}{(1 - \epsilon^2/2)(\alpha_n - \alpha_1 + \eta)} \right]^T \\
&\equiv \frac{1 - \epsilon^2/2}{\eta} \cdot (e_1 + \hat{d}_{i+1}).
\end{aligned}$$

To bound $\|\hat{d}_{i+1}\|_2$, we note that we can bound each denominator using $|\alpha_j - \alpha_1 + \eta| \geq \mathrm{gap}(1, \Lambda) - |\eta|$, so using (5.12) as well we get

$$\|\hat{d}_{i+1}\|_2 \leq \frac{\|d_i\|_2 |\eta|}{(1 - \epsilon^2/2)(\mathrm{gap}(1, \Lambda) - |\eta|)} \leq \frac{2\|\Lambda\|_2 \epsilon^3}{(1 - \epsilon^2/2)(\mathrm{gap}(1, \Lambda) - 2\|\Lambda\|\epsilon^2)}$$

or $\|\hat{d}_{i+1}\|_2 = O(\epsilon^3)$. Finally, since $x_{i+1} = e_1 + d_{i+1} = (e_1 + \hat{d}_{i+1})/\|e_1 + \hat{d}_{i+1}\|_2$, we see $\|d_{i+1}\|_2 = O(\epsilon^3)$ as well. $\square$

5.3.3. Divide-and-Conquer

This method is the fastest now available if you want all eigenvalues and eigenvectors of a tridiagonal matrix whose dimension is larger than about 25. (The exact threshold depends on the computer.) It is quite subtle to implement in a numerically stable way. Indeed, although this method was first introduced in 1981 [59], the "right" implementation was not discovered until 1992 [127, 131]). This routine is available as LAPACK routines `ssyevd` for dense matrices and `sstevd` for tridiagonal matrices. This routine uses divide-and-conquer for matrices of dimension larger than 25 and automatically switches to QR iteration for smaller matrices (or if eigenvalues only are desired).

We first discuss the overall structure of the algorithm, and leave numerical details for later. Let

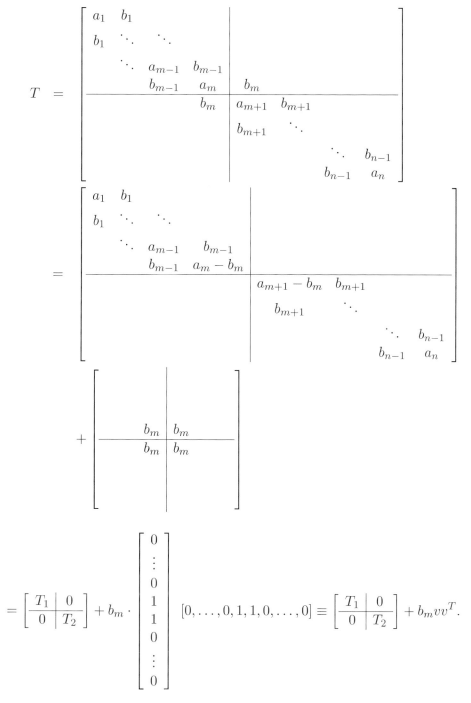

$$= \left[\begin{array}{c|c} T_1 & 0 \\ \hline 0 & T_2 \end{array}\right] + b_m \cdot \begin{bmatrix} 0 \\ \vdots \\ 0 \\ 1 \\ 1 \\ 0 \\ \vdots \\ 0 \end{bmatrix} [0,\ldots,0,1,1,0,\ldots,0] \equiv \left[\begin{array}{c|c} T_1 & 0 \\ \hline 0 & T_2 \end{array}\right] + b_m v v^T.$$

Suppose that we have the eigendecompositions of T_1 and T_2: $T_i = Q_i \Lambda_i Q_i^T$. These will be computed recursively by this same algorithm. We relate the eigenvalues of T to those of T_1 and T_2 as follows:

$$
\begin{aligned}
T &= \begin{bmatrix} T_1 & 0 \\ 0 & T_2 \end{bmatrix} + b_m v v^T \\
&= \begin{bmatrix} Q_1 \Lambda_1 Q_1^T & 0 \\ 0 & Q_2 \Lambda_2 Q_2^T \end{bmatrix} + b_m v v^T \\
&= \begin{bmatrix} Q_1 & 0 \\ 0 & Q_2 \end{bmatrix} \left(\begin{bmatrix} \Lambda_1 & \\ & \Lambda_2 \end{bmatrix} + b_m u u^T \right) \begin{bmatrix} Q_1^T & 0 \\ 0 & Q_2^T \end{bmatrix},
\end{aligned}
$$

where

$$
u = \begin{bmatrix} Q_1^T & 0 \\ 0 & Q_2^T \end{bmatrix} \quad v = \begin{bmatrix} \text{last column of } Q_1^T \\ \text{first column of } Q_2^T \end{bmatrix}
$$

since $v = [0, \ldots, 0, 1, 1, 0, \ldots, 0]^T$. Therefore, the eigenvalues of T are the same as those of the similar matrix $D + \rho u u^T$ where $D = \begin{bmatrix} \Lambda_1 & 0 \\ 0 & \Lambda_2 \end{bmatrix}$ is diagonal, $\rho = b_m$ is a scalar, and u is a vector. Henceforth we will assume without loss of generality that the diagonal $d_1, \ldots, d_n$ of D is sorted: $d_n \leq \cdots \leq d_1$.

To find the eigenvalues of $D + \rho u u^T$, assume first that $D - \lambda I$ is nonsingular, and compute the characteristic polynomial as follows:

$$
\det(D + \rho u u^T - \lambda I) = \det((D - \lambda I)(I + \rho(D - \lambda I)^{-1} u u^T)). \tag{5.13}
$$

Since $D - \lambda I$ is nonsingular, $\det(I + \rho(D - \lambda I)^{-1} u u^T) = 0$ whenever λ is an eigenvalue. Note that $I + \rho(D - \lambda I)^{-1} u u^T$ is the identity plus a rank-1 matrix; the determinant of such a matrix is easy to compute:

LEMMA 5.1. *If x and y are vectors, $\det(I + x y^T) = 1 + y^T x$.*

The proof is left to Question 5.14.
Therefore

$$
\begin{aligned}
\det(I + \rho(D - \lambda I)^{-1} u u^T) &= 1 + \rho u^T (D - \lambda I)^{-1} u \tag{5.14} \\
&= 1 + \rho \sum_{i=1}^n \frac{u_i^2}{d_i - \lambda} \equiv f(\lambda),
\end{aligned}
$$

and the eigenvalues of T are the roots of the so-called *secular equation* $f(\lambda) = 0$. If all d_i are distinct and all $u_i \neq 0$ (the generic case), the function $f(\lambda)$ has the graph shown in Figure 5.2 (for $n = 4$ and $\rho > 0$).

As we can see, the line $y = 1$ is a horizontal asymptote, and the lines $\lambda = d_i$ are vertical asymptotes. Since $f'(\lambda) = \rho \sum_{i=1}^n \frac{u_i^2}{(d_i - \lambda)^2} > 0$, the function is strictly increasing except at $\lambda = d_i$. Thus the roots of $f(\lambda)$ are interlaced by the d_i, and there is one more root to the right of d_1 ($d_1 = 4$ in Figure 5.2). (If $\rho < 0$, then $f(\lambda)$ is decreasing and there is one more root to the left of d_n.)

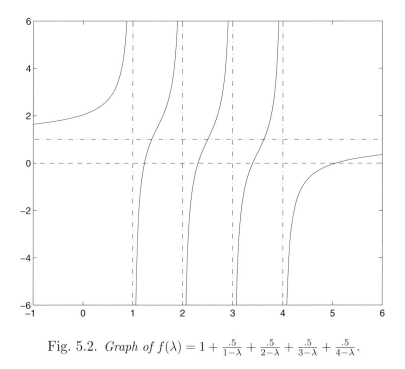

Fig. 5.2. *Graph of* $f(\lambda) = 1 + \frac{.5}{1-\lambda} + \frac{.5}{2-\lambda} + \frac{.5}{3-\lambda} + \frac{.5}{4-\lambda}$.

Since $f(\lambda)$ is monotonic and smooth on the intervals (d_i, d_{i+1}), it is possible to find a version of Newton's method that converges fast and monotonically to each root, given a starting point in (d_i, d_{i+1}). We discuss details later in this section. All we need to know here is that in practice Newton converges in a bounded number of steps per eigenvalue. Since evaluating $f(\lambda)$ and $f'(\lambda)$ costs $O(n)$ flops, finding one eigenvalue costs $O(n)$ flops, and so finding all n eigenvalues of $D + \rho u u^T$ costs $O(n^2)$ flops.

It is also easy to derive an expression for the eigenvectors of $D + u u^T$.

LEMMA 5.2. *If α is an eigenvalue of $D + \rho u u^T$, then $(D - \alpha I)^{-1} u$ is its eigenvector. Since $D - \alpha I$ is diagonal, this costs $O(n)$ flops to compute.*

Proof.

$$
\begin{aligned}
(D + \rho u u^T)[(D - \alpha I)^{-1} u] &= (D - \alpha I + \alpha I + \rho u u^T)(D - \alpha I)^{-1} u \\
&= u + \alpha (D - \alpha I)^{-1} u + u[\rho u^T (D - \alpha I)^{-1} u] \\
&= u + \alpha (D - \alpha I)^{-1} u - u \\
&\qquad \text{since } \rho u^T (D - \alpha I)^{-1} u + 1 = f(\alpha) = 0 \\
&= \alpha[(D - \alpha I)^{-1} u] \quad \text{as desired.} \quad \square
\end{aligned}
$$

Evaluating this formula for all n eigenvectors costs $O(n^2)$ flops. Unfortunately, this simple formula for the eigenvectors is not numerically stable, because two very close values of α_i can result in nonorthogonal computed eigenvectors u_i. Finding a stable alternative took over a decade from the original formulation of this algorithm. We discuss details later in this section.

The overall algorithm is recursive.

ALGORITHM 5.2. *Finding eigenvalues and eigenvectors of a symmetric tridiagonal matrix using divide-and-conquer:*

*proc dc_eig (T, Q, Λ) ..... from input T compute
    outputs Q and Λ where $T = Q\Lambda Q^T$*

if T is 1-by-1
        return $Q = 1, \Lambda = T$
else
        form $T = \begin{bmatrix} T_1 & 0 \\ 0 & T_2 \end{bmatrix} + b_m v v^T$
        call dc_eig (T_1, Q_1, Λ_1)
        call dc_eig (T_2, Q_2, Λ_2)
        form $D + \rho u u^T$ from $\Lambda_1, \Lambda_2, Q_1, Q_2$
        find eigenvalues Λ and eigenvectors Q' of $D + \rho u u^T$
        form $Q = \begin{bmatrix} Q_1 & 0 \\ 0 & Q_2 \end{bmatrix} \cdot Q'$ = eigenvectors of T
        return Q and Λ
endif

We analyze the complexity of Algorithm 5.2 as follows. Let $t(n)$ be the number of flops to run dc_eig on an n-by-n matrix. Then

$$
\begin{aligned}
t(n) \;=\; & 2t(n/2) && \text{for the 2 recursive calls to dc\_eig}(T_i, Q_i, \Lambda_i) \\
& +O(n^2) && \text{to find the eigenvalues of } D + \rho u u^T \\
& +O(n^2) && \text{to find the eigenvectors of } D + \rho u u^T \\
& +c \cdot n^3 && \text{to multiply } Q = \begin{bmatrix} Q_1 & 0 \\ 0 & Q_2 \end{bmatrix} \cdot Q'.
\end{aligned}
$$

If we treat Q_1, Q_2, and Q' as dense matrices and use the standard matrix multiplication algorithm, the constant in the last line is $c = 1$. Thus we see that the major cost in the algorithm is the matrix multiplication in the last line. Ignoring the $O(n^2)$ terms, we get $t(n) = 2t(n/2) + cn^3$. This geometric sum can be evaluated, yielding $t(n) \approx c\frac{4}{3}n^3$ (see Question 5.15). In practice, c is usually much less than 1, because a phenomenon called *deflation* makes Q' quite sparse.

After discussing deflation in the next section, we discuss details of solving the secular equation, and computing the eigenvectors stably. Finally, we discuss how to accelerate the method by exploiting FMM techniques used in electrostatic particle simulation [124]. These sections may be skipped on a first reading.

Deflation

So far in our presentation we have assumed that the d_i are distinct, and the u_i nonzero. When this is not the case, the secular equation $f(\lambda) = 0$ will have $k < n$ vertical asymptotes, and so $k < n$ roots. But it turns out that the remaining $n - k$ eigenvalues are available very cheaply: If $d_i = d_{i+1}$, or if $u_i = 0$, one can easily show that d_i is also an eigenvalue of $D + \rho u u^T$ (see Question 5.16). This process is called *deflation*. In practice we use a threshold and deflate d_i either if it is close enough to d_{i+1} or if u_i is small enough.

In practice, deflation happens quite frequently: In experiments with random dense matrices with uniformly distributed eigenvalues, over 15% of the eigenvalues of the largest $D + \rho u u^T$ deflated, and in experiments with random dense matrices with eigenvalues approaching 0 geometrically, over 85% deflated! It is essential to take advantage of this behavior to make the algorithm fast [59, 210].

The payoff in deflation is not in making the solution of the secular equation faster; this costs only $O(n^2)$ anyway. The payoff is in making the matrix multiplication in the last step of the algorithm fast. For if $u_i = 0$, then the corresponding eigenvector is e_i, the ith column of the identity matrix (see Question 5.16). This means that the ith column of Q' is e_i, so no work is needed to compute the ith column of Q in the two multiplications by Q_1 and Q_2. There is a similar simplification when $d_i = d_{i+1}$. When many eigenvalues deflate, much of the work in the matrix multiplication can be eliminated. This is borne out in the numerical experiments presented in section 5.3.6.

Solving the Secular Equation

When some u_i is small but too large to deflate, a problem arises when trying to use Newton's method to solve the secular equation. Recall that the principle of Newton's method for updating an approximate solution λ_j of $f(\lambda) = 0$ is

1. to approximate the function $f(\lambda)$ near $\lambda = \lambda_j$ with a linear function $l(\lambda)$, whose graph is a straight line tangent to the graph of $f(\lambda)$ at $\lambda = \lambda_j$,

2. to let λ_{j+1} be the zero of this linear approximation: $l(\lambda_{j+1}) = 0$.

The graph in Figure 5.2 offers no apparent difficulties to Newton's method, because the function $f(\lambda)$ appears to be reasonably well approximated by straight lines near each zero. But now consider the graph in Figure 5.3, which differs from Figure 5.2 only by changing u_i^2 from .5 to .001, which is not nearly small enough to deflate. The graph of $f(\lambda)$ in the left-hand figure is visually indistinguishable from its vertical and horizontal asymptotes, so in the right-hand figure we blow it up around one of the vertical asymptotes, $\lambda = 2$. We see that the graph of $f(\lambda)$ "turns the corner" very rapidly and is nearly horizontal for most values of λ. Thus, if we started Newton's method from almost any λ_0, the linear approximation $l(\lambda)$ would also be nearly horizontal

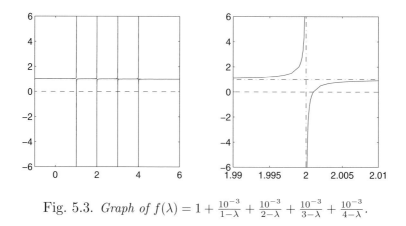

Fig. 5.3. *Graph of* $f(\lambda) = 1 + \frac{10^{-3}}{1-\lambda} + \frac{10^{-3}}{2-\lambda} + \frac{10^{-3}}{3-\lambda} + \frac{10^{-3}}{4-\lambda}$.

with a slightly positive slope, so λ_1 would be an enormous negative number, a useless approximation to the true zero.

Newton's method can be modified to deal with this situation as follows. Since $f(\lambda)$ is not well approximated by a straight line $l(x)$, we approximate it by another simple function $h(x)$. There is nothing special about straight lines; any approximation $h(\lambda)$ that is both easy to compute and has zeros that are easy to compute can be used in place of $l(x)$ in Newton's method. Since $f(\lambda)$ has poles at d_i and d_{i+1} and these poles dominate the behavior of $f(\lambda)$ near them, it is natural when seeking the root in (d_i, d_{i+1}) to choose $h(\lambda)$ to have these poles as well, i.e.,

$$h(\lambda) = \frac{c_1}{d_i - \lambda} + \frac{c_2}{d_{i+1} - \lambda} + c_3.$$

There are several ways to choose the constants c_1, c_2, and c_3 so that $h(\lambda)$ approximates $f(\lambda)$; we present a slightly simplified version of the one used in the LAPACK routine `slaed4` [172, 45]. Assuming for a moment that we have chosen c_1, c_2, and c_3, we can easily solve $h(\lambda) = 0$ for λ by solving the equivalent quadratic equation

$$c_1(d_{i+1} - \lambda) + c_2(d_i - \lambda) + c_3(d_i - \lambda)(d_{i+1} - \lambda) = 0.$$

Given the approximate zero λ_j, here is how we compute c_1, c_2, and c_3 so that for λ near λ_j

$$\frac{c_1}{d_i - \lambda} + \frac{c_2}{d_{i+1} - \lambda} + c_3 = h(\lambda) \approx f(\lambda) = 1 + \rho \sum_{k=1}^{n} \frac{u_k^2}{d_k - \lambda}.$$

Write

$$f(\lambda) = 1 + \sum_{k=1}^{i} \frac{u_k^2}{d_k - \lambda} + \sum_{k=i+1}^{n} \frac{u_k^2}{d_k - \lambda} \equiv 1 + \psi_1(\lambda) + \psi_2(\lambda).$$

For $\lambda \in (d_i, d_{i+1})$, $\psi_1(\lambda)$ is a sum of positive terms and $\psi_2(\lambda)$ is a sum of negative terms. Thus both $\psi_1(\lambda)$ and $\psi_2(\lambda)$ can be computed accurately, whereas adding them together would likely result in cancellation and loss of relative accuracy in the sum. We now choose c_1 and $\hat{c}_1$ so that

$$h_1(\lambda) \equiv \hat{c}_1 + \frac{c_1}{d_i - \lambda} \quad \text{satisfies}$$
$$h_1(\lambda_j) = \psi_1(\lambda_j) \quad \text{and} \quad h_1'(\lambda_j) = \psi_1'(\lambda_j). \tag{5.15}$$

This means that the graph of $h_1(\lambda)$ (a hyperbola) is tangent to the graph of $\psi_1(\lambda)$ at $\lambda = \lambda_j$. The two conditions in equation (5.15) are the usual conditions in Newton's method, except instead of using a straight line approximation, we use a hyperbola. It is easy to verify that $c_1 = \psi_1'(\lambda_j)(d_i - \lambda_j)^2$ and $\hat{c}_1 = \psi_1(\lambda_j) - \psi_1'(\lambda_j)(d_i - \lambda_j)$. (See Question 5.17.)

Similarly, we choose c_2 and $\hat{c}_2$ so that

$$h_2(\lambda) \equiv \hat{c}_2 + \frac{c_2}{d_{i+1} - \lambda} \quad \text{satisfies}$$
$$h_2(\lambda_j) = \psi_2(\lambda_j) \quad \text{and} \quad h_2'(\lambda_j) = \psi_2'(\lambda_j). \tag{5.16}$$

Finally, we set

$$\begin{aligned}
h(\lambda) &= 1 + h_1(\lambda) + h_2(\lambda) \\
&= (1 + \hat{c}_1 + \hat{c}_2) + \frac{c_1}{d_i - \lambda} + \frac{c_2}{d_{i+1} - \lambda} \\
&\equiv c_3 + \frac{c_1}{d_i - \lambda} + \frac{c_2}{d_{i+1} - \lambda}.
\end{aligned}$$

EXAMPLE 5.8. For example, in the example in Figure 5.3, if we start with $\lambda_0 = 2.5$, then

$$h(\lambda) = \frac{1.1111 \cdot 10^{-3}}{2 - \lambda} + \frac{1.1111 \cdot 10^{-3}}{3 - \lambda} + 1,$$

and its graph is visually indistinguishable from the graph of $f(\lambda)$ in the right-hand figure. Solving $h(\lambda_1) = 0$, we get $\lambda_1 = 2.0011$, which is accurate to 4 decimal digits. Continuing, λ_2 is accurate to 11 digits, and λ_3 is accurate to all 16 digits. $\diamond$

The algorithm used in LAPACK routine `slaed4` is a slight variation on the one described here (the one here is called *the Middle Way* in [172]). The LAPACK routine averages two or three iterations per eigenvalue to converge to full machine precision, and never took more than seven steps in extensive numerical tests.

Computing the Eigenvectors Stably

Once we have solved the secular equation to get the eigenvalues α_i of $D+\rho u u^T$, Lemma 5.2 provides a simple formula for the eigenvectors: $(D - \alpha_i I)^{-1}u$. Unfortunately, the formula can be unstable [59, 90, 234], in particular when two eigenvalues α_i and α_{i+1} are very close together. Intuitively, the problem is that $(D - \alpha_i I)^{-1}u$ and $(D - \alpha_{i+1}I)^{-1}u$ are "very close" formulas yet are supposed to yield orthogonal eigenvectors. More precisely, when α_i and α_{i+1} are very close, they must also be close to the d_i between them. Therefore, there is a great deal of cancellation, either when evaluating $d_i - \alpha_i$ and $d_i - \alpha_{i+1}$ or when evaluating the secular equation during Newton iteration. Either way, $d_i - \alpha_i$ and $d_i - \alpha_{i+1}$ may contain large relative errors, so the computed eigenvectors $(D - \alpha_i)^{-1}u$ and $(D - \alpha_{i+1})^{-1}u$ are quite inaccurate and far from orthogonal.

Early attempts to address this problem [90, 234] used double precision arithmetic (when the input data was single precision) to solve the secular equation to high accuracy so that $d_i - \alpha_i$ and $d_i - \alpha_{i+1}$ could be computed to high accuracy. But when the input data are already in double precision, this means quadruple precision would be needed, and this is not available in many machines and languages, or at least not cheaply. As described in section 1.5, it is possible to simulate quadruple precision using double precision [234, 204]. This can be done portably and relatively efficiently, as long as the underlying floating point arithmetic rounds sufficiently accurately. In particular, these simulations require that $\text{fl}(a \pm b) = (a \pm b)(1 + \delta)$ with $|\delta| = O(\varepsilon)$, barring overflow or underflow (see section 1.5 and Question 1.18). Unfortunately, the Cray 2, YMP, and C90 do not round accurately enough to use these efficient algorithms.

Finally, an alternative formula was found that makes simulating high precision arithmetic unnecessary. It is based on the following theorem of Löwner [129, 179].

THEOREM 5.10. *Löwner. Let* $D = \text{diag}(d_1, \ldots, d_n)$ *be diagonal with* $d_n < \cdots < d_1$. *Let* $\alpha_n < \cdots < \alpha_1$ *be given, satisfying the interlacing property*

$$d_n < \alpha_n < \cdots < d_{i+1} < \alpha_{i+1} < d_i < \alpha_i < \cdots < d_1 < \alpha_1.$$

Then there is a vector $\hat{u}$ *such that the* α_i *are the exact eigenvalues of* $\hat{D} \equiv D + \hat{u}\hat{u}^T$. *The entries of* $\hat{u}$ *are given by*

$$|\hat{u}_i| = \left[\frac{\prod_{j=1}^{n}(\alpha_j - d_i)}{\prod_{j=1, \, j\neq i}^{n}(d_j - d_i)} \right]^{1/2}.$$

Proof. The characteristic polynomial of $\hat{D}$ can be written both as $\det(\hat{D} - $

$\lambda I) = \prod_{j=1}^{n}(\alpha_j - \lambda)$ and (using equations (5.13) and (5.14)) as

$$
\begin{aligned}
\det(\hat{D} - \lambda I) &= \left[\prod_{j=1}^{n}(d_j - \lambda)\right] \cdot \left(1 + \sum_{j=1}^{n} \frac{\hat{u}_j^2}{d_j - \lambda}\right) \\
&= \left[\prod_{j=1}^{n}(d_j - \lambda)\right] \cdot \left(1 + \sum_{\substack{j=1 \\ j \neq i}}^{n} \frac{\hat{u}_j^2}{d_j - \lambda}\right) \\
&\quad + \left[\prod_{\substack{j=1 \\ j \neq i}}^{n}(d_j - \lambda)\right] \cdot \hat{u}_i^2.
\end{aligned}
$$

Setting $\lambda = d_i$ and equating both expressions for $\det(\hat{D} - \lambda I)$ yield

$$
\prod_{j=1}^{n}(\alpha_j - d_i) = \hat{u}_i^2 \cdot \prod_{\substack{j=1 \\ j \neq i}}^{n}(d_j - d_i)
$$

or

$$
\hat{u}_i^2 = \frac{\prod_{j=1}^{n}(\alpha_j - d_i)}{\prod_{j=1, \, j \neq i}^{n}(d_j - d_i)}.
$$

Using the interlacing property, we can show that the fraction on the right is positive, so we can take its square root to get the desired expression for $\hat{u}_i$. $\square$

Here is the stable algorithm for computing the eigenvalues and eigenvectors (where we assume for simplicity of presentation that $\rho = 1$).

ALGORITHM 5.3. *Compute the eigenvalues and eigenvectors of $D + uu^T$.*

> *Solve the secular equation $1 + \sum_{i=1}^{n} \frac{u_i^2}{d_i - \lambda} = 0$ to get the eigenvalues*
> *α_i of $D + uu^T$.*
> *Use Löwner's theorem to compute $\hat{u}$ so that the α_i are "exact"*
> *eigenvalues of $D + \hat{u}\hat{u}^T$.*
> *Use Lemma 5.2 to compute the eigenvectors of $D + \hat{u}\hat{u}^T$.*

Here is a sketch of why this algorithm is numerically stable. By analyzing the stopping criterion in the secular equation solver,[19] one can show that $\|uu^T - \hat{u}\hat{u}^T\|_2 \leq O(\varepsilon)(\|D\|_2 + \|uu^T\|_2)$; this means that $D + uu^T$ and $D + \hat{u}\hat{u}^T$ are so close together that the eigenvalues and eigenvectors of $D + \hat{u}\hat{u}^T$ are stable approximations of the eigenvalues and eigenvectors of $D + uu^T$. Next

---

[19]In more detail, the secular equation solver must solve for $\alpha_i - d_i$ or $d_{i+1} - \alpha_i$ (whichever is smaller), *not* α_i, to attain this accuracy.

note that the formula for $\hat{u}_i$ in Löwner's theorem requires only differences of floating point numbers $d_j - d_i$ and $\alpha_j - d_i$, products and quotients of these differences, and a square root. *Provided that the floating point arithmetic is accurate enough* that $\mathrm{fl}(a \odot b) = (a \odot b)(1 + \delta)$ for all $\odot \in \{+, -, \times, /\}$ and $\mathrm{sqrt}(a) = \sqrt{a} \cdot (1 + \delta)$ with $|\delta| = O(\varepsilon)$, this formula can be evaluated to high relative accuracy. In particular, we can easily show that

$$\mathrm{fl}\left(\left[\frac{\prod_{j=1}^{n}(\alpha_j - d_i)}{\prod_{j=1,\, j \neq i}^{n}(d_j - d_i)}\right]^{1/2}\right) = (1 + (4n - 2)\delta) \cdot \left[\frac{\prod_{j=1}^{n}(\alpha_j - d_i)}{\prod_{j=1,\, j \neq i}^{n}(d_j - d_i)}\right]^{1/2}$$

with $|\delta| = O(\varepsilon)$, barring overflow or underflow. Similarly, the formula in Lemma 5.2 can also be evaluated to high relative accuracy, so we can compute the eigenvectors of $D + \hat{u}\hat{u}^T$ to high relative accuracy. In particular, they are very accurately orthogonal.

In summary, provided the floating point arithmetic is accurate enough, Algorithm 5.3 computes very accurate eigenvalues and eigenvectors of a matrix $D + \hat{u}\hat{u}^T$ that differs only slightly from the original matrix $D + uu^T$. This means that it is numerically stable.

The reader should note that our need for sufficiently accurate floating point arithmetic is precisely what prevented the simulation of quadruple precision proposed in [234, 204] from working on some Cray machines. So we have not yet succeeded in providing an algorithm that works reliably on these machines. One more trick is necessary: The only operations that fail to be accurate enough on some Cray machines are addition and subtraction, because of the lack of a so-called *guard digit* in the floating point hardware. This means that the bottom-most bit of an operand may be treated as 0 during addition or subtraction, even if it is 1. If most higher-order bits cancel, this "lost bit" becomes significant. For example, subtracting 1 from the next smaller floating point number, in which case all leading bits cancel, results in a number twice too large on the Cray C90 and in 0 on the Cray 2. But if the bottom bit is already 0, no harm is done. So the trick is to deliberately set all the bottom bits of the d_i to 0 before applying Löwner's theorem or Lemma 5.2 in Algorithm 5.3. This modification causes only a small relative change in the d_i and α_i, and so the algorithm is still stable.[20]

This algorithm is described in more detail in [129, 131] and implemented in LAPACK routine `slaed3`.

---

[20]To set the bottom bit of a floating point number β to 0 on a Cray, one can show that it is necessary only to set $\beta := (\beta + \beta) - \beta$. This inexpensive computation does not change β at all on a machine with accurate binary arithmetic (barring overflow, which is easily avoided). But on a Cray machine it sets the bottom bit to 0. The reader familiar with Cray arithmetic is invited to prove this. The only remaining difficulty is preventing an optimizing compiler from removing this line of code entirely, which some overzealous optimizers might do; this is accomplished (for the current generation of compilers) by computing $(\beta + \beta)$ with a function call to a function stored in a separate file from the main routine. We hope that by the time compilers become clever enough to optimize even this situation, Cray arithmetic will have died out.

Accelerating Divide-and-Conquer Using the FMM

The FMM [124] was originally invented for the completely different problem of computing the mutual forces on n electrically charged particles or the mutual gravitational forces on n masses. We only sketch how these problems are related to finding eigenvalues and eigenvectors, leaving details to [131].

Let d_1 through d_n be the three-dimensional position vectors of n particles with charges $z_i \cdot u_i$. Let α_1 through α_n be the position vectors of n other particles with unit positive charges. Then the inverse-square law tells us that the force on the particle at α_j due to the particles at d_1 through d_n is proportional to

$$f_j = \sum_{i=1}^{n} \frac{z_i u_i (d_i - \alpha_j)}{\|d_i - \alpha_j\|_2^3}.$$

If we are modeling electrostatics in two dimensions instead of three, the force law changes to the inverse-first-power law[21]

$$f_j = \sum_{i=1}^{n} \frac{z_i u_i (d_i - \alpha_j)}{\|d_i - \alpha_j\|_2^2}.$$

Since d_i and α_j are vectors in $\mathbb{R}^2$, we can also consider them to be complex variables. In this case

$$f_j = \sum_{i=1}^{n} \frac{z_i u_i}{\bar{d}_i - \bar{\alpha}_j},$$

where $\bar{d}_i$ and $\bar{\alpha}_j$ are the complex conjugates of d_i and α_j, respectively. If d_i and α_j happen to be real numbers, this simplifies further to

$$f_j = \sum_{i=1}^{n} \frac{z_i u_i}{d_i - \alpha_j}.$$

Now consider performing a matrix-vector multiplication $f^T = z^T Q'$, where Q' is the eigenvector matrix of $D + uu^T$. From Lemma 5.2, $Q'_{ij} = u_i s_j / (d_i - \alpha_j)$, where s_j is a scale factor chosen so that column j is a unit vector. Then the jth entry of $f^T = z^T Q'$ is

$$f_j = \sum_{i=1}^{n} z_i Q'_{ij} = s_j \sum_{i=1}^{n} \frac{z_i u_i}{d_i - \alpha_j},$$

which is the same sum as for the electrostatic force, except for the scale factor s_j. Thus, the most expensive part of the divide-and-conquer algorithm, the matrix multiplication in the last line of Algorithm 5.2, is equivalent to evaluating electrostatic forces.

---

[21]Technically, this means the potential function satisfies Poisson's equation in two space coordinates rather than three.

Evaluating this sum for $j = 1, \ldots, n$ appears to require $O(n^2)$ flops. The FMM and others like it [124, 23] can be used to approximately (but very accurately) evaluate this sum in $O(n \cdot \log n)$ time (or even $O(n)$) time instead. (See the lectures on "Fast Hierarchical Methods for the N-body Problem" at PARALLEL_HOMEPAGE for details.)

But this idea alone is not enough to reduce the cost of divide-and-conquer to $O(n \cdot \log^p n)$. After all, the output eigenvector matrix Q has n^2 entries, which appears to mean that the complexity should be at least n^2. So we must represent Q using fewer than n^2 independent numbers. This is possible, because an n-by-n tridiagonal matrix has only $2n - 1$ "degrees of freedom" (the diagonal and superdiagonal entries), of which n can be represented by the eigenvalues, leaving $n-1$ for the orthogonal matrix Q. In other words, not every orthogonal matrix can be the eigenvector matrix of a symmetric tridiagonal T; only an $(n - 1)$-dimensional subset of the entire $(n(n - 1)/2)$-dimensional set of orthogonal matrices can be such eigenvector matrices.

We will represent Q using the divide-and-conquer tree computed by Algorithm 5.2. Rather than accumulating $Q = \begin{bmatrix} Q_1 & 0 \\ 0 & Q_2 \end{bmatrix} \cdot Q'$, we will store all the Q' matrices, one at each node in the tree. And we will not store Q' explicitly but rather just store D, ρ, u, and the eigenvalues α_i of $D + \rho u u^T$. We can do this since this is all we need to use the FMM to multiply by Q'. This reduces the storage needed for Q from n^2 to $O(n \cdot \log n)$. Thus, the output of the algorithm is a "factored" form of Q consising of all the Q' factors at the nodes of the tree. This is an adequate representation of Q, because we can use the FMM to multiply any vector by Q in $O(n \cdot \log^p n)$ time.

5.3.4. Bisection and Inverse Iteration

The Bisection algorithm exploits Sylvester's inertia theorem (Theorem 5.3) to find only those k eigenvalues that one wants, at cost $O(nk)$. Recall that Inertia$(A) = (\nu, \zeta, \pi)$, where ν, ζ, and π are the number of negative, zero, and positive eigenvalues of A, respectively. Suppose that X is nonsingular; Sylvester's inertia theorem asserts that Inertia$(A) = $ Inertia$(X^T A X)$.

Now suppose that one uses Gaussian elimination to factorize $A - zI = LDL^T$, where L is nonsingular and D diagonal. Then Inertia$(A - zI) = $ Inertia(D). Since D is diagonal, its inertia is trivial to compute. (In what follows, we use notation such as "# $d_{ii} < 0$" to mean "the number of values of d_{ii} that are less than zero.")

$$
\begin{aligned}
\text{Inertia}(A - zI) \quad &= \quad (\#\ d_{ii} < 0,\ \#\ d_{ii} = 0,\ \#\ d_{ii} > 0) \\
&= \quad (\#\ \text{negative eigenvalues of } A - zI, \\
&\qquad \#\ \text{zero eigenvalues of } A - zI, \\
&\qquad \#\ \text{positive eigenvalues of } A - zI) \\
&= \quad (\#\ \text{eigenvalues of } A < z, \\
&\qquad \#\ \text{eigenvalues of } A = z, \\
&\qquad \#\ \text{eigenvalues of } A > z).
\end{aligned}
$$

Suppose $z_1 < z_2$ and we compute Inertia $(A - z_1 I)$ and Inertia $(A - z_2 I)$. Then the number of eigenvalues in the interval $[z_1, z_2)$ equals (# eigenvalues of $A < z_2$) – (# eigenvalues of $A < z_1$).

To make this observation into an algorithm, define

$$\text{Negcount}(A, z) = \text{# eigenvalues of } A < z.$$

ALGORITHM 5.4. *Bisection: Find all eigenvalues of A inside $[a, b)$ to a given error tolerance* tol:

> $n_a = \text{Negcount}(A, a)$
> $n_b = \text{Negcount}(A, b)$
> *if* $n_a = n_b$, *quit* ... *because there are no eigenvalues in* $[a, b)$
> *put* $[a, n_a, b, n_b]$ *onto Worklist*
>   /* *Worklist contains a list of intervals* $[a, b)$ *containing*
>   *eigenvalues* $n - n_a$ *through* $n - n_b + 1$, *which the algorithm*
>   *will repeatedly bisect until they are narrower than* tol. */
> *while Worklist is not empty do*
>   *remove* $[low, n_{\text{low}}, up, n_{\text{up}}]$ *from Worklist*
>   *if* $(up - low < $ tol$)$ *then*
>   *print* "*there are* $n_{\text{up}} - n_{\text{low}}$ *eigenvalues in* $[low, up)$"
>   *else*
> $mid = (low + up)/2$
> $n_{\text{mid}} = \text{Negcount }(A, mid)$
>   *if* $n_{\text{mid}} > n_{\text{low}}$ *then* ... *there are eigenvalues in* $[low, mid)$
>   *put* $[low, n_{\text{low}}, mid, n_{\text{mid}}]$ *onto Worklist*
>   *end if*
>   *if* $n_{\text{up}} > n_{\text{mid}}$ *then* ... *there are eigenvalues in* $[mid, up)$
>   *put* $[mid, n_{\text{mid}}, up, n_{\text{up}}]$ *onto Worklist*
>   *end if*
>   *end if*
> *end while*

If $\alpha_1 \geq \cdots \geq \alpha_n$ are eigenvalues, the same idea can be used to compute α_j for $j = j_0, j_0 + 1, \ldots, j_1$. This is because we know $\alpha_{n - n_{\text{low}}}$ through $\alpha_{n - n_{\text{up}} + 1}$ lie in the interval $[low, up)$.

If A were dense, we could implement Negcount(A, z) by doing symmetric Gaussian elimination with pivoting as described in section 2.7.2. But this would cost $O(n^3)$ flops per evaluation and so not be cost effective. On the other hand, Negcount(A, z) is quite simple to compute for tridiagonal A, provided that we do not pivot:

$$A - zI \equiv \begin{bmatrix} a_1 - z & b_1 & & & \\ b_1 & a_2 - z & \ddots & & \\ & \ddots & \ddots & b_{n-1} \\ & & b_{n-1} & a_n \end{bmatrix} = LDL^T$$

$$
\equiv
\begin{bmatrix}
1 & & & \\
l_1 & \ddots & & \\
& \ddots & \ddots & \\
& & l_{n-1} & 1
\end{bmatrix}
\cdot
\begin{bmatrix}
d_1 & & & \\
& \ddots & & \\
& & \ddots & \\
& & & d_n
\end{bmatrix}
\cdot
\begin{bmatrix}
1 & l_1 & & \\
& \ddots & \ddots & \\
& & \ddots & l_{n-1} \\
& & & 1
\end{bmatrix},
$$

so $a_1 - z = d_1, d_1 l_1 = b_1$ and thereafter $l_{i-1}^2 d_{i-1} + d_i = a_i - z$, $d_i l_i = b_i$. Substituting $l_i = b_i/d_i$ into $l_{i-1}^2 d_{i-1} + d_i = a_i - z$ yields the simple recurrence

$$
d_i = (a_i - z) - \frac{b_{i-1}^2}{d_{i-1}}. \tag{5.17}
$$

Notice that we are not pivoting, so you might think that this is dangerously unstable, especially when d_{i-1} is small. In fact, since $A - zI$ is tridiagonal, (5.17) can be shown to be very stable [73, 74, 156].

LEMMA 5.3. *The d_i computed in floating point arithmetic, using (5.17), have the same signs (and so compute the same Inertia) as the $\hat{d}_i$ computed exactly from $\hat{A}$, where $\hat{A}$ is very close to A:*

$$
(\hat{A})_{ii} \equiv \hat{a}_i = a_i \quad and \quad (\hat{A})_{i,i+1} \equiv \hat{b}_i = b_i(1 + \epsilon_i), \quad where \ |\epsilon_i| \leq 2.5\varepsilon + O(\varepsilon^2).
$$

Proof. Let $\tilde{d}_i$ denote the quantities computed using equation (5.17) including rounding errors:

$$
\tilde{d}_i = \left[(a_i - z)(1 + \epsilon_{-,1,i}) - \frac{b_{i-1}^2(1 + \epsilon_{*,i})}{\tilde{d}_{i-1}} \cdot (1 + \epsilon_{/,i}) \right] (1 + \epsilon_{-,2,i}), \tag{5.18}
$$

where all the ϵ's are bounded by machine roundoff ε in magnitude, and their subscripts indicate which floating point operation they come from (for example, $\epsilon_{-,2,i}$ is from the second subtraction when computing $\tilde{d}_i$). Define the new variables

$$
\hat{d}_i = \frac{\tilde{d}_i}{(1 + \epsilon_{-,1,i})(1 + \epsilon_{-,2,i})}, \tag{5.19}
$$

$$
\hat{b}_{i-1} = b_{i-1} \left[\frac{(1 + \epsilon_{*,i})(1 + \epsilon_{/,i})}{(1 + \epsilon_{-,1,i})(1 + \epsilon_{-,1,i-1})(1 + \epsilon_{-,2,i-1})} \right]^{1/2} \equiv b_{i-1}(1 + \epsilon_i).
$$

Note that $\hat{d}_i$ and $\tilde{d}_i$ have the same signs, and $|\epsilon_i| \leq 2.5\varepsilon + O(\varepsilon^2)$. Substituting (5.19) into (5.18) yields

$$
\hat{d}_i = a_i - z - \frac{\hat{b}_{i-1}^2}{\hat{d}_{i-1}},
$$

completing the proof. $\square$

A complete analysis must take the possibility of overflow or underflow into account. Indeed, using the exception handling facilities of IEEE arithmetic, one can safely compute even when some d_{i-1} is exactly zero! For in this case

$d_i = -\infty$, $d_{i+1} = a_{i+1} - z$, and the computation continues unexceptionally [73, 81].

The cost of a single call to Negcount on a tridiagonal matrix is at most $4n$ flops. Therefore the overall cost to find k eigenvalues is $O(kn)$. This is implemented in LAPACK routine sstebz.

Note that Bisection converges linearly, with one more bit of accuracy for each bisection of an interval. There are many ways to accelerate convergence, using algorithms like Newton's method and its relatives, to find zeros of the characteristic polynomial (which may be computed by multiplying all the d_i's together) [173, 174, 175, 176, 178, 269].

To compute eigenvectors once we have computed (selected) eigenvalues, we can use inverse iteration (Algorithm 4.2); this is available in LAPACK routine sstein. Since we can use accurate eigenvalues as shifts, convergence usually takes one or two iterations. In this case the cost is $O(n)$ flops per eigenvector, since one step of inverse iteration requires us only to solve a tridiagonal system of equations (see section 2.7.3). When several computed eigenvalues $\hat{\alpha}_i, \ldots, \hat{\alpha}_j$ are close together, their corresponding computed eigenvectors $\hat{q}_i, \ldots, \hat{q}_j$ may not be orthogonal. In this case the algorithm *reorthogonalizes* the computed eigenvectors, computing the QR decomposition $[\hat{q}_i, \ldots, \hat{q}_j] = QR$ and replacing each $\hat{q}_k$ with the kth column of Q; this guarantees that the $\hat{q}_k$ are orthonormal. This QR decomposition is usually computed using the MGS orthogonalization process (Algorithm 3.1); i.e., each computed eigenvector has any components in the directions of previously computed eigenvectors explicitly subtracted out. When the cluster size k is small, the cost $O(k^2 n)$ of this reorthogonalization is small, so in principle all the eigenvalues and all the eigenvectors could be computed by Bisection followed by inverse iteration in just $O(n^2)$ flops total. This is much faster than the $O(n^3)$ cost of QR iteration or divide-and-conquer (in the worst case). The obstacle to obtaining this speedup reliably is that if the cluster size k is large, i.e., a sizable fraction of n, then the total cost rises to $O(n^3)$ again. Worse, there is no guarantee that the computed eigenvectors are accurate or orthogonal. (The trouble is that after reorthogonalizing a set of nearly dependent $\hat{q}_k$, cancellation may mean some computed eigenvectors consist of little more than roundoff errors.)

There has been recent progress on this problem, however [105, 83, 201, 203], and it now appears possible that inverse iteration may be "repaired" to provide accurate, orthogonal eigenvectors without spending more than $O(n)$ flops per eigenvector. This would make Bisection and "repaired" inverse iteration the algorithm of choice in all cases, no matter how many eigenvalues and eigenvectors are desired. We look forward to describing this algorithm in a future edition.

Note that Bisection and inverse iteration are "embarrassingly parallel," since each eigenvalue and later eigenvector may be found independently of the others. (This presumes that inverse iteration has been repaired so that reorthogonalization with many other eigenvectors is no longer necessary.) This makes these algorithms very attractive for parallel computers [76].

5.3.5. Jacobi's Method

Jacobi's method does not start by reducing A to tridiagonal from as do the previous methods but instead works on the original dense matrix. Jacobi's method is usually much slower than the previous methods and remains of interest only because it can sometimes compute tiny eigenvalues and their eigenvectors with much higher accuracy than the previous methods and can be easily parallelized. Here we describe only the basic implementation of Jacobi's method and defer the discussion of high accuracy to section 5.4.3.

Given a symmetric matrix $A = A_0$, Jacobi's method produces a sequence $A_1, A_2, \ldots$ of orthogonally similar matrices, which eventually converge to a diagonal matrix with the eigenvalues on the diagonal. A_{i+1} is obtained from A_i by the formula $A_{i+1} = J_i^T A_i J_i$, where J_i is an orthogonal matrix called a *Jacobi rotation*. Thus

$$
\begin{aligned}
A_m &= J_{m-1}^T A_{m-1} J_{m-1} \\
&= J_{m-1}^T J_{m-2}^T A_{m-2} J_{m-2} J_{m-1} = \cdots \\
&= J_{m-1}^T \cdots J_0^T A_0 J_0 \cdots J_{m-1} \\
&\equiv J^T A J.
\end{aligned}
$$

If we choose each J_i appropriately, A_m approaches a diagonal matrix Λ for large m. Thus we can write $\Lambda \approx J^T A J$ or $J\Lambda J^T \approx A$. Therefore, the columns of J are approximate eigenvectors.

We will make $J^T A J$ nearly diagonal by iteratively choosing J_i to make *one* pair of offdiagonal entries of $A_{i+1} = J_i^T A_i J_i$ zero at a time. We will do this by choosing J_i to be a Givens rotation,

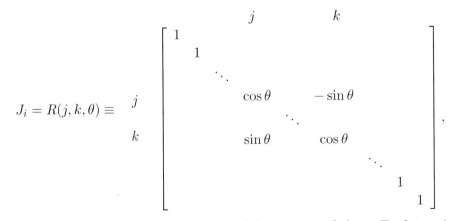

$$
J_i = R(j, k, \theta) \equiv
$$

where θ is chosen to zero out the j, k and k, j entries of A_{i+1}. To determine θ (or actually $\cos\theta$ and $\sin\theta$), write

$$
\begin{bmatrix} a_{jj}^{(i+1)} & a_{jk}^{(i+1)} \\ a_{kj}^{(i+1)} & a_{kk}^{(i+1)} \end{bmatrix} = \begin{bmatrix} \cos\theta & -\sin\theta \\ \sin\theta & \cos\theta \end{bmatrix}^T \begin{bmatrix} a_{jj}^{(i)} & a_{jk}^{(i)} \\ a_{kj}^{(i)} & a_{kk}^{(i)} \end{bmatrix} \begin{bmatrix} \cos\theta & -\sin\theta \\ \sin\theta & \cos\theta \end{bmatrix}
$$

$$= \begin{bmatrix} \lambda_1 & 0 \\ 0 & \lambda_2 \end{bmatrix},$$

where λ_1 and λ_2 are the eigenvalues of

$$\begin{bmatrix} a_{jj}^{(i)} & a_{jk}^{(i)} \\ a_{kj}^{(i)} & a_{kk}^{(i)} \end{bmatrix}.$$

It is easy to compute $\cos\theta$ and $\sin\theta$: Multiplying out the last expression, using symmetry, abbreviating $c \equiv \cos\theta$ and $s \equiv \sin\theta$, and dropping the superscript (i) for simplicity yield

$$\begin{bmatrix} \lambda_1 & 0 \\ 0 & \lambda_2 \end{bmatrix} = \begin{bmatrix} a_{jj}c^2 + a_{kk}s^2 + 2sca_{jk} & sc(a_{kk} - a_{jj}) + a_{jk}(c^2 - s^2) \\ sc(a_{kk} - a_{jj}) + a_{jk}(c^2 - s^2) & a_{jj}s^2 + a_{kk}c^2 - 2sca_{jk} \end{bmatrix}.$$

Setting the offdiagonals to 0 and solving for θ we get $0 = sc(a_{kk} - a_{jj}) + a_{jk}(c^2 - s^2)$, or

$$\frac{a_{jj} - a_{kk}}{2a_{jk}} = \frac{c^2 - s^2}{2sc} = \frac{\cos 2\theta}{\sin 2\theta} = \cot 2\theta \equiv \tau.$$

We now let $t = \frac{s}{c} = \tan\theta$ and note that $t^2 + 2\tau t - 1 = 0$ to get (via the quadratic formula) $t = \frac{\text{sign}(\tau)}{|\tau| + \sqrt{1+\tau^2}}$, $c = \frac{1}{\sqrt{1+t^2}}$ and $s = t \cdot c$. We summarize this derivation in the following algorithm.

ALGORITHM 5.5. *Compute and apply a Jacobi rotation to A in coordinates j, k:*

proc Jacobi-Rotation (A, j, k)
    if $|a_{jk}|$ is not too small
        $\tau = (a_{jj} - a_{kk})/(2 \cdot a_{jk})$
        $t = \text{sign}(\tau)/(|\tau| + \sqrt{1 + \tau^2})$
        $c = 1/\sqrt{1 + t^2}$
        $s = c \cdot t$
        $A = R^T(j, k, \theta) \cdot A \cdot R(j, k, \theta)$ *... where $c = \cos\theta$ and $s = \sin\theta$*
        if eigenvectors are desired
            $J = J \cdot R(j, k, \theta)$
        end if
    end if

The cost of applying $R(j, k, \theta)$ to A (or J) is only $O(n)$ flops, because only rows and columns j and k of A (and columns j and k of J) are modified. The overall Jacobi algorithm is then as follows.

ALGORITHM 5.6. *Jacobi's method to find the eigenvalues of a symmetric matrix:*

repeat
    choose a j, k pair
    call Jacobi-Rotation(A, j, k)
until A is sufficiently diagonal

We still need to decide how to pick j, k pairs. There are several possibilities. To measure progress to convergence and describe these possibilities, we define

$$\text{off}(A) \equiv \sqrt{\sum_{1 \leq j < k \leq n} a_{jk}^2}.$$

Thus $\text{off}(A)$ is the root-sum-of-squares of the (upper) offdiagonal entries of A, so A is diagonal if and only if $\text{off}(A) = 0$. Our goal is to make $\text{off}(A)$ approach 0 quickly. The next lemma tells us that $\text{off}(A)$ decreases monotonically with every Jacobi rotation.

LEMMA 5.4. *Let A' be the matrix after calling Jacobi-Rotation(A, j, k) for any $j \neq k$. Then $\text{off}^2(A') = \text{off}^2(A) - a_{jk}^2$.*

Proof. Note that $A' = A$ except in rows and columns j and k. Write

$$\text{off}^2(A) = \left(\sum_{\substack{1 \leq j' < k' \leq n \\ j' \neq j \text{ or } k' \neq k}} a_{j'k'}^2 \right) + a_{jk}^2 \equiv S^2 + a_{jk}^2$$

and similarly $\text{off}^2(A') = S'^2 + a_{jk}'^2 = S'^2$, since $a_{jk}' = 0$ after calling Jacobi-Rotation(A, j, k). Since $\|X\|_F = \|QX\|_F$ and $\|X\|_F = \|XQ\|_F$ for any X and any orthogonal Q, we can show $S^2 = S'^2$. Thus $\text{off}^2(A') = \text{off}^2(A) - a_{jk}^2$ as desired. $\square$

The next algorithm was the original version of the algorithm (from Jacobi in 1846), and it has an attractive analysis although it is too slow to use.

ALGORITHM 5.7. *Classical Jacobi's algorithm:*

    while $\text{off}(A) > \text{tol}$ (where tol is the stopping criterion set by user)
        choose j and k so a_{jk} is the largest offdiagonal entry in magnitude
        call Jacobi-Rotation (A, j, k)
    end while

THEOREM 5.11. *After one Jacobi rotation in the classical Jacobi's algorithm, we have $\text{off}(A') \leq \sqrt{1 - \frac{1}{N}} \, \text{off}(A)$, where $N = \frac{n(n-1)}{2} =$ the number of superdiagonal entries of A. After k Jacobi-Rotations $\text{off}(\cdot)$ is no more than $\left(1 - \frac{1}{N}\right)^{k/2} \text{off}(A)$.*

Proof. By Lemma 5.4, after one step, $\text{off}^2(A') = \text{off}^2(A) - a_{jk}^2$, where a_{jk} is the largest offdiagonal entry. Thus $\text{off}^2(A) \leq \frac{n(n-1)}{2} a_{jk}^2$, or $a_{jk}^2 \geq \frac{1}{n(n-1)/2} \text{off}^2(A)$, so $\text{off}^2(A) - a_{jk}^2 \leq \left(1 - \frac{1}{N}\right) \text{off}^2(A)$ as desired. $\square$

So the classical Jacobi's algorithm converges at least linearly with the error (measured by $\text{off}(A)$) decreasing by a factor of at least $\sqrt{1 - \frac{1}{N}}$ at a time. In fact, it eventually converges quadratically.

THEOREM 5.12. *Jacobi's method is locally quadratically convergent after N steps (i.e., enough steps to choose each a_{jk} once). This means that for i large enough*

$$\text{off}(A_{i+N}) = O(\text{off}^2(A_i)).$$

In practice, we do not use the classical Jacobi's algorithm because searching for the largest entry is too slow: We would need to search $\frac{n^2-n}{2}$ entries for every Jacobi rotation, which costs only $O(n)$ flops to perform, and so for large n the search time would dominate. Instead, we use the following simple method to choose j and k.

ALGORITHM 5.8. *Cyclic-by-row-Jacobi: Sweep through the offdiagonals of A rowwise.*

> *repeat*
> > *for $j = 1$ to $n-1$*
> > > *for $k = j+1$ to n*
> > > > *call Jacobi-Rotation(A, j, k)*
> > > *end for*
> > *end for*
> *until A is sufficiently diagonal*

A no longer changes when Jacobi-Rotation(A, j, k) chooses only $c = 1$ and $s = 0$ for an entire pass through the inner loop. The cyclic Jacobi's algorithm is also asymptotically quadratically convergent like the classical Jacobi's algorithm [262, p. 270].

The cost of one Jacobi "sweep" (where each j, k pair is selected once) is approximately half the cost of reduction to tridiagonal form and the computation of eigenvalues and eigenvectors using QR iteration, and more than the cost using divide-and-conquer. Since Jacobi's method often takes 5–10 sweeps to converge, it is much slower than the competition.

5.3.6. Performance Comparison

In this section we analyze the performance of the three fastest algorithms for the symmetric eigenproblem: QR iteration, Bisection with inverse iteration, and divide-and-conquer. More details may be found in [10, chap. 3] or NETLIB/lapack/lug/lapack_lug.html.

We begin by discussing the fastest algorithm and later compare the others. We used the LAPACK routine `ssyevd`. The algorithm to find only eigenvalues is reduction to tridiagonal form followed by QR iteration, for an operation count of $\frac{4}{3}n^3 + O(n^2)$ flops. The algorithm to find eigenvalues and eigenvectors is tridiagonal reduction followed by divide-and-conquer. We timed `ssyevd` on an IBM RS6000/590, a workstation with a peak speed of 266 Mflops, although optimized matrix-multiplication runs at only 233 Mflops for 100-by-100 matrices and 256 Mflops for 1000-by-1000 matrices. The actual performance is given in the table below. The "Mflop rate" is the actual speed of the code in Mflops, and "Time / Time(Matmul)" is the time to solve the eigenproblem divided by the time to multiply two square matrices of the same size. We see that for large enough matrices, matrix-multiplication and finding only the eigenvalues of a symmetric matrix are about equally expensive. (In contrast, the nonsymmetric eigenproblem is least 16 times more costly [10].) Finding the eigenvectors as well is a little under three times as expensive as matrix-multiplication.

Dimension	Eigenvalues only		Eigenvalues and eigenvectors	
	Mflop rate	Time / Time(Matmul)	Mflop rate	Time / Time(Matmul)
100	72	3.1	72	9.3
1000	160	1.1	174	2.8

Now we compare the relative performance of QR iteration, Bisection with inverse iteration, and divide-and-conquer. In Figures 5.4 and 5.5 these are labeled QR, BZ (for the LAPACK routine `sstebz`, which implements Bisection), and DC, respectively. The horizontal axis in these plots is matrix dimension, and the vertical axis is time divided by the time for DC. Therefore, the DC curve is a horizontal line at 1, and the other curves measure how many times slower BZ and QR are than DC. Figure 5.4 shows only the time for the tridiagonal eigenproblem, whereas Figure 5.5 shows the entire time, starting from a dense matrix.

In the top graph in Figure 5.5 the matrices tested were random symmetric matrices; in Figure 5.4, the tridiagonal matrices were obtained by reducing these dense matrices to tridiagonal form. Such random matrices have well-separated eigenvalues on average, so inverse iteration requires little or no expensive reorthogonalization. Therefore BZ was comparable in performance to DC, although QR was significantly slower, up to 15 times slower in the tridiagonal phase on large matrices.

In the bottom two graphs, the dense symmetric matrices had eigenvalues 1, .5, .25, ..., $.5^{n-1}$. In other words, there were many eigenvalues clustered near zero, so inverse iteration had a lot of reorthogonalization to do. Thus the tridiagonal part of BZ was over 70 times slower than DC. QR was up to 54 times slower than DC, too, because DC actually *speeds up* when there is a large cluster of eigenvalues; this is because of deflation.

The distinction in speeds among QR, BZ, and DC is less noticeable in Figure 5.5 than in Figure 5.4, because Figure 5.5 includes the common $O(n^3)$ overhead of reduction to tridiagonal form and transforming the eigenvalues of the tridiagonal matrix to eigenvalues of the original dense matrix; this common overhead is labeled TRD. Since DC is so close to TRD in Figure 5.5, this means that any further acceleration of DC will make little difference in the overall speed of the dense algorithm.

5.4.   Algorithms for the Singular Value Decomposition

In Theorem 3.3, we showed that the SVD of the general matrix G is closely related to the eigendecompositions of the symmetric matrices $G^T G$, GG^T and $\begin{bmatrix} 0 & G^T \\ G & 0 \end{bmatrix}$. Using these facts, the algorithms in the previous section can be transformed into algorithms for the SVD. The transformations are not straightforward, however, because the added structure of the SVD can often be exploited to make the algorithms more efficient or more accurate [120, 80, 67].

All the algorithms for the eigendecomposition of a symmetric matrix A, except Jacobi's method, have the following structure:

1. Reduce A to tridiagonal form T with an orthogonal matrix Q_1: $A = Q_1 T Q_1^T$.

2. Find the eigendecomposition of T: $T = Q_2 \Lambda Q_2^T$, where Λ is the diagonal matrix of eigenvalues and Q_2 is the orthogonal matrix whose columns are eigenvectors.

3. Combine these decompositions to get $A = (Q_1 Q_2)\Lambda(Q_1 Q_2)^T$. The columns of $Q = Q_1 Q_2$ are the eigenvectors of A.

All the algorithms for the SVD of a general matrix G, except Jacobi's method, have an analogous structure:

1. Reduce G to bidiagonal form B with orthogonal matrices U_1 and V_1: $G = U_1 B V_1^T$. This means B is nonzero only on the main diagonal and first superdiagonal.

2. Find the SVD of B: $B = U_2 \Sigma V_2^T$, where Σ is the diagonal matrix of singular values, and U_2 and V_2 are orthogonal matrices whose columns are the left and right singular vectors, respectively.

3. Combine these decompositions to get $G = (U_1 U_2)\Sigma(V_1 V_2)^T$. The columns of $U = U_1 U_2$ and $V = V_1 V_2$ are the left and right singular vectors of G, respectively.

Reduction to bidiagonal form is accomplished by the algorithm in section 4.4.7. Recall from the discussion there that it costs $\frac{8}{3}n^3 + O(n^2)$ flops to compute B;

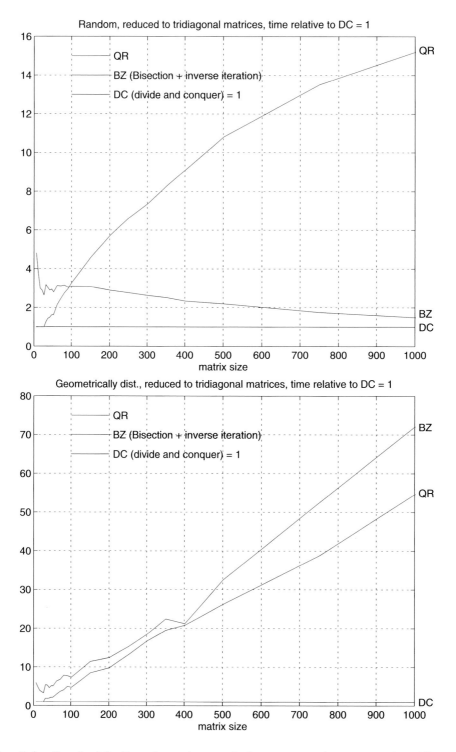

Fig. 5.4. *Speed of finding eigenvalues and eigenvectors of a symmetric tridiagonal matrix, relative to divide-and-conquer.*

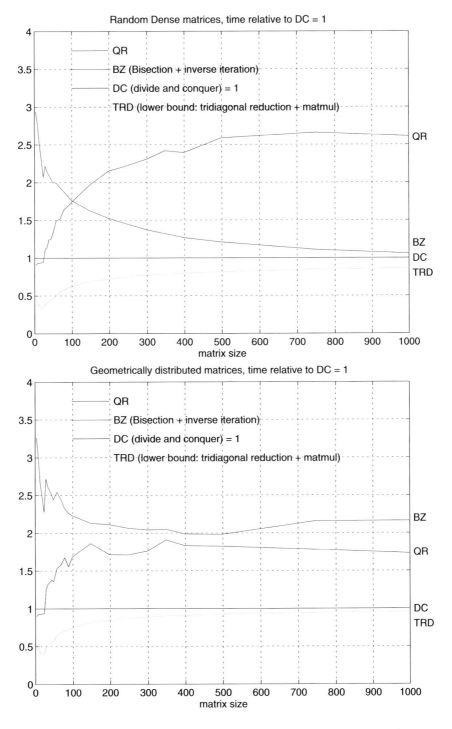

Fig. 5.5. *Speed of finding eigenvalues and eigenvectors of a symmetric dense matrix, relative to divide-and-conquer.*

this is all that is needed if only the singular values Σ are to be computed. It costs another $4n^3 + O(n^2)$ flops to compute U_1 and V_1, which are needed to compute the singular vectors as well.

The following simple lemma shows how to convert the problem of finding the SVD of the bidiagonal matrix B into the eigendecomposition of a symmetric tridiagonal matrix T.

LEMMA 5.5. *Let B be an n-by-n bidiagonal matrix, with diagonal $a_1, \ldots, a_n$ and superdiagonal $b_1, \ldots, b_{n-1}$. There are three ways to convert the problem of finding the SVD of B to finding the eigenvalues and eigenvectors of a symmetric tridiagonal matrix.*

1. *Let $A = \begin{bmatrix} 0 & B^T \\ B & 0 \end{bmatrix}$. Let P be the permutation matrix $P = [e_1, e_{n+1}, e_2, e_{n+2}, \ldots, e_n, e_{2n}]$, where e_i is the ith column of the 2n-by-2n identity matrix. Then $T_{ps} \equiv P^T A P$ is symmetric tridiagonal. The subscript "ps" stands for perfect shuffle, because multiplying P times a vector x "shuffles" the entries of x like a deck of cards. One can show that T_{ps} has all zeros on its main diagonal, and its superdiagonal and subdiagonal is $a_1, b_1, a_2, b_2, \ldots, b_{n-1}, a_n$. If $T_{ps} x_i = \alpha_i x_i$ is an eigenpair for T_{ps}, with x_i a unit vector, then $\alpha_i = \pm \sigma_i$, where σ_i is a singular value of B, and $P x_i = \frac{1}{\sqrt{2}} \begin{bmatrix} v_i \\ \pm u_i \end{bmatrix}$, where u_i and v_i are left and right singular vectors of B, respectively.*

2. *Let $T_{BB^T} \equiv BB^T$. Then T_{BB^T} is symmetric tridiagonal with diagonal $a_1^2 + b_1^2, a_2^2 + b_2^2, \ldots, a_{n-1}^2 + b_{n-1}^2, a_n^2$, and superdiagonal and subdiagonal $a_2 b_1, a_3 b_2, \ldots, a_n b_{n-1}$. The singular values of B are the square roots of the eigenvalues of T_{BB^T}, and the left singular vectors of B are the eigenvectors of T_{BB^T}.*

3. *Let $T_{B^T B} \equiv B^T B$. Then $T_{B^T B}$ is symmetric tridiagonal with diagonal $a_1^2, a_2^2 + b_1^2, a_3^2 + b_2^2, \ldots, a_n^2 + b_{n-1}^2$ and superdiagonal and subdiagonal $a_1 b_1, a_2 b_2, \ldots, a_{n-1} b_{n-1}$. The singular values of B are the square roots of the eigenvalues of $T_{B^T B}$, and the right singular vectors of B are the eigenvectors of $T_{B^T B}$. $T_{B^T B}$ contains no information about the left singular vectors of B.*

For a proof, see Question 5.19.

Thus, we could in principle apply any of QR iteration, divide-and-conquer, or Bisection with inverse iteration to one of the tridiagonal matrices from Lemma 5.5 and then extract the singular and (perhaps only left or right) singular vectors from the resulting eigendecomposition. However, this simple approach would sacrifice both speed and accuracy by ignoring the special properties of the underlying SVD problem. We give two illustrations of this.

First, it would be inefficient to run symmetric tridiagonal QR iteration or divide-and-conquer on T_{ps}. This is because these algorithms both compute all the eigenvalues (and perhaps eigenvectors) of T_{ps}, whereas Lemma 5.5 tells us we only need the nonnegative eigenvalues (and perhaps eigenvectors). There are some accuracy difficulties with singular vectors for tiny singular values too.

Second, explicitly forming either T_{BB^T} or T_{B^TB} is numerically unstable. In fact one can lose half the accuracy in the small singular values of B. For example, let $\eta = \varepsilon/2$, so $1 + \eta$ rounds to 1 in floating point arithmetic. Let $B = \begin{bmatrix} 1 & 1 \\ 0 & \sqrt{\eta} \end{bmatrix}$, which has singular values near $\sqrt{2}$ and $\sqrt{\eta/2}$. Then $B^TB = \begin{bmatrix} 1 & 1 \\ 1 & 1+\eta \end{bmatrix}$ rounds to $T_{B^TB} = \begin{bmatrix} 1 & 1 \\ 1 & 1 \end{bmatrix}$, an exactly singular matrix. Thus, rounding $1 + \eta$ to 1 changes the smaller computed singular value from its true value near $\sqrt{\eta/2} = \sqrt{\varepsilon}/2$ to 0. In contrast, a backward stable algorithm should change the singular values by no more than $O(\varepsilon)\|B\|_2 = O(\varepsilon)$. In IEEE double precision floating point arithmetic, $\varepsilon \approx 10^{-16}$ and $\sqrt{\varepsilon}/2 \approx 10^{-8}$, so the error introduced by forming B^TB is 10^8 times larger than roundoff, a much larger change. The same loss of accuracy can occur by explicitly forming T_{BB^T}.

Because of the instability caused by computing T_{BB^T} or T_{B^TB}, good SVD algorithms work directly on B or possibly T_{ps}.

In summary, we describe the practical algorithms used for computing the SVD.

1. *QR iteration and its variations.* Properly implemented [104], this is the fastest algorithm for finding all the singular values of a bidiagonal matrix. Furthermore, it finds all the singular values to high relative accuracy, as discussed in section 5.2.1. This means that all the digits of all the singular values are correct, even the tiniest ones. In contrast, symmetric tridiagonal QR iteration may compute tiny eigenvalues with no relative accuracy at all. A different variation of QR iteration [80] is used to compute the singular vectors as well: by using QR iteration with a zero shift to compute the smallest singular vectors, this variation computes the singular values nearly as accurately, as well as getting singular vectors as accurately as described in section 5.2.1. But this is only the fastest algorithm for small matrices, up to about dimension $n = 25$. This routine is available in LAPACK subroutine sbdsqr.

2. *Divide-and-conquer.* This is currently the fastest method to find all singular values and singular vectors for matrices larger than $n = 25$. (The implementation in LAPACK, sbdsdc, defaults to sbdsqr for small matrices.) However, divide-and-conquer does not guarantee that the tiny singular values are computed to high relative accuracy. Instead, it guarantees only the same error bound as in the symmetric eigenproblem: the error in singular value σ_j is at most $O(\varepsilon)\sigma_1$ rather than $O(\varepsilon)\sigma_j$. This is sufficiently accurate for most applications.

3. *Bisection and inverse iteration.* One can apply Bisection and inverse iteration to T_{ps} of part 1 of Lemma 5.5 to find only the singular values in

a desired interval. This algorithm is guaranteed to find the singular values to high relative accuracy, although the singular vectors may occasionally suffer loss of orthogonality as described in section 5.3.4.

4. *Jacobi's method.* We may compute the SVD of a dense matrix G by applying Jacobi's method of section 5.3.5 *implicitly* to GG^T or $G^T G$, i.e., without explicitly forming either one and so possibly losing stability. For some classes of G, i.e., those to which we can profitably apply the relative perturbation theory of section 5.2.1, we can show that Jacobi's method computes the singular values and singular vectors to high relative accuracy, as described in section 5.2.1.

The following sections describe some of the above algorithms in more detail, notably QR iteration and its variation dqds in section 5.4.1; the proof of high accuracy of dqds and Bisection in section 5.4.2; and Jacobi's method in section 5.4.3. We omit divide-and-conquer because of its overall similarity to the algorithm discussed in section 5.3.3, and refer the reader to [130] for details.

5.4.1. QR Iteration and Its Variations for the Bidiagonal SVD

There is a long history of variations on QR iteration for the SVD, designed to be as efficient and accurate as possible; see [200] for a good survey. The algorithm in the LAPACK routine `sbdsqr` was originally based on [80] and later updated to use the algorithm in [104] in the case when singular values only are desired. This latter algorithm, called dqds for historical reasons,[22] is elegant, fast, and accurate, so we will present it.

To derive dqds, we begin with an algorithm that predates QR iteration, called LR iteration, specialized to symmetric positive definite matrices.

ALGORITHM 5.9. *LR iteration: Let T_0 be any symmetric positive definite matrix. The following algorithm produces a sequence of similar symmetric positive definite matrices T_i:*

> $i = 0$
> *repeat*
> > *Choose a shift τ_i^2 smaller than the smallest eigenvalue of T_i.*
> > *Compute the Cholesky factorization $T_i - \tau_i^2 I = B_i^T B_i$*
> > > *(B_i is an upper triangular matrix with positive diagonal.)*
> >
> > $T_{i+1} = B_i B_i^T + \tau_i^2 I$
> > $i = i + 1$
> *until convergence*

---

[22]dqds is short for "differential quotient-difference algorithm with shifts" [209].

LR iteration is very similar in structure to QR iteration: We compute a factorization, and multiply the factors in reverse order to get the next iterate T_{i+1}. It is easy to see that T_{i+1} and T_i are similar: $T_{i+1} = B_i B_i^T + \tau_i^2 I = B_i^{-T} B_i^T B_i B_i^T + \tau_i^2 B_i^{-T} B_i^T = B_i^{-T} T_i B_i^T$.

In fact, when the shift $\tau_i^2 = 0$, we can show that two steps of LR iteration produce the same T_2 as one step of QR iteration.

LEMMA 5.6. *Let T_2 be the matrix produced by two steps of Algorithm 5.9 using $\tau_i^2 = 0$, and let T' be the matrix produced by one step of QR iteration ($QR = T_0$, $T' = RQ$). Then $T_2 = T'$.*

Proof. Since T_0 is symmetric, we can factorize T_0^2 in two ways: First, $T_0^2 = T_0^T T_0 = (QR)^T QR = R^T R$. We assume without loss of generality that $R_{ii} > 0$. This is a factorization of T_0^2 into a lower triangular matrix R^T times its transpose; since the Cholesky factorization is unique, this must in fact be the Cholesky factorization. The second factorization is $T_0^2 = B_0^T B_0 B_0^T B_0$. Now by Algorithm 5.9, $T_1 = B_0 B_0^T = B_1^T B_1$, so we can rewrite $T_0^2 = B_0^T B_0 B_0^T B_0 = B_0^T (B_1^T B_1) B_0 = (B_1 B_0)^T B_1 B_0$. This is also a factorization of T_0^2 into a lower triangular matrix $(B_1 B_0)^T$ times its transpose, so this must again be the Cholesky factorization. By uniqueness of the Cholesky factorization, we conclude $R = B_1 B_0$, thus relating two steps of LR iteration to one step of QR iteration. We exploit this relationship as follows: $T_0 = QR$ implies

$$
\begin{aligned}
T' &= RQ = RQ(RR^{-1}) = R(QR)R^{-1} = RT_0 R^{-1} \quad \text{because } T_0 = QR \\
&= (B_1 B_0)(B_0^T B_0)(B_1 B_0)^{-1} \quad \text{because } R = B_1 B_0 \text{ and } T_0 = B_0^T B_0 \\
&= B_1 B_0 B_0^T B_0 B_0^{-1} B_1^{-1} = B_1 (B_0 B_0^T) B_1^{-1} \\
&= B_1 (B_1^T B_1) B_1^{-1} \quad \text{because } B_0 B_0^T = T_1 = B_1^T B_1 \\
&= B_1 B_1^T \\
&= T_2 \text{ as desired.} \quad \square
\end{aligned}
$$

Neither Algorithm 5.9 nor Lemma 5.6 depends on T_0 being tridiagonal, just symmetric positive definite. Using the relationship between LR iteration and QR iteration in Lemma 5.6, one can show that much of the convergence analysis of QR iteration goes over to LR iteration; we will not explore this here.

Our ultimate algorithm, dqds, is mathematically equivalent to LR iteration. But it is *not* implemented as described in Algorithm 5.9, because this would involve explicitly forming $T_{i+1} = B_i B_i^T + \tau_i^2 I$, which in section 5.4 we showed could be numerically unstable. Instead, we will form B_{i+1} *directly* from B_i, without ever forming the intermediate matrix T_{i+1}.

To simplify notation, let B_i have diagonal $a_1, \ldots, a_n$ and superdiagonal $b_1, \ldots, b_{n-1}$, and B_{i+1} have diagonal $\hat{a}_1, \ldots, \hat{a}_n$ and superdiagonal $\hat{b}_1, \ldots, \hat{b}_{n-1}$. We use the convention $b_0 = \hat{b}_0 = b_n = \hat{b}_n = 0$. We relate B_i to B_{i+1} by

$$
B_{i+1}^T B_{i+1} + \tau_{i+1}^2 I = T_{i+1} = B_i B_i^T + \tau_i^2 I. \tag{5.20}
$$

Equating the j, j entries of the left and right sides of equation (5.20) for $j < n$ yields

$$\hat{a}_j^2 + \hat{b}_{j-1}^2 + \tau_{i+1}^2 = a_j^2 + b_j^2 + \tau_i^2 \quad \text{or} \quad \hat{a}_j^2 = a_j^2 + b_j^2 - \hat{b}_{j-1}^2 - \delta, \qquad (5.21)$$

where $\delta = \tau_{i+1}^2 - \tau_i^2$. Since τ_i^2 must be chosen to approach the smallest eigenvalue of T from below (to keep T_i positive definite and the algorithm well defined), $\delta \geq 0$. Equating the squares of the $j, j+1$ entries of the left and right sides of equation (5.20) yields

$$\hat{a}_j^2 \hat{b}_j^2 = a_{j+1}^2 b_j^2 \quad \text{or} \quad \hat{b}_j^2 = a_{j+1}^2 b_j^2 / \hat{a}_j^2. \qquad (5.22)$$

Combining equations (5.21) and (5.22) yields the not-yet-final algorithm

for $j = 1$ to $n - 1$
$\quad \hat{a}_j^2 = a_j^2 + b_j^2 - \hat{b}_{j-1}^2 - \delta$
$\quad \hat{b}_j^2 = b_j^2 \cdot (a_{j+1}^2 / \hat{a}_j^2)$
end for
$\hat{a}_n^2 = a_n^2 - \hat{b}_{n-1}^2 - \delta$

This version of the algorithm has only five floating point operations in the inner loop, which is quite inexpensive. It maps directly from the *squares* of the entries of B_i to the *squares* of the entries of B_{i+1}. There is no reason to take square roots until the very end of the algorithm. Indeed, square roots, along with divisions, can take 10 to 30 times longer than additions, subtractions, or multiplications on modern computers, so we should avoid as many of them as possible. To emphasize that we are computing squares of entries, we change variables to $q_j \equiv a_j^2$ and $e_j \equiv b_j^2$, yielding the penultimate algorithm qds (again, the name is for historical reasons that do not concern us [209]).

ALGORITHM 5.10. *One step of the qds algorithm:*

for $j = 1$ to $n - 1$
$\quad \hat{q}_j = q_j + e_j - \hat{e}_{j-1} - \delta$
$\quad \hat{e}_j = e_j \cdot (q_{j+1} / \hat{q}_j)$
end for
$\hat{q}_n = q_n - \hat{e}_{n-1} - \delta$

The final algorithm, dqds, will do about the same amount of work as qds but will be significantly more accurate, as will be shown in section 5.4.2. We take the subexpression $q_j - \hat{e}_{j-1} - \delta$ from the first line of Algorithm 5.10 and rewrite it as follows:

$$\begin{aligned} d_j &\equiv q_j - \hat{e}_{j-1} - \delta \\ &= q_j - \frac{q_j e_{j-1}}{\hat{q}_{j-1}} - \delta \quad \text{from} \quad (5.22) \end{aligned}$$

$$= q_j \cdot \left[\frac{\hat{q}_{j-1} - e_{j-1}}{\hat{q}_{j-1}} \right] - \delta$$

$$= q_j \cdot \left[\frac{q_{j-1} - \hat{e}_{j-2} - \delta}{\hat{q}_{j-1}} \right] - \delta \quad \text{from} \quad (5.21)$$

$$= \frac{q_j}{\hat{q}_{j-1}} \cdot d_{j-1} - \delta.$$

This lets us rewrite the inner loop of Algorithm 5.10 as

$$\hat{q}_j = d_j + e_j$$
$$\hat{e}_j = e_j \cdot (q_{j+1}/\hat{q}_j)$$
$$d_{j+1} = d_j \cdot (q_{j+1}/\hat{q}_j) - \delta$$

Finally, we note that d_{j+1} can overwrite d_j and that $t = q_{j+1}/\hat{q}_j$ need be computed only once to get the final dqds algorithm.

ALGORITHM 5.11. *One step of the dqds algorithm:*

$$d = q_1 - \delta$$
$$for\ j = 1\ to\ n - 1$$
$$\quad \hat{q}_j = d + e_j$$
$$\quad t = (q_{j+1}/\hat{q}_j)$$
$$\quad \hat{e}_j = e_j \cdot t$$
$$\quad d = d \cdot t - \delta$$
$$end\ for$$
$$\hat{q}_n = d$$

The dqds algorithm has the same number of floating point operations in its inner loop as qds but trades a subtraction for a multiplication. This modification pays off handsomely in guaranteed high relative accuracy, as described in the next section.

There are two important issues we have not discussed: choosing a shift $\delta = \tau_{i+1}^2 - \tau_i^2$ and detecting convergence. These are discussed in detail in [104].

5.4.2. Computing the Bidiagonal SVD to High Relative Accuracy

This section, which depends on section 5.2.1, may be skipped on a first reading.

Our ability to compute the SVD of a bidiagonal matrix B to high relative accuracy (as defined in section 5.2.1) depends on Theorem 5.13 below, which says that small relative changes in the entries of B cause only small relative changes in the singular values.

LEMMA 5.7. *Let B be a bidiagonal matrix, with diagonal entries $a_1, \ldots, a_n$ and superdiagonal entries $b_1, \ldots, b_{n-1}$. Let $\hat{B}$ be another bidiagonal matrix* with diagonal entries $\hat{a}_i = a_i\chi_i$ and superdiagonal entries $\hat{b}_i = b_i\zeta_i$. Then $\hat{B} = D_1 B D_2$, where

$$D_1 = \mathrm{diag}\left(\chi_1, \frac{\chi_2\chi_1}{\zeta_1}, \frac{\chi_3\chi_2\chi_1}{\zeta_2\zeta_1}, \dots, \frac{\chi_n\cdots\chi_1}{\zeta_{n-1}\cdots\zeta_1}\right),$$

$$D_2 = \mathrm{diag}\left(1, \frac{\zeta_1}{\chi_1}, \frac{\zeta_2\zeta_1}{\chi_2\chi_1}, \dots, \frac{\zeta_{n-1}\cdots\zeta_1}{\chi_{n-1}\cdots\chi_1}\right).$$

The proof of this lemma is a simple computation (see Question 5.20). We can now apply Corollary 5.2 to conclude the following.

THEOREM 5.13. *Let B and $\hat{B}$ be defined as in Lemma 5.7. Suppose that there is a $\tau \geq 1$ such that $\tau^{-1} \leq \chi_i \leq \tau$ and $\tau^{-1} \leq \zeta_i \leq \tau$. In other words $\epsilon \equiv \tau - 1$ is a bound on the relative difference between each entry of B and the corresponding entry of $\hat{B}$. Let $\sigma_n \leq \cdots \leq \sigma_1$ be the singular values of B and $\hat{\sigma}_n \leq \cdots \leq \hat{\sigma}_1$ be the singular values of $\hat{B}$. Then $|\hat{\sigma}_i - \sigma_i| \leq \sigma_i(\tau^{4n-2} - 1)$. If $\sigma_i \neq 0$ and $\tau - 1 = \epsilon \ll 1$, then we can write*

$$\frac{|\hat{\sigma}_i - \sigma_i|}{\sigma_i} \leq \tau^{4n-2} - 1 = (4n-2)\epsilon + O(\epsilon^2).$$

Thus, the relative change in the singular values $|\hat{\sigma}_i - \sigma_i|/\sigma_i$ is bounded by $4n-2$ times the relative change ϵ in the matrix entries. With a little more work, the factor $4n - 2$ can be improved to $2n - 1$ (see Question 5.21). The singular vectors can also be shown to be determined quite accurately, proportional to the reciprocal of the relative gap, as defined in section 5.2.1.

We will show that both Bisection (Algorithm 5.4 applied to T_{ps} from Lemma 5.5) and dqds (Algorithm 5.11) can be used to find the singular values of a bidiagonal matrix to high relative accuracy. First we consider Bisection. Recall that the eigenvalues of the symmetric tridiagonal matrix T_{ps} are the singular values of B and their negatives. Lemma 5.3 implies that the inertia of $T_{ps} - \lambda I$ computed using equation (5.17) is the exact inertia of some $\hat{B}$, where the relative difference of corresponding entries of $\hat{B}$ and B is at most about 2.5ε. Therefore, by Theorem 5.13, the relative difference between the computed singular values (the singular values of $\hat{B}$) and the true singular values is at most about $(10n - 5)\varepsilon$.

Now we consider Algorithm 5.11. We will use Theorem 5.13 to prove that the singular values of B (the input to Algorithm 5.11) and the singular values of $\hat{B}$ (the output from Algorithm 5.11) agree to high relative accuracy. This fact implies that after many steps of dqds, when $\hat{B}$ is nearly diagonal with its singular values on the diagonal, these singular values match the singular values of the original input matrix to high relative accuracy.

The simplest situation to understand is when the shift $\delta = 0$. In this case, the only operations in dqds are additions of positive numbers, multiplications, and divisions; no cancellation occurs. Roughly speaking, any sequence of expressions built of these basic operations is guaranteed to compute each output to high relative accuracy. Therefore, $\hat{B}$ is computed to high relative accuracy, and so by Theorem 5.13, the singular values of B and $\hat{B}$ agree to high relative accuracy. The general case, where $\delta > 0$, is trickier [104].

THEOREM 5.14. *One step of Algorithm 5.11 in floating point arithmetic, applied to B and yielding $\hat{B}$, is equivalent to the following sequence of operations:*

1. *Make a small relative change (by at most 1.5ε) in each entry of B, getting $\tilde{B}$.*

2. *Apply one step of Algorithm 5.11 in exact arithmetic to $\tilde{B}$, getting $\check{B}$.*

3. *Make a small relative change (by at most ε) in each entry of $\check{B}$, getting $\hat{B}$.*

Steps 1 and 3 above make only small relative changes in the singular values of the bidiagonal matrix, so by Theorem 5.13 the singular values of B and $\hat{B}$ agree to high relative accuracy.

Proof. Let us write the inner loop of Algorithm 5.11 as follows, introducing subscripts on the d and t variables to let us keep track of them in different iterations and including subscripted $1 + \epsilon$ terms for the roundoff errors:

$$\hat{q}_j = (d_j + e_j)(1 + \epsilon_{j,+})$$
$$t_j = (q_{j+1}/\hat{q}_j)(1 + \epsilon_{j,/})$$
$$\hat{e}_j = e_j \cdot t_j(1 + \epsilon_{j,*1})$$
$$d_{j+1} = (d_j \cdot t_j(1 + \epsilon_{j,*2}) - \delta)(1 + \epsilon_{j,-})$$

Substituting the first line into the second line yields

$$t_j = \frac{q_{j+1}}{d_j + e_j} \cdot \frac{1 + \epsilon_{j,/}}{1 + \epsilon_{j,+}}.$$

Substituting this expression for t_j into the last line of the algorithm and dividing through by $1 + \epsilon_{j,-}$ yield

$$\frac{d_{j+1}}{1 + \epsilon_{j,-}} = \frac{d_j q_{j+1}}{d_j + e_j} \cdot \frac{(1 + \epsilon_{j,/})(1 + \epsilon_{j,*2})}{1 + \epsilon_{j,+}} - \delta. \qquad (5.23)$$

This tells us how to define $\tilde{B}$: Let

$$\tilde{d}_{j+1} = \frac{d_{j+1}}{1 + \epsilon_{j,-}},$$
$$\tilde{e}_j = \frac{e_j}{1 + \epsilon_{j-1,-}}, \qquad (5.24)$$
$$\tilde{q}_{j+1} = q_{j+1}\frac{(1 + \epsilon_{j,/})(1 + \epsilon_{j,*2})}{1 + \epsilon_{j,+}},$$

so (5.23) becomes

$$\tilde{d}_{j+1} = \frac{\tilde{d}_j \tilde{q}_{j+1}}{\tilde{d}_j + \tilde{e}_j} - \delta.$$

Note from (5.24) that $\tilde{B}$ differs from B by a relative change of at most 1.5ε in each entry (from the three $1 + \epsilon$ factors in $\tilde{q}_{j+1} = \tilde{a}_{j+1,j+1}^2$).

Now we can define $\check{q}_j$ and $\check{e}_j$ in $\check{B}$ by

$$\check{q}_j = \tilde{d}_j + \tilde{e}_j,$$
$$\check{t}_j = (\tilde{q}_{j+1}/\check{q}_j),$$
$$\check{e}_j = \tilde{e}_j \cdot \check{t}_j,$$
$$\check{d}_{j+1} = \tilde{d}_j \cdot \check{t}_j - \delta.$$

This is one step of the dqds algorithm applied exactly to $\tilde{B}$, getting $\check{B}$. To finally show that $\check{B}$ differs from $\hat{B}$ by a relative change of at most ε in each entry, note that

$$\begin{aligned}
\check{q}_j &= \tilde{d}_j + \tilde{e}_j \\
&= \frac{d_j}{1 + \epsilon_{j-1,-}} + \frac{e_j}{1 + \epsilon_{j-1,-}} \\
&= (d_j + e_j)(1 + \epsilon_{j,+}) \cdot \frac{1}{(1 + \epsilon_{j,+})(1 + \epsilon_{j-1,-})} \\
&= \hat{q}_j \cdot \left[\frac{1}{(1 + \epsilon_{j,+})(1 + \epsilon_{j-1,-})} \right]
\end{aligned}$$

and

$$\begin{aligned}
\check{e}_j &= \tilde{e}_j \cdot \check{t}_j \\
&= \frac{e_j}{1 + \epsilon_{j-1,-}} \cdot \frac{\tilde{q}_{j+1}}{\check{q}_j} \\
&= \frac{e_j}{1 + \epsilon_{j-1,-}} \cdot t_j(1 + \epsilon_{j,*2})(1 + \epsilon_{j-1,-}) \\
&= e_j t_j(1 + \epsilon_{j,*1}) \frac{1 + \epsilon_{j,*2}}{1 + \epsilon_{j,*1}} \\
&= \hat{e}_j \cdot \left[\frac{1 + \epsilon_{j,*2}}{1 + \epsilon_{j,*1}} \right]. \quad \square
\end{aligned}$$

5.4.3. Jacobi's Method for the SVD

In section 5.3.5 we discussed Jacobi's method for finding the eigenvalues and eigenvectors of a dense symmetric matrix A, and said it was the slowest available method for this problem. In this section we will show how to apply Jacobi's method to find the SVD of a dense matrix G by *implicitly* applying Algorithm 5.8 of section 5.3.5 to the symmetric matrix $A = G^T G$. This implies that the convergence properties of this method are nearly the same as those of Algorithm 5.8, and in particular Jacobi's method is also the slowest method available for the SVD.

Jacobi's method is still interesting, however, because for some kinds of matrices G, it can compute the singular values and singular vectors much more accurately than the other algorithms we have discussed. For these G, Jacobi's method computes the singular values and singular vectors to high relative accuracy, as described in section 5.2.1.

After describing the implicit Jacobi's method for the SVD of G, we will show that it computes the SVD to high relative accuracy when G can be written in the form $G = DX$, where D is diagonal and X is well conditioned. (This means that G is ill conditioned if and only if D has both large and small diagonal entries.) More generally, we benefit as long as X is significantly better conditioned than G. We will illustrate this with a matrix where any algorithm involving reduction to bidiagonal form necessarily loses all significant digits in all but the largest singular value, whereas Jacobi's method computes all singular values to full machine precision. Then we survey other classes of matrices G for which Jacobi's method is also significantly more accurate than methods using bidiagonalization.

Note that if G is bidiagonal, then we showed in section 5.4.2 that we could use either Bisection or the dqds algorithm (section 5.4.1) to compute its SVD to high relative accuracy. The trouble is that reducing a matrix from dense to bidiagonal form can introduce errors that are large enough to destroy high relative accuracy, as our example will show. Since Jacobi's method operates on the original matrix without first reducing it to bidiagonal form, it can achieve high relative accuracy in many more situations.

The implicit Jacobi's method is mathematically equivalent to applying Algorithm 5.8 to $A = G^T G$. In other words, at each step we compute a Jacobi rotation J and implicitly update $G^T G$ to $J^T G^T G J$, where J is chosen so that two offdiagonal entries of $G^T G$ are set to zero in $J^T G^T G J$. But instead of computing $G^T G$ or $J^T G^T G J$ explicitly, we instead only compute GJ. For this reason, we call our algorithm *one-sided Jacobi rotation*.

ALGORITHM 5.12. *Compute and apply a one-sided Jacobi rotation to G in coordinates j, k:*

proc One-Sided-Jacobi-Rotation (G, j, k)
    Compute $a_{jj} = (G^T G)_{jj}$, $a_{jk} = (G^T G)_{jk}$, and $a_{kk} = (G^T G)_{kk}$
    if $|a_{jk}|$ is not too small
        $\tau = (a_{jj} - a_{kk})/(2 \cdot a_{jk})$
        $t = \text{sign}(\tau)/(|\tau| + \sqrt{1 + \tau^2})$
        $c = 1/\sqrt{1 + t^2}$
        $s = c \cdot t$
        $G = G \cdot R(j, k, \theta)$ *... where $c = \cos\theta$ and $s = \sin\theta$*
        if right singular vectors are desired

$$J = J \cdot R(j, k, \theta)$$
    end if
  end if

Note that the jj, jk, and kk entries of $A = G^T G$ are computed by procedure One-Sided-Jacobi-Rotation, after which it computes the Jacobi rotation $R(j, k, \theta)$ in the same way as procedure Jacobi-Rotation (Algorithm 5.5).

ALGORITHM 5.13. *One-sided Jacobi: Assume that G is n-by-n. The outputs are the singular values σ_i, the left singular vector matrix U, and the right singular vector matrix V so that $G = U\Sigma V^T$, where $\Sigma = \text{diag}(\sigma_i)$.*

  repeat
      for $j = 1$ to $n - 1$
          for $k = j + 1$ to n
              call One-Sided-Jacobi-Rotation(G, j, k)
          end for
      end for
  until $G^T G$ is diagonal enough
  Let $\sigma_i = \|G(:, i)\|_2$ (the 2-norm of column i of G)
  Let $U = [u_1, \ldots, u_n]$, where $u_i = G(:, i)/\sigma_i$
  let $V = J$, the accumulated product of Jacobi rotations

Question 5.22 asks for a proof that the matrices Σ, U, and V computed by one-sided Jacobi do indeed form the SVD of G.

The following theorem shows that one-sided Jacobi can compute the SVD to high relative accuracy, despite roundoff, provided that we can write $G = DX$, where D is diagonal and X is well-conditioned.

THEOREM 5.15. *Let $G = DX$ be an n-by-n matrix, where D is diagonal and nonsingular, and X is nonsingular. Let $\hat{G}$ be the matrix after calling One-Sided-Jacobi-Rotation(G, j, k) m times in floating point arithmetic. Let $\sigma_1 \geq \cdots \geq \sigma_n$ be the singular values of G, and let $\hat{\sigma}_1 \geq \cdots \geq \hat{\sigma}_n$ be the singular values of $\hat{G}$. Then*

$$\frac{|\sigma_i - \hat{\sigma}_i|}{\sigma_i} \leq O(m\varepsilon)\kappa(X), \tag{5.25}$$

where $\kappa(X) = \|X\| \cdot \|X^{-1}\|$ is the condition number of X. In other words, the relative error in the singular values is small if the condition number of X is small.

Proof.   We first consider $m = 1$; i.e., we apply only a single Jacobi rotation and later generalize to larger m.

Examining One-Sided-Jacobi-Rotation(G, j, k), we see that $\hat{G} = \text{fl}(G \cdot \tilde{R})$, where $\tilde{R}$ is a floating point Givens rotation. By construction, $\tilde{R}$ differs from some exact Givens rotation R by $O(\varepsilon)$ in norm. (It is not important or necessarily true that $\tilde{R}$ differs by $O(\varepsilon)$ from the "true" Jacobi rotation, the one that One-Sided-Jacobi-Rotation(G, j, k) would have computed in exact arithmetic. It is necessary only that it differs from *some* rotation by $O(\varepsilon)$. This requires only that $c^2 + s^2 = 1 + O(\varepsilon)$, which is easy to verify.)

Our goal is to show that $\hat{G} = GR(I + E)$ for some E that is small in norm: $\|E\|_2 = O(\varepsilon)\kappa(X)$. If E were zero, then $\hat{G}$ and GR would have the same singular values, since R is exactly orthogonal. When E is less than one in norm, we can use Corollary 5.2 to bound the relative difference in singular values by

$$\frac{|\sigma_i - \hat{\sigma}_i|}{\sigma_i} \leq \|(I + E)(I + E)^T - I\|_2 = \|E + E^T + EE^T\|_2 \leq 3\|E\|_2$$
$$= O(\varepsilon)\kappa(X) \tag{5.26}$$

as desired.

Now we construct E. Since $\tilde{R}$ multiplies G on the right, each row of $\hat{G}$ depends only on the corresponding row of G; write this in Matlab notation as $\hat{G}(i,:) = \text{fl}(G(i,:) \cdot \tilde{R})$. Let $F = \hat{G} - GR$. Then by Lemma 3.1 and the fact that $G = DX$,

$$\|F(i,:)\|_2 = \|\hat{G}(i,:) - G(i,:)R\|_2 = O(\varepsilon)\|G(i,:)\|_2 = O(\varepsilon)\|d_{ii}X(i,:)\|_2$$

and so $\|d_{ii}^{-1}F(i,:)\|_2 = O(\varepsilon)\|X(i,:)\|_2$, or $\|D^{-1}F\|_2 = O(\varepsilon)\|X\|_2$. Therefore, since $R^{-1} = R^T$ and $G^{-1} = (DX)^{-1} = X^{-1}D^{-1}$,

$$\hat{G} = GR + F = GR(I + R^TG^{-1}F) = GR(I + R^TX^{-1}D^{-1}F) \equiv GR(I + E)$$

where

$$\|E\|_2 \leq \|R^T\|_2\|X^{-1}\|_2\|D^{-1}F\|_2 = O(\varepsilon)\|X\|_2\|X^{-1}\|_2 = O(\varepsilon)\kappa(X)$$

as desired.

To extend this result to $m > 1$ rotations, note that in exact arithmetic we would have $\hat{G} = GR = DXR = D\hat{X}$, with $\kappa(\hat{X}) = \kappa(X)$, so that the bound (5.26) would apply at each of the m steps, yielding bound (5.25). Because of roundoff, $\kappa(\hat{X})$ could grow by as much as $\kappa(I+E) \leq (1+O(\varepsilon)\kappa(X))$ at each step, a factor very close to 1, which we absorb into the $O(m\varepsilon)$ term. $\square$

To complete the algorithm, we need to be careful about the stopping criterion, i.e., how to implement the statement "if $|a_{jk}|$ is not too small" in Algorithm 5.12, One-Sided-Jacobi-Rotation. The appropriate criterion

$$|a_{jk}| \geq \varepsilon\sqrt{a_{jj}a_{kk}}$$

is discussed further in Question 5.24.

EXAMPLE 5.9. We consider an extreme example $G = DX$ where Jacobi's method computes all singular values to full machine precision; any method relying on bidiagonalization computes only the largest one, $\sqrt{3}$, to full machine precision; and all the others with no accuracy at all (although it still computes them with errors $\pm O(\varepsilon) \cdot \sqrt{3}$, as expected from a backward stable algorithm). In this example $\varepsilon = 2^{-53} \approx 10^{-16}$ (IEEE double precision) and $\eta = 10^{-20}$ (nearly any value of $\eta < \varepsilon$ will do). We define

$$
G \equiv
\begin{bmatrix}
\eta & 1 & 1 & 1 \\
\eta & \eta & 0 & 0 \\
\eta & 0 & \eta & 0 \\
\eta & 0 & 0 & \eta
\end{bmatrix}
=
\begin{bmatrix}
1 & & & \\
 & \eta & & \\
 & & \eta & \\
 & & & \eta
\end{bmatrix}
\cdot
\begin{bmatrix}
\eta & 1 & 1 & 1 \\
1 & 1 & 0 & 0 \\
1 & 0 & 1 & 0 \\
1 & 0 & 0 & 1
\end{bmatrix}
\equiv D \cdot X.
$$

To at least 16 digits, the singular values of G are $\sqrt{3}$, $\sqrt{3} \cdot \eta$, η, and η. To see how accuracy is lost by reducing G to bidiagonal form, we consider just the first step of the algorithm in section 4.4.7: After step 1, premultiplication by a Householder transformation to zero out $G(2:4,1)$, G in exact arithmetic would be

$$
\begin{bmatrix}
-2\eta & -.5 - \frac{\eta}{2} & -.5 - \frac{\eta}{2} & -.5 - \frac{\eta}{2} \\
0 & -.5 + \frac{5\eta}{6} & -.5 - \frac{\eta}{6} & -.5 - \frac{\eta}{6} \\
0 & -.5 - \frac{\eta}{6} & -.5 + \frac{5\eta}{6} & -.5 - \frac{\eta}{6} \\
0 & -.5 - \frac{\eta}{6} & -.5 - \frac{\eta}{6} & -.5 + \frac{5\eta}{6}
\end{bmatrix},
$$

but since η is so small, this rounds to

$$
G_1 =
\begin{bmatrix}
-2\eta & -.5 & -.5 & -.5 \\
0 & -.5 & -.5 & -.5 \\
0 & -.5 & -.5 & -.5 \\
0 & -.5 & -.5 & -.5
\end{bmatrix}.
$$

Note that all information about η has been "lost" from the last three columns of G_1. Since the last three columns of G_1 are identical, G_1 is exactly singular and indeed of rank 2. Thus the two smallest singular values have been changed from η to 0, a complete loss of relative accuracy. If we made no further rounding errors, we would reduce G_1 to the bidiagonal form

$$
B =
\begin{bmatrix}
-2\eta & \sqrt{.75} & & \\
 & 1.5 & 0 & \\
 & & 0 & 0 \\
 & & & 0
\end{bmatrix}
$$

with singular values $\sqrt{3}$, $\sqrt{3}\eta$, 0, and 0, the larger two of which are accurate singular values of G. But as the algorithm proceeds to reduce G_1 to bidiagonal form, roundoff introduces nonzero quantities of $O(\varepsilon)$ into the zero entries of B, making all three small singular values inaccurate. The two smallest nonzero computed singular values are accidents of roundoff and proportional to ε.

One-sided Jacobi's method has no difficulty with this matrix, converging in three sweeps to $G = U\Sigma V^T$, where to machine precision

$$
U = \begin{bmatrix} 0 & -\frac{\eta^2}{\sqrt{2}} & 1 & 0 \\ \frac{1}{\sqrt{3}} & \frac{1}{\sqrt{2}} & \frac{\eta}{3} & \frac{-1}{\sqrt{6}} \\ \frac{1}{\sqrt{3}} & -\frac{1}{\sqrt{2}} & \frac{\eta}{3} & \frac{-1}{\sqrt{6}} \\ \frac{1}{\sqrt{3}} & \frac{\eta}{\sqrt{2}} & \frac{\eta}{3} & \frac{2}{\sqrt{6}} \end{bmatrix}, \quad V = \begin{bmatrix} 1 & \frac{\eta}{\sqrt{2}} & \frac{\eta}{\sqrt{3}} & \frac{-\eta}{\sqrt{6}} \\ -\eta & \frac{1}{\sqrt{2}} & \frac{1}{\sqrt{3}} & \frac{-1}{\sqrt{6}} \\ 0 & \frac{-1}{\sqrt{2}} & \frac{1}{\sqrt{3}} & \frac{-1}{\sqrt{6}} \\ 0 & \frac{-\eta^3}{\sqrt{2}} & \frac{1}{\sqrt{3}} & \frac{2}{\sqrt{6}} \end{bmatrix},
$$

and $\Sigma = \mathrm{diag}(\sqrt{3}\eta, \eta, \sqrt{3}, \eta)$. (Jacobi does not automatically sort the singular values; this can be done as a postprocessing step.) ⋄

Here are some other examples where versions of Jacobi's method can be shown to guarantee high relative accuracy in the SVD (or symmetric eigendecomposition), whereas methods relying on bidiagonalization (or tridiagonalization) may lose all significant digits in the smallest singular value (or eigenvalues). Many other examples appear in [75].

1. If $A = LL^T$ is the Cholesky decomposition of a symmetric positive definite matrix, then the SVD of $L = U\Sigma V^T$ provides the eigendecomposition of $A = U\Sigma^2 U^T$. If $L = DX$, where X is well-conditioned and D is diagonal, then Theorem 5.15 tells us that we can use Jacobi's method to compute the singular values σ_i of L to high relative accuracy, with relative errors bounded by $O(\varepsilon)\kappa(X)$. But we also have to account for the roundoff errors in computing the Cholesky factor L: using Cholesky's backward error bound (2.16) (along with Theorem 5.6) one can bound the relative error in the singular values introduced by roundoff during Cholesky by $O(\varepsilon)\kappa^2(X)$. So if X is well-conditioned, all the eigenvalues of A will be computed to high relative accuracy (see Question 5.23 and [82, 92, 183]).

EXAMPLE 5.10. As in Example 5.9, we choose an extreme case where any algorithm relying on initially reducing A to tridiagonal form is guaranteed to lose all relative accuracy in the smallest eigenvalue, whereas Cholesky followed by one-sided Jacobi's method on the Cholesky factor computes all eigenvalues to nearly full machine precision. As in that example, let $\eta = 10^{-20}$ (any $\eta < \varepsilon/120$ will do), and let

$$
A = \begin{bmatrix} 1 & \sqrt{\eta} & \sqrt{\eta} \\ \sqrt{\eta} & 1 & 10\eta \\ \sqrt{\eta} & 10\eta & 100\eta \end{bmatrix} = \begin{bmatrix} 1 & 10^{-10} & 10^{-10} \\ 10^{-10} & 1 & 10^{-19} \\ 10^{-10} & 10^{-19} & 10^{-20} \end{bmatrix}.
$$

If we reduce A to tridiagonal form T exactly, then

$$
T = \begin{bmatrix} 1 & \sqrt{2\eta} & \\ \sqrt{2\eta} & .5 + 60\eta & .5 - 50\eta \\ & .5 - 50\eta & .5 + 40\eta \end{bmatrix},
$$

but since η is so small, this rounds to

$$\hat{T} = \begin{bmatrix} 1 & \sqrt{2\eta} & \\ \sqrt{2\eta} & .5 & .5 \\ & .5 & .5 \end{bmatrix},$$

which is not even positive definite, since the bottom right 2-by-2 submatrix is exactly singular. Thus, the smallest eigenvalues of $\hat{T}$ is nonpositive, and so tridiagonal reduction has lost all relative accuracy in the smallest eigenvalue. In contrast, one-sided Jacobi's method has no trouble computing the correct square roots of eigenvalues of A, namely, $1 + \sqrt{\eta} = 1 + 10^{-10}$, $1 - \sqrt{\eta} = 1 - 10^{-10}$, and $.99\eta = .99 \cdot 10^{-20}$, to nearly full machine precision. $\diamond$

2. The most general situation in which we understand how to compute the SVD of A to high relative accuracy is when we can accurately compute *any* factorization $A = YDX$, where X and Y are well-conditioned but otherwise arbitrary and D is diagonal. In the last example we had $L = DX$; i.e., Y was the identity matrix. Gaussian elimination with complete pivoting is another source of such factorizations (with Y lower triangular and X upper triangular). For details, see [74]. For applications of this idea to indefinite symmetric eigenproblems, see [228, 250], and for generalized symmetric eigenvalue problems, see [66, 92]

5.5. Differential Equations and Eigenvalue Problems

We seek our motivation for this section from conservation laws in physics. We consider once again the mass-spring system introduced in Example 4.1 and reexamined in Example 5.1. We start with the simplest case of one spring and one mass, without friction:

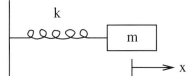

We let x denote horizontal displacement from equilibrium. Then Newton's law $F = ma$ becomes $m\ddot{x}(t) + kx(t) = 0$. Let $E(t) = \frac{1}{2}m\dot{x}^2(t) + \frac{1}{2}kx^2(t) =$ "kinetic energy" + "potential energy." Conservation of energy tells us that $\frac{d}{dt}E(t)$ should be zero. We can confirm this is true by computing $\frac{d}{dt}E(t) = m\dot{x}(t)\ddot{x}(t) + kx(t)\dot{x}(t) = \dot{x}(t)(m\ddot{x}(t) + kx(t)) = 0$ as desired.

More generally we have $M\ddot{x}(t) + Kx(t) = 0$, where M is the *mass matrix* and K is the *stiffness matrix*. The energy is defined to be $E(t) = \frac{1}{2}\dot{x}^T(t)M\dot{x}(t) + \frac{1}{2}x^T(t)Kx(t)$. That this is the correct definition is confirmed by verifying that it is conserved:

$$
\begin{aligned}
\frac{d}{dt}E(t) &= \frac{d}{dt}\left(\frac{1}{2}\dot{x}^T(t)M\dot{x}(t) + \frac{1}{2}x^T(t)Kx(t)\right) \\
&= \frac{1}{2}(\ddot{x}^T(t)M\dot{x}(t) + \dot{x}^T(t)M\ddot{x}(t) + \dot{x}^T(t)Kx(t) + x^T(t)K\dot{x}(t)) \\
&= \dot{x}^T(t)M\ddot{x}(t) + \dot{x}^T(t)Kx(t) \\
&= \dot{x}^T(t)(M\ddot{x}(t) + Kx(t)) = 0,
\end{aligned}
$$

where we have used the symmetry of M and K.

The differential equations $M\ddot{x}(t) + Kx(t) = 0$ are linear. It is a remarkable fact that some nonlinear differential equations also conserve quantities such as "energy."

5.5.1. The Toda Lattice

For ease of notation, we will write $\dot{x}$ instead of $\dot{x}(t)$ when the argument is clear from context.

The *Toda lattice* is also a mass-spring system, but the force from the spring is an exponentially decaying function of its stretch, instead of a linear function:

$$\ddot{x}_i = e^{-(x_i - x_{i-1})} - e^{-(x_{i+1} - x_i)}.$$

We use the boundary conditions $e^{-(x_1 - x_0)} = 0$ (i.e., $x_0 = -\infty$) and $e^{-(x_{n+1} - x_n)} = 0$ (i.e., $x_{n+1} = +\infty$). More simply, these boundary conditions mean there are no walls at the left or right (see Figure 4.1).

Now we change variables to $b_k = \frac{1}{2}e^{(x_k - x_{k+1})/2}$ and $a_k = -\frac{1}{2}\dot{x}_k$. This yields the differential equations

$$
\begin{aligned}
\dot{b}_k &= \frac{1}{2}e^{(x_k - x_{k+1})/2} \cdot \frac{1}{2}(\dot{x}_k - \dot{x}_{k+1}) = b_k(a_{k+1} - a_k), \\
\dot{a}_k &= -\frac{1}{2}\ddot{x}_k = 2(b_k^2 - b_{k-1}^2)
\end{aligned}
\qquad (5.27)
$$

with $b_0 \equiv 0$ and $b_n \equiv 0$. Now define the two tridiagonal matrices

$$
T = \begin{bmatrix} a_1 & b_1 & & \\ b_1 & \ddots & \ddots & \\ & \ddots & \ddots & b_{n-1} \\ & & b_{n-1} & a_n \end{bmatrix}
\quad \text{and} \quad
B = \begin{bmatrix} 0 & b_1 & & \\ -b_1 & \ddots & \ddots & \\ & \ddots & \ddots & b_{n-1} \\ & & -b_{n-1} & 0 \end{bmatrix},
$$

where $B = -B^T$. Then one can easily confirm that equation (5.27) is the same as $\frac{dT}{dt} = BT - TB$. This is called the *Toda flow*.

THEOREM 5.16. *$T(t)$ has the same eigenvalues as $T(0)$ for all t. In other words, the eigenvalues, like "energy," are conserved by the differential equation.*

Proof. Define $\frac{d}{dt}U = BU$, $U(0) = I$. We claim that $U(t)$ is orthogonal for all t. To prove this, it suffices to show $\frac{d}{dt}U^T U = 0$ since $U^T U(0) = I$:

$$\frac{d}{dt}U^T U = \dot{U}^T U + U^T \dot{U} = U^T B^T U + U^T BU = -U^T BU + U^T BU = 0$$

since B is skew symmetric.

Now we claim that $T(t) = U(t)T(0)U^T(t)$ satisfies the Toda flow $\frac{dT}{dt} = BT - TB$, implying each $T(t)$ is orthogonally similar to $T(0)$ and so has the same eigenvalues:

$$\begin{aligned}
\frac{d}{dt}T(t) &= \dot{U}(t)T(0)U^T(t) + U(t)T(0)\dot{U}^T(t) \\
&= B(t)U(t)T(0)U^T(t) + U(t)T(0)U^T(t)B^T(t) \\
&= B(t)T(t) - T(t)B(t)
\end{aligned}$$

as desired. $\square$

Note that the only property of B used was skew symmetry, so if $\frac{d}{dt}T = BT - TB$ and $B^T = -B$, then $T(t)$ has the same eigenvalues for all t.

THEOREM 5.17. *As $t \to +\infty$ or $t \to -\infty$, $T(t)$ converges to a diagonal matrix with the eigenvalues on the diagonal.*

Proof. We want to show $b_i(t) \to 0$ as $t \to \pm\infty$. We begin by showing $\int_{-\infty}^{\infty} \sum_{i=1}^{n-1} b_i^2(t)dt < \infty$. We use induction to show $\int_{-\infty}^{\infty}(b_j^2(t) + b_{n-j}^2(t))dt < \infty$ and then add these inequalities for all j. When $j = 0$, we get $\int_{-\infty}^{\infty}(b_0^2(t) + b_n^2(t))dt$, which is 0 by assumption.

Now let $\varphi(t) = a_j(t) - a_{n-j+1}(t)$. $\varphi(t)$ is bounded by $2\|T(t)\|_2 = 2\|T(0)\|_2$ for all t. Then

$$\begin{aligned}
\dot{\varphi}(t) &= \dot{a}_j(t) - \dot{a}_{n-j+1}(t) \\
&= 2(b_j^2(t) - b_{j-1}^2(t)) - 2(b_{n-j+1}^2(t) - b_{n-j}^2(t)) \\
&= 2(b_j^2(t) + b_{n-j}^2(t)) - 2(b_{j-1}^2(t) - b_{n-j+1}^2(t))
\end{aligned}$$

and so

$$\begin{aligned}
\varphi(\tau) - \varphi(-\tau) &= \int_{-\tau}^{\tau} \dot{\varphi}(t)dt \\
&= 2\int_{-\tau}^{\tau}(b_j^2(t) + b_{n-j}^2(t))dt - 2\int_{-\tau}^{\tau}(b_{j-1}^2(t) + b_{n-j+1}^2(t))dt.
\end{aligned}$$

The last integral is bounded for all τ by the induction hypothesis, and $\varphi(\tau) - \varphi(-\tau)$ is also bounded for all τ, so $\int_{-\infty}^{\infty}(b_j^2(t) + b_{n-j}^2(t))dt$ must be bounded as desired.

Let $p(t) = \sum_{i=1}^{n-1} b_i^2(t)$. We now know that $\int_{-\infty}^{\infty} p(t)dt < \infty$, and since $p(t) \geq 0$ we want to conclude that $\lim_{t \to \pm\infty} p(t) = 0$. But we need to exclude the possibility that $p(t)$ has narrow spikes as $t \to \pm\infty$, in which case $\int_{-\infty}^{\infty} p(t)dt$ could be finite without $p(t)$ approaching 0. We show $p(t)$ has no spikes by showing its derivative is bounded:

$$|\dot{p}(t)| = \left| \sum_{i=1}^{n-1} 2\dot{b}_i(t)b_i(t) \right| = \left| \sum_{i=1}^{n-1} 2b_i^2(t)(a_{i+1}(t) - a_i(t)) \right| \leq 4(n-1)\|T\|_2^2. \quad \square$$

Thus, in principle, one could use an ODE solver on the Toda flow to solve the eigenvalue problem, but this is no faster than other existing methods. The interest in the Toda flow lies in its close relationship with with QR algorithm.

DEFINITION 5.5. *Let X_- denote the strictly lower triangle of X, and $\pi_0(X) = X_- - X_-^T$.*

Note that $\pi_0(X)$ is skew symmetric and that if X is already skew symmetric, then $\pi_0(X) = X$. Thus π_0 *projects* onto skew symmetric matrices.

Consider the differential equation

$$\frac{d}{dt}T = BT - TB, \tag{5.28}$$

where $B = -\pi_0(F(T))$ and F is any smooth function from the real numbers to the real numbers. Since $B = -B^T$, Theorem 5.16 shows that $T(t)$ has the same eigenvalues for all t. Choosing $F(x) = x$ corresponds to the Toda flow that we just studied, since in this case

$$-\pi_0(F(T)) = -\pi_0(T) = \begin{bmatrix} 0 & b_1 & & \\ -b_1 & \ddots & \ddots & \\ & \ddots & \ddots & b_{n-1} \\ & & -b_{n-1} & 0 \end{bmatrix} = B.$$

The next theorem relates the QR decomposition to the solution of differential equation (5.28).

THEOREM 5.18. *Let $F(T(0)) = F_0$. Let $e^{tF_0} = Q(t)R(t)$ be the QR decomposition. Then $T(t) = Q^T(t)T(0)Q(t)$ solves equation (5.28).*

We delay the proof of the theorem until later. If we choose the function F correctly, it turns out that the iterates computed by QR iteration (Algorithm 4.4) are *identical* to the solutions of the differential equation.

DEFINITION 5.6. *Choosing $F(x) = \log x$ in equation (5.28) yields a differential equation called the* QR flow.

COROLLARY 5.3. *Let $F(x) = \log x$. Suppose that $T(0)$ is positive definite, so $\log T(0)$ is real. Let $T_0 \equiv T(0) = QR$, $T_1 = RQ$, etc. be the sequence of matrices produced by the unshifted QR iteration. Then $T(i) = T_i$. Thus the QR algorithm gives solutions to the QR flow at integer times t.*[23]

---

[23]Note that since the QR decomposition is not completely unique (Q can be replaced by QS and R can be replaced by SR, where S is a diagonal matrix with diagonal entries ± 1). T_i and $T(i)$ could actually differ by a similarity $T_i = ST(i)S^{-1}$. For simplicity we will assume here, and in Corollary 5.4, that S has been chosen so that $T_i = T(i)$.

Proof of Corollary. At $t = 1$, we get $e^{t \log T_0} = T_0 = Q(1)R(1)$, the QR decomposition of T_0, and $T(1) = Q^T(1)T_0Q(1) = R(1)Q(1) = T_1$ as desired. Since the solution of the ODE is unique, this extends to show $T(i) = T_i$ for larger i. $\square$

The following figure illustrates this corollary graphically. The curve represents the solution of the differential equation. The dots represent the solutions $T(i)$ at the integer times $t = 0, 1, 2, \ldots$ and indicate that they are equal to the QR iterates T_i.

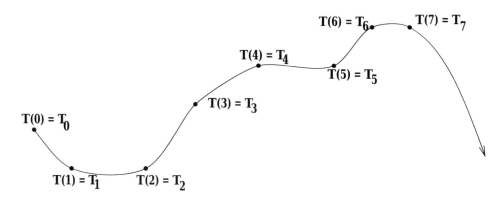

Proof of Theorem 5.18. Differentiate $e^{tF_0} = QR$ to get

$$
\begin{aligned}
F_0 e^{tF_0} &= \dot{Q}R + Q\dot{R} \\
\text{or } \dot{Q} &= F_0 e^{tF_0} R^{-1} - Q\dot{R}R^{-1} \\
\text{or } Q^T\dot{Q} &= Q^T F_0 e^{tF_0} R^{-1} - \dot{R}R^{-1} \\
&= Q^T F_0 (QR)R^{-1} - \dot{R}R^{-1} \quad \text{because } e^{tF_0} = QR \\
&= Q^T F(T(0))Q - \dot{R}R^{-1} \quad \text{because } F_0 = F(T(0)) \\
&= F(Q^T T(0)Q) - \dot{R}R^{-1} \\
&= F(T) - \dot{R}R^{-1}.
\end{aligned}
$$

Now $I = Q^TQ$ implies that $0 = \frac{d}{dt}Q^TQ = \dot{Q}^TQ + Q^T\dot{Q} = (Q^T\dot{Q})^T + (Q^T\dot{Q})$. This means $Q^T\dot{Q}$ is skew symmetric, and so $\pi_0(Q^T\dot{Q}) = Q^T\dot{Q} = \pi_0(F(T) - \dot{R}R^{-1})$. Since $\dot{R}R^{-1}$ is upper triangular, it doesn't affect π_0 and so finally $Q^T\dot{Q} = \pi_0(F(T))$. Now

$$
\begin{aligned}
\frac{d}{dt}T(t) &= \frac{d}{dt}Q^T(t)T(0)Q(t) \\
&= \dot{Q}^T T(0)Q + Q^T T(0)\dot{Q} \\
&= \dot{Q}^T(QQ^T)T(0)Q + Q^T T(0)(QQ^T)\dot{Q} \\
&= \dot{Q}^T QT(t) + T(t)Q^T\dot{Q} \\
&= -Q^T\dot{Q}T(t) + T(t)Q^T\dot{Q} \\
&= -\pi_0(F(T(t)))T(t) + T(t)\pi_0(F(T(t)))
\end{aligned}
$$

as desired. $\square$

The next corollary explains the phenomenon observed in Question 4.15, where QR could be made to "run backward" and return to its starting matrix. See also Question 5.25.

COROLLARY 5.4. *Suppose that we obtain T_4 from the positive definite matrix T_0 by the following steps:*

1. *Do m steps of the unshifted QR algorithm on T_0 to get T_1.*

2. *Let $T_2 =$ "flipped T_1" $= JT_1J$, where J equals the identity matrix with its columns in reverse order.*

3. *Do m steps of unshifted QR on T_2 to get T_3.*

4. *Let $T_4 = JT_3J$.*

Then $T_4 = T_0$.

Proof. If $X = X^T$, it is easy to verify that $\pi_0(JXJ) = -J\pi_0(X)J$ so $T_J(t) \equiv JT(t)J$ satisfies

$$
\begin{aligned}
\tfrac{d}{dt}T_J(t) &= J\tfrac{d}{dt}T(t)J \\
&= J[-\pi_0(F(T))T + T\pi_0(F(T))]J \\
&= -J\pi_0(F(T))J(JTJ) + (JTJ)J\pi_0(F(T))J \quad \text{since } J^2 = I \\
&= \pi_0(JF(T)J)T_J - T_J\pi_0(JF(T)J) \\
&= \pi_0(F(JTJ))T_J - T_J\pi_0(F(JTJ)) \\
&= \pi_0(F(T_J))T_J - T_J\pi_0(F(T_J)).
\end{aligned}
$$

This is nearly the same equation as $T(t)$. In fact, it satisfies *exactly* the same equation as $T(-t)$:

$$
\frac{d}{dt}T(-t) = -\frac{d}{dt}T|_{-t} = -[\pi_0(F(T))T + T\pi_0(F(T))]_{-t}.
$$

So with the same initial conditions T_2, $T_J(t)$, and $T(-t)$ must be equal. Integrating for time m, $T(-t)$ takes $T_2 = JT_1J$ back to JT_0J, the initial state, so $T_3 = JT_0J$ and $T_4 = JT_3J = T_0$ as desired.   □

5.5.2. The Connection to Partial Differential Equations

This section may be skipped on a first reading.

Let $T(t) = -\frac{\partial^2}{\partial x^2} + q(x,t)$ and $B(t) = -4\frac{\partial^3}{\partial x^3} + 3(q(x,t)\frac{\partial}{\partial x} + \frac{\partial}{\partial x}q(x,t))$. Both $T(t)$ and $B(t)$ are linear operators on functions, i.e., generalizations of matrices.

Substituting into $\frac{dT}{dt} = BT - TB$ yields

$$
q_t = 6qq_x - q_{xxx}, \tag{5.29}
$$

provided that we choose the correct boundary conditions for q. (B must be skew symmetric, and T symmetric.) Equation (5.29) is called the *Korteweg–de Vries equation* and describes water flow in a shallow channel. One can rigorously show that (5.29) preserves the eigenvalues of $T(t)$ for all t in the sense that the ODE

$$\left(-\frac{\partial^2}{\partial x^2} + q(x,t)\right) h(x) = \lambda h(x)$$

has some infinite set of eigenvalues $\lambda_1, \lambda_2, \ldots$ for all t. In other words, there is an infinite sequence of energylike quantities conserved by the Korteweg–de Vries equation. This is important for both theoretical and numerical reasons.

For more details on the Toda flow, see [144, 170, 67, 68, 239] and papers by Kruskal [166], Flaschka [106], and Moser [187] in [188].

5.6. References and Other Topics for Chapter 5

An excellent general reference for the symmetric eigenproblem is [197]. The material on relative perturbation theory can be found in [75, 82, 101]; section 5.2.1 was based on the latter of these references. Related work is found in [66, 92, 228, 250]. A classical text on perturbation theory for general linear operators is [161]. For a survey of parallel algorithms for the symmetric eigenproblem, see [76]. The QR algorithm for finding the SVD of bidiagonal matrices is discussed in [80, 67, 120], and the dqds algorithm is in [104, 200, 209]. For an error analysis of the Bisection algorithm, see [73, 74, 156], and for recent attempts to accelerate Bisection see [105, 203, 201, 176, 173, 175, 269]. Current work in improving inverse iteration appears in [105, 83, 201, 203]. The divide-and-conquer eigenroutine was introduced in [59] and further developed in [13, 90, 127, 131, 153, 172, 210, 234]. The possibility of high-accuracy eigenvalues obtained from Jacobi is discussed in [66, 75, 82, 92, 183, 228]. The Toda flow and related phenomena are discussed in [67, 68, 106, 144, 166, 170, 187, 188, 239].

5.7. Questions for Chapter 5

QUESTION 5.1. *(Easy; Z. Bai)* Show that $A = B + iC$ is Hermitian if and only if

$$M = \left[\begin{array}{cc} B & -C \\ C & B \end{array}\right]$$

is symmetric. Express the eigenvalues and eigenvectors of M in terms of those of A.

QUESTION 5.2. *(Medium)* Prove Corollary 5.1, using Weyl's theorem (Theorem 5.1) and part 4 of Theorem 3.3.

QUESTION 5.3. *(Medium)* Consider Figure 5.1. Consider the corresponding contour plot for an arbitrary 3-by-3 matrix A with eigenvalues $\alpha_3 \leq \alpha_2 \leq \alpha_1$. Let C_1 and C_2 be the two great circles along which $\rho(u, A) = \alpha_2$. At what angle do they intersect?

QUESTION 5.4. *(Hard)* Use the Courant–Fischer minimax theorem (Theorem 5.2) to prove the *Cauchy interlace theorem*:

- Suppose that $A = \begin{bmatrix} H & b \\ b^T & u \end{bmatrix}$ is an n-by-n symmetric matrix and H is $(n-1)$-by-$(n-1)$. Let $\alpha_n \leq \cdots \leq \alpha_1$ be the eigenvalues of A and $\theta_{n-1} \leq \cdots \leq \theta_1$ be the eigenvalues of H. Show that these two sets of eigenvalues *interlace*:

$$\alpha_n \leq \theta_{n-1} \leq \cdots \leq \theta_i \leq \alpha_i \leq \theta_{i-1} \leq \alpha_{i-1} \leq \cdots \leq \theta_1 \leq \alpha_1.$$

- Let $A = \begin{bmatrix} H & B \\ B^T & U \end{bmatrix}$ be n-by-n and H be m-by-m, with eigenvalues $\theta_m \leq \cdots \leq \theta_1$. Show that the eigenvalues of A and H interlace in the sense that $\alpha_{j+(n-m)} \leq \theta_j \leq \alpha_j$ (or equivalently $\alpha_j \leq \theta_{j-(n-m)} \leq \alpha_{j-(n-m)}$).

QUESTION 5.5. *(Medium)* Let $A = A^T$ with eigenvalues $\alpha_1 \geq \cdots \geq \alpha_n$. Let $H = H^T$ with eigenvalues $\theta_1 \geq \cdots \geq \theta_n$. Let $A + H$ have eigenvalues $\lambda_1 \geq \cdots \geq \lambda_n$. Use the Courant–Fischer minimax theorem (Theorem 5.2) to show that $\alpha_j + \theta_n \leq \lambda_j \leq \alpha_j + \theta_1$. If H is positive definite, conclude that $\lambda_j > \alpha_j$. In other words, adding a symmetric positive definite matrix H to another symmetric matrix A can only increase its eigenvalues.

This result will be used in the proof of Theorem 7.1.

QUESTION 5.6. *(Medium)* Let $A = [A_1, A_2]$ be n-by-n, where A_1 is n-by-m and A_2 is n-by-$(n-m)$. Let $\sigma_1 \geq \cdots \geq \sigma_n$ be the singular values of A and $\tau_1 \geq \cdots \geq \tau_m$ be the singular values of A_1. Use the Cauchy interlace theorem from Question 5.4 and part 4 of Theorem 3.3 to prove that $\sigma_j \geq \tau_j \geq \sigma_{j+n-m}$.

QUESTION 5.7. *(Medium)* Let q be a unit vector and d be any vector orthogonal to q. Show that $\|(q+d)q^T - I\|_2 = \|q+d\|_2$. (This result is used in the proof of Theorem 5.4.)

QUESTION 5.8. *(Hard)* Formulate and prove a theorem for singular vectors analogous to Theorem 5.4.

QUESTION 5.9. *(Hard)* Prove bound (5.6) from Theorem 5.5.

QUESTION 5.10. *(Harder)* Prove bound (5.7) from Theorem 5.5.

QUESTION 5.11. *(Easy)* Suppose $\theta = \theta_1 + \theta_2$, where all three angles lie between 0 and $\pi/2$. Prove that $\frac{1}{2}\sin 2\theta \leq \frac{1}{2}\sin 2\theta_1 + \frac{1}{2}\sin 2\theta_2$. This result is used in the proof of Theorem 5.7.

QUESTION 5.12. *(Hard)* Prove Corollary 5.2. Hint: Use part 4 of Theorem 3.3.

QUESTION 5.13. *(Medium)* Let A be a symmetric matrix. Consider running shifted QR iteration (Algorithm 4.5) with a Rayleigh quotient shift ($\sigma_i = a_{nn}$) at every iteration, yielding a sequence $\sigma_1, \sigma_2, \ldots$ of shifts. Also run Rayleigh quotient iteration (Algorithm 5.1), starting with $x_0 = [0, \ldots, 0, 1]^T$, yielding a sequence of Rayleigh quotients $\rho_1, \rho_2, \ldots$. Show that these sequences are identical: $\sigma_i = \rho_i$ for all i. This justifies the claim in section 5.3.2 that shifted QR iteration enjoys local cubic convergence.

QUESTION 5.14. *(Easy)* Prove Lemma 5.1.

QUESTION 5.15. *(Easy)* Prove that if $t(n) = 2t(n/2) + cn^3 + O(n^2)$, then $t(n) \approx c\frac{4}{3}n^3$. This justifies the complexity analysis of the divide-and-conquer algorithm (Algorithm 5.2).

QUESTION 5.16. *(Easy)* Let $A = D + \rho uu^T$, where $D = \text{diag}(d_1, \ldots, d_n)$ and $u = [u_1, \ldots, u_n]^T$. Show that if $d_i = d_{i+1}$ or $u_i = 0$, then d_i is an eigenvalue of A. If $u_i = 0$, show that the eigenvector corresponding to d_i is e_i, the ith column of the identity matrix. Derive a similarly simple expression when $d_i = d_{i+1}$. This shows how to handle deflation in the divide-and-conquer algorithm, Algorithm 5.2.

QUESTION 5.17. *(Easy)* Let ψ and ψ' be given scalars. Show how to compute scalars c and $\hat{c}$ in the function definition $h(\lambda) = \hat{c} + \frac{c}{d-\lambda}$ so that at $\lambda = \xi$, $h(\xi) = \psi$ and $h'(\xi) = \psi'$. This result is needed to derive the secular equation solver in section 5.3.3.

QUESTION 5.18. *(Easy; Z. Bai)* Use the SVD to show that if A is an m-by-n real matrix with $m \geq n$, then there exists an m-by-n matrix Q with orthonormal columns ($Q^T Q = I$) and an n-by-n positive semidefinite matrix P such that $A = QP$. This decomposition is called the *polar decomposition of A*, because it is analogous to the polar form of a complex number $z = e^{i\arg(z)} \cdot |z|$.) Show that if A is nonsingular, then the polar decomposition is unique.

QUESTION 5.19. *(Easy)* Prove Lemma 5.5.

QUESTION 5.20. *(Easy)* Prove Lemma 5.7.

QUESTION 5.21. *(Hard)* Prove Theorem 5.13. Also, reduce the exponent $4n - 2$ in Theorem 5.13 to $2n - 1$. Hint: In Lemma 5.7, multiply D_1 and divide D_2 by an appropriately chosen constant.

QUESTION 5.22. *(Medium)* Prove that Algorithm 5.13 computes the SVD of G, assuming that $G^T G$ converges to a diagonal matrix.

QUESTION 5.23. *(Harder)* Let A be an n-by-n symmetric positive definite matrix with Cholesky decomposition $A = LL^T$, and let $\hat{L}$ be the Cholesky factor computed in floating point arithmetic. In this question we will bound the relative error in the (squared) singular values of $\hat{L}$ as approximations of the eigenvalues of A. Show that A can be written $A = D\bar{A}D$, where $D = \text{diag}(a_{11}^{1/2}, \ldots, a_{nn}^{1/2})$ and $\bar{a}_{ii} = 1$ for all i. Write $L = DX$. Show that $\kappa^2(X) = \kappa(\bar{A})$. Using bound (2.16) for the backward error δA of Cholesky $A + \delta A = \hat{L}\hat{L}^T$, show that one can write $\hat{L}^T\hat{L} = Y^TL^TLY$, where $\|Y^TY - I\|_2 \leq O(\varepsilon)\kappa(\bar{A})$. Use Theorem 5.6 to conclude that the eigenvalues of $\hat{L}^T\hat{L}$ and of L^TL differ relatively by at most $O(\varepsilon)\kappa(\bar{A})$. Then show that this is also true of the eigenvalues of $\hat{L}\hat{L}^T$ and LL^T. This means that the squares of the singular values of $\hat{L}$ differ relatively from the eigenvalues of A by at most $O(\varepsilon)\kappa(\bar{A}) = O(\varepsilon)\kappa^2(L)$.

QUESTION 5.24. *(Harder)* This question justifies the stopping criterion for one-sided Jacobi's method for the SVD (Algorithm 5.13). Let $A = G^TG$, where G and A are n-by-n. Suppose that $|a_{jk}| \leq \epsilon\sqrt{a_{jj}a_{kk}}$ for all $j \neq k$. Let $\sigma_n \leq \cdots \leq \sigma_1$ be the singular values of G, and $\alpha_n^2 \leq \cdots \leq \alpha_1^2$ be the sorted diagonal entries of A. Prove that $|\sigma_i - \alpha_i| \leq n\epsilon|\alpha_i|$ so that the α_i equal the singular values to high relative accuracy. Hint: Use Corollary 5.2.

QUESTION 5.25. *(Harder)* In Question 4.15, you "noticed" that running QR for m steps on a symmetric matrix, "flipping" the rows and columns, running for another m steps, and flipping again got you back to the original matrix. (Flipping X means replacing X by JXJ, where J is the identity matrix with its row in reverse order.) In this exercise we will prove this for symmetric positive definite matrices T, using an approach different from Corollary 5.4.

Consider LR iteration (Algorithm 5.9) with a zero shift, applied to the symmetric positive definite matrix T (which is not necessarily tridiagonal): Let $T = T_0 = B_0^TB_0$ be the Cholesky decomposition, $T_1 = B_0B_0^T = B_1^TB_1$, and more generally $T_i = B_{i-1}B_{i-1}^T = B_i^TB_i$. Let $\hat{T}_i$ denote the matrix obtained from T_0 after i steps of unshifted QR iteration; i.e., if $\hat{T}_i = Q_iR_i$ is the QR decomposition, then $\hat{T}_{i+1} = R_iQ_i$. In Lemma 5.6 we showed that $\hat{T}_i = T_{2i}$; i.e., one step of QR is the same as two steps of LR.

1. Show that $T_i = (B_{i-1}B_{i-2}\cdots B_0)^{-T}T_0(B_{i-1}B_{i-2}\cdots B_0)^T$.

2. Show that $T_i = (B_{i-1}B_{i-2}\cdots B_0)T_0(B_{i-1}B_{i-2}\cdots B_0)^{-1}$.

3. Show that $T_0^i = (B_iB_{i-1}\cdots B_0)^T(B_iB_{i-1}\cdots B_0)$ is the Cholesky decomposition of T_0^i.

4. Show that $T_0^i = (Q_0\cdots Q_{i-2}Q_{i-1}) \cdot (R_{i-1}R_{i-2}\cdots R_0)$ is the QR decomposition of T_0^i.

5. Show that $T_0^{2i} = (R_{2i-1}R_{2i-2}\cdots R_0)^T(R_{2i-1}R_{2i-2}\cdots R_0)$ is the Cholesky decomposition of T_0^{2i}.

6. Show that the result after m steps of QR, flipping, m steps of QR, and flipping is the same as the original matrix. Hint: Use the fact that the Cholesky factorization is unique.

QUESTION 5.26. *(Hard; Z. Bai)* Suppose that x is an n-vector. Define the matrix C by $c_{ij} = |x_i| + |x_j| - |x_i - x_j|$. Show that $C(x)$ is positive semidefinite.

QUESTION 5.27. *(Easy; Z. Bai)* Let

$$A = \begin{pmatrix} I & B \\ B^H & I \end{pmatrix}$$

with $\|B\|_2 < 1$. Show that

$$\|A\|_2 \|A^{-1}\|_2 = \frac{1 + \|B\|_2}{1 - \|B\|_2}.$$

QUESTION 5.28. *(Medium; Z. Bai)* A square matrix A is said to be *skew Hermitian* if $A^* = -A$. Prove that

1. the eigenvalues of a skew Hermitian are purely imaginary.

2. $I - A$ is nonsingular.

3. $C = (I - A)^{-1}(I + A)$ is unitary. C is called the *Cayley transform* of A.

6

Iterative Methods for Linear Systems

6.1. Introduction

Iterative algorithms for solving $Ax = b$ are used when methods such as Gaussian elimination require too much time or too much space. Methods such as Gaussian elimination, which compute the exact answers after a finite number of steps (in the absence of roundoff!), are called *direct methods*. In contrast to direct methods, iterative methods generally do not produce the exact answer after a finite number of steps but decrease the error by some fraction after each step. Iteration ceases when the error is less than a user-supplied threshold. The final error depends on how many iterations one does as well as on properties of the method and the linear system. Our overall goal is to develop methods that decrease the error by a large amount at each iteration and do as little work per iteration as possible.

Much of the activity in this field involves exploiting the underlying mathematical or physical problem that gives rise to the linear system in order to design better iterative methods. The underlying problems are often finite difference or finite element models of physical systems, usually involving a differential equation. There are many kinds of physical systems, differential equations, and finite difference and finite element models, and so many methods. We cannot hope to cover all or even most interesting situations, so we will limit ourselves to a *model problem*, the standard finite difference approximation to Poisson's equation on a square. Poisson's equation and its close relation, Laplace's equation, arise in many applications, including electromagnetics, fluid mechanics, heat flow, diffusion, and quantum mechanics, to name a few. In addition to describing how each method works on Poisson's equation, we will indicate how generally applicable it is, and describe common variations.

The rest of this chapter is organized as follows. Section 6.2 describes on-line help and software for iterative methods discussed in this chapter. Section 6.3 describes the formulation of the model problem in detail. Section 6.4 summarizes and compares the performance of (nearly) all the iterative methods in this chapter for solving the model problem.

The next five sections describe methods in roughly increasing order of their effectiveness on the model problem. Section 6.5 describes the most basic iterative methods: Jacobi's, Gauss–Seidel, successive overrelaxation, and their variations. Section 6.6 describes Krylov subspace methods, concentrating on the conjugate gradient method. Section 6.7 describes the fast Fourier transform and how to use it to solve the model problem. Section 6.8 describes block cyclic reduction. Finally, section 6.9 discusses multigrid, our fastest algorithm for the model problem. Multigrid requires only $O(1)$ work per unknown, which is optimal.

Section 6.10 describes domain decomposition, a family of techniques for combining the simpler methods described in earlier sections to solve more complicated problems than the model problem.

6.2.   On-line Help for Iterative Methods

For Poisson's equation, there will be a short list of numerical methods that are clearly superior to all the others we discuss. But for other linear systems it is not always clear which method is best (which is why we talk about so many!). To help users select the best method for solving their linear systems among the many available, on-line help is available at NETLIB/templates. This directory contains a short book [24] and software for most of the iterative methods discussed in this chapter. The book is available in both PostScript (NETLIB/templates/templates.ps) and Hypertext Markup Language (NETLIB/templates/Templates.html). The software is available in Matlab, Fortran, and C++.

The word *template* is used to describe this book and the software, because the implementations separate the details of matrix representations from the algorithm itself. In particular, the *Krylov subspace methods* (see section 6.6) require only the ability to multiply the matrix A by an arbitrary vector z. The best way to do this depends on how A is represented but does not otherwise affect the organization of the algorithm. In other words, matrix-vector multiplication is a "black-box" called by the template. It is the user's responsibility to supply an implementation of this black-box.

An analogous templates project for eigenvalue problems is underway. Other recent textbooks on iterative methods are [15, 136, 214].

For the most challenging practical problems arising from differential equations more challenging than our model problem, the linear system $Ax = b$ must be "preconditioned," or replaced with the equivalent systems $M^{-1}Ax = M^{-1}b$, which is somehow easier to solve. This is discussed at length in sections 6.6.5 and 6.10. Implementations, including parallel ones, of many of these techniques are available on-line in the package PETSc, or Portable Extensible Toolkit for Scientific computing, at http://www.mcs.anl.gov/petsc/petsc.html [232].

6.3. Poisson's Equation

6.3.1. Poisson's Equation in One Dimension

We begin with a one-dimensional version of Poisson's equation,

$$-\frac{d^2 v(x)}{dx^2} = f(x), \quad 0 < x < 1, \tag{6.1}$$

where $f(x)$ is a given function and $v(x)$ is the unknown function that we want to compute. $v(x)$ must also satisfy the boundary conditions[24] $v(0) = v(1) = 0$. We *discretize* the problem by trying to compute an approximate solution at $N + 2$ evenly spaced points x_i between 0 and 1: $x_i = ih$, where $h = \frac{1}{N+1}$ and $0 \le i \le N + 1$. We abbreviate $v_i = v(x_i)$ and $f_i = f(x_i)$. To convert differential equation (6.1) into a linear equation for the unknowns $v_1, \ldots, v_N$, we use *finite differences* to approximate

$$\left. \frac{dv(x)}{dx} \right|_{x=(i-.5)h} \approx \frac{v_i - v_{i-1}}{h},$$

$$\left. \frac{dv(x)}{dx} \right|_{x=(i+.5)h} \approx \frac{v_{i+1} - v_i}{h}.$$

Subtracting these approximations and dividing by h yield the *centered difference approximation*

$$\left. -\frac{d^2 v(x)}{dx^2} \right|_{x=x_i} = \frac{2v_i - v_{i-1} - v_{i+1}}{h^2} - \tau_i, \tag{6.2}$$

where τ_i, the so-called *truncation error*, can be shown to be $O(h^2 \cdot \|\frac{d^4 v}{dx^4}\|_\infty)$. We may now rewrite equation (6.1) at $x = x_i$ as

$$-v_{i-1} + 2v_i - v_{i+1} = h^2 f_i + h^2 \tau_i,$$

where $0 < i < N+1$. Since the boundary conditions imply that $v_0 = v_{N+1} = 0$, we have N equations in N unknowns $v_1, \ldots, v_N$:

$$T_N \cdot \begin{bmatrix} v_1 \\ \vdots \\ \vdots \\ v_N \end{bmatrix} \equiv \begin{bmatrix} 2 & -1 & & 0 \\ -1 & \ddots & \ddots & \\ & \ddots & \ddots & -1 \\ 0 & & -1 & 2 \end{bmatrix} \cdot \begin{bmatrix} v_1 \\ \vdots \\ \vdots \\ v_N \end{bmatrix}$$

$$= h^2 \begin{bmatrix} f_1 \\ \vdots \\ \vdots \\ f_N \end{bmatrix} + h^2 \begin{bmatrix} \tau_1 \\ \vdots \\ \vdots \\ \tau_N \end{bmatrix} \tag{6.3}$$

---

[24] These are called *Dirichlet boundary conditions*. Other kinds of boundary conditions are also possible.

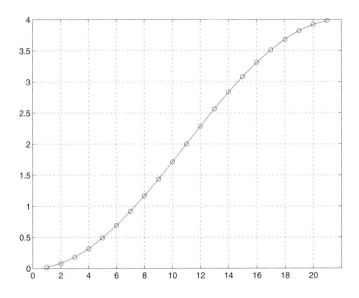

Fig. 6.1. *Eigenvalues of T_{21}.*

or

$$T_N v = h^2 f + h^2 \bar{\tau}. \tag{6.4}$$

To solve this equation, we will ignore $\bar{\tau}$, since it is small compared to f, to get

$$T_N \hat{v} = h^2 f. \tag{6.5}$$

(We bound the error $v - \hat{v}$ later.)

The coefficient matrix T_N plays a central role in all that follows, so we will examine it in some detail. First, we will compute its eigenvalues and eigenvectors. One can easily use trigonometric identities to confirm the following lemma (see Question 6.1).

LEMMA 6.1. *The eigenvalues of T_N are $\lambda_j = 2(1 - \cos \frac{\pi j}{N+1})$. The eigenvectors are z_j, where $z_j(k) = \sqrt{\frac{2}{N+1}} \sin(jk\pi/(N+1))$. z_j has unit two-norm. Let $Z = [z_1, \ldots, z_n]$ be the orthogonal matrix whose columns are the eigenvectors, and $\Lambda = \mathrm{diag}(\lambda_1, \ldots, \lambda_n)$, so we can write $T_N = Z\Lambda Z^T$.*

Figure 6.1 is a plot of the eigenvalues of T_N for $N = 21$.

The largest eigenvalue is $\lambda_N = 2(1 - \cos \pi \frac{N}{N+1}) \approx 4$. The smallest eigenvalue[25] is λ_1, where for small i

$$\lambda_i = 2\left(1 - \cos \frac{i\pi}{N+1}\right) \approx 2\left(1 - \left(1 - \frac{i^2\pi^2}{2(N+1)^2}\right)\right) = \left(\frac{i\pi}{N+1}\right)^2.$$

---

[25] Note that λ_N is the largest eigenvalue and λ_1 is the smallest eigenvalue, the opposite of the convention of Chapter 5.

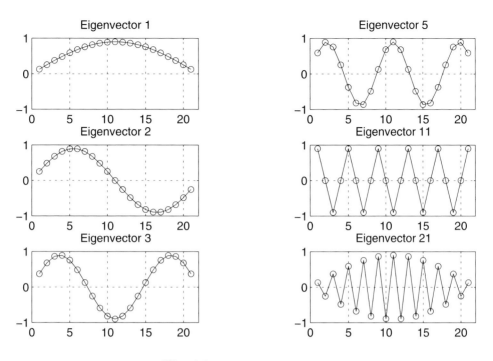

Fig. 6.2. *Eigenvectors of T_{21}.*

Thus T_N is positive definite with condition number $\lambda_N/\lambda_1 \approx 4(N+1)^2/\pi^2$ for large N. The eigenvectors are sinusoids with lowest frequency at $j = 1$ and highest at $j = N$, shown in Figure 6.2 for $N = 21$.

Now we know enough to bound the error, i.e., the difference between the solution of $T_N \hat{v} = h^2 f$ and the true solution v of the differential equation: Subtract equation (6.5) from equation (6.4) to get $v - \hat{v} = h^2 T_N^{-1} \bar{\tau}$. Taking norms yields

$$\|v - \hat{v}\|_2 \leq h^2 \|T_N^{-1}\|_2 \|\bar{\tau}\|_2 \approx h^2 \frac{(N+1)^2}{\pi^2} \|\bar{\tau}\|_2 = O(\|\bar{\tau}\|_2) = O\left(h^2 \left\| \frac{d^4 v}{dx^4} \right\|_\infty \right),$$

so the error $v - \hat{v}$ goes to zero proportionally to h^2, provided that the solution is smooth enough. ($\|\frac{d^4 v}{dx^4}\|_\infty$ is bounded.)

From now on we will not distinguish between v and its approximation $\hat{v}$ and so will simplify notation by letting $T_N v = h^2 f$.

In addition to the solution of the linear system $h^{-2} T_N v = f$ approximating the solution of the differential equation (6.1), it turns out that the eigenvalues and eigenvectors of $h^{-2} T_N$ also approximate the eigenvalues and *eigenfunctions* of the differential equation: We say that $\hat{\lambda}_i$ is an eigenvalue and $\hat{z}_i(x)$ is an eigenfunction of the differential equation if

$$-\frac{d^2 \hat{z}_i(x)}{dx^2} = \hat{\lambda}_i \hat{z}_i(x) \quad \text{with} \quad \hat{z}_i(0) = \hat{z}_i(1) = 0.$$

Let us solve for $\hat{\lambda}_i$ and $\hat{z}_i(x)$: It is easy to see that $\hat{z}_i(x)$ must equal $\alpha \sin(\sqrt{\hat{\lambda}_i}x) + \beta \cos(\sqrt{\hat{\lambda}_i}x)$ for some constants α and β. The boundary condition $\hat{z}_i(0) = 0$ implies $\beta = 0$, and the boundary condition $\hat{z}_i(1) = 0$ implies that $\sqrt{\hat{\lambda}_i}$ is an integer multiple of π, which we can take to be $i\pi$. Thus $\hat{\lambda}_i = i^2\pi^2$ and $\hat{z}_i(x) = \alpha \sin(i\pi x)$ for any nonzero constant α (which we can set to 1). Thus the eigenvector z_i is *precisely* equal to the eigenfunction $\hat{z}_i(x)$ evaluated at the sample points $x_j = jh$ (when scaled by $\sqrt{\frac{2}{N+1}}$). And when i is small, $\hat{\lambda}_i = i^2\pi^2$ is well approximated by $h^{-2}\cdot\lambda_i = (N+1)^2\cdot 2(1-\cos\frac{i\pi}{N+1}) = i^2\pi^2 + O((N+1)^{-2})$.

Thus we see there is a close correspondence between T_N (or $h^{-2}T_N$) and the second derivative operator $-\frac{d^2}{dx^2}$. This correspondence will be the motivation for the design and analysis of later algorithms.

It is also possible to write down simple formulas for the Cholesky and LU factors of T_N; see Question 6.2 for details.

6.3.2. Poisson's Equation in Two Dimensions

Now we turn to Poisson's equation in two dimensions:

$$-\frac{\partial^2 v(x,y)}{\partial x^2} - \frac{\partial^2 v(x,y)}{\partial y^2} = f(x,y) \tag{6.6}$$

on the unit square $\{(x,y) : 0 < x,y < 1\}$, with boundary condition $v = 0$ on the boundary of the square. We discretize at the grid points in the square which are at (x_i, y_j) with $x_i = ih$ and $y_j = jh$, with $h = \frac{1}{N+1}$. We abbreviate $v_{ij} = v(ih, jh)$ and $f_{ij} = f(ih, jh)$, as shown below for $N = 3$:

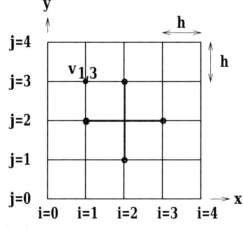

From equation (6.2), we know that we can approximate

$$-\frac{\partial^2 v(x,y)}{\partial x^2}\bigg|_{x=x_i,y=y_j} \approx \frac{2v_{i,j} - v_{i-1,j} - v_{i+1,j}}{h^2} \quad \text{and} \tag{6.7}$$

$$-\frac{\partial^2 v(x,y)}{\partial y^2}\bigg|_{x=x_i,y=y_j} \approx \frac{2v_{i,j} - v_{i,j-1} - v_{i,j+1}}{h^2}. \tag{6.8}$$

Adding these approximations lets us write

$$-\frac{\partial^2 v(x,y)}{\partial x^2} - \frac{\partial^2 v(x,y)}{\partial y^2}\bigg|_{x=x_i, y=y_j}$$

$$= \frac{4v_{ij} - v_{i-1,j} - v_{i+1,j} - v_{i,j-1} - v_{i,j+1}}{h^2} - \tau_{ij}, \tag{6.9}$$

where τ_{ij} is again a truncation error bounded by $O(h^2)$. The heavy (blue) cross in the middle of the above figure is called the (5-*point*) *stencil* of this equation, because it connects all (5) values of v present in equation (6.9). From the boundary conditions we know $v_{0j} = v_{N+1,j} = v_{i,0} = v_{i,N+1} = 0$ so that equation (6.9) defines a set of $n = N^2$ linear equations in the n unknowns v_{ij} for $1 \le i, j \le N$:

$$4v_{ij} - v_{i-1,j} - v_{i+1,j} - v_{i,j-1} - v_{i,j+1} = h^2 f_{ij}. \tag{6.10}$$

There are two ways to rewrite the n equations represented by (6.10) as a single matrix equation, both of which we will use later.

The first way is to think of the unknowns v_{ij} as occupying an N-by-N matrix V with entries v_{ij} and the right-hand sides $h^2 f_{ij}$ as similarly occupying an N-by-N matrix $h^2 F$. The trick is to write the matrix with i, j entry $4v_{ij} - v_{i-1,j} - v_{i+1,j} - v_{i,j-1} - v_{i,j+1}$ in a simple way in terms of V and T_N: Simply note that

$$\begin{aligned}
2v_{ij} - v_{i-1,j} - v_{i+1,j} &= (T_N \cdot V)_{ij}, \\
2v_{ij} - v_{i,j-1} - v_{i,j+1} &= (V \cdot T_N)_{ij},
\end{aligned}$$

so adding these two equations yields

$$(T_N \cdot V + V \cdot T_N)_{ij} = 4v_{ij} - v_{i-1,j} - v_{i+1,j} - v_{i,j-1} - v_{i,j+1} = h^2 f_{ij} = (h^2 F)_{ij}$$

or

$$T_N \cdot V + V \cdot T_N = h^2 F. \tag{6.11}$$

This is a linear system of equations for the unknown entries of the matrix V, even though it is not written in the usual "$Ax = b$" format, with the unknowns forming a vector x. (We will write the "$Ax = b$" format below.) Still, it is enough to tell us what the eigenvalues and eigenvectors of the underlying matrix A are, because "$Ax = \lambda x$" is the same as "$T_N V + V T_N = \lambda V$." Now suppose that $T_N z_i = \lambda_i z_i$ and $T_N z_j = \lambda_j z_j$ are any two eigenpairs of T_N, and let $V = z_i z_j^T$. Then

$$\begin{aligned}
T_N V + V T_N &= (T_N z_i) z_j^T + z_i (z_j^T T_N) \\
&= (\lambda_i z_i) z_j^T + z_i (z_j^T \lambda_j) \\
&= (\lambda_i + \lambda_j) z_i z_j^T \\
&= (\lambda_i + \lambda_j) V, \tag{6.12}
\end{aligned}$$

so $V = z_i z_j^T$ is an "eigenvector" and $\lambda_i + \lambda_j$ is an eigenvalue. Since V has N^2 entries, we expect N^2 eigenvalues and eigenvectors, one for each pair of eigenvalues λ_i and λ_j of T_N. In particular, the smallest eigenvalue is $2\lambda_1$ and the largest eigenvalue is $2\lambda_N$, so the condition number is the same as in the one-dimensional case. We rederive this result below using the "$Ax = b$" format. See Figure 6.3 for plots of some eigenvectors, represented as surfaces defined by the matrix entries of $z_i z_j^T$.

Just as the eigenvalues and eigenvectors of $h^{-2}T_N$ were good approximations to the eigenvalues and eigenfunctions of one-dimensional Poisson's equation, the same is true of two-dimensional Poisson's equation, whose eigenvalues and eigenfunctions are as follows (see Question 6.3):

$$\left(-\frac{\partial^2}{\partial x^2} - \frac{\partial^2}{\partial y^2}\right)\sin(i\pi x)\sin(j\pi y)$$
$$= (i^2\pi^2 + j^2\pi^2)\sin(i\pi x)\sin(j\pi y). \tag{6.13}$$

The second way to write the n equations represented by equation (6.10) as a single matrix equation is to write the unknowns v_{ij} in a single long N^2-by-1 vector. This requires us to choose an order for them, and we (somewhat arbitrarily) choose to number them as shown in Figure 6.4, columnwise from the upper left to the lower right.

For example, when $N = 3$ one gets a column vector $v \equiv [v_1, \ldots, v_9]^T$. If we number f accordingly, we can transform equation (6.10) to get

$$T_{3\times 3} \cdot \begin{bmatrix} v_1 \\ v_2 \\ \vdots \\ \vdots \\ \vdots \\ v_9 \end{bmatrix} \equiv \begin{bmatrix} 4 & -1 & & -1 & & & & & \\ -1 & 4 & -1 & & -1 & & & & \\ & -1 & 4 & & & -1 & & & \\ -1 & & & 4 & -1 & & -1 & & \\ & -1 & & -1 & 4 & -1 & & -1 & \\ & & -1 & & -1 & 4 & & & -1 \\ & & & -1 & & & 4 & -1 & \\ & & & & -1 & & -1 & 4 & -1 \\ & & & & & -1 & & -1 & 4 \end{bmatrix} \begin{bmatrix} v_1 \\ v_2 \\ \vdots \\ \vdots \\ \vdots \\ v_9 \end{bmatrix}$$

$$= h^2 \begin{bmatrix} f_1 \\ f_2 \\ \vdots \\ \vdots \\ \vdots \\ f_9 \end{bmatrix}. \tag{6.14}$$

The -1's immediately next to the diagonal correspond to subtracting the top and bottom neighbors $-v_{i,j-1} - v_{i,j+1}$. The -1's farther away away from the diagonal correspond to subtracting the left and right neighbors $-v_{i-1,j} - v_{i+1,j}$. For general N, we confirm in the next section that we get an N^2-by-N^2 linear system

$$T_{N\times N} \cdot v = h^2 f, \tag{6.15}$$

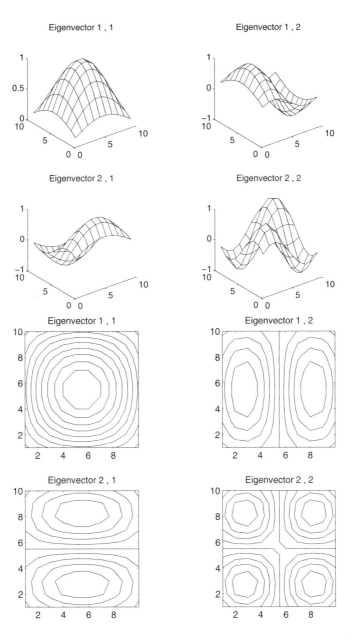

Fig. 6.3. *Three-dimensional and contour plots of first four eigenvectors of the 10-by-10 Poisson equation.*

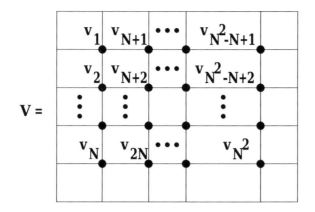

Fig. 6.4. *Numbering the unknowns in Poisson's equation.*

where $T_{N \times N}$ has N N-by-N blocks of the form $T_N + 2I_N$ on its diagonal and $-I_N$ blocks on its offdiagonals:

$$
T_{N \times N} =
\begin{bmatrix}
T_N + 2I_N & -I_N & & \\
-I_N & \ddots & \ddots & \\
& \ddots & \ddots & -I_N \\
& & -I_N & T_N + 2I_N
\end{bmatrix}. \tag{6.16}
$$

6.3.3. Expressing Poisson's Equation with Kronecker Products

Here is a systematic way to derive equations (6.15) and (6.16) as well as to compute the eigenvalues and eigenvectors of $T_{N \times N}$. The method works equally well for Poisson's equation in three or more dimensions.

DEFINITION 6.1. *Let X be m-by-n. Then $\text{vec}(X)$ is defined to be a column vector of size $m \cdot n$ made of the columns of X stacked atop one another from left to right.*

Note that N^2-by-1 vector v defined in Figure 6.4 can also be written $v = \text{vec}(V)$.

To express $T_{N \times N}$ as well as compute its eigenvalues and eigenvectors, we need to introduce *Kronecker products*.

DEFINITION 6.2. *Let A be an m-by-n matrix and B be a p-by-q matrix. Then $A \otimes B$, the* Kronecker product *of A and B, is the $(m \cdot p)$-by-$(n \cdot q)$ matrix*

$$
\begin{bmatrix}
a_{1,1} \cdot B & \cdots & a_{1,n} \cdot B \\
\vdots & & \vdots \\
a_{m,1} \cdot B & \cdots & a_{m,n} \cdot B
\end{bmatrix}.
$$

The following lemma tells us how to rewrite the Poisson equation in terms of Kronecker products and the $\text{vec}(\cdot)$ operator.

LEMMA 6.2. *Let A be m-by-m, B be n-by-n, and X and C be m-by-n. Then the following properties hold:*

1. $\text{vec}(AX) = (I_n \otimes A) \cdot \text{vec}(X)$.

2. $\text{vec}(XB) = (B^T \otimes I_m) \cdot \text{vec}(X)$.

3. *The Poisson equation $T_N V + V T_N = h^2 F$ is equivalent to*

$$T_{N \times N} \cdot \text{vec}(V) \equiv (I_N \otimes T_N + T_N \otimes I_N) \cdot \text{vec}(V) = h^2 \text{vec}(F). \quad (6.17)$$

Proof. We prove only part 3, leaving the other parts to Question 6.4. We start with the Poisson equation $T_N V + V T_N = h^2 F$ as expressed in equation (6.11), which is clearly equivalent to

$$\text{vec}(T_N V + V T_N) = \text{vec}(T_N V) + \text{vec}(V T_N) = \text{vec}(h^2 F).$$

By part 1 of the lemma

$$\text{vec}(T_N V) = (I_N \otimes T_N)\text{vec}(V).$$

By part 2 of the lemma and the symmetry of T_N,

$$\text{vec}(V T_N) = (T_N^T \otimes I_N)\text{vec}(V) = (T_N \otimes I_N)\text{vec}(V).$$

Adding the last two expressions completes the proof of part 3. $\square$

The reader can confirm that the expression

$$
\begin{aligned}
T_{N \times N} \;=\;& I_N \otimes T_N + T_N \otimes I_N \\[4pt]
=\;& \begin{bmatrix} T_N & & & \\ & \ddots & & \\ & & \ddots & \\ & & & T_N \end{bmatrix} + \begin{bmatrix} 2I_N & -I_N & & \\ -I_N & \ddots & \ddots & \\ & \ddots & \ddots & -I_N \\ & & -I_N & 2I_N \end{bmatrix}
\end{aligned}
$$

from equation (6.17) agrees with equation (6.16).[26]

To compute the eigenvalues of matrices defined by Kronecker products, like $T_{N \times N}$, we need the following lemma, whose proof is also part of Question 6.4.

LEMMA 6.3. *The following facts about Kronecker products hold:*

1. *Assume that the products $A \cdot C$ and $B \cdot D$ are well defined. Then $(A \otimes B) \cdot (C \otimes D) = (A \cdot C) \otimes (B \cdot D)$.*

---

[26] We can use this formula to compute $T_{N \times N}$ in two lines of Matlab:

```
TN = 2*eye(N) - diag(ones(N-1,1),1) - diag(ones(N-1,1),-1);
TNxN = kron(eye(N),TN) + kron(TN,eye(N));
```

2. *If A and B are invertible, then $(A \otimes B)^{-1} = A^{-1} \otimes B^{-1}$.*

3. $(A \otimes B)^T = A^T \otimes B^T$.

PROPOSITION 6.1. *Let $T_N = Z \Lambda Z^T$ be the eigendecomposition of T_N, with $Z = [z_1, \ldots, z_N]$ the orthogonal matrix whose columns are eigenvectors, and $\Lambda = \mathrm{diag}(\lambda_1, \ldots, \lambda_N)$. Then the eigendecomposition of $T_{N \times N} = I \otimes T_N + T_N \otimes I$ is*

$$I \otimes T_N + T_N \otimes I = (Z \otimes Z) \cdot (I \otimes \Lambda + \Lambda \otimes I) \cdot (Z \otimes Z)^T. \qquad (6.18)$$

$I \otimes \Lambda + \Lambda \otimes I$ *is a diagonal matrix whose $(iN + j)$th diagonal entry, the (i,j)th eigenvalue of $T_{N \times N}$, is $\lambda_{i,j} = \lambda_i + \lambda_j$. $Z \otimes Z$ is an orthogonal matrix whose $(iN + j)$th column, the corresponding eigenvector, is $z_i \otimes z_j$.*

Proof. From parts 1 and 3 of Lemma 6.3, it is easy to verify that $Z \otimes Z$ is orthogonal, since $(Z \otimes Z)(Z \otimes Z)^T = (Z \otimes Z)(Z^T \otimes Z^T) = (Z \cdot Z^T) \otimes (Z \cdot Z^T) = I \otimes I = I$. We can now verify equation (6.18):

$$
\begin{aligned}
&(Z \otimes Z) \cdot (I \otimes \Lambda + \Lambda \otimes I) \cdot (Z \otimes Z)^T \\
&= (Z \otimes Z) \cdot (I \otimes \Lambda + \Lambda \otimes I) \cdot (Z^T \otimes Z^T) \\
&\qquad \text{by part 3 of Lemma 6.3} \\
&= (Z \cdot I \cdot Z^T) \otimes (Z \cdot \Lambda \cdot Z^T) + (Z \cdot \Lambda \cdot Z^T) \otimes (Z \cdot I \cdot Z^T) \\
&\qquad \text{by part 1 of Lemma 6.3} \\
&= (I) \otimes (T_N) + (T_N) \otimes (I) \\
&= T_{N \times N}.
\end{aligned}
$$

Also, it is easy to verify that $I \otimes \Lambda + \Lambda \otimes I$ is diagonal, with diagonal entry $(iN + j)$ given by $\lambda_j + \lambda_i$, so that equation (6.18) really is the eigendecomposition of $T_{N \times N}$. Finally, from the definition of Kronecker product, one can see that column $iN + j$ of $Z \otimes Z$ is $z_i \otimes z_j$. $\square$

The reader can confirm that the eigenvector $z_i \otimes z_j = \mathrm{vec}(z_j z_i^T)$, thus matching the expression for an eigenvector in equation (6.12).

For a generalization of Proposition 6.1 to the matrix $A \otimes I + B^T \otimes I$, which arises when solving the Sylvester equation $AX - XB = C$, see Question 6.5 (and Question 4.6).

Similarly, Poisson's equation in three dimensions leads to

$$T_{N \times N \times N} \equiv T_N \otimes I_N \otimes I_N + I_N \otimes T_N \otimes I_N + I_N \otimes I_N \otimes T_N,$$

with eigenvalues all possible triple sums of eigenvalues of T_N, and eigenvector matrix $Z \otimes Z \otimes Z$. Poisson's equation in higher dimensions is represented analogously.

Method	Serial Time	Space	Direct or Iterative	Section
Dense Cholesky	n^3	n^2	D	2.7.1
Explicit inverse	n^2	n^2	D	
Band Cholesky	n^2	$n^{3/2}$	D	2.7.3
Jacobi's	n^2	n	I	6.5
Gauss–Seidel	n^2	n	I	6.5
Sparse Cholesky	$n^{3/2}$	$n \cdot \log n$	D	2.7.4
Conjugate gradients	$n^{3/2}$	n	I	6.6
Successive overrelaxation	$n^{3/2}$	n	I	6.5
SSOR with Chebyshev accel.	$n^{5/4}$	n	I	6.5
Fast Fourier transform	$n \cdot \log n$	n	D	6.7
Block cyclic reduction	$n \cdot \log n$	n	D	6.8
Multigrid	n	n	I	6.9
Lower bound	n	n		

Table 6.1. *Order of complexity of solving Poisson's equation on an N-by-N grid* $(n = N^2)$.

6.4. Summary of Methods for Solving Poisson's Equation

Table 6.1 lists the costs of various direct and iterative methods for solving the model problem on an N-by-N grid. The variable $n = N^2$, the number of unknowns. Since direct methods provide the exact answer (in the absence of roundoff), whereas iterative methods provide only approximate answers, we must be careful when comparing their costs, since a low-accuracy answer can be computed more cheaply by an iterative method than a high-accuracy answer. Therefore, we compare costs, assuming that the iterative methods iterate often enough to make the error at most some fixed small value[27] (say, 10^{-6}).

The second and third columns of Table 6.1 give the number of arithmetic operations (or time) and space required on a serial machine. Column 4 indicates whether the method is direct (D) or iterative (I). All entries are meant in the $O(\cdot)$ sense; the constants depend on implementation details and the stopping criterion for the iterative methods (say, 10^{-6}). For example, the entry for Cholesky also applies to Gaussian elimination, since this changes the constant only by a factor of two. The last column indicates where the algorithm is discussed in the text.

The methods are listed in increasing order of speed, from slowest (dense

---

[27]Alternatively, we could iterate until the error is $O(h^2) = O((N+1)^{-2})$, the size of the truncation error. One can show that this would increase the costs of the iterative methods in Table 6.1 by a factor of $O(\log n)$.

Cholesky) to fastest (multigrid), ending with a lower bound applying to any method. The lower bound is n because at least one operation is required per solution component, since otherwise they could not all be different and also depend on the input. The methods are also, roughly speaking, in order of decreasing generality, with dense Cholesky applicable to any symmetric positive definite matrix and later algorithms applicable (or at least provably convergent) only for limited classes of matrices. In later sections we will describe the applicability of various methods in more detail.

The "explicit inverse" algorithm refers to precomputing the explicit inverse of $T_{N \times N}$, and computing $v = T_{N \times N}^{-1} f$ by a single matrix-vector multiplication (and not counting the flops to precompute $T_{N \times N}^{-1}$). Along with dense Cholesky, it uses n^2 space, vastly more than the other methods. It is not a good method. Band Cholesky was discussed in section 2.7.3; this is just Cholesky taking advantage of the fact that there are no entries to compute or store outside a band of $2N + 1$ diagonals.

Jacobi's and Gauss–Seidel are classical iterative methods and not particularly fast, but they form the basis for other faster methods: successive overrelaxation, symmetric successive overrelaxation, and multigrid, our fastest algorithm. So we will study them in some detail in section 6.5.

Sparse Cholesky refers to the algorithm discussed in section 2.7.4: it is an implementation of Cholesky that avoids storing or operating on the zero entries of $T_{N \times N}$ or its Cholesky factor. Furthermore, we are assuming the rows and columns of $T_{N \times N}$ have been "optimally ordered" to minimize work and storage (using nested dissection [112, 113]). While sparse Cholesky is reasonably fast on Poisson's equation in two dimensions, it it significantly worse in three dimensions (using $O(N^6) = O(n^2)$ time and $O(N^4) = O(n^{4/3})$ space), because there is more "fill-in" of zero entries during the algorithm.

Conjugate gradients are a representative of a much larger class of methods, called *Krylov subspace* methods, which are very widely applicable both for linear system solving and finding eigenvalues of sparse matrices. We will discuss these methods in more detail in section 6.6.

The fastest methods are block cyclic reduction, the fast Fourier transform (FFT), and multigrid. In particular, multigrid does only $O(1)$ operations per solution component, which is asymptotically optimal.

A final warning is that this table does not give a complete picture, since the constants are missing. For a particular size problem on a particular machine, one cannot immediately deduce which method is fastest. Still, it is clear that iterative methods such as Jacobi's, Gauss–Seidel, conjugate gradients, and successive overrelaxation are inferior to the FFT, block cyclic reduction, and multigrid for large enough n. But they remain of interest because they are building blocks for some of the faster methods and because they apply to larger classes of problems than the faster methods.

All of these algorithms can be implemented in parallel; see the lectures on PARALLEL_HOMEPAGE for details. It is interesting that, depending on the parallel machine, multigrid may no longer be fastest. This is because on a parallel machine the time required for separate processors to communicate data to one another may be as costly as the floating point operations, and other algorithms may communicate less than multigrid.

6.5. Basic Iterative Methods

In this section we will talk about the most basic iterative methods:

> Jacobi's,
> Gauss–Seidel,
> successive overrelaxation (SOR(ω)),
> Chebyshev acceleration with symmetric successive overrelaxation
>   (SSOR(ω)).

These methods are also discussed and their implementations are provided at NETLIB/templates.

Given x_0, these methods generate a sequence x_m converging to the solution $A^{-1}b$ of $Ax = b$, where x_{m+1} is cheap to compute from x_m.

DEFINITION 6.3. *A* splitting *of A is a decomposition* $A = M - K$, *with M nonsingular.*

A splitting yields an iterative method as follows: $Ax = Mx - Kx = b$ implies $Mx = Kx + b$ or $x = M^{-1}Kx + M^{-1}b \equiv Rx + c$. So we can take $x_{m+1} = Rx_m + c$ as our iterative method. Let us see when it converges.

LEMMA 6.4. *Let $\|\cdot\|$ be any operator norm ($\|R\| \equiv \max_{x \neq 0} \frac{\|Rx\|}{\|x\|}$). If $\|R\| < 1$, then $x_{m+1} = Rx_m + c$ converges for any x_0.*

Proof. Subtract $x = Rx + c$ from $x_{m+1} = Rx_m + c$ to get $x_{m+1} - x = R(x_m - x)$. Thus $\|x_{m+1} - x\| \leq \|R\| \cdot \|x_m - x\| \leq \|R\|^{m+1} \cdot \|x_0 - x\|$, which converges to 0 since $\|R\| < 1$. □

Our ultimate convergence criterion will depend on the following property of R.

DEFINITION 6.4. *The* spectral radius *of R is $\rho(R) \equiv \max |\lambda|$, where the maximum is taken over all eigenvalues λ of R.*

LEMMA 6.5. *For all operator norms $\rho(R) \leq \|R\|$. For all R and for all $\epsilon > 0$ there is an operator norm $\|\cdot\|_\star$ such that $\|R\|_\star \leq \rho(R) + \epsilon$. The norm $\|\cdot\|_\star$ depends on both R and ϵ.*

Proof. To show $\rho(R) \leq \|R\|$ for any operator norm, let x be an eigenvector for λ, where $\rho(R) = |\lambda|$ and so $\|R\| = \max_{y \neq 0} \frac{\|Ry\|}{\|y\|} \geq \frac{\|Rx\|}{\|x\|} = \frac{\|\lambda x\|}{\|x\|} = |\lambda|$.

To construct an operator norm $\|\cdot\|_\star$ such that $\|R\|_\star \le \rho(R)+\epsilon$, let $S^{-1}RS = J$ be in Jordan form. Let $D_\epsilon = \mathrm{diag}(1, \epsilon, \epsilon^2, \ldots, \epsilon^{n-1})$. Then

$$(SD_\epsilon)^{-1}R(SD_\epsilon) \;=\; D_\epsilon^{-1}JD_\epsilon$$

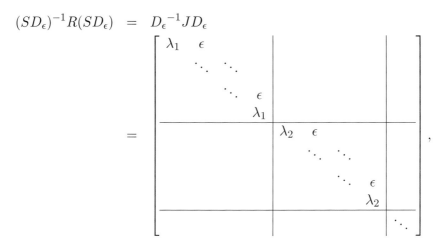

i.e., a "Jordan form" with ϵ's above the diagonal. Now use the vector norm $\|x\|_\star \equiv \|(SD_\epsilon)^{-1}x\|_\infty$ to generate the operator norm

$$
\begin{aligned}
\|R\|_\star &\equiv \max_{x\neq 0}\frac{\|Rx\|_\star}{\|x\|_\star}\\
&= \max_{x\neq 0}\frac{\|(SD_\epsilon)^{-1}Rx\|_\infty}{\|(SD_\epsilon)^{-1}x\|_\infty}\\
&= \max_{y\neq 0}\frac{\|(SD_\epsilon)^{-1}R(SD_\epsilon)y\|_\infty}{\|y\|_\infty}\\
&= \|(SD_\epsilon)^{-1}R(SD_\epsilon)\|_\infty\\
&= \max_i |\lambda_i| + \epsilon\\
&= \rho(R) + \epsilon. \quad \square
\end{aligned}
$$

THEOREM 6.1. *The iteration* $x_{m+1} = Rx_m + c$ *converges to the solution of* $Ax = b$ *for all starting vectors* x_0 *and for all* b *if and only if* $\rho(R) < 1$.

Proof. If $\rho(R) \ge 1$, choose $x_0 - x$ to be an eigenvector of R with eigenvalue λ where $|\lambda| = \rho(R)$. Then

$$(x_{m+1} - x) = R(x_m - x) = \cdots = R^{m+1}(x_0 - x) = \lambda^{m+1}(x_0 - x)$$

will not approach 0. If $\rho(R) < 1$, use Lemma 6.5 to choose an operator norm so $\|R\|_\star < 1$ and then apply Lemma 6.4 to conclude that the method converges. $\square$

DEFINITION 6.5. *The* rate of convergence *of* $x_{m+1} = Rx_m + c$ *is* $r(R) \equiv -\log_{10}\rho(R)$.

$r(R)$ is the increase in the number of correct decimal places in the solution per iteration, since $\log_{10}\|x_m - x\|_\star - \log_{10}\|x_{m+1} - x\|_\star \geq r(R) + O(\epsilon)$. The smaller is $\rho(R)$, the higher is the rate of convergence, i.e., the greater is the number of correct decimal places computed per iteration.

Our goal is now to choose a splitting $A = M - K$ so that both

(1) $Rx = M^{-1}Kx$ and $c = M^{-1}b$ are easy to evaluate,

(2) $\rho(R)$ is small.

We will need to balance these conflicting goals. For example, choosing $M = I$ is good for goal (1) but may not make $\rho(R) < 1$. On the other hand, choosing $M = A$ and $K = 0$ is good for goal (2) but probably bad for goal (1).

The splittings for the methods discussed in this section all share the following notation. When A has no zeros on its diagonal, we write

$$A = D - \tilde{L} - \tilde{U} = D(I - L - U), \tag{6.19}$$

where D is the diagonal of A, $-\tilde{L}$ is the strictly lower triangular part of A, $DL = \tilde{L}$, $-\tilde{U}$ is the strictly upper triangular part of A, and $DU = \tilde{U}$.

6.5.1. Jacobi's Method

Jacobi's method can be described as repeatedly looping through the equations, changing variable j so that equation j is satisfied exactly. Using the notation of equation (6.19), the splitting for Jacobi's method is $A = D - (\tilde{L} + \tilde{U})$; we denote $R_J \equiv D^{-1}(\tilde{L} + \tilde{U}) = L + U$ and $c_J \equiv D^{-1}b$, so we can write one step of Jacobi's method as $x_{m+1} = R_J x_m + c_J$. To see that this formula corresponds to our first description of Jacobi's method, note that it implies $Dx_{m+1} = (\tilde{L} + \tilde{U})x_m + b$, or $a_{jj}x_{m+1,j} = -\sum_{k \neq j} a_{jk}x_{m,k} + b_j$, or $a_{jj}x_{m+1,j} + \sum_{k \neq j} a_{jk}x_{m,k} = b_j$.

ALGORITHM 6.1. *One step of Jacobi's method:*

> *for $j = 1$ to n*
> $\qquad x_{m+1,j} = \frac{1}{a_{jj}}(b_j - \sum_{k \neq j} a_{jk}x_{m,k})$
> *end for*

In the special case of the model problem, the implementation of Jacobi's algorithm simplifies as follows. Working directly from equation (6.10) and letting $v_{m,i,j}$ denote the mth value of the solution at grid point i, j, Jacobi's method becomes the following.

ALGORITHM 6.2. *One step of Jacobi's method for two-dimensional Poisson's equation:*

> *for $i = 1$ to N*
> $\qquad$ *for $j = 1$ to N*
> $\qquad\qquad v_{m+1,i,j} = (v_{m,i-1,j} + v_{m,i+1,j} + v_{m,i,j-1} + v_{m,i,j+1} + h^2 f_{ij})/4$

end for
  end for

In other words, at each step the new value of v_{ij} is obtained by "averaging" its neighbors with $h^2 f_{ij}$. Note that all new values $v_{m+1,i,j}$ may be computed independently of one another. Indeed, Algorithm 6.2 can be implemented in one line of Matlab if the $v_{m+1,i,j}$ are stored in a square array $\hat{V}$ that includes an extra first and last row of zeros and first and last column of zeros (see Question 6.6).

6.5.2. Gauss–Seidel Method

The motivation for this method is that at the jth step of the loop for Jacobi's method, we have improved values of the first $j-1$ components of the solution, so we should use them in the sum.

ALGORITHM 6.3. *One step of the Gauss–Seidel method:*

for $j = 1$ to n

$$x_{m+1,j} = \frac{1}{a_{jj}} \left(b_j - \underbrace{\sum_{k=1}^{j-1} a_{jk} x_{m+1,k}}_{\text{updated } x\text{'s}} - \underbrace{\sum_{k=j+1}^{n} a_{jk} x_{m,k}}_{\text{older } x\text{'s}} \right)$$

  end for

For the purpose of later analysis, we want to write this algorithm in the form $x_{m+1} = R_{GS} x_m + c_{GS}$. To this end, note that it can first be rewritten as

$$\sum_{k=1}^{j} a_{jk} x_{m+1,k} = -\sum_{k=j+1}^{n} a_{jk} x_{m,k} + b_j. \tag{6.20}$$

Then using the notation of equation (6.19), we can rewrite equation (6.20) as $(D - \tilde{L}) x_{m+1} = \tilde{U} x_m + b$ or

$$
\begin{aligned}
x_{m+1} &= (D - \tilde{L})^{-1} \tilde{U} x_m + (D - \tilde{L})^{-1} b \\
&= (I - L)^{-1} U x_m + (I - L)^{-1} D^{-1} b \\
&\equiv R_{GS} x_m + c_{GS}.
\end{aligned}
$$

As with Jacobi's method, we consider how to implement the Gauss–Seidel method for our model problem. In principle it is quite similar, except that we have to keep track of which variables are new (numbered $m + 1$) and which are old (numbered m). But depending on the order in which we loop through the grid points i, j, we will get different (and valid) implementations of the

Gauss–Seidel method. This is unlike Jacobi's method, in which the order in which we update the variables is irrelevant. For example, if we update $v_{m,1,1}$ first (before any other $v_{m,i,j}$), then all its neighboring values are necessarily old. But if we update $v_{m,1,1}$ last, then all its neighboring values are necessarily new, so we get a different value for $v_{m,1,1}$. Indeed, there are as many possible implementations of the Gauss–Seidel method as there are ways to order N^2 variables (namely, $N^2!$). But of all these orderings, two are of most interest. The first is the ordering shown in Figure 6.4; this is called the *natural ordering*.

The second ordering is called *red-black ordering*. It is important because our best convergence results in sections 6.5.4 and 6.5.5 depend on it. To explain red-black ordering, consider the chessboard-like coloring of the grid of unknowns below; the **B** nodes correspond to the black squares on a chessboard, and the **R** nodes correspond to the red squares.

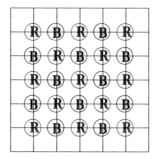

The red-black ordering is to order the red nodes before the black nodes. Note that red nodes are adjacent to only black nodes. So if we update all the red nodes first, they will use only old data from the black nodes. Then when we update the black nodes, which are only adjacent to red nodes, they will use only new data from the red nodes. Thus the algorithm becomes the following.

ALGORITHM 6.4. *One step of the Gauss–Seidel method on two-dimensional Poisson's equation with red-black ordering:*

for all nodes i, j that are red (**R** *)*
$$v_{m+1,i,j} = (v_{m,i-1,j} + v_{m,i+1,j} + v_{m,i,j-1} + v_{m,i,j+1} + h^2 f_{ij})/4$$
end for

for all nodes i, j that are black (**B** *)*
$$v_{m+1,i,j} = (v_{m+1,i-1,j} + v_{m+1,i+1,j} + v_{m+1,i,j-1} + v_{m+1,i,j+1} + h^2 f_{ij})/4$$
end for

6.5.3. Successive Overrelaxation

We refer to this method as SOR(ω), where ω is the *relaxation parameter*. The motivation is to improve the Gauss–Seidel loop by taking an appropriate weighted average of the $x_{m+1,j}$ and $x_{m,j}$:

$$\text{SOR's} \quad x_{m+1,j} = (1-\omega)x_{m,j} + \omega x_{m+1,j},$$

yielding the following algorithm.

ALGORITHM 6.5. *SOR:*

for $j = 1$ to n
$$x_{m+1,j} = (1-\omega)x_{m,j} + \frac{\omega}{a_{jj}}\left[b_j - \sum_{k=1}^{j-1} a_{jk}x_{m+1,k} - \sum_{k=j+1}^{n} a_{jk}x_{m,k}\right]$$
end for

We may rearrange this to get, for $j = 1$ to n,

$$a_{jj}x_{m+1,j} + \omega\sum_{k=1}^{j-1} a_{jk}x_{m+1,k} = (1-\omega)a_{jj}x_{m,j} - \omega\sum_{k=j+1}^{n} a_{jk}x_{m,k} + \omega b_j$$

or, again using the notation of equation (6.19),

$$(D - \omega\tilde{L})x_{m+1} = ((1-\omega)D + \omega\tilde{U})x_m + \omega b$$

or

$$
\begin{aligned}
x_{m+1} &= (D - \omega\tilde{L})^{-1}((1-\omega)D + \omega\tilde{U})x_m + \omega(D - \omega\tilde{L})^{-1}b \\
&= (I - \omega L)^{-1}((1-\omega)I + \omega U)x_m + \omega(I - \omega L)^{-1}D^{-1}b \\
&\equiv R_{SOR(\omega)}x_m + c_{SOR(\omega)}. \quad (6.21)
\end{aligned}
$$

We distinguish three cases, depending on the values of ω: $\omega = 1$ is equivalent to the Gauss–Seidel method, $\omega < 1$ is called *underrelaxation*, and $\omega > 1$ is called *overrelaxation*. A somewhat superficial motivation for overrelaxation is that if the direction from x_m to x_{m+1} is a good direction in which to move the solution, then moving $\omega > 1$ times as far in that direction is better.

In the next two sections, we will show how to pick the optimal ω for the model problem. This optimality depends on using red-black ordering.

ALGORITHM 6.6. *One step of SOR(ω) on two-dimensional Poisson's equation with red-black ordering:*

for all nodes i, j that are red ($\textbf{R}$)
$$v_{m+1,i,j} = (1-\omega)v_{m,i,j} +$$
$$\omega(v_{m,i-1,j} + v_{m,i+1,j} + v_{m,i,j-1} + v_{m,i,j+1} + h^2 f_{ij})/4$$
end for
for all nodes i, j that are black ($\textbf{B}$)
$$v_{m+1,i,j} = (1-\omega)v_{m,i,j} +$$
$$\omega(v_{m+1,i-1,j} + v_{m+1,i+1,j} + v_{m+1,i,j-1} + v_{m+1,i,j+1} + h^2 f_{ij})/4$$
end for

6.5.4.   Convergence of Jacobi's, Gauss–Seidel, and SOR(ω) Methods on the Model Problem

It is easy to compute how fast Jacobi's method converges on the model problem, since the corresponding splitting is $T_{N \times N} = 4I - (4I - T_{N \times N})$, and so $R_J = (4I)^{-1}(4I - T_{N \times N}) = I - T_{N \times N}/4$. Thus the eigenvalues of R_J are $1 - \lambda_{i,j}/4$, where the $\lambda_{i,j}$ are the eigenvalues of $T_{N \times N}$:

$$\lambda_{i,j} = \lambda_i + \lambda_j = 4 - 2\left(\cos \frac{\pi i}{N+1} + \cos \frac{\pi j}{N+1}\right).$$

$\rho(R_J)$ is the largest of $|1 - \lambda_{i,j}/4|$, namely,

$$\rho(R_J) = |1 - \lambda_{1,1}/4| = |1 - \lambda_{N,N}/4| = \cos \frac{\pi}{N+1} \approx 1 - \frac{\pi^2}{2(N+1)^2}.$$

Note that as N grows and T becomes more ill-conditioned, the spectral radius $\rho(R_J)$ approaches 1. Since the error is multiplied by the spectral radius at each step, convergence slows down. To estimate the speed of convergence more precisely, let us compute the number m of Jacobi iterations required to decrease the error by $e^{-1} = \exp(-1)$. Then m must satisfy $(\rho(R_J))^m = e^{-1}$, or $(1 - \frac{\pi^2}{2(N+1)^2})^m = e^{-1}$, or $m \approx \frac{2(N+1)^2}{\pi^2} = O(N^2) = O(n)$. Thus the number of iterations is proportional to the number of unknowns. Since one step of Jacobi costs $O(1)$ to update each solution component or $O(n)$ to update all of them, it costs $O(n^2)$ to decrease the error by e^{-1} (or by any constant factor less than 1). This explains the entry for Jacobi's method in Table 6.1.

This is a common phenomenon: the more ill-conditioned the original problem, the more slowly most iterative methods converge. There are important exceptions, such as multigrid and domain decomposition, which we discuss later.

In the next section we will show, provided that the variables in Poisson's equation are updated in red-black order (see Algorithm 6.4 and Corollary 6.1), that $\rho(R_{GS}) = \rho(R_J)^2 = \cos^2 \frac{\pi}{N+1}$. In other words, one Gauss–Seidel step decreases the error as much as two Jacobi steps. This is a general phenomenon for matrices arising from approximating differential equations with certain finite difference approximations. This also explains the entry for the Gauss–Seidel method in Table 6.1; since it is only twice as fast as Jacobi, it still has the same complexity in the $O(\cdot)$ sense.

For the same red-black update order (see Algorithm 6.6 and Theorem 6.7), we will also show that for the relaxation parameter $1 < \omega = 2/(1 + \sin \frac{\pi}{N+1}) < 2$

$$\rho(R_{SOR(\omega)}) = \frac{\cos^2 \frac{\pi}{N+1}}{(1 + \sin \frac{\pi}{N+1})^2} \approx 1 - \frac{2\pi}{N+1} \text{ for large } N.$$

This is in contrast to $\rho(R) = 1 - O(\frac{1}{N^2})$ for R_J and R_{GS}. This is the optimal value for ω; i.e., it minimizes $R_{SOR(\omega)}$. With this choice of ω, SOR(ω) is approximately N times faster than Jacobi's or the Gauss–Seidel method, since if $\text{SOR}(\omega)$ takes j steps to decrease the error as much as k steps of Jacobi's or the Gauss–Seidel method, then $(1 - \frac{1}{N^2})^k \approx (1 - \frac{1}{N})^j$, implying $1 - \frac{k}{N^2} \approx 1 - \frac{j}{N}$ or $k \approx j \cdot N$. This lowers the complexity of $\text{SOR}(\omega)$ from $O(n^2)$ to $O(n^{3/2})$, as shown in Table 6.1.

In the next section we will show generally for certain finite difference matrices how to choose ω to minimize $\rho(R_{SOR(\omega)})$.

6.5.5. Detailed Convergence Criteria for Jacobi's, Gauss–Seidel, and SOR(ω) Methods

We will give a sequence of criteria that guarantee the convergence of these methods. The first criterion is simple to evaluate but is not always applicable, in particular not to the model problem. Then we give several more complicated criteria, which place stronger conditions on the matrix A but in return give more information about convergence. These more complicated criteria are tailored to fit the matrices arising from discretizing certain kinds of partial differential equations such as Poisson's equation.

Here is a summary of the results of this section:

1. If A is strictly row diagonally dominant (Definition 6.6), then Jacobi's and Gauss–Seidel methods both converge, and the Gauss–Seidel method is faster (Theorem 6.2). Strict row diagonal dominance means that each diagonal entry of A is larger in magnitude than the sum of the magnitudes of the other entries in its row.

2. Since our model problem is not strictly row diagonally dominant, the last result does not apply. So we ask for a weaker form of diagonal dominance (Definition 6.11) but impose a condition called *irreducibility* on the pattern of nonzero entries of A (Definition 6.7) to prove convergence of Jacobi's and Gauss–Seidel methods. The Gauss–Seidel method again converges faster than Jacobi's method (Theorem 6.3). This result applies to the model problem.

3. Turning to $\text{SOR}(\omega)$, we show that $0 < \omega < 2$ is necessary for convergence (Theorem 6.4). If A is also positive definite (like the model problem), $0 < \omega < 2$ is also sufficient for convergence (Theorem 6.5).

4. To quantitatively compare Jacobi's, Gauss–Seidel, and SOR(ω) methods, we make one more assumption about the pattern of nonzero entries of A. This property is called *property A* (Definition 6.12) and is equivalent to saying that the *graph of the matrix* is *bipartite*. Property A essentially says that we can update the variables using red-black ordering. Given property A there is a simple algebraic formula relating the eigenvalues of R_J, R_{GS}, and $R_{SOR(\omega)}$ (Theorem 6.6), which lets us compare their rates

of convergence. This formula also lets us compute the optimal ω that makes SOR(ω) converge as fast as possible (Theorem 6.7).

DEFINITION 6.6. *A is* strictly row diagonally dominant *if* $|a_{ii}| > \sum_{j \neq i} |a_{ij}|$ *for all i.*

THEOREM 6.2. *If A is strictly row diagonally dominant, both Jacobi's and Gauss–Seidel methods converge. In fact $\|R_{GS}\|_\infty \leq \|R_J\|_\infty < 1$.*

The inequality $\|R_{GS}\|_\infty \leq \|R_J\|_\infty$ implies that one step of the worst problem for the Gauss–Seidel method converges at least as fast as one step of the worst problem for Jacobi's method. It does *not* guarantee that for any particular $Ax = b$, the Gauss–Seidel method will be faster than Jacobi's method; Jacobi's method could "accidentally" have a smaller error at some step.

Proof.   Again using the notation of equation (6.19), we write $R_J = L + U$ and $R_{GS} = (I - L)^{-1}U$. We want to prove

$$\|R_{GS}\|_\infty = \||R_{GS}|e\|_\infty \leq \||R_J|e\|_\infty = \|R_J\|_\infty, \tag{6.22}$$

where $e = [1, \ldots, 1]^T$ is the vector of all ones. Inequality (6.22) will be true if can prove the stronger componentwise inequality

$$|(I - L)^{-1}U| \cdot e = |R_{GS}| \cdot e \leq |R_J| \cdot e = (|L| + |U|) \cdot e. \tag{6.23}$$

Since

$$
\begin{aligned}
|(I - L)^{-1}U| \cdot e &\leq |(I - L)^{-1}| \cdot |U| \cdot e && \text{by the triangle inequality} \\
&= \left| \sum_{i=0}^{n-1} L^i \right| \cdot |U| \cdot e && \text{since } L^n = 0 \\
&\leq \sum_{i=0}^{n-1} |L|^i \cdot |U| \cdot e && \text{by the triangle inequality} \\
&= (I - |L|)^{-1} \cdot |U| \cdot e && \text{since } |L|^n = 0,
\end{aligned}
$$

inequality (6.23) will be true if can prove the even stronger componentwise inequality

$$(I - |L|)^{-1} \cdot |U| \cdot e \leq (|L| + |U|) \cdot e. \tag{6.24}$$

Since all entries of $(I - |L|)^{-1} = \sum_{i=0}^{n-1} |L|^i$ are nonnegative, inequality (6.24) will be true if we can prove

$$|U| \cdot e \leq (I - |L|) \cdot (|L| + |U|) \cdot e = (|L| + |U| - |L|^2 - |L| \cdot |U|) \cdot e$$

or

$$0 \leq (|L| - |L|^2 - |L| \cdot |U|) \cdot e = |L| \cdot (I - |L| - |U|) \cdot e. \tag{6.25}$$

Since all entries of $|L|$ are nonnegative, inequality (6.25) will be true if we can prove

$$0 \le (I - |L| - |U|) \cdot e \quad \text{or} \quad |R_J| \cdot e = (|L| + |U|)e \le e. \tag{6.26}$$

Finally, inequality (6.26) is true because by assumption $\||R_J| \cdot e\|_\infty = \|R_J\|_\infty = \rho < 1$. $\square$

An analogous result holds when A is strictly column diagonally dominant (i.e., A^T is strictly row diagonally dominant).

The reader may easily confirm that this simple criterion does not apply to the model problem, so we need to weaken the assumption of strict diagonal dominance. Doing so requires looking at the *graph properties* of a matrix.

DEFINITION 6.7. *A is an* irreducible matrix *if there is no permutation matrix P such that*

$$PAP^T = \left[\begin{array}{c|c} A_{11} & A_{12} \\ \hline 0 & A_{22} \end{array} \right].$$

We connect this definition to *graph theory* as follows.

DEFINITION 6.8. *A* directed graph *is a finite collection of* nodes *connected by a finite collection of* directed edges, *i.e., arrows from one node to another. A* path *in a directed graph is a sequence of nodes $n_1, \ldots, n_m$ with an edge from each n_i to n_{i+1}. A* self edge *is an edge from a node to itself.*

DEFINITION 6.9. *The* directed graph of A, $G(A)$, *is a graph with nodes $1, 2, \ldots, n$ and an edge from node i to node j if and only if $a_{ij} \ne 0$.*

EXAMPLE 6.1. The matrix

$$A = \left[\begin{array}{cccc} 2 & -1 & & \\ -1 & 2 & -1 & \\ & -1 & 2 & -1 \\ & & -1 & 2 \end{array} \right]$$

has the directed graph

$\diamond$

DEFINITION 6.10. *A directed graph is called* strongly connected *if there exists a path from every node i to every node j. A* strongly connected component *of a directed graph is a subgraph (a subset of the nodes with all edges connecting them) which is strongly connected and cannot be made larger yet still be strongly connected.*

EXAMPLE 6.2. The graph in Example 6.1 is strongly connected.   ◇

EXAMPLE 6.3. Let

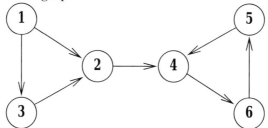

$$A = \begin{bmatrix} 1 & 1 & & & \\ & & 1 & & \\ 1 & & & & \\ \hline & & & & 1 \\ & 1 & & & \\ & & & 1 & \end{bmatrix},$$

which has the directed graph

   This graph is not strongly connected, since there is no path to node 1 from anywhere else. Nodes 4, 5, and 6 form a strongly connected component, since there is a path from any one of them to any other.   ◇

EXAMPLE 6.4. The graph of the model problem is strongly connected. The graph is essentially

except that each edge in the grid represents two edges (one in each direction), and the self edges are not shown.   ◇

LEMMA 6.6. *A is irreducible if and only if $G(A)$ is strongly connected.*

Proof. If $A = \begin{bmatrix} A_{11} & A_{12} \\ 0 & A_{22} \end{bmatrix}$ is reducible, then there is clearly no way to get from the nodes corresponding to A_{22} back to the ones corresponding to A_{11}; i.e., $G(A)$ is not strongly connected. Similarly, if $G(A)$ is not strongly connected, renumber the rows (and columns) so that all the nodes in a particular strongly connected component come first; then the matrix PAP^T will be block upper triangular.   □

EXAMPLE 6.5. The matrix A in Example 6.3 is reducible.

DEFINITION 6.11. *A is* weakly row diagonally dominant *if for all i, $|a_{ii}| \geq \sum_{k \neq i} |a_{ik}|$ with strict inequality at least once.*

THEOREM 6.3. *If A is irreducible and weakly row diagonally dominant, then both Jacobi's and Gauss–Seidel methods converge, and $\rho(R_{GS}) < \rho(R_J) < 1$.*

For a proof of this theorem, see [249].

EXAMPLE 6.6. The model problem is weakly diagonally dominant and irreducible but not strongly diagonally dominant. (The diagonal is 4, and the offdiagonal sums are either 2, 3, or 4.) So Jacobi's and Gauss–Seidel methods converge on the model problem. ◇

Despite the above results showing that under certain conditions the Gauss–Seidel method is faster than Jacobi's method, no such general result holds. This is because there are nonsymmetric matrices for which Jacobi's method converges and the Gauss–Seidel method diverges, as well as matrices for which the Gauss–Seidel method converges and Jacobi's method diverges [249].

Now we consider the convergence of $SOR(\omega)$ [249]. Recall its definition:

$$R_{SOR(\omega)} = (I - \omega L)^{-1}((1 - \omega)I + \omega U).$$

THEOREM 6.4. *$\rho(R_{SOR(\omega)}) \geq |\omega - 1|$. Therefore $0 < \omega < 2$ is required for convergence.*

Proof. Write the characteristic polynomial of $R_{SOR(\omega)}$ as $\varphi(\lambda) = \det(\lambda I - R_{SOR(\omega)}) = \det((I - \omega L)(\lambda I - R_{SOR(\omega)})) = \det((\lambda + \omega - 1)I - \omega \lambda L - \omega U)$ so that

$$\varphi(0) = \pm \prod_{i=1}^{n} \lambda_i(R_{SOR(\omega)}) = \pm \det((\omega - 1)I) = \pm(\omega - 1)^n,$$

implying $\max_i |\lambda_i(R_{SOR(\omega)})| \geq |\omega - 1|$. ☐

THEOREM 6.5. *If A is symmetric positive definite, then $\rho(R_{SOR(\omega)}) < 1$ for all $0 < \omega < 2$, so $SOR(\omega)$ converges for all $0 < \omega < 2$. Taking $\omega = 1$, we see that the Gauss–Seidel method also converges.*

Proof. There are two steps. We abbreviate $R_{SOR(\omega)} = R$. Using the notation of equation (6.19), let $M = \omega^{-1}(D - \omega \tilde{L})$. Then we

  (1) define $Q = A^{-1}(2M - A)$ and show $\Re\lambda_i(Q) > 0$ for all i,

  (2) show that $R = (Q - I)(Q + I)^{-1}$, implying $|\lambda_i(R)| < 1$ for all i.

For (1), note that $Qx = \lambda x$ implies $(2M - A)x = \lambda Ax$ or $x^*(2M - A)x = \lambda x^* Ax$. Add this last equation to its conjugate transpose to get $x^*(M + M^* - A)x = (\Re\lambda)(x^* Ax)$. So $\Re\lambda = x^*(M + M^* - A)x/x^* Ax = x^*(\frac{2}{\omega} - 1)Dx/x^* Ax > 0$ since A and $(\frac{2}{\omega} - 1)D$ are positive definite.

To prove (2), note that $(Q - I)(Q + I)^{-1} = (2A^{-1}M - 2I)(2A^{-1}M)^{-1} = I - M^{-1}A = R$, so by the spectral mapping theorem (Question 4.5)

$$|\lambda(R)| = \left| \frac{\lambda(Q) - 1}{\lambda(Q) + 1} \right| = \left| \frac{(\Re\lambda(Q) - 1)^2 + (\Im\lambda(Q))^2}{(\Re\lambda(Q) + 1)^2 + (\Im\lambda(Q))^2} \right|^{1/2} < 1. \quad \square$$

Together, Theorems 6.4 and 6.5 imply that if A is symmetric positive definite, then $\text{SOR}(\omega)$ converges if and only if $0 < \omega < 2$.

EXAMPLE 6.7. The model problem is symmetric positive definite, so $\text{SOR}(\omega)$ converges for $0 < \omega < 2$. $\diamond$

For the final comparison of the costs of Jacobi's, Gauss–Seidel, and $\text{SOR}(\omega)$ methods on the model problem we impose another graph theoretic condition on A that often arises from certain discretized partial differential equations, such as Poisson's equation. This condition will let us compute $\rho(R_{GS})$ and $\rho(R_{SOR(\omega)})$ explicitly in terms of $\rho(R_J)$.

DEFINITION 6.12. *A matrix T has* property A *if there exists a permutation P such that*

$$PTP^T = \left[\begin{array}{c|c} T_{11} & T_{12} \\ \hline T_{21} & T_{22} \end{array} \right],$$

where T_{11} and T_{22} are diagonal. In other words in the graph $G(A)$ the nodes divide into two sets $S_1 \cup S_2$, where there are no edges between two nodes both in S_1 or both in S_2 (ignoring self edges); such a graph is called bipartite.

EXAMPLE 6.8. *Red-black ordering* for the model problem. This was introduced in section 6.5.2, using the following chessboard-like depiction of the graph of the model problem: The black $\textcircled{B}$ nodes are in S_1, and the red $\textcircled{R}$ nodes are in S_2.

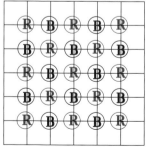

As described in section 6.5.2, each equation in the model problem relates the value at a grid point to the values at its left, right, top, and bottom neighbors, which are colored differently from the grid point in the middle. In other words, there is no direct connection from an $\textcircled{R}$ node to an $\textcircled{R}$ node or from a $\textcircled{B}$ node to a $\textcircled{B}$ node. So if we number the red nodes before the black nodes, the matrix will be in the form demanded by Definition 6.12. For example, in the case of a 3-by-3 grid, we get the following:

$$P \begin{bmatrix} 4 & -1 & & -1 & & & & & \\ -1 & 4 & -1 & & -1 & & & & \\ & -1 & 4 & & & -1 & & & \\ -1 & & & 4 & -1 & & -1 & & \\ & -1 & & -1 & 4 & -1 & & -1 & \\ & & -1 & & -1 & 4 & & & -1 \\ & & & -1 & & & 4 & -1 & \\ & & & & -1 & & -1 & 4 & -1 \\ & & & & & -1 & & -1 & 4 \end{bmatrix} P^T$$

$$= \begin{bmatrix} 4 & & & & & -1 & -1 & & \\ & 4 & & & & -1 & & -1 & \\ & & 4 & & & -1 & -1 & -1 & -1 \\ & & & 4 & & & -1 & & -1 \\ & & & & 4 & & & -1 & -1 \\ -1 & -1 & -1 & & & 4 & & & \\ -1 & & -1 & -1 & & & 4 & & \\ & -1 & -1 & & -1 & & & 4 & \\ & & -1 & -1 & -1 & & & & 4 \end{bmatrix}. \quad \diamond$$

Now suppose that T has property A, so we can write (where $D_i = T_{ii}$ is diagonal)

$$PTP^T = \begin{bmatrix} D_1 & T_{12} \\ T_{21} & D_2 \end{bmatrix} = \begin{bmatrix} D_1 & \\ & D_2 \end{bmatrix} - \begin{bmatrix} 0 & 0 \\ -T_{21} & 0 \end{bmatrix} - \begin{bmatrix} 0 & -T_{12} \\ 0 & 0 \end{bmatrix}$$
$$= D - \tilde{L} - \tilde{U}.$$

DEFINITION 6.13. *Let $R_J(\alpha) = \alpha L + \frac{1}{\alpha}U$. Then $R_J(1) = R_J$ is the iteration matrix for Jacobi's method.*

PROPOSITION 6.2. *The eigenvalues of $R_J(\alpha)$ are independent of α.*

Proof.

$$R_J(\alpha) = -\begin{bmatrix} 0 & \frac{1}{\alpha}D_1^{-1}T_{12} \\ \alpha D_2^{-1}T_{21} & 0 \end{bmatrix}$$

has the same eigenvalues as the similar matrix

$$\begin{bmatrix} I & \\ & \alpha I \end{bmatrix}^{-1} R_J(\alpha) \begin{bmatrix} I & \\ & \alpha I \end{bmatrix} = -\begin{bmatrix} 0 & D_1^{-1}T_{12} \\ D_2^{-1}T_{21} & 0 \end{bmatrix} = R_J(1). \quad \square$$

DEFINITION 6.14. *Let T be any matrix, with $T = D - \tilde{L} - \tilde{U}$ and $R_J(\alpha) = \alpha D^{-1}\tilde{L} + \frac{1}{\alpha}D^{-1}\tilde{U}$. If $R_J(\alpha)$'s eigenvalues are independent of α, then T is called* consistent ordering.

It is an easy fact that if T has property A, such as the model problem, then PTP^T is consistently ordered for the permutation P that makes $PTP^T = \begin{bmatrix} T_{11} & T_{12} \\ T_{21} & T_{22} \end{bmatrix}$ have diagonal T_{11} and T_{22}. It is not true that consistent ordering implies a matrix has property A.

EXAMPLE 6.9. *Any block tridiagonal matrix*

$$\begin{bmatrix} D_1 & A_1 & & \\ B_1 & \ddots & \ddots & \\ & \ddots & \ddots & A_{n-1} \\ & & B_{n-1} & D_n \end{bmatrix}$$

is consistently ordered when the D_i are diagonal. $\diamond$

Consistent ordering implies that there are simple formulas relating the eigenvalues of R_J, R_{GS}, and $R_{SOR(\omega)}$ [249].

THEOREM 6.6. *If A is consistently ordered and $\omega \neq 0$, then the following are true:*

1) *The eigenvalues of R_J appear in $\pm$ pairs.*

2) *If μ is an eigenvalue of R_J and*

$$(\lambda + \omega - 1)^2 = \lambda \omega^2 \mu^2, \tag{6.27}$$

then λ is an eigenvalue of $R_{SOR(\omega)}$.

3) *Conversely, if $\lambda \neq 0$ is an eigenvalue of $R_{SOR(\omega)}$, then μ in equation (6.27) is an eigenvalue of R_J.*

Proof.

1) Consistent ordering implies that the eigenvalues of $R_J(\alpha)$ are independent of α, so $R_J = R_J(1)$ and $R_J(-1) = -R_J(1)$ have same eigenvalues; hence they appear in $\pm$ pairs.

2) If $\lambda = 0$ and equation (6.27) holds, then $\omega = 1$ and 0 is indeed an eigenvalue of $R_{SOR(1)} = R_{GS} = (I - L)^{-1}U$ since R_{GS} is singular. Otherwise

$$\begin{aligned}
0 &= \det(\lambda I - R_{SOR(\omega)}) \\
&= \det((I - \omega L)(\lambda I - R_{SOR(\omega)})) \\
&= \det((\lambda + \omega - 1)I - \omega \lambda L - \omega U) \\
&= \det\left(\sqrt{\lambda}\omega\left(\left(\frac{\lambda + \omega - 1}{\sqrt{\lambda}\omega}\right)I - \sqrt{\lambda}L - \frac{1}{\sqrt{\lambda}}U\right)\right) \\
&= \det\left(\left(\frac{\lambda + \omega - 1}{\sqrt{\lambda}\omega}\right)I - L - U\right)(\sqrt{\lambda}\omega)^n,
\end{aligned}$$

where the last equality is true because of Proposition 6.2. Therefore $\frac{\lambda + \omega - 1}{\sqrt{\lambda}\omega} = \mu$, an eigenvalue of $L + U = R_J$, and $(\lambda + \omega - 1)^2 = \mu^2 \omega^2 \lambda$.

3) If $\lambda \neq 0$, the last set of equalities works in the opposite direction.   □

COROLLARY 6.1. *If A is consistently ordered, then $\rho(R_{GS}) = (\rho(R_J))^2$. This means that the Gauss–Seidel method is twice as fast as Jacobi's method.*

Proof.   The choice $\omega = 1$ is equivalent to the Gauss–Seidel method, so $\lambda^2 = \lambda\mu^2$ or $\lambda = \mu^2$.   □

To get the most benefit from overrelaxation, we would like to find ω_{opt} minimizing $\rho(R_{SOR(\omega)})$ [249].

THEOREM 6.7. *Suppose that A is consistently ordered, R_J has real eigenvalues, and $\mu = \rho(R_J) < 1$. Then*

$$\omega_{opt} = \frac{2}{1 + \sqrt{1 - \mu^2}},$$

$$\rho(R_{SOR(\omega_{opt})}) = \omega_{opt} - 1 = \frac{\mu^2}{[1 + \sqrt{1 - \mu^2}]^2},$$

$$\rho(R_{SOR(\omega)}) = \begin{cases} \omega - 1, & \omega_{opt} \leq \omega \leq 2, \\ 1 - \omega + \frac{1}{2}\omega^2\mu^2 + \omega\mu\sqrt{1 - \omega + \frac{1}{4}\omega^2\mu^2}, & 0 < \omega \leq \omega_{opt}. \end{cases}$$

Proof.   Solve $(\lambda + \omega - 1)^2 = \lambda\omega^2\mu^2$ for λ.   □

EXAMPLE 6.10. The model problem is an example: R_J is symmetric, so it has real eigenvalues. Figure 6.5 shows a plot of $\rho(R_{SOR(\omega)})$ versus ω, along with $\rho(R_{GS})$ and $\rho(R_J)$, for the model problem on an N-by-N grid with $N = 16$ and $N = 64$. The plots on the left are of $\rho(R)$, and the plots on the right are semilogarithmic plots of $1 - \rho(R)$. The main conclusion that we can draw is that the graph of $\rho(R_{SOR(\omega)})$ has a vary narrow minimum, so if ω is even slightly different from ω_{opt}, the convergence will slow down significantly. The second conclusion is that if you have to guess ω_{opt}, a large value (near 2) is a better guess than a small value.   ◇

6.5.6. Chebyshev Acceleration and Symmetric SOR (SSOR)

Of the methods we have discussed so far, Jacobi's and Gauss–Seidel methods require no information about the matrix to execute them (although proving that they converge requires some information). $SOR(\omega)$ depends on a parameter ω, which can be chosen depending on $\rho(R_J)$ to accelerate convergence. Chebyshev acceleration is useful when we know even more about the spectrum of R_J than just $\rho(R_J)$ and lets us further accelerate convergence.

Suppose that we convert $Ax = b$ to the iteration $x_{i+1} = Rx_i + c$, using some method (Jacobi's, Gauss–Seidel, or $SOR(\omega)$). Then we get a sequence $\{x_i\}$ where $x_i \to x$ as $i \to \infty$ if $\rho(R) < 1$.

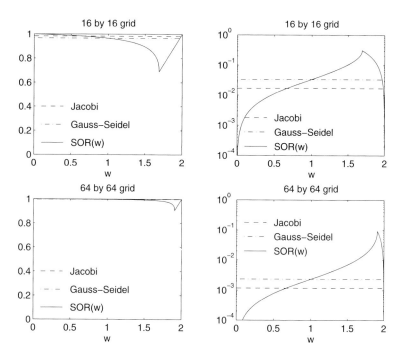

Fig. 6.5. *Convergence of Jacobi's, Gauss–Seidel, and SOR(ω) methods versus ω on the model problem on a 16-by-16 grid and a 64-by-64 grid. The spectral radius $\rho(R)$ of each method ($\rho(R_J)$, $\rho(R_{GS})$, and $\rho(R_{SOR(\omega)})$) is plotted on the left, and $1 - \rho(R)$ on the right.*

Given all these approximations x_i, it is natural to ask whether some linear combination of them, $y_m = \sum_{i=1}^{m} \gamma_{mi} x_i$, is an even better approximation of the solution x. Note that the scalars γ_{mi} must satisfy $\sum_{i=0}^{m} \gamma_{mi} = 1$, since if $x_0 = x_1 = \cdots = x$, we want $y_m = x$, too. So we can write the error in y_m as

$$y_m - x = \sum_{i=0}^{m} \gamma_{mi} x_i - x$$
$$= \sum_{i=0}^{m} \gamma_{mi} (x_i - x)$$
$$= \sum_{i=0}^{m} \gamma_{mi} R^i (x_0 - x)$$
$$= p_m(R)(x_0 - x), \tag{6.28}$$

where $p_m(R) = \sum_{i=0}^{m} \gamma_{mi} R^i$ is a polynomial of degree m with $p_m(1) = \sum_{i=0}^{m} \gamma_{mi} = 1$.

EXAMPLE 6.11. If we could choose p_m to be the characteristic polynomial of R, then $p_m(R) = 0$ by the Cayley–Hamilton theorem, and we would converge in m steps. But this is not practical, because we seldom know the eigenvalues

of R and we want to converge much faster than in $m = \dim(R)$ steps anyway.
$\diamond$

Instead of seeking a polynomial such that $p_m(R)$ is zero, we will settle for making the spectral radius of $p_m(R)$ as small as we can. Suppose that we knew

- the eigenvalues of R were real, and

- the eigenvalues of R lay in an interval $[-\rho, \rho]$ not containing 1.

Then we could try to choose a polynomial p_m where

1) $p_m(1) = 1$, and

2) $\max_{-\rho \le x \le \rho} |p_m(x)|$ is as small as possible.

Since the eigenvalues of $p_m(R)$ are $p_m(\lambda(R))$ (see Problem 4.5), these eigenvalues would be small and so the spectral radius (the largest eigenvalue in absolute value) would be small.

Finding a polynomial p_m to satisfy conditions 1) and 2) above is a classical problem in approximation theory whose solution is based on *Chebyshev polynomials*.

DEFINITION 6.15. *The mth Chebyshev polynomial is defined by the recurrence* $T_m(x) \equiv 2xT_{m-1}(x) - T_{m-2}(x)$, *where* $T_0(x) = 1$ *and* $T_1(x) = x$.

Chebyshev polynomials have many interesting properties [240]. Here are a few, which are easy to prove from the definition (see Question 6.7).

LEMMA 6.7. *Chebyshev polynomials have the following properties:*

- $T_m(1) = 1$.

- $T_m(x) = 2^{m-1}x^m + O(x^{m-1})$.

- $T_m(x) = \begin{cases} \cos(m \cdot \arccos x) & \text{if } |x| \le 1, \\ \cosh(m \cdot \text{arccosh} x) & \text{if } |x| \ge 1. \end{cases}$

- $|T_m(x)| \le 1$ *if* $|x| \le 1$.

- *The zeros of* $T_m(x)$ *are* $x_i = \cos((2i-1)\pi/(2m))$ *for* $i = 1, \dots, m$.

- $T_m(x) = \frac{1}{2}[(x + \sqrt{x^2 - 1})^m + (x + \sqrt{x^2 - 1})^{-m}]$ *if* $|x| > 1$.

- $T_m(1 + \epsilon) \ge .5(1 + m\sqrt{2\epsilon})$ *if* $\epsilon > 0$.

Here is a table of values of $T_m(1 + \epsilon)$. Note how fast it grows as m grows, even when ϵ is tiny (see Figure 6.6).

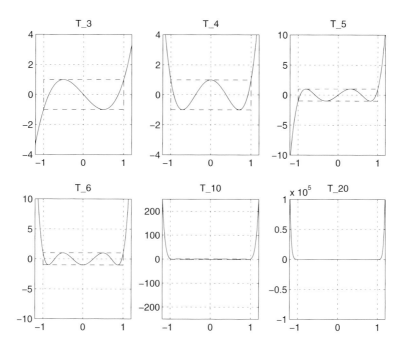

Fig. 6.6. *Graph of $T_m(x)$ versus x. The dotted lines indicate that $|T_m(x)| \leq 1$ for $|x| \leq 1$.*

m	ϵ		
	10^{-4}	10^{-3}	10^{-2}
10	1.0	1.1	2.2
100	2.2	44	$6.9 \cdot 10^5$
200	8.5	$3.8 \cdot 10^3$	$9.4 \cdot 10^{11}$
1000	$6.9 \cdot 10^5$	$1.3 \cdot 10^{19}$	$1.2 \cdot 10^{61}$

A polynomial with the properties we want is $p_m(x) = T_m(x/\rho)/T_m(1/\rho)$. To see why, note that $p_m(1) = 1$ and that if $x \in [-\rho, \rho]$, then $|p_m(x)| \leq 1/T_m(1/\rho)$. For example, if $\rho = 1/(1 + \epsilon)$, then $|p_m(x)| \leq 1/T_m(1 + \epsilon)$. As we have just seen, this bound is tiny for small ϵ and modest m.

To implement this cheaply, we use the three-term recurrence $T_m(x) = 2xT_{m-1}(x) - T_{m-2}(x)$ used to define Chebyshev polynomials. This means that we need to save and combine only three vectors y_m, y_{m-1}, and y_{m-2}, not all the previous x_m. To see how this works, let $\mu_m \equiv 1/T_m(1/\rho)$, so $p_m(R) = \mu_m T_m(R/\rho)$ and $\dfrac{1}{\mu_m} = \dfrac{2}{\rho\mu_{m-1}} - \dfrac{1}{\mu_{m-2}}$ by the three-term recurrence in Definition 6.15. Then

$$
\begin{aligned}
y_m - x &= p_m(R)(x_0 - x) \quad \text{by equation (6.28)} \\
&= \mu_m T_m\left(\frac{R}{\rho}\right)(x_0 - x) \\
&= \mu_m \left[2 \cdot \frac{R}{\rho} \cdot T_{m-1}\left(\frac{R}{\rho}\right)(x_0 - x) - T_{m-2}\left(\frac{R}{\rho}\right)(x_0 - x)\right]
\end{aligned}
$$

by Definition 6.15

$$= \mu_m \left[2 \cdot \frac{R}{\rho} \cdot \frac{p_{m-1}(\frac{R}{\rho})(x_0 - x)}{\mu_{m-1}} - \frac{p_{m-2}(\frac{R}{\rho})(x_0 - x)}{\mu_{m-2}} \right]$$

$$= \mu_m \left[2 \cdot \frac{R}{\rho} \cdot \frac{y_{m-1} - x}{\mu_{m-1}} - \frac{y_{m-2} - x}{\mu_{m-2}} \right] \quad \text{by equation (6.28)}$$

or

$$y_m = \frac{2\mu_m}{\mu_{m-1}} \frac{R}{\rho} y_{m-1} - \frac{\mu_m}{\mu_{m-2}} y_{m-2} + d_m,$$

where

$$\begin{aligned}
d_m &= x - \frac{2\mu_m}{\mu_{m-1}} \left(\frac{R}{\rho} \right) x + \frac{\mu_m}{\mu_{m-2}} x \\
&= x - \frac{2\mu_m}{\mu_{m-1}} \left(\frac{x - c}{\rho} \right) + \frac{\mu_m}{\mu_{m-2}} x \quad \text{since } x = Rx + c \\
&= \mu_m \left(\frac{1}{\mu_m} - \frac{2}{\rho\mu_{m-1}} + \frac{1}{\mu_{m-2}} \right) x + \frac{2\mu_m}{\rho\mu_{m-1}} c \\
&= \frac{2\mu_m}{\rho\mu_{m-1}} c \quad \text{by the definition of } \mu_m.
\end{aligned}$$

This yields the algorithm.

ALGORITHM 6.7. *Chebyshev acceleration of $x_{i+1} = Rx_i + c$:*

> $\mu_0 = 1; \; \mu_1 = \rho; \; y_0 = x_0; \; y_1 = Rx_0 + c$
> *for* $m = 2, 3, \ldots$
> $\mu_m = 1 / \left(\frac{2}{\rho\mu_{m-1}} - \frac{1}{\mu_{m-2}} \right)$
> $y_m = \frac{2\mu_m}{\rho\mu_{m-1}} Ry_{m-1} - \frac{\mu_m}{\mu_{m-2}} y_{m-2} + \frac{2\mu_m}{\rho\mu_{m-1}} c$
> *end for*

Note that each iteration takes just one application of R, so if this is significantly more expensive than the other scalar and vector operations, this algorithm is no more expensive per step than the original iteration $x_{m+1} = Rx_m + c$.

Unfortunately, we cannot apply this directly to SOR(ω) for solving $Ax = b$, because $R_{SOR(\omega)}$ generally has complex eigenvalues, and Chebyshev acceleration requires that R have real eigenvalues in the interval $[-\rho, \rho]$. But we can fix this by using the following algorithm.

ALGORITHM 6.8. *Symmetric SOR (SSOR):*

1. *Take one step of SOR(ω) computing the components of x in the usual increasing order:* $x_{i,1}, x_{i,2}, \ldots, x_{i,n}$,

2. *Take one step of SOR(ω) computing backwards:* $x_{i,n}, x_{i,n-1}, \ldots, x_{i,1}$.

We will reexpress this algorithm as $x_{i+1} = E_\omega x_i + c_\omega$ and show that E_ω has real eigenvalues, so we can use Chebyschev acceleration.

Suppose A is symmetric as in the model problem and again write $A = D - \tilde{L} - \tilde{U} = D(I - L - U)$ as in equation (6.19). Since $A = A^T$, $U = L^T$. Use equation (6.21) to rewrite the two steps of SSOR as

1. $x_{i+1/2} = (I - \omega L)^{-1}((1 - \omega)I + \omega U)x_i + c_{1/2} \equiv L_\omega x_i + c_{1/2}$,
2. $x_i = (I - \omega U)^{-1}((1 - \omega)I + \omega L)x_{i+1/2} + c_1 \equiv U_\omega x_{i+1/2} + c_1$.

Eliminating $x_{i+1/2}$ yields $x_{i+1} = E_\omega x_i + \hat{c}$, where

$$
\begin{aligned}
E_\omega &= U_\omega L_\omega \\
&= I + (\omega - 2)^2 (I - \omega U)^{-1}(I - \omega L)^{-1} + (\omega - 2)(I - \omega U)^{-1} \\
&\quad + (\omega - 2)(I - \omega U)^{-1}(I - \omega L)^{-1}(I - \omega U).
\end{aligned}
$$

We claim that E_ω has real eigenvalues, since it has the same eigenvalues as the similar matrix

$$
\begin{aligned}
(I - \omega U)&E_\omega(I - \omega U)^{-1} \\
&= I + (2 - \omega)^2 (I - \omega L)^{-1}(I - \omega U)^{-1} + (\omega - 2)(I - \omega U)^{-1} \\
&\quad + (\omega - 2)(I - \omega L)^{-1} \\
&= I + (2 - \omega)^2 (I - \omega L)^{-1}(I - \omega L^T)^{-1} + (\omega - 2)(I - \omega L^T)^{-1} \\
&\quad + (\omega - 2)(I - \omega L)^{-1},
\end{aligned}
$$

which is clearly symmetric and so must have real eigenvalues.

EXAMPLE 6.12. Let us apply SSOR(ω) with Chebyshev acceleration to the model problem. We need to both choose ω and estimate the spectral radius $\rho = \rho(E_\omega)$. The optimal ω that minimizes ρ is not known but Young [267, 137] has shown that the choice $\omega = \frac{2}{1 + [2(1 - \rho(R_J))]^{1/2}}$ is a good one, yielding $\rho(E_\omega) \approx 1 - \frac{\pi}{2N}$. With Chebyshev acceleration the error is multiplied by $\mu_m \approx \frac{1}{T_m(1 + \frac{\pi}{2N})} \leq 2/(1 + m\sqrt{\frac{\pi}{N}})$ at step m. Therefore, to decrease the error by a fixed factor < 1 requires $m = O(N^{1/2}) = O(n^{1/4})$ iterations. Since each iteration has the same cost as an iteration of SOR(ω), $O(n)$, the overall cost is $O(n^{5/4})$. This explains the entry for SSOR with Chebyshev acceleration in Table 6.1.

In contrast, after m steps of SOR(ω_{opt}), the error would decrease only by $(1 - \frac{\pi}{N})^m$. For example, consider $N = 1000$. Then SOR(ω_{opt}) requires $m = 220$ iterations to cut the error in half, whereas SSOR(ω_{opt}) with Chebyshev acceleration requires only $m = 17$ iterations. $\diamond$

6.6. Krylov Subspace Methods

These methods are used both to solve $Ax = b$ and to find eigenvalues of A. They assume that A is accessible only via a "black-box" subroutine that returns $y = Az$ given any z (and perhaps $y = A^T z$ if A is nonsymmetric). In other words, no direct access or manipulation of matrix entries is used. This is a reasonable assumption for several reasons. First, the cheapest nontrivial operation that one can perform on a (sparse) matrix is to multiply it by a vector; if A has m nonzero entries, matrix-vector multiplication costs m multiplications and (at most) m additions. Second, A may not be represented explicitly as a matrix but may be available only as a subroutine for computing Ax.

EXAMPLE 6.13. Suppose that we have a physical device whose behavior is modeled by a program that takes a vector x of input parameters and produces a vector y of output parameters describing the device's behavior. The output y may be an arbitrarily complicated function $y = f(x)$, perhaps requiring the solution of nonlinear differential equations. For example, x could be parameters describing the shape of a wing and $f(x)$ could be the drag on the wing, computed by solving the Navier–Stokes equations for the airflow over the wing. A common engineering design problem is to pick the input x to optimize the device behavior $f(x)$, where for concreteness we assume that this means making $f(x)$ as small as possible. Our problem is then to try to solve $f(x) = 0$ as nearly as we can. Assume for illustration that x and y are vectors of equal dimension. Then Newton's method is an obvious candidate, yielding the iteration $x^{(m+1)} = x^{(m)} - (\nabla f(x^{(m)}))^{-1} f(x^{(m)})$, where $\nabla f(x^{(m)})$ is the Jacobian of f at $x^{(m)}$. We can rewrite this as solving the linear system $(\nabla f(x^{(m)})) \cdot \delta^{(m)} = f(x^{(m)})$ for $\delta^{(m)}$ and then computing $x^{(m+1)} = x^{(m)} - \delta^{(m)}$. But how do we solve this linear system with coefficient matrix $\nabla f(x^{(m)})$ when computing $f(x^{(m)})$ is already complicated? It turns out that we can compute the matrix-vector product $(\nabla f(x)) \cdot z$ for an arbitrary vector z so that we can use Krylov subspace methods to solve the linear system. One way to compute $(\nabla f(x)) \cdot z$ is with *divided differences* or by using a Taylor expansion to see that $[f(x + hz) - f(x)]/h \approx (\nabla f(x)) \cdot z$. Thus, computing $(\nabla f(x)) \cdot z$ requires two calls to the subroutine that computes $f(\cdot)$, once with argument x and once with $x + hz$. However, sometimes it is difficult to choose h to get an accurate approximation of the derivative (choosing h too small results in a loss of accuracy due to roundoff). Another way to compute $(\nabla f(x)) \cdot z$ is to actually differentiate the function f. If f is simple enough, this can be done by hand. For complicated f, compiler tools can take a (nearly) arbitrary subroutine for computing $f(x)$ and automatically produce another subroutine for computing $(\nabla f(x)) \cdot z$ [29]. This can also be done by using the operator overloading facilities of C++ or Fortran 90, although this is less efficient. ◇

A variety of different Krylov subspace methods exist. Some are suitable for nonsymmetric matrices, and others assume symmetry or positive-definiteness. Some methods for nonsymmetric matrices assume that $A^T z$ can be computed as well as Az; depending on how A is represented, $A^T z$ may or may not be available (see Example 6.13). The most efficient and best understood method, the conjugate gradient method (CG), is suitable only for symmetric positive definite matrices, including the model problem. We will concentrate on CG in this chapter.

Given a matrix that is not symmetric positive definite, it can be difficult to pick the best method from the many available. In section 6.6.6 we will give a short summary of the other methods available, besides CG, along with advice on which method to use in which situation. We also refer the reader to the more comprehensive on-line help at NETLIB/templates, which includes a book [24] and implementations in Matlab, Fortran, and C++. For a survey of current research in Krylov subspace methods, see [15, 107, 136, 214].

In Chapter 7, we will also discuss Krylov subspace methods for finding eigenvalues.

6.6.1. Extracting Information about A via Matrix-Vector Multiplication

Given a vector b and a subroutine for computing $A \cdot x$, what can we deduce about A? The most obvious thing that we can do is compute the sequence of matrix-vector products $y_1 = b$, $y_2 = Ay_1$, $y_3 = Ay_2 = A^2 y_1$, ..., $y_n = Ay_{n-1} = A^{n-1} y_1$, where A is n-by-n. Let $K = [y_1, y_2, \ldots, y_n]$. Then we can write

$$A \cdot K = [Ay_1, \ldots, Ay_{n-1}, Ay_n] = [y_2, \ldots, y_n, A^n y_1]. \qquad (6.29)$$

Note that the leading $n - 1$ columns of $A \cdot K$ are the same as the trailing $n - 1$ columns of K, shifted left by one. Assume for the moment that K is nonsingular, so we can compute $c = -K^{-1} A^n y_1$. Then

$$A \cdot K = K \cdot [e_2, e_3, \ldots, e_n, -c] \equiv K \cdot C,$$

where e_i is the ith column of the identity matrix, or

$$K^{-1} A K = C = \begin{bmatrix} 0 & 0 & \cdots & 0 & -c_1 \\ 1 & 0 & \cdots & 0 & -c_2 \\ 0 & 1 & \cdots & \vdots & \vdots \\ \vdots & 0 & \cdots & \vdots & \vdots \\ \vdots & \vdots & \cdots & 0 & \vdots \\ \vdots & \vdots & \cdots & 1 & -c_n \end{bmatrix}.$$

Note that C is upper Hessenberg. In fact, it is a *companion matrix* (see section 4.5.3), which means that its characteristic polynomial is $p(x) = x^n + \sum_{i=1}^{n} c_i x^{i-1}$. Thus, just by matrix-vector multiplication, we have reduced A to a very simple form, and in principle we could now find the eigenvalues of A by finding the zeros of $p(x)$.

However, this simple form is not useful in practice, for the following reasons:

1. Finding c requires $n - 1$ matrix-vector multiplications by A and then solving a linear system with K. Even if A is sparse, K is likely to be dense, so there is no reason to expect solving a linear system with K will be any easier than solving the original problem $Ax = b$.

2. K is likely to be very ill-conditioned, so c would be very inaccurately computed. This is because the algorithm is performing the power method (Algorithm 4.1) to get the columns y_i of K, so that y_i is converging to an eigenvector corresponding to the largest eigenvalue of A. Thus, the columns of K tend to get more and more parallel.

We will overcome these problems as follows: We will replace K with an orthogonal matrix Q such that for all k, the leading k columns of K and Q span the same the same space. This space is called a *Krylov subspace*. In contrast to K, Q is well conditioned and easy to invert. Furthermore, we will compute only as many leading columns of Q as needed to get an accurate solution (for $Ax = b$ or $Ax = \lambda x$). In practice we usually need very few columns compared to the matrix dimension n.

We proceed by writing $K = QR$, the QR decomposition of K. Then

$$K^{-1}AK = (R^{-1}Q^T)A(QR) = C,$$

implying

$$Q^T AQ = RCR^{-1} \equiv H.$$

Since R and R^{-1} are both upper triangular and C is upper Hessenberg, it is easy to confirm that $H = RCR^{-1}$ is also upper Hessenberg (see Question 6.11). In other words, we have reduced A to upper Hessenberg form by an orthogonal transformation Q. (This is the first step of the algorithm for finding eigenvalues of nonsymmetric matrices discussed in section 4.4.6.) Note that if A is symmetric, so is $Q^T AQ = H$, and a symmetric matrix which is upper Hessenberg must also be lower Hessenberg, i.e., tridiagonal. In this case we write $Q^T AQ = T$.

We still need to show how to compute the columns of Q one at a time, rather than all of them: Let $Q = [q_1, \ldots, q_n]$. Since $Q^T AQ = H$ implies $AQ = QH$, we can equate column j on both sides of $AQ = QH$, yielding

$$Aq_j = \sum_{i=1}^{j+1} h_{i,j}q_i.$$

Since the q_i are orthonormal, we can multiply both sides of this last equality by q_m^T to get

$$q_m^T Aq_j = \sum_{i=1}^{j+1} h_{i,j}q_m^T q_i = h_{m,j} \quad \text{for } 1 \le m \le j$$

and so

$$h_{j+1,j} q_{j+1} = A q_j - \sum_{i=1}^{j} h_{i,j} q_i.$$

This justifies the following algorithm.

ALGORITHM 6.9. *The Arnoldi algorithm for (partial) reduction to Hessenberg form:*

$q_1 = b / \|b\|_2$
/* k is the number of columns of Q and H to compute */
for $j = 1$ to k
    $z = A q_j$
    for $i = 1$ to j
        $h_{i,j} = q_i^T z$
        $z = z - h_{i,j} q_i$
    end for
    $h_{j+1,j} = \|z\|_2$
    if $h_{j+1,j} = 0$, quit
    $q_{j+1} = z / h_{j+1,j}$
end for

The q_j computed by Arnoldi's algorithm are often called *Arnoldi vectors*. The loop over i updating z can be also be described as applying the *modified Gram–Schmidt algorithm* (Algorithm 3.1) to subtract the components in the directions q_1 through q_j away from z, leaving z orthogonal to them. Computing q_1 through q_k costs k matrix-vector multiplications by A, plus $O(k^2 n)$ other work. If we stop the algorithm here, what have we learned about A? Let us write $Q = [Q_k, Q_u]$, where $Q_k = [q_1, \dots, q_k]$ and $Q_u = [q_{k+1}, \dots, q_n]$. Note that we have computed only Q_k and q_{k+1}; the other columns of Q_u are unknown. Then

$$
\begin{aligned}
H &= Q^T A Q = [Q_k, Q_u]^T A [Q_k, Q_u] = \begin{bmatrix} Q_k^T A Q_k & Q_k^T A Q_u \\ Q_u^T A Q_k & Q_u^T A Q_u \end{bmatrix} \\[2mm]
&\qquad\qquad\qquad\qquad\;\; k \qquad\;\; n-k \\[1mm]
&\equiv \begin{array}{c} k \\ n-k \end{array} \begin{pmatrix} H_k & H_{uk} \\ H_{ku} & H_u \end{pmatrix}.
\end{aligned}
\tag{6.30}
$$

Note that H_k is upper Hessenberg, because H has the same property. For the same reason, H_{ku} has a single (possibly) nonzero entry in its upper right corner, namely, $h_{k+1,k}$. Thus, H_u and H_{uk} are unknown; we know only H_k and H_{ku}.

When A is symmetric, $H = T$ is symmetric and tridiagonal, and the Arnoldi algorithm simplifies considerably, because most of the $h_{i,j}$ are zero: Write

$$T = \begin{bmatrix} \alpha_1 & \beta_1 & & \\ \beta_1 & \ddots & \ddots & \\ & \ddots & \ddots & \beta_{n-1} \\ & & \beta_{n-1} & \alpha_n \end{bmatrix}.$$

Equating column j on both sides of $AQ = QT$ yields

$$Aq_j = \beta_{j-1}q_{j-1} + \alpha_j q_j + \beta_j q_{j+1}.$$

Since the columns of Q are orthonormal, multiplying both sides of this equation by q_j yields $q_j A q_j = \alpha_j$. This justifies the following version of the Arnoldi algorithm, called the *Lanczos algorithm*.

ALGORITHM 6.10. *The Lanczos algorithm for (partial) reduction to symmetric tridiagonal form.*

$q_1 = b/\|b\|_2$, $\beta_0 = 0$, $q_0 = 0$
$for\ j = 1\ to\ k$
$\quad z = Aq_j$
$\quad \alpha_j = q_j^T z$
$\quad z = z - \alpha_j q_j - \beta_{j-1}q_{j-1}$
$\quad \beta_j = \|z\|_2$
$\quad if\ \beta_j = 0,\ quit$
$\quad q_{j+1} = z/\beta_j$
$end\ for$

The q_j computed by the Lanczos algorithm are often called *Lanczos vectors*. After k steps of the Lanczos algorithm, here is what we have learned about A:

$$\begin{aligned} T & = Q^T AQ = [Q_k, Q_u]^T A [Q_k, Q_u]^T \\ & = \begin{bmatrix} Q_k^T AQ_k & Q_k^T AQ_u \\ Q_u^T AQ_k & Q_u^T AQ_u \end{bmatrix} \\ & \qquad\qquad\quad k \qquad n-k \\ & \equiv \begin{matrix} k \\ n-k \end{matrix} \begin{pmatrix} T_k & T_{uk} \\ T_{ku} & T_u \end{pmatrix} \\ & = \begin{bmatrix} T_k & T_{ku}^T \\ T_{ku} & T_u \end{bmatrix}. \end{aligned} \qquad (6.31)$$

Because A is symmetric, we know T_k and $T_{ku} = T_{uk}^T$ but not T_u. T_{ku} has a single (possibly) nonzero entry in its upper right corner, namely, β_k. Note that β_k is nonnegative, because it is computed as the norm of z.

We define some standard notation associated with the partial factorization of A computed by the Arnoldi and Lanczos algorithms.

DEFINITION 6.16. *The* Krylov subspace $\mathcal{K}_k(A, b)$ *is* span$[b, Ab, A^2b, \ldots, A^{k-1}b]$.

We will write $\mathcal{K}_k$ instead of $\mathcal{K}_k(A, b)$ if A and b are implicit from the context. Provided that the algorithm does not quit because $z = 0$, the vectors Q_k computed by the Arnoldi or Lanczos algorithms form an orthonormal basis of the Krylov subspace $\mathcal{K}_k$. (One can show that $\mathcal{K}_k$ has dimension k if and only if the Arnoldi or Lanczos algorithm can compute q_k without quitting first; see Question 6.12.) We also call H_k (or T_k) the *projection* of A onto the Krylov subspace $\mathcal{K}_k$.

Our goal is to design algorithms to solve $Ax = b$ using only the information computed by k steps of the Arnoldi or the Lanczos algorithm. We hope that k can be much smaller than n, so the algorithms are efficient.

(In Chapter 7 we will use this same information for find eigenvalues of A. We can already sketch how we will do this: Note that if $h_{k+1,k}$ happens to be zero, then H (or T) is block upper triangular and so all the eigenvalues of H_k are also eigenvalues of H, and therefore also of A, since A and H are similar. The (right) eigenvectors of H_k are eigenvectors of H, and if we multiply them by Q_k, we get eigenvectors of A. When $h_{k+1,k}$ is nonzero but small, we expect the eigenvalues and eigenvectors of H_k to provide good approximations to the eigenvalues and eigenvectors of A.)

We finish this introduction by noting that roundoff error causes a number of the algorithms that we discuss to behave *entirely differently* from how they would in exact arithmetic. In particular, the vectors q_i computed by the Lanczos algorithm can quickly lose orthogonality and in fact often become linearly dependent. This apparently disastrous numerical instability led researchers to abandon these algorithms for several years after their discovery. But eventually researchers learned either how to stabilize the algorithms or that convergence occurred despite instability! We return to these points in section 6.6.4, where we analyze the convergence of the conjugate gradient method for solving $Ax = b$ (which is "unstable" but converges anyway), and in Chapter 7, especially in sections 7.4 and 7.5, where we show how to compute eigenvalues (and the basic algorithm is modified to ensure stability).

6.6.2. Solving $Ax = b$ Using the Krylov Subspace $\mathcal{K}_k$

How do we solve $Ax = b$, given only the information available from k steps of either the Arnoldi or the Lanczos algorithm?

Since the only vectors we know are the columns of Q_k, the only place to "look" for an approximate solution is in the Krylov subspace $\mathcal{K}_k$ spanned by these vectors. In other words, we see the "best" approximate solution of the form

$$x_k = \sum_{j=1}^{k} z_k q_k = Q_k \cdot z, \quad \text{where} \quad z = [z_1, \ldots, z_k]^T.$$

Now we have to define "best." There are several natural but different definitions, leading to different algorithms. We let $x = A^{-1}b$ denote the true solution and $r_k = b - Ax_k$ denote the residual.

1. The "best" x_k minimizes $\|x_k - x\|_2$. Unfortunately, we do not have enough information in our Krylov subspace to compute this x_k.

2. The "best" x_k minimizes $\|r_k\|_2$. This is implementable, and the corresponding algorithms are called MINRES (for *minimum residual*) when A is symmetric [194] and GMRES (for *generalized minimum residual*) when A is nonsymmetric [215].

3. The "best" x_k makes $r_k \perp \mathcal{K}_k$, i.e., $Q_k^T r_k = 0$. This is sometimes called the *orthogonal residual* property, or a *Galerkin condition*, by analogy to a similar condition in the theory of finite elements. When A is symmetric, the corresponding algorithm is called SYMMLQ [194]. When A is nonsymmetric, a variation of GMRES works [211].

4. When A is symmetric and positive definite, it defines a norm $\|r\|_{A^{-1}} = (r^T A^{-1} r)^{1/2}$ (see Lemma 1.3). We say the "best" x_k minimizes $\|r_k\|_{A^{-1}}$. This norm is the same as $\|x_k - x\|_A$. The algorithm is called the conjugate gradient algorithm [145].

When A is symmetric positive definite, the last two definitions of "best" also turn out to be equivalent.

THEOREM 6.8. *Let A be symmetric, $T_k = Q_k^T A Q_k$, and $r_k = b - Ax_k$, where $x_k \in \mathcal{K}_k$. If T_k is nonsingular and $x_k = Q_k T_k^{-1} e_1 \|b\|_2$, where $e_1^{k \times 1} = [1, 0, \ldots, 0]^T$, then $Q_k^T r_k = 0$. If A is also positive definite, then T_k must be nonsingular, and this choice of x_k also minimizes $\|r_k\|_{A^{-1}}$ over all $x_k \in \mathcal{K}_k$. We also have that $r_k = \pm \|r_k\|_2 q_{k+1}$.*

Proof. We drop the subscripts k for ease of notation. Let $x = QT^{-1}e_1\|b\|_2$ and $r = b - Ax$, and assume that $T = Q^T AQ$ is nonsingular. We confirm that $Q^T r = 0$ by computing

$$
\begin{aligned}
Q^T r = Q^T (b - Ax) &= Q^T b - Q^T Ax \\
&= e_1 \|b\|_2 - Q^T A(QT^{-1} e_1 \|b\|_2) \\
&\qquad \text{because the first column of } Q \text{ is } b/\|b\|_2 \\
&\qquad \text{and its other columns are orthogonal to } b \\
&= e_1 \|b\|_2 - (Q^T AQ) T^{-1} e_1 \|b\|_2 \\
&= e_1 \|b\|_2 - (T) T^{-1} e_1 \|b\|_2 \quad \text{because } Q^T AQ = T \\
&= 0.
\end{aligned}
$$

Now assume that A is also positive definite. Then T must be positive definite and thus nonsingular too (see Question 6.13). Let $\hat{x} = x + Qz$ be another candidate solution in $\mathcal{K}$, and let $\hat{r} = b - A\hat{x}$. We need to show that $\|\hat{r}\|_{A^{-1}}$ is minimized when $z = 0$. But

$$
\begin{aligned}
\|\hat{r}\|_{A^{-1}}^2 &= \hat{r}^T A^{-1} \hat{r} \qquad \text{by definition} \\
&= (r - AQz)^T A^{-1} (r - AQz) \\
&\qquad \text{since } \hat{r} = b - A\hat{x} = b - A(x + Qz) = r - AQz \\
&= r^T A^{-1} r^T - 2(AQz)^T A^{-1} r + (AQz)^T A^{-1} (AQz) \\
&= \|r\|_{A^{-1}}^2 - 2z^T Q^T r + \|AQz\|_{A^{-1}}^2 \\
&\qquad \text{since } (AQz)^T A^{-1} r = z^T Q^T A A^{-1} r = z^T Q^T r \\
&= \|r\|_{A^{-1}}^2 + \|AQz\|_{A^{-1}}^2 \qquad \text{since } Q^T r = 0,
\end{aligned}
$$

so $\|\hat{r}\|_{A^{-1}}$ is minimized if and only if $AQz = 0$. But $AQz = 0$ if and only if $z = 0$ since A is nonsingular and Q has full column rank.

To show that $r_k = \pm\|r_k\|_2 q_{k+1}$, we reintroduce subscripts. Since $x_k \in \mathcal{K}_k$, we must have $r_k = b - Ax_k \in \mathcal{K}_{k+1}$, so r_k is a linear combination of the columns of Q_{k+1}, since these columns span $\mathcal{K}_{k+1}$. But since $Q_k^T r_k = 0$, the only column of Q_{k+1} to which r_k is not orthogonal is q_{k+1}. $\square$

6.6.3. Conjugate Gradient Method

The algorithm of choice for symmetric positive definite matrices is CG. Theorem 6.8 characterizes the solution x_k computed by CG. While MINRES might seem more natural than CG because it minimizes $\|r_k\|_2$ instead of $\|r_k\|_{A^{-1}}$, it turns out that MINRES requires more work to implement, is more susceptible to numerical instabilities, and thus often produces less accurate answers than CG. We will see that CG has the particularly attractive property that it can be implemented by keeping only four vectors in memory at one time, and not k (q_1 through q_k). Furthermore, the work in the inner loop, beyond the matrix-vector product, is limited to two dot products, three "saxpy" operations (adding a multiple of one vector to another), and a handful of scalar operations. This is a very small amount of work and storage.

Now we derive CG. There are several ways to do this. We will start with the Lanczos algorithm (Algorithm 6.10), which computes the columns of the orthogonal matrix Q_k and the entries of the tridiagonal matrix T_k, along with the formula $x_k = Q_k T_k^{-1} e_1 \|b\|_2$ from Theorem 6.8. We will show how to compute x_k directly via recurrences for three sets of vectors. We will keep only the most recent vector from each set in memory at one time, overwriting the old ones. The first set of vectors are the approximate solutions x_k. The second set of vectors are the residuals $r_k = b - Ax_k$, which Theorem 6.8 showed were parallel to the Lanczos vectors q_{k+1}. The third set of vectors are the *conjugate gradients* p_k. The p_k are called *gradients* because a single step of CG can be interpreted as choosing a scalar ν so that the new solution $x_k = x_{k-1} + \nu p_k$ minimizes the residual norm $\|r_k\|_{A^{-1}} = (r_k^T A^{-1} r_k)^{1/2}$. In other words, the p_k are used as *gradient search directions*. The p_k are called *conjugate*, or more precisely *A-conjugate*, because $p_k^T A p_j = 0$ if $j \neq k$. In other words, the p_k are orthogonal with respect to the inner product defined by A (see Lemma 1.3).

Since A is symmetric positive definite, so is $T_k = Q_k^T A Q_k$ (see Question 6.13). This means we can perform Cholesky on T_k to get $T_k = \hat{L}_k \hat{L}_k^T = L_k D_k L_k^T$, where L_k is unit lower bidiagonal and D_k is diagonal. Then using the formula for x_k from Theorem 6.8, we get

$$
\begin{aligned}
x_k &= Q_k T_k^{-1} e_1 \|b\|_2 \\
&= Q_k (L_k^{-T} D_k^{-1} L_k^{-1}) e_1 \|b\|_2 \\
&= (Q_k L_k^{-T})(D_k^{-1} L_k^{-1} e_1 \|b\|_2) \\
&\equiv (\tilde{P}_k)(y_k),
\end{aligned}
$$

where $\tilde{P}_k \equiv Q_k L_k^{-T}$ and $y_k \equiv D_k^{-1} L_k^{-1} e_1 \|b\|_2$. Write $\tilde{P}_k = [\tilde{p}_1, \ldots, \tilde{p}_k]$. The conjugate gradients p_i will turn out to be parallel to the columns $\tilde{p}_i$ of $\tilde{P}_k$. We know enough to prove the following lemma.

LEMMA 6.8. *The columns $\tilde{p}_i$ of $\tilde{P}_k$ are A-conjugate. In other words, $\tilde{P}_k^T A \tilde{P}_k$ is diagonal.*

Proof. We compute

$$
\begin{aligned}
\tilde{P}_k^T A \tilde{P}_k &= (Q_k L_k^{-T})^T A (Q_k L_k^{-T}) = L_k^{-1}(Q_k^T A Q_k) L_k^{-T} = L_k^{-1}(T_k) L_k^{-T} \\
&= L_k^{-1}(L_k D_k L_k^T) L_k^{-T} = D_k. \quad \square
\end{aligned}
$$

Now we derive simple recurrences for the columns of $\tilde{P}_k$ and entries of y_k. We will show that $y_{k-1} \equiv [\eta_1, \ldots, \eta_{k-1}]^T$ is identical to the leading $k-1$ entries of $y_k = [\eta_1, \ldots, \eta_{k-1}, \eta_k]^T$ and that $\tilde{P}_{k-1}$ is identical to the leading $k-1$ columns of $\tilde{P}_k$. Therefore we can let

$$
x_k = \tilde{P}_k \cdot y_k = [\tilde{P}_{k-1}, \tilde{p}_k] \cdot \begin{bmatrix} y_{k-1} \\ \eta_k \end{bmatrix} = \tilde{P}_{k-1} y_{k-1} + \tilde{p}_k \eta_k = x_{k-1} + \tilde{p}_k \eta_k \quad (6.32)
$$

be our recurrence for x_k.

The recurrence for the η_k is derived as follows. Since T_{k-1} is the leading $(k-1)$-by-$(k-1)$ submatrix of T_k, L_{k-1} and D_{k-1} are also the leading $(k-1)$-by-$(k-1)$ submatrices of L_k and D_k, respectively:

$$
\begin{aligned}
T_k &= \begin{bmatrix} \alpha_1 & \beta_1 & & \\ \beta_1 & \ddots & \ddots & \\ & \ddots & \ddots & \beta_{k-1} \\ & & \beta_{k-1} & \alpha_k \end{bmatrix} \\
&= L_k D_k L_k^T
\end{aligned}
$$

$$= \begin{bmatrix} 1 & & & \\ l_1 & \ddots & & \\ & \ddots & \ddots & \\ & & l_{k-1} & 1 \end{bmatrix} \cdot \begin{bmatrix} d_1 & & & \\ & \ddots & & \\ & & d_{k-1} & \\ & & & d_k \end{bmatrix} \cdot \begin{bmatrix} 1 & & & \\ l_1 & \ddots & & \\ & \ddots & \ddots & \\ & & l_{k-1} & 1 \end{bmatrix}^T$$

$$= \begin{bmatrix} L_{k-1} & \\ l_{k-1}\hat{e}_{k-1}^T & 1 \end{bmatrix} \cdot \mathrm{diag}(D_{k-1}, d_k) \cdot \begin{bmatrix} L_{k-1} & \\ l_{k-1}\hat{e}_{k-1}^T & 1 \end{bmatrix}^T ,$$

where $\hat{e}_{k-1}^T = [0, \ldots, 0, 1]$ has dimension $k - 1$. Similarly, D_{k-1}^{-1} and L_{k-1}^{-1} are also the leading $(k - 1)$-by-$(k - 1)$ submatrices of $D_k^{-1} = \mathrm{diag}(D_{k-1}^{-1}, d_k^{-1})$ and

$$L_k^{-1} = \begin{bmatrix} L_{k-1}^{-1} & \\ \star & 1 \end{bmatrix} ,$$

respectively, where the details of the last row $\star$ do not concern us. This means that $y_{k-1} = D_{k-1}^{-1} L_{k-1}^{-1} \hat{e}_1 \|b\|_2$, where $\hat{e}_1$ has dimension $k - 1$, is identical to the leading $k - 1$ components of

$$y_k = D_k^{-1} L_k^{-1} e_1 \|b\|_2 = \begin{bmatrix} D_{k-1}^{-1} & \\ & d_k^{-1} \end{bmatrix} \cdot \begin{bmatrix} L_{k-1}^{-1} & \\ \star & 1 \end{bmatrix} \cdot e_1 \|b\|_2$$

$$= \begin{bmatrix} D_{k-1}^{-1} L_{k-1}^{-1} \hat{e}_1 \|b\|_2 \\ \eta_k \end{bmatrix} = \begin{bmatrix} y_{k-1} \\ \eta_k \end{bmatrix} .$$

Now we need a recurrence for the columns of $\tilde{P}_k = [\tilde{p}_1, \ldots, \tilde{p}_k]$. Since L_{k-1}^T is upper triangular, so is L_{k-1}^{-T}, and it forms the leading $(k - 1)$-by-$(k - 1)$ submatrix of L_k^{-T}. Therefore $\tilde{P}_{k-1}$ is identical to the leading $k - 1$ columns of

$$\tilde{P}_k = Q_k L_k^{-T} = [Q_{k-1}, q_k] \begin{bmatrix} L_{k-1}^{-T} & \star \\ 0 & 1 \end{bmatrix} = [Q_{k-1} L_{k-1}^{-T}, \tilde{p}_k] = [\tilde{P}_{k-1}, \tilde{p}_k].$$

From $\tilde{P}_k = Q_k L_k^{-T}$, we get $\tilde{P}_k L_k^T = Q_k$ or, equating the kth column on both sides, the recurrence

$$\tilde{p}_k = q_k - l_{k-1}\tilde{p}_{k-1}. \tag{6.33}$$

Altogether, we have recursions for q_k (from the Lanczos algorithm), for $\tilde{p}_k$ (from equation (6.33)), and for the approximate solution x_k (from equation (6.32)). All these recursions are *short*; i.e., they require only the previous iterate or two to implement. Thus, they together provide the means to compute x_k while storing a small number of vectors and doing a small number of dot products, saxpys, and scalar work in the inner loop.

We still have to simplify these recursions slightly to get the ultimate CG algorithm. Since Theorem 6.8 tells us that r_k and q_{k+1} are parallel, we can replace the Lanczos recurrence for q_{k+1} with the recurrence $r_k = b - Ax_k$ or equivalently $r_k = r_{k-1} - \eta_k A\tilde{p}_k$ (gotten from multiplying the recurrence

$x_k = x_{k-1} + \eta_k \tilde{p}_k$ by A and subtracting from $b = b$). This yields the three vector recurrences

$$r_k = r_{k-1} - \eta_k A\tilde{p}_k, \tag{6.34}$$

$$x_k = x_{k-1} + \eta_k \tilde{p}_k \quad \text{from equation (6.32)}, \tag{6.35}$$

$$\tilde{p}_k = q_k - l_{k-1}\tilde{p}_{k-1} \quad \text{from equation (6.33)}. \tag{6.36}$$

In order to eliminate q_k, substitute $q_k = r_{k-1}/\|r_{k-1}\|_2$ and $p_k \equiv \|r_{k-1}\|_2 \tilde{p}_k$ into the above recurrences to get

$$
\begin{aligned}
r_k &= r_{k-1} - \frac{\eta_k}{\|r_{k-1}\|_2} A p_k \\
&\equiv r_{k-1} - \nu_k A p_k, \tag{6.37} \\
x_k &= x_{k-1} + \frac{\eta_k}{\|r_{k-1}\|_2} p_k \\
&\equiv x_{k-1} + \nu_k p_k, \tag{6.38} \\
p_k &= r_{k-1} - \frac{\|r_{k-1}\|_2 l_{k-1}}{\|r_{k-2}\|_2} \cdot p_{k-1} \\
&\equiv r_{k-1} + \mu_k \cdot p_{k-1}. \tag{6.39}
\end{aligned}
$$

We still need formulas for the scalars ν_k and μ_k. As we will see, there are several equivalent mathematical expression for them in terms of dot products of vectors computed by the algorithm. Our ultimate formulas are chosen to minimize the number of dot products needed and because they are more stable than the alternatives.

To get a formula for ν_k, first we multiply both sides of equation (6.39) on the left by $p_k^T A$ and use the fact that p_k and p_{k-1} are A-conjugate (Lemma 6.8) to get

$$p_k^T A p_k = p_k^T A r_{k-1} + 0 = r_{k-1}^T A p_k. \tag{6.40}$$

Then, multiply both sides of equation (6.37) on the left by r_{k-1}^T and use the fact that $r_{k-1}^T r_k = 0$ (since the r_i are parallel to the columns of the orthogonal matrix Q) to get

$$
\begin{aligned}
\nu_k &= \frac{r_{k-1}^T r_{k-1}}{r_{k-1}^T A p_k} \\
&= \frac{r_{k-1}^T r_{k-1}}{p_k^T A p_k} \quad \text{by equation (6.40)}. \tag{6.41}
\end{aligned}
$$

(Equation (6.41) can also be derived from a property of ν_k in Theorem 6.8, namely, that it minimizes the residual norm

$$
\begin{aligned}
\|r_k\|_{A^{-1}}^2 &= r_k^T A^{-1} r_k \\
&= (r_{k-1} - \nu_k A p_k)^T A^{-1}(r_{k-1} - \nu_k A p_k) \quad \text{by equation (6.37)} \\
&= r_{k-1} A^{-1} r_{k-1}^T - 2\nu_k p_k^T r_{k-1} + \nu_k^2 p_k^T A p_k.
\end{aligned}
$$

This expression is a quadratic function of ν_k, so it can be easily minimized by setting its derivative with respect to ν_k to zero and solving for ν_k. This yields

$$
\begin{aligned}
\nu_k &= \frac{p_k^T r_{k-1}}{p_k^T A p_k} \\
&= \frac{(r_{k-1} + \mu_k \cdot p_{k-1})^T r_{k-1}}{p_k^T A p_k} \quad \text{by equation (6.39)} \\
&= \frac{r_{k-1}^T r_{k-1}}{p_k^T A p_k},
\end{aligned}
$$

where we have used the fact that $p_{k-1}^T r_{k-1} = 0$, which holds since r_{k-1} is orthogonal to all vectors in $\mathcal{K}_{k-1}$, including p_{k-1}.)

To get a formula for μ_k, multiply both sides of equation (6.39) on the left by $p_{k-1}^T A$ and use the fact that p_k and p_{k-1} are A-conjugate (Lemma 6.8) to get

$$
\mu_k = -\frac{p_{k-1}^T A r_{k-1}}{p_{k-1}^T A p_{k-1}}. \tag{6.42}
$$

The trouble with this formula for μ_k is that it requires another dot product, $p_{k-1}^T A r_{k-1}$, besides the two required for ν_k. So we will derive another formula requiring no new dot products.

We do this by deriving an alternate formula for ν_k: Multiply both sides of equation (6.37) on the left by r_k^T, again use the fact that $r_{k-1}^T r_k = 0$, and solve for ν_k to get

$$
\nu_k = -\frac{r_k^T r_k}{r_k^T A p_k}. \tag{6.43}
$$

Equating the two expressions (6.41) and (6.43) for ν_{k-1} (note that we have subtracted 1 from the subscript), rearranging, and comparing to equation (6.42) yield our ultimate formula for μ_k:

$$
\begin{aligned}
\mu_k &= -\frac{p_{k-1}^T A r_{k-1}}{p_{k-1}^T A p_{k-1}} \\
&= \frac{r_{k-1}^T r_{k-1}}{r_{k-2}^T r_{k-2}}. \tag{6.44}
\end{aligned}
$$

Combining recurrences (6.37), (6.38), and (6.39) and formulas (6.41) and (6.44) yields our final implementation of the conjugate gradient algorithm.

ALGORITHM 6.11. *Conjugate gradient algorithm:*

$k = 0;\ x_0 = 0;\ r_0 = b;\ p_1 = b;$
repeat
    $k = k + 1$

$$z = A \cdot p_k$$
$$\nu_k = (r_{k-1}^T r_{k-1})/(p_k^T z)$$
$$x_k = x_{k-1} + \nu_k p_k$$
$$r_k = r_{k-1} - \nu_k z$$
$$\mu_{k+1} = (r_k^T r_k)/(r_{k-1}^T r_{k-1})$$
$$p_{k+1} = r_k + \mu_{k+1} p_k$$
$$until \ \|r_k\|_2 \ is \ small \ enough$$

The cost of the inner loop for CG is one matrix-vector product $z = A \cdot p_k$, two inner products (by saving the value of $r_k^T r_k$ from one loop iteration to the next), three saxpys, and a few scalar operations. The only vectors that need to be stored are the current values of r, x, p, and $z = Ap$. For more implementation details, including how to decide if "$\|r_k\|_2$ is small enough," see NETLIB/templates/Templates.html.

6.6.4. Convergence Analysis of the Conjugate Gradient Method

We begin with a convergence analysis of CG that depends only on the condition number of A. This analysis will show that the number of CG iterations needed to reduce the error by a fixed factor less than 1 is proportional to the square root of the condition number. This worst-case analysis is a good estimate for the speed of convergence on our model problem, Poisson's equation. But it severely *underestimates* the speed of convergence in many other cases. After presenting the bound based on the condition number, we describe when we can expect faster convergence.

We start with the initial approximate solution $x_0 = 0$. Recall that x_k minimizes the A^{-1}-norm of the residual $r_k = b - Ax_k$ over all possible solutions $x_k \in \mathcal{K}_k(A, b)$. This means x_k minimizes

$$\|b - Az\|_{A^{-1}}^2 \equiv f(z) = (b - Az)^T A^{-1}(b - Az) = (x - z)^T A(x - z)$$

over all $z \in \mathcal{K}_k = \mathrm{span}[b, Ab, A^2 b, \ldots, A^{k-1} b]$. Any $z \in \mathcal{K}_k(A, b)$ may be written $z = \sum_{j=0}^{k-1} \alpha_j A^j b = p_{k-1}(A) b = p_{k-1}(A) Ax$, where $p_{k-1}(\xi) = \sum_{j=0}^{k-1} \alpha_j \xi^j$ is a polynomial of degree $k - 1$. Therefore,

$$
\begin{aligned}
f(z) &= [(I - p_{k-1}(A)A)x]^T A[(I - p_{k-1}(A)A)x] \\
&\equiv (q_k(A)x)^T A(q_k(A)x) \\
&= x^T q_k(A) A q_k(A) x,
\end{aligned}
$$

where $q_k(\xi) \equiv 1 - p_{k-1}(\xi) \cdot \xi$ is a degree-k polynomial with $q_k(0) = 1$. Note that $(q_k(A))^T = q_k(A)$ because $A = A^T$. Letting $\mathcal{Q}_k$ be the set of all degree-k polynomials which take the value 1 at 0, this means

$$f(x_k) = \min_{z \in \mathcal{K}_k} f(z) = \min_{q_k \in \mathcal{Q}_k} x^T q_k(A) A q_k(A) x. \tag{6.45}$$

To simplify this expression, write the eigendecomposition $A = Q\Lambda Q^T$ and let $Q^T x = y$ so that

$$
\begin{aligned}
f(x_k) = \min_{z \in \mathcal{K}_k} f(z) &= \min_{q_k \in \mathcal{Q}_k} x^T (q_k(Q\Lambda Q^T))(Q\Lambda Q^T)(q_k(Q\Lambda Q^T))x \\
&= \min_{q_k \in \mathcal{Q}_k} x^T (Q q_k(\Lambda)Q^T)(Q\Lambda Q^T)(Q q_k(\Lambda)Q^T)x \\
&= \min_{q_k \in \mathcal{Q}_k} y^T q_k(\Lambda)\Lambda q_k(\Lambda)y \\
&= \min_{q_k \in \mathcal{Q}_k} y^T \cdot \mathrm{diag}(q_k(\lambda_i)\lambda_i q_k(\lambda_i)) \cdot y \\
&= \min_{q_k \in \mathcal{Q}_k} \sum_{i=1}^{n} y_i{}^2 \lambda_i (q_k(\lambda_i))^2 \\
&\leq \min_{q_k \in \mathcal{Q}_k} \left(\max_{\lambda_i \in \lambda(A)} (q_k(\lambda_i))^2 \right) \sum_{i=1}^{n} y_i{}^2 \lambda_i \\
&= \min_{q_k \in \mathcal{Q}_k} \left(\max_{\lambda_i \in \lambda(A)} (q_k(\lambda_i))^2 \right) f(x_0)
\end{aligned}
$$

since $x_0 = 0$ implies $f(x_0) = x^T A x = y^T \Lambda y = \sum_{i=1}^{n} y_i{}^2 \lambda_i$. Therefore,

$$
\frac{\|r_k\|_{A^{-1}}^2}{\|r_0\|_{A^{-1}}^2} = \frac{f(x_k)}{f(x_0)} \leq \min_{q_k \in \mathcal{Q}} \max_{\lambda_i \in \lambda(A)} (q_k(\lambda_i))^2
$$

or

$$
\frac{\|r_k\|_{A^{-1}}}{\|r_0\|_{A^{-1}}} \leq \min_{q_k \in \mathcal{Q}} \max_{\lambda_i \in \lambda(A)} |q_k(\lambda_i)|.
$$

We have thus reduced the question of how fast CG converges to a question about polynomials: How small can a degree-k polynomial $q_k(\xi)$ be when ξ ranges over the eigenvalues of A, while simultaneously satisfying $q_k(0) = 1$? Since A is positive definite, its eigenvalues lie in the interval $[\lambda_{\min}, \lambda_{\max}]$, where $0 < \lambda_{\min} \leq \lambda_{\max}$, so to get a simple upper bound we will instead seek a degree-k polynomial $\hat{q}_k(\xi)$ that is small on the whole interval $[\lambda_{\min}, \lambda_{\max}]$ and 1 at 0. A polynomial $\hat{q}_k(\xi)$ that has this property is easily constructed from the Chebyshev polynomials $T_k(\xi)$ discussed in section 6.5.6. Recall that $|T_k(\xi)| \leq 1$ when $|\xi| \leq 1$ and increases rapidly when $|\xi| > 1$ (see Figure 6.6). Now let

$$
\hat{q}_k(\xi) = T_k \left(\frac{\lambda_{\max} + \lambda_{\min} - 2\xi}{\lambda_{\max} - \lambda_{\min}} \right) \Big/ T_k \left(\frac{\lambda_{\max} + \lambda_{\min}}{\lambda_{\max} - \lambda_{\min}} \right).
$$

It is easy to see that $\hat{q}(0) = 1$, and if $\xi \in [\lambda_{\min}, \lambda_{\max}]$, then

$$
\left| \frac{\lambda_{\max} + \lambda_{\min} - 2\xi}{\lambda_{\max} - \lambda_{\min}} \right| \leq 1,
$$

so

$$
\begin{aligned}
\frac{\|r_k\|_{A^{-1}}}{\|r_0\|_{A^{-1}}} &\leq \min_{q_k \in \mathcal{Q}} \max_{\lambda_i \in \lambda(A)} |q_k(\lambda_i)| \\
&\leq \frac{1}{T_k(\frac{\lambda_{\max}+\lambda_{\min}}{\lambda_{\max}-\lambda_{\min}})} = \frac{1}{T_k(\frac{\kappa+1}{\kappa-1})} = \frac{1}{T_k(1 + \frac{2}{\kappa-1})},
\end{aligned} \tag{6.46}
$$

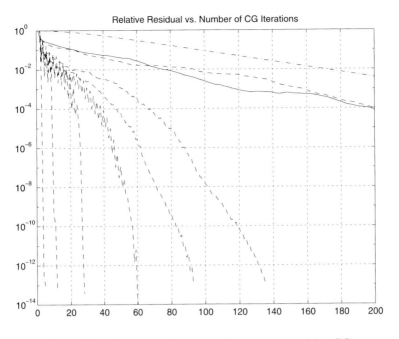

Fig. 6.7. *Graph of relative residuals computed by CG.*

where $\kappa = \lambda_{\max}/\lambda_{\min}$ is the condition number of A.

If the condition number κ is near 1, $1 + 2/(\kappa - 1)$ is large, $1/T_k(1 + \frac{2}{\kappa-1})$ is small, and convergence is rapid. If κ is large, convergence slows down, with the A^{-1}-norm of the residual r_k going to zero like

$$\frac{1}{T_k(1 + \frac{2}{\kappa-1})} \leq \frac{2}{1 + \frac{2k}{\sqrt{\kappa-1}}}.$$

EXAMPLE 6.14. For the N-by-N model problem, $\kappa = O(N^2)$, so after k steps of CG the residual is multiplied by about $(1 - O(N^{-1}))^k$, the same as SOR with optimal overrelaxation parameter ω. In other words, CG takes $O(N) = O(n^{1/2})$ iterations to converge. Since each iteration costs $O(n)$, the overall cost is $O(n^{3/2})$. This explains the entry for CG in Table 6.1. $\diamond$

This analysis using the condition number does not explain all the important convergence behavior of CG. The next example shows that the entire distribution of eigenvalues of A is important, not just the ratio of the largest to the smallest one.

EXAMPLE 6.15. Let us consider Figure 6.7, which plots the relative residual $\|r_k\|_2/\|r_0\|_2$ at each CG step for eight different linear systems. The relative residual $\|r_k\|_2/\|r_0\|_2$ measures the speed of convergence; our implementation of CG terminates when this ratio sinks below 10^{-13}, or after $k = 200$ steps, whichever comes first.

All eight linear systems shown have the same dimension $n = 10^4$ and the same condition number $\kappa \approx 4134$, yet their convergence behaviors are radically different. The uppermost (dash-dot) line is $1/T_k(1 + \frac{2}{\kappa-1})$, which inequality (6.46) tells us is an upper bound on $\|r_k\|_{A^{-1}}/\|r_0\|_{A^{-1}}$. It turns out the graphs of $\|r_k\|_2/\|r_0\|_2$ and the graphs of $\|r_k\|_{A^{-1}}/\|r_0\|_{A^{-1}}$ are nearly the same, so we plot only the former, which are easier to interpret.

The solid line is $\|r_k\|_2/\|r_0\|_2$ for Poisson's equation on a 100-by-100 grid with a random right-hand side b. We see that the upper bound captures its general convergence behavior. The seven dashed lines are plots of $\|r_k\|_2/\|r_0\|_2$ for seven diagonal linear systems $D_i x = b$, numbered from D_1 on the left to D_7 on the right. Each D_i has the same dimension and condition number as Poisson's equation, so we need to study them more closely to understand their differing convergence behaviors.

We have constructed each D_i so that its smallest m_i and largest m_i eigenvalues are identical to those of Poisson's equation, with the remaining $n - 2m_i$ eigenvalues equal to the geometric mean of the largest and smallest eigenvalues. In other words, D_i has only $d_i = 2m_i + 1$ distinct eigenvalues. We let k_i denote the number of CG iterations it takes for the solution of $D_i x = b$ to reach $\|r_k\|_2/\|r_0\|_2 \leq 10^{-13}$. The convergence properties are summarized in the following table:

Example number	i	1	2	3	4	5	6	7
Number of distinct eigenvalues	d_i	3	11	41	81	201	401	5000
Number of steps to converge	k_i	3	11	27	59	94	134	> 200

We see that the number k_i of steps required to converge grows with the number d_i of distinct eigenvalues. D_7 has the same spectrum as Poisson's equation, and converges about as slowly.

In the absence of roundoff, we claim that CG would take *exactly* $k_i = d_i$ steps to converge. The reason is that we can find a polynomial $q_{d_i}(\xi)$ of degree d_i that is zero at the eigenvalues α_j of A, while $q_{d_i}(0) = 1$, namely,

$$q_{d_i}(\xi) = \frac{\prod_{j=1}^{d_i}(\alpha_j - \xi)}{\prod_{j=1}^{d_i}(\alpha_j)}.$$

Equation (6.45) tells us that after d_i steps, CG minimizes $\|r_{d_i}\|_{A^{-1}}^2 = f(x_{d_i})$ over all possible degree-d_i polynomials equaling 1 at 0. Since q_{d_i} is one of those polynomials and $q_{d_i}(A) = 0$, we must have $\|r_{d_i}\|_{A^{-1}}^2 = 0$, or $r_{d_i} = 0$. ◇

One lesson of Example 6.15 is that if the largest and smallest eigenvalues of A are few in number (or clustered closely together), then CG will converge much more quickly than an analysis based just on A's condition number would indicate.

Another lesson is that the behavior of CG in floating point arithmetic can differ significantly from its behavior in exact arithmetic. We saw this because the number d_i of distinct eigenvalues frequently differed from the number k_i of steps required to converge, although in theory we showed that they should be identical. Still, d_i and k_i were of the same order of magnitude.

Indeed, if one were to perform CG in exact arithmetic and compare the computed solutions and residuals with those computed in floating point arithmetic, they would very probably diverge and soon be quite different. Still, as long as A is not too ill-conditioned, the floating point result will eventually converge to the desired solution of $Ax = b$, and so CG is still very useful. The fact that the exact and floating point results can differ dramatically is interesting but does not prevent the practical use of CG.

When CG was discovered, it was proven that in exact arithmetic it would provide the exact answer after n steps, since then r_{n+1} would be orthogonal to n other orthogonal vectors r_1 through r_n, and so must be zero. In other words, CG was thought of as a *direct method* rather than an *iterative method*. When convergence after n steps did not occur in practice, CG was considered unstable and then abandoned for many years. Eventually it was recognized as a perfectly good iterative method, often providing quite accurate answers after $k \ll n$ steps.

Recently, a subtle backward error analysis was devised to explain the observed behavior of CG in floating point and explain how it can differ from exact arithmetic [123]. This behavior can also include long "plateaus" in the convergence, with $\|r_k\|_2$ decreasing little for many iterations, interspersed with periods of rapid convergence. This behavior can be explained by showing that CG applied to $Ax = b$ in floating point arithmetic behaves *exactly* like CG applied to $\tilde{A}\tilde{x} = \tilde{b}$ in exact arithmetic, where $\tilde{A}$ is close to A in the following sense: $\tilde{A}$ has a much larger dimension than A, but $\tilde{A}$'s eigenvalues all lie in narrow clusters around the eigenvalues of A. Thus the plateaus in convergence correspond to the polynomial q_k underlying CG developing more and more zeros near the eigenvalues of $\tilde{A}$ lying in a cluster.

6.6.5. Preconditioning

In the previous section we saw that the convergence rate of CG depended on the condition number of A, or more generally the distribution of A's eigenvalues. Other Krylov subspace methods have the same property. *Preconditioning* means replacing the system $Ax = b$ with the system $M^{-1}Ax = M^{-1}b$, where M is an approximation to A with the properties that

1. M is symmetric and positive definite,

2. $M^{-1}A$ is well conditioned or has few extreme eigenvalues,

3. $Mx = b$ is easy to solve.

A careful, problem-dependent choice of M can often make the condition number of $M^{-1}A$ much smaller than the condition number of A and thus accelerate convergence dramatically. Indeed, a good preconditioner is often necessary for an iterative method to converge at all, and much current research in iterative methods is directed at finding better preconditioners (see also section 6.10).

We cannot apply CG directly to the system $M^{-1}Ax = M^{-1}b$, because $M^{-1}A$ is generally not symmetric. We derive the *preconditioned conjugate gradient method* as follows. Let $M = Q\Lambda Q^T$ be the eigendecomposition of M, and define $M^{1/2} \equiv Q\Lambda^{1/2}Q^T$. Note that $M^{1/2}$ is also symmetric positive definite, and $(M^{1/2})^2 = M$. Now multiply $M^{-1}Ax = M^{-1}b$ through by $M^{1/2}$ to get the new symmetric positive definite system $(M^{-1/2}AM^{-1/2})(M^{1/2}x) = M^{-1/2}b$, or $\hat{A}\hat{x} = \hat{b}$. Note that $\hat{A}$ and $M^{-1}A$ have the same eigenvalues since they are similar ($M^{-1}A = M^{-1/2}\hat{A}M^{1/2}$). We now apply CG *implicitly* to the system $\hat{A}\hat{x} = \hat{b}$ in such a way that avoids the need to multiply by $M^{-1/2}$. This yields the following algorithm.

ALGORITHM 6.12. *Preconditioned CG algorithm:*

    $k = 0;\ x_0 = 0;\ r_0 = b;\ p_1 = M^{-1}b;\ y_0 = M^{-1}r_0$
    repeat
        $k = k + 1$
        $z = A \cdot p_k$
        $\nu_k = (y_{k-1}^T r_{k-1})/(p_k^T z)$
        $x_k = x_{k-1} + \nu_k p_k$
        $r_k = r_{k-1} - \nu_k z$
        $y_k = M^{-1}r_k$
        $\mu_{k+1} = (y_k^T r_k)/(y_{k-1}^T r_{k-1})$
        $p_{k+1} = y_k + \mu_{k+1}p_k$
    until $\|r_k\|_2$ *is small enough*

THEOREM 6.9. *Let A and M be symmetric positive definite, $\hat{A} = M^{-1/2}AM^{1/2}$, and $\hat{b} = M^{-1/2}b$. The CG algorithm applied to $\hat{A}\hat{x} = \hat{b}$,*

    $k = 0;\ \hat{x}_0 = 0;\ \hat{r}_0 = \hat{b};\ \hat{p}_1 = \hat{b};$
    repeat
        $k = k + 1$
        $\hat{z} = \hat{A} \cdot \hat{p}_k$
        $\hat{\nu}_k = (\hat{r}_{k-1}^T \hat{r}_{k-1})/(\hat{p}_k^T \hat{z})$
        $\hat{x}_k = \hat{x}_{k-1} + \hat{\nu}_k \hat{p}_k$
        $\hat{r}_k = \hat{r}_{k-1} - \hat{\nu}_k \hat{z}$
        $\hat{\mu}_{k+1} = (\hat{r}_k^T \hat{r}_k)/(\hat{r}_{k-1}^T \hat{r}_{k-1})$
        $\hat{p}_{k+1} = \hat{r}_k + \hat{\mu}_{k+1}\hat{p}_k$
    until $\|\hat{r}_k\|_2$ *is small enough*

and Algorithm 6.12 are related as follows:

$$
\begin{aligned}
\hat{\mu}_k &= \mu_k, \\
\hat{\nu}_k &= \nu_k,
\end{aligned}
$$

$$
\begin{aligned}
\hat{z} &= M^{-1/2}z, \\
\hat{x}_k &= M^{1/2}x_k, \\
\hat{r}_k &= M^{-1/2}r_k, \\
\hat{p}_k &= M^{1/2}p_k.
\end{aligned}
$$

Therefore, x_k converges to $M^{-1/2}$ times the solution of $\hat{A}\hat{x} = \hat{b}$, i.e., to $M^{-1/2}\hat{A}^{-1}\hat{b} = A^{-1}b$.

For a proof, see Question 6.14.

Now we describe some common preconditioners. Note that our twin goals of minimizing the condition number of $M^{-1}A$ and keeping $Mx = b$ easy to solve are in conflict with one another: Choosing $M = A$ minimizes the condition number of $M^{-1}A$ but leaves $Mx = b$ as hard to solve as the original problem. Choosing $M = I$ makes solving $Mx = b$ trivial but leaves the condition number of $M^{-1}A$ unchanged. Since we need to solve $Mx = b$ in the inner loop of the algorithm, we restrict our discussion to those M for which solving $Mx = b$ is easy, and describe when they are likely to decrease the condition number of $M^{-1}A$.

- If A has widely varying diagonal entries, we may use the simple *diagonal preconditioner* $M = \text{diag}(a_{11}, \ldots, a_{nn})$. One can show that among all possible diagonal preconditioners, this choice reduces the condition number of $M^{-1}A$ to within a factor of n of its minimum value [244]. This is also called *Jacobi preconditioning*.

- As a generalization of the first preconditioner, let

$$
A = \begin{bmatrix} A_{11} & \cdots & A_{1k} \\ \vdots & \ddots & \vdots \\ A_{k1} & \cdots & A_{kk} \end{bmatrix}
$$

  be a block matrix, where the diagonal blocks A_{ii} are square. Then among all block diagonal preconditioners

$$
M = \begin{bmatrix} M_{11} & & \\ & \ddots & \\ & & M_{kk} \end{bmatrix},
$$

  where M_{ii} and A_{ii} have the same dimensions, the choice $M_{ii} = A_{ii}$ minimizes the condition number of $M^{-1/2}AM^{-1/2}$ to within a factor of k [69]. This is also called *block Jacobi preconditioning*.

- Like Jacobi, SSOR can also be used to create a (block) preconditioner.

- An *incomplete Cholesky factorization* LL^T of A is an approximation $A \approx LL^T$, where L is limited to a particular sparsity pattern, such as the original pattern of A. In other words, no fill-in is allowed during Cholesky. Then $M = LL^T$ is used. (For nonsymmetric problems, there is a corresponding *incomplete LU preconditioner*.)

- *Domain decomposition* is used when A represents an equation (such as Poisson's equation) on a physical region Ω. So far, for Poisson's equation, we have let Ω be the unit square. More generally, the region Ω may be broken up into disjoint (or slightly overlapping) subregions $\Omega = \cup_j \Omega_j$, and the equation may be solved on each subregion independently. For example, if we are solving Poisson's equation and if the subregions are squares or rectangles, these subproblems can be solved very quickly using FFTs (see section 6.7). Solving these subproblem corresponds to a block diagonal M (if the subregions are disjoint) or a product of block diagonal M (if the subregions overlap). This is discussed in more detail in section 6.10.

A number of these preconditioners have been implemented in the software packages PETSc [232] and PARPRE (NETLIB/scalapack/parpre.tar.gz).

6.6.6. Other Krylov Subspace Algorithms for Solving $Ax = b$

So far we have concentrated on the symmetric positive definite linear systems and minimized the A^{-1}-norm of the residual. In this section we describe methods for other kinds of linear systems and offer advice on which method to use, based on simple properties of the matrix. See Figure 6.8 for a summary [15, 107, 136, 214] and NETLIB/templates for details, in particular for more comprehensive advice on choosing a method, along with software.

Any system $Ax = b$ can be changed to a symmetric positive definite system by solving the normal equations $A^T A x = A^T b$ (or $AA^T y = b$, $x = A^T y$). This includes the least squares problem $\min_x \|Ax - b\|_2$. This lets us use CG, provided that we can multiply vectors both by A and A^T. Since the condition number of $A^T A$ or AA^T is the square of the condition number of A, this method can lead to slow convergence if A is ill-conditioned but is fast if A is well-conditioned (or $A^T A$ has a "good" distribution of eigenvalues, as discussed in section 6.6.4).

We can minimize the two-norm of the residual instead of the A^{-1}-norm when A is symmetric positive definite. This is called the minimum residual algorithm, or MINRES [194]. Since MINRES is more expensive than CG and is often less accurate because of numerical instabilities, it is not used for positive definite systems. But MINRES can be used when the matrix is symmetric indefinite, whereas CG cannot. In this case, we can also use the SYMMLQ algorithm of Paige and Saunders [194], which produces a residual $r_k \perp \mathcal{K}_k(A, b)$ at each step.

Unfortunately, there are few matrices other than symmetric matrices where algorithms like CG exist that simultaneously

1. either minimize the residual $\|r_k\|_2$ or keep it orthogonal $r_k \perp \mathcal{K}_k$,

2. require a fixed number of dot products and saxpy's in the inner loop, independent of k.

Essentially, algorithms satisfying these two properties exist only for matrices of the form $e^{i\theta}(T + \sigma I)$, where $T = T^T$ (or $TH = (HT)^T$ for some symmetric positive definite H), θ is real, and σ is complex [102, 251]. For these symmetric and special nonsymmetric A, it turns out we can find a short recurrence, as in the Lanczos algorithm, for computing an orthogonal basis $[q_1, \ldots, q_k]$ of $\mathcal{K}_k(A, b)$. The fact that there are just a few terms in the recurrence for updating q_k means that it can be computed very efficiently.

This existence of short recurrences no longer holds for general nonsymmetric A. In this case, we can use Arnoldi's algorithm. So instead of the tridiagonal matrix $T_k = Q_k^T A Q_k$, we get a fully upper Hessenberg matrix $H_k = Q_k^T A Q_k$. The *GMRES algorithm (generalized minimum residual)* uses this decomposition to choose $x_k = Q_k y_k \in \mathcal{K}_k(A, b)$ to minimize the residual

$$
\begin{aligned}
\|r_k\|_2 &= \|b - A x_k\|_2 \\
&= \|b - A Q_k y_k\|_2 \\
&= \|b - (QHQ^T)Q_k y_k\|_2 \quad \text{by equation (6.30)} \\
&= \|Q^T b - H Q^T Q_k y_k\|_2 \quad \text{since } Q \text{ is orthogonal} \\
&= \left\| e_1 \|b\|_2 - \begin{bmatrix} H_k & H_{uk} \\ H_{ku} & H_u \end{bmatrix} \cdot \begin{bmatrix} y_k \\ 0 \end{bmatrix} \right\|_2 \\
&\qquad \text{by equation (6.30) and since the first column of} \\
&\qquad Q = [Q_k, Q_u] \text{ is } b/\|b\|_2 \\
&= \left\| e_1 \|b\|_2 - \begin{bmatrix} H_k \\ H_{ku} \end{bmatrix} y_k \right\|_2.
\end{aligned}
$$

Since only the first row of H_{ku} is nonzero, this is a $(k+1)$-by-k upper Hessenberg least squares problem for the entries of y_k. Since it is upper Hessenberg, the QR decomposition needed to solve it can be accomplished with k Givens rotations, at a cost of $O(k^2)$ instead of $O(k^3)$. Also, the storage required is $O(kn)$, since Q_k must be stored. One way to limit the growth in cost and storage is to *restart* GMRES, i.e., taking the answer x_k computed after k steps, restarting GMRES to solve the linear system $Ad = r_k = b - A x_k$, and updating the solution to get $x_k + d$; this is called GMRES(k). Still, even GMRES(k) is more expensive than CG, where the cost of the inner loop does not depend on k at all.

Another approach to nonsymmetric linear systems is to abandon computing an orthonormal basis of $\mathcal{K}_k(A, b)$ and compute a nonorthonormal basis that again reduces A to (nonsymmetric) tridiagonal form. This is called the *non-symmetric Lanczos method* and requires matrix-vector multiplication by both A and A^T. This is important because $A^T z$ is sometimes harder (or impossible)

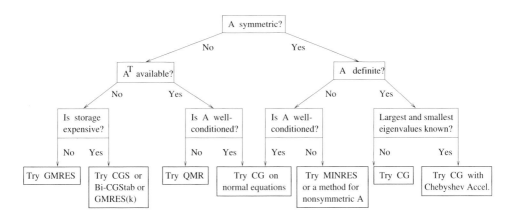

Fig. 6.8. *Decision tree for choosing an iterative algorithm for $Ax = b$. Bi-CGStab = bi-conjugate gradient stabilized. QMR = quasi-minimum residuals. CGS = CG squared.*

to compute (see Example 6.13). The advantage of tridiagonal form is that it is much easier to solve with a tridiagonal matrix than a Hessenberg one. The disadvantage is that the basis vectors may be very ill-conditioned and may in fact fail to exist at all, a phenomenon called *breakdown*. The potential efficiency has led to a great deal of research on avoiding or alleviating this instability (*look-ahead Lanczos*) and to competing methods, including *biconjugate gradients* and *quasi-minimum residuals*. There are also some versions that do not require multiplication by A^T, including *conjugate gradients squared* and *bi-conjugate gradient stabilized*. No one method is best in all cases.

Figure 6.8 shows a decision tree giving simple advice on which method to try first, assuming that we have no other deep knowledge of the matrix A (such as that it arises from the Poisson equation).

6.7. Fast Fourier Transform

In this section i will always denote $\sqrt{-1}$.

We begin by showing how to solve the two-dimensional Poisson's equation in a way requiring multiplication by the matrix of eigenvectors of T_N. A straightforward implementation of this matrix-matrix multiplication would cost $O(N^3) = O(n^{3/2})$ operations, which is expensive. Then we show how this multiplication can be implemented using the FFT in only $O(N^2 \log N) = O(n \log n)$ operations, which is within a factor of $\log n$ of optimal.

This solution is a discrete analogue of the Fourier series solution of the original differential equation (6.1) or (6.6). Later we will make this analogy more precise.

Let $T_N = Z\Lambda Z^T$ be the eigendecomposition of T_N, as defined in Lemma 6.1. We begin with the formulation of the two-dimensional Poisson's equation in equation (6.11):

$$T_N V + V T_N = h^2 F.$$

Substitute $T_N = Z \Lambda Z^T$ and multiply by the Z^T on the left and Z on the right to get

$$Z^T (Z \Lambda Z^T) V Z + Z^T V (Z \Lambda Z^T) Z = Z^T (h^2 F) Z$$

or

$$\Lambda V' + V' \Lambda = h^2 F',$$

where $V' = Z^T V Z$ and $F' = Z^T F Z$. The (j, k)th entry of this last equation is

$$(\Lambda V' + V' \Lambda)_{jk} = \lambda_j v'_{jk} + v'_{jk} \lambda_k = h^2 f'_{jk},$$

which can be solved for v'_{jk} to get

$$v'_{jk} = \frac{h^2 f'_{jk}}{\lambda_j + \lambda_k}.$$

This yields the first version of our algorithm.

ALGORITHM 6.13. *Solving the two-dimensional Poisson's equation using the eigendecomposition $T_N = Z \Lambda Z^T$:*

1) $F' = Z^T F Z$
2) *For all j and k,* $v'_{jk} = \frac{h^2 f'_{jk}}{\lambda_j + \lambda_k}$
3) $V = Z V' Z^T$

The cost of step 2 is $3N^2 = 3n$ operations, and the cost of steps 1 and 3 is 4 matrix-matrix multiplications by Z and $Z^T = Z$, which is $8N^3 = 8n^{3/2}$ operations using a conventional algorithm. In the next section we show how multiplication by Z is essentially the same as computing a *discrete Fourier transform*, which can be done in $O(N^2 \log N) = O(n \log n)$ operations using the FFT.

(Using the language of Kronecker products introduced in section 6.3.3, and in particular the eigendecomposition of $T_{N \times N}$ from Proposition 6.1,

$$T_{N \times N} = I \otimes T_N + T_N \otimes I = (Z \otimes Z) \cdot (I \otimes \Lambda + \Lambda \otimes I) \cdot (Z \otimes Z)^T,$$

we can rewrite the formula justifying Algorithm 6.13 as follows:

$$
\begin{aligned}
\mathrm{vec}(V) &= (T_{N \times N})^{-1} \cdot \mathrm{vec}(h^2 F) \\
&= ((Z \otimes Z) \cdot (I \otimes \Lambda + \Lambda \otimes I) \cdot (Z \otimes Z)^T)^{-1} \cdot \mathrm{vec}(h^2 F) \\
&= (Z \otimes Z)^{-T} \cdot (I \otimes \Lambda + \Lambda \otimes I)^{-1} \cdot (Z \otimes Z)^{-1} \cdot \mathrm{vec}(h^2 F) \\
&= (Z \otimes Z) \cdot (I \otimes \Lambda + \Lambda \otimes I)^{-1} \cdot (Z^T \otimes Z^T) \cdot \mathrm{vec}(h^2 F). \quad (6.47)
\end{aligned}
$$

We claim that doing the indicated matrix-vector multiplications from right to left is mathematically the same as Algorithm 6.13; see Question 6.9. This also shows how to extend the algorithm to Poisson's equation in higher dimensions.)

6.7.1. The Discrete Fourier Transform

In this subsection, we will number the rows and columns of matrices from 0 to $N - 1$ instead of from 1 to N.

DEFINITION 6.17. *The* discrete Fourier transform (DFT) *of an N-vector x is the vector $y = \Phi x$, where Φ is an N-by-N matrix defined as follows. Let $\omega = e^{\frac{-2\pi i}{N}} = \cos\frac{2\pi}{N} - i \cdot \sin\frac{2\pi}{N}$, a principal Nth root of unity. Then $\phi_{jk} = \omega^{jk}$. The* inverse discrete Fourier transform (IDFT) *of y is the vector $x = \Phi^{-1}y$.*

LEMMA 6.9. $\frac{1}{\sqrt{N}}\Phi$ *is a symmetric unitary matrix, so* $\Phi^{-1} = \frac{1}{N}\Phi^* = \frac{1}{N}\bar{\Phi}$.

Proof.   Clearly $\Phi = \Phi^T$, so $\bar{\Phi} = \Phi^*$, and we need only show $\Phi \cdot \bar{\Phi} = N \cdot I$. Compute $(\Phi\bar{\Phi})_{lj} = \sum_{k=0}^{N-1} \phi_{lk}\bar{\phi}_{kj} = \sum_{k=0}^{N-1} \omega^{lk}\bar{\omega}^{kj} = \sum_{k=0}^{N-1} \omega^{k(l-j)}$, since $\bar{\omega} = \omega^{-1}$. If $l = j$, this sum is clearly N. If $l \neq j$, it is a geometric sum with value $\frac{1-\omega^{N(l-j)}}{1-\omega^{l-j}} = 0$, since $\omega^N = 1$.   □

Thus, both the DFT and IDFT are just matrix-vector multiplications and can be straightforwardly implemented in $2N^2$ flops. This operation is called a DFT because of its close mathematical relationship to two other kinds of Fourier analyses:

the Fourier transform	$F(\zeta) = \int_{-\infty}^{\infty} e^{-2\pi i \zeta x} f(x)dx$
and its inverse	$f(x) = \int_{-\infty}^{\infty} e^{+2\pi i \zeta x} F(\zeta)d\zeta$
the Fourier series	$c_j = \int_0^1 e^{-2\pi i j x} f(x)dx$
where f is periodic on $[0, 1]$	
and its inverse	$f(x) = \sum_{j=-\infty}^{\infty} e^{+2\pi i j x} c_j$
the DFT	$y_j = (\Phi x)_j = \sum_{k=0}^{N-1} e^{-2\pi i j k/N} x_k$
and its inverse	$x_k = (\Phi^{-1}y)_k = \frac{1}{N}\sum_{j=0}^{N-1} e^{+2\pi i j k/N} y_j$

We will make this close relationship more concrete in two ways. First, we will show how to solve the model problem using the DFT and then the original Poisson's equation (6.1) using Fourier series. This example will motivate us to find a fast way to multiply by Φ, because this will give us a fast way to solve the model problem. This fast way is called the *fast Fourier transform* or *FFT*. Instead of $2N^2$ flops, it will require only about $\frac{3}{2}N \log_2 N$ flops, which is much less. We will derive the FFT by stressing a second mathematical relationship shared among the different kinds of Fourier analyses: reducing convolution to multiplication.

In Algorithm 6.13 we showed that to solve the discrete Poisson equation $T_N V + V T_N = h^2 F$ for V required the ability to multiply by the N-by-N matrix Z, where

$$z_{jk} = \sqrt{\frac{2}{N+1}} \sin \frac{\pi(j+1)(k+1)}{N+1}.$$

(Recall that we number rows and columns from 0 to $N - 1$ in this section.) Now consider the $(2N + 2)$-by-$(2N + 2)$ DFT matrix Φ, whose j, k entry is

$$\exp\left(\frac{-2\pi ijk}{2N + 2}\right) = \exp\left(\frac{-\pi ijk}{N + 1}\right) = \cos\frac{\pi jk}{N + 1} - i \cdot \sin\frac{\pi jk}{N + 1}.$$

Thus the N-by-N matrix Z consists of $-\sqrt{\frac{2}{N+1}}$ times the imaginary part of the second through $(N + 1)$st rows and columns of Φ. So if we can multiply efficiently by Φ using the FFT, then we can multiply efficiently by Z. (To be most efficient, one modifies the FFT algorithm, which we describe below, to multiply by Z directly; this is called the *fast sine transform*. But one can also just use the FFT.) Thus, multiplying ZF quickly requires an FFT-like operation on each column of F, and multiplying FZ requires the same operation on each row. (In three dimensions, we would let V be an N-by-N-by-N array of unknowns and apply the same operation to each of the $3N^2$ sections parallel to the coordinate axes.)

6.7.2. Solving the Continuous Model Problem Using Fourier Series

We now return to numbering rows and columns of matrices from 1 to N.

In this section we show how the algorithm for solving the discrete model problem is a natural analogue of using Fourier series to solve the original differential equation (6.1). We will do this for the one-dimensional model problem.

Recall that Poisson's equation on $[0, 1]$ is $-\frac{d^2 v}{dx^2} = f(x)$ with boundary conditions $v(0) = v(1)$. To solve this, we will expand $v(x)$ in a Fourier series: $v(x) = \sum_{j=1}^{\infty} \alpha_j \sin(j\pi x)$. (The boundary condition $v(1) = 0$ tells us that no cosine terms appear.) Plugging $v(x)$ into Poisson's equation yields

$$\sum_{j=1}^{\infty} \alpha_j (j^2 \pi^2) \sin(j\pi x) = f(x).$$

Multiply both sides by $\sin(k\pi x)$, integrate from 0 to 1, and use the fact that $\int_0^1 \sin(j\pi x) \sin(k\pi x) dx = 0$ if $j \neq k$ and $1/2$ if $j = k$ to get

$$\alpha_k = \frac{2}{k^2 \pi^2} \int_0^1 \sin(k\pi x) f(x) dx$$

and finally

$$v(x) = \sum_{j=1}^{\infty} \left(\frac{2}{j^2 \pi^2} \int_0^1 \sin(j\pi y) f(y) dy\right) \sin(j\pi x). \tag{6.48}$$

Now consider the discrete model problem $T_N v = h^2 f$. Since $T_N = Z\Lambda Z^T$, we can write $v = T_N^{-1} h^2 f = Z\Lambda^{-1} Z^T h^2 f$, so

$$v_k = \sum_{j=1}^{N} z_{kj} \frac{h^2}{\lambda_j} (Z^T f)_j = \sum_{j=1}^{N} \sin\frac{\pi jk}{N + 1} \left(\frac{h^2}{\lambda_j} \sqrt{\frac{2}{N + 1}} (Z^T f)_j\right), \tag{6.49}$$

where

$$\sqrt{\frac{2}{N+1}}(Z^T f)_j \;=\; \sqrt{\frac{2}{N+1}}\sum_{l=1}^{N}\sqrt{\frac{2}{N+1}}\sin\left(\frac{\pi jl}{N+1}\right)f_l$$

$$=\; 2\sum_{l=1}^{N}\frac{1}{N+1}\sin\left(\frac{\pi jl}{N+1}\right)f_l$$

$$\approx\; 2\int_{0}^{1}\sin(\pi jy)f(y)dy,$$

since the last sum is just a Riemann sum approximation of the integral. Furthermore, for small j, recall that $\frac{h^2}{\lambda_j}\approx\frac{1}{j^2\pi^2}$. So we see how the solution of the discrete problem (6.49) approximates the solution of the continuous problem (6.48), with multiplication by Z^T corresponding to multiplication by $\sin(j\pi x)$ and integration, and multiplication by Z corresponding to summing the different Fourier components.

6.7.3. Convolutions

The *convolution* is an important operation in Fourier analysis, whose definition depends on whether we are doing Fourier transforms, Fourier series, or the DFT:

Fourier transform	$(f * g)(x) \equiv \int_{-\infty}^{\infty} f(x-y)g(y)dy$
Fourier series	$(f * g)(x) \equiv \int_{0}^{1} f(x-y)g(y)dy$
DFT	If $a = [a_0,\dots,a_{N-1},0,\dots,0]^T$ and $b = [b_0,\dots,b_{N-1},0,\dots,0]^T$ are $2N$-vectors, then $a * b \equiv c = [c_0,\dots,c_{2N-1}]^T$, where $c_k = \sum_{j=0}^{k} a_j b_{k-j}$

To illustrate the use of the discrete convolution, consider polynomial multiplication. Let $a(x) = \sum_{k=0}^{N-1} a_k x^k$ and $b(x) = \sum_{k=0}^{N-1} b_k x^k$ be degree-$(N-1)$ polynomials. Then their product $c(x) \equiv a(x)\cdot b(x) = \sum_{k=0}^{2N-1} c_k x^k$, where the coefficients $c_0,\dots,c_{2N-1}$ are given by the discrete convolution.

One purpose of the Fourier transform, Fourier series, or DFT is to convert convolution into multiplication. In the case of the Fourier transform, $\mathcal{F}(f * g) = \mathcal{F}(f)\cdot\mathcal{F}(g)$; i.e., the Fourier transform of the convolution is the product of the Fourier transforms. In the case of Fourier series, $c_j(f * g) = c_j(f)\cdot c_j(g)$; i.e., the Fourier coefficients of the convolution are the product of the Fourier coefficients. The same is true of the discrete convolution.

THEOREM 6.10. *Let $a = [a_0,\dots,a_{N-1},0,\dots,0]^T$ and $b = [b_0,\dots,b_{N-1},0,\dots,0]^T$ be vectors of dimension $2N$, and let $c = a * b = [c_0,\dots,c_{2N-1}]^T$. Then $(\Phi c)_k = (\Phi a)_k \cdot (\Phi b)_k$.*

Proof.   If $a' = \Phi a$, then $a'_k = \sum_{j=0}^{2N-1} a_j \omega^{kj}$, the value of the polynomial $a(x) \equiv \sum_{j=0}^{N-1} a_j x^j$ at $x = \omega^k$. Similarly $b' = \Phi b$ means $b'_k = \sum_{j=0}^{N-1} b_j \omega^{kj} = b(\omega^k)$ and $c' = \Phi c$ means $c'_k = \sum_{j=0}^{2N-1} c_j \omega^{kj} = c(\omega^k)$. Therefore

$$a'_k \cdot b'_k = a(\omega^k) \cdot b(\omega^k) = c(\omega^k) = c'_k$$

as desired. $\square$

In other words, the DFT is polynomial evaluation at the points $\omega^0, \dots, \omega^{N-1}$, and conversely the IDFT is polynomial interpolation, producing the coefficients of a polynomial given its values at $\omega^0, \dots, \omega^{N-1}$.

6.7.4.   Computing the Fast Fourier Transform

We will derive the FFT via its interpretation as polynomial evaluation just discussed. The goal is to evaluate $a(x) = \sum_{k=0}^{N-1} a_k x^k$ at $x = \omega^j$ for $0 \leq j \leq N - 1$. For simplicity we will assume $N = 2^m$. Now write

$$
\begin{aligned}
a(x) &= a_0 + a_1 x + a_2 x^2 + \cdots + a_{N-1} x^{N-1} \\
&= (a_0 + a_2 x^2 + a_4 x^4 + \cdots) + x(a_1 + a_3 x^2 + a_5 x^4 + \cdots) \\
&\equiv a_{even}(x^2) + x \cdot a_{odd}(x^2).
\end{aligned}
$$

Thus, we need to evaluate two polynomials a_{even} and a_{odd} of degree $\frac{N}{2} - 1$ at $(\omega^j)^2$, $0 \leq j \leq N - 1$. But this is really just $\frac{N}{2}$ points ω^{2j} for $0 \leq j \leq \frac{N}{2} - 1$, since $\omega^{2j} = \omega^{2(j + \frac{N}{2})}$.

Thus evaluating a polynomial of degree $N - 1 = 2^m - 1$ at all N Nth roots of unity is the same as evaluating two polynomials of degree $\frac{N}{2} - 1$ at all $\frac{N}{2}$ $\frac{N}{2}$th roots of unity and then combining the results with N multiplications and additions. This can be done recursively.

ALGORITHM 6.14. *FFT (recursive version):*

> *function FFT(a, N)*
> > *if $N = 1$*
> > > *return a*
> > *else*
> > > $a'_{even} = FFT(a_{even}, N/2)$
> > > $a'_{odd} = FFT(a_{odd}, N/2)$
> > > $\omega = e^{-2\pi i/N}$
> > > $w = [\omega^0, \dots, \omega^{N/2-1}]$
> > > *return $a' = [a'_{even} + w. * a'_{odd}, a'_{even} - w. * a'_{odd}]$*
> > *endif*

Here .* means componentwise multiplication of arrays (as in Matlab), and we have used the fact that $\omega^{j+N/2} = -\omega^j$.

Let the cost of this algorithm be denoted $C(N)$. Then we see that $C(N)$ satisfies the recurrence $C(N) = 2C(N/2) + 3N/2$ (assuming that the powers of ω are precomputed and stored in tables). To solve this recurrence write

$$
\begin{aligned}
C(N) &= 2C\left(\frac{N}{2}\right) + \frac{3N}{2} = 4C\left(\frac{N}{4}\right) + 2\cdot\frac{3N}{2} = 8C\left(\frac{N}{8}\right) + 3\cdot\frac{3N}{2} \\
&= \cdots \\
&= \log_2 N \cdot \frac{3N}{2}.
\end{aligned}
$$

To compute the FFT of each column (or each row) of an N-by-N matrix therefore costs $\log_2 N \cdot \frac{3N^2}{2}$. This complexity analysis justifies the entry for the FFT in Table 6.1.

In practice, implementations of the FFT use simple nested loops rather than recursion in order to be as efficient as possible; see NETLIB/fftpack. In addition, these implementations sometimes return the components in *bit-reversed* order: This means that instead of returning $y_0, y_1, \ldots, y_{N-1}$, where $y = \Phi x$, the subscripts j are reordered so that the bit patterns are reversed. For example, if $N = 8$, the subscripts run from $0 = 000_2$ to $7 = 111_2$. The following table shows the normal order and the bit-reversed order:

normal increasing order	bit-reversed order
$0 = 000_2$	$0 = 000_2$
$1 = 001_2$	$4 = 100_2$
$2 = 010_2$	$2 = 010_2$
$3 = 011_2$	$6 = 110_2$
$4 = 100_2$	$1 = 001_2$
$5 = 101_2$	$5 = 101_2$
$6 = 110_2$	$3 = 011_2$
$7 = 111_2$	$7 = 111_2$

The inverse FFT undoes this reordering and returns the results in their original order. Therefore, these algorithms can be used for solving the model problem, provided that we divide by the appropriate eigenvalues, whose subscripts correspond to bit-reversed order. (Note that Matlab always returns results in normal increasing order.)

6.8. Block Cyclic Reduction

Block cyclic reduction is another fast $(O(N^2 \log_2 N))$ method for the model problem but is slightly more generally applicable than the FFT-based solution. The fastest algorithms for the model problem on vector computers are often a hybrid of block cyclic reduction and FFT.

First we describe a simple but numerically unstable version version of the algorithm; then we say a little about how to stabilize it. Write the model problem as

$$
\begin{bmatrix} A & -I & & \\ -I & \ddots & \ddots & \\ & \ddots & \ddots & -I \\ & & -I & A \end{bmatrix}
\begin{bmatrix} x_1 \\ \vdots \\ x_N \end{bmatrix} =
\begin{bmatrix} b_1 \\ \vdots \\ b_N \end{bmatrix},
$$

where we assume that N, the dimension of $A = T_N + 2I_N$, is odd. Note also that x_i and b_i are N-vectors.

We use block Gaussian elimination to combine three consecutive sets of equations,

$$
\begin{array}{lllllll}
+ & [& -x_{j-2} & +Ax_{j-1} & -x_j & & = b_{j-1} &], \\
+A* & [& & -x_{j-1} & +Ax_j & -x_{j+1} & = b_j &], \\
+ & [& & & -x_j & +Ax_{j+1} & -x_{j+2} & = b_{j+1} &],
\end{array}
$$

thus eliminating x_{j-1} and x_{j+1}:

$$
-x_{j-2} + (A^2 - 2I)x_j - x_{j+2} = b_{j-1} + Ab_j + b_{j+1}.
$$

Doing this for every set of three consecutive equations yields two sets of equations: one for the x_j with j even,

$$
\begin{bmatrix} B & -I & & & \\ -I & B & -I & & \\ & -I & \ddots & \ddots & \\ & & \ddots & \ddots & -I \\ & & & -I & B \end{bmatrix}
\begin{bmatrix} x_2 \\ x_4 \\ \vdots \\ x_{N-1} \end{bmatrix} =
\begin{bmatrix} b_1 + Ab_2 + b_3 \\ b_3 + Ab_4 + b_5 \\ \vdots \\ b_{N-2} + Ab_{N-1} + b_N \end{bmatrix}, \quad (6.50)
$$

where $B = A^2 - 2I$, and one set of equations for the x_j with j odd, which we can solve after solving equation (6.50) for the odd x_j:

$$
\begin{bmatrix} A & & & \\ & A & & \\ & & \ddots & \\ & & & A \end{bmatrix}
\begin{bmatrix} x_1 \\ x_3 \\ \vdots \\ x_N \end{bmatrix} =
\begin{bmatrix} b_1 + x_2 \\ b_3 + x_2 + x_4 \\ \vdots \\ b_N + x_{N-1} \end{bmatrix}.
$$

Note that equation (6.50) has the same form as the original problem, so we may repeat this process recursively. For example, at the next step we get

$$
\begin{bmatrix} C & -I & & & \\ -I & C & -I & & \\ & -I & \ddots & \ddots & \\ & & \ddots & \ddots & -I \\ & & & -I & C \end{bmatrix}
\begin{bmatrix} x_4 \\ x_8 \\ \vdots \end{bmatrix} =
\begin{bmatrix} \vdots \\ \vdots \\ \vdots \end{bmatrix}, \quad \text{where } C = B^2 - 2I,
$$

and

$$\begin{bmatrix} B & & & \\ & B & & \\ & & \ddots & \\ & & & B \end{bmatrix} \begin{bmatrix} x_2 \\ x_6 \\ \vdots \end{bmatrix} = \begin{bmatrix} \vdots \\ \vdots \\ \vdots \\ \vdots \end{bmatrix}.$$

We repeat this until only one equation is left, which we solve another way.

We formalize this algorithm as follows: Assume $N = N_0 = 2^{k+1} - 1$, and let $N_r = 2^{k+1-r} - 1$. Let $A^{(0)} = A$ and $b_j{}^{(0)} = b_j$ for $j = 1, \ldots, N$.

ALGORITHM 6.15. *Block cyclic reduction:*

1) *Reduce:*

> *for* $r = 0$ *to* $k - 1$
> $\quad A^{(r+1)} = (A^{(r)})^2 - 2I$
> $\quad$ *for* $j = 1$ *to* N_{r+1}
> $\quad\quad b_j{}^{(r+1)} = b_{2j-1}{}^{(r)} + A^{(r)} b_{2j}{}^{(r)} + b_{2j+1}{}^{(r)}$
> $\quad$ *end for*
> *end for*

> *Comment: at the rth step the problem is reduced to*

$$\begin{bmatrix} A^{(r)} & -I & & \\ -I & \ddots & \ddots & \\ & \ddots & \ddots & -I \\ & & -I & A^{(r)} \end{bmatrix} \begin{bmatrix} x_1{}^{(r)} \\ \vdots \\ x_{N_r}{}^{(r)} \end{bmatrix} = \begin{bmatrix} b_1{}^{(r)} \\ \vdots \\ b_{N_r}{}^{(r)} \end{bmatrix}$$

2) $A^{(k)} x^{(k)} = b^{(k)}$ *is solved another way.*

3) *Backsolve:*

> *for* $r = k - 1, \ldots, 0$
> $\quad$ *for* $j = 1$ *to* N_{r+1}
> $\quad\quad x_{2j}{}^{(r)} = x_j{}^{(r+1)}$
> $\quad$ *end for*
> $\quad$ *for* $j = 1$ *to* N_r *step* 2
> $\quad\quad$ *solve* $A^{(r)} x_j{}^{(r)} = b_j{}^{(r)} + x_{j-1}{}^{(r)} + x_{j+1}{}^{(r)}$ *for* $x_j{}^{(r)}$
> $\quad\quad$ (*we take* $x_0{}^{(r)} = x_{N_r+1}{}^{(r)} \equiv 0$)
> $\quad$ *end for*
> *end for*

> *Finally,* $x = x^{(0)}$ *is the desired result.*

This simple approach has two drawbacks:

1) It is numerically unstable because $A^{(r)}$ grows quickly: $\|A^{(r)}\| \sim \|A^{(r-1)}\|^2$ $\approx 4^{2^r}$, so in computing $b_j^{(r+1)}$, the $b_{2j \pm 1}^{(r)}$ are lost in roundoff.

2) $A^{(r)}$ has bandwidth $2^r + 1$ if A is tridiagonal, so it soon becomes dense and thus expensive to multiply or solve.

Here is a fix for the second drawback. Note that $A^{(r)}$ is a polynomial $p_r(A)$ of degree 2^r:

$$p_0(A) = A \text{ and } p_{r+1}(A) = (p_r(A))^2 - 2I.$$

LEMMA 6.10. *Let* $t = 2\cos\theta$. *Then* $p_r(t) = p_r(2\cos\theta) = 2\cos(2^r\theta)$.

Proof. This is a simple trigonometric identity. □

Note that $p_r(t) = 2\cos(2^r \arccos(\frac{t}{2})) = 2T_{2^r}(\frac{t}{2})$, where $T_{2^r}(\cdot)$ is a *Chebyshev polynomial* (see section 6.5.6).

LEMMA 6.11. $p_r(t) = \prod_{j=1}^{2^r}(t - t_j)$, *where* $t_j = 2\cos(\pi \frac{2j-1}{2^r})$.

Proof. The zeros of the Chebyshev polynomials are given in Lemma 6.7. □

Thus $A^{(r)} = \prod_{j=1}^{2^r}(A - 2\cos(\pi \frac{2j-1}{2^r}))$, so solving $A^{(r)}z = c$ is equivalent to solving 2^r tridiagonal systems with tridiagonal coefficient matrices $A + 2\cos(\pi \frac{2j-1}{2^r})$, each of which costs $O(N)$ via tridiagonal Gaussian elimination or Cholesky.

More changes are needed to have a numerically stable algorithm. The final algorithm is due to Buneman and described in [47, 46].

We analyze the cost of the simple algorithm as follows; the stable algorithm is analogous. Multiplying by a tridiagonal matrix or solving a tridiagonal system of size N costs $O(N)$ flops. Therefore multiplying by $A^{(r)}$ or solving a system with $A^{(r)}$ costs $O(2^r N)$ flops, since $A^{(r)}$ is the product of 2^r tridiagonal matrices. The inner loop of step 1) of the algorithm therefore costs $\frac{N}{2^{r+1}} \cdot O(2^r N) = O(N^2)$ flops to update the $N_{r+1} \approx \frac{N}{2^{r+1}}$ vectors $b_j^{(r+1)}$. $A^{(r+1)}$ is not computed explicitly. Since the loop in step 1) is executed $k \approx \log_2 N$ times, the total cost of step 1) is $O(N^2 \log_2 N)$. For similar reasons, step 2) costs $O(2^k N) = O(N^2)$ flops, and step 3) costs $O(N^2 \log_2 N)$ flops, for a total cost of $O(N^2 \log_2 N)$ flops. This justifies the entry for block cyclic reduction in Table 6.1.

This algorithm generalizes to any block tridiagonal matrix with a symmetric matrix A repeated along the diagonal and a symmetric matrix F that commutes with A ($FA = AF$) repeated along the offdiagonals. See also Question 6.10. This is a common situation when solving linear systems arising from discretized differential equations such as Poisson's equation.

6.9.   Multigrid

Multigrid methods were invented for partial differential equations such as Poisson's equation, but they work on a wider class of problems too. In contrast to other iterative schemes that we have discussed so far, multigrid's convergence rate is *independent* of the problem size N, instead of slowing down for larger problems. As a consequence, it can solve problems with n unknowns in $O(n)$ time or for a constant amount of work per unknown. This is optimal, modulo the (modest) constant hidden inside the $O(\cdot)$.

Here is why the other iterative algorithms that we have discussed *cannot* be optimal for the model problem. In fact, this is true of *any* iterative algorithm that computes approximation x_{m+1} by averaging values of x_m and the right-hand side b from neighboring grid points. This includes Jacobi's, Gauss–Seidel, SOR(ω), SSOR with Chebyshev acceleration (the last three with red-black ordering), and any Krylov subspace method based on matrix-vector multiplication with the matrix $T_{N\times N}$; this is because multiplying a vector by $T_{N\times N}$ is also equivalent to averaging neighboring grid point values. Suppose that we start with a right-hand side b on a 31-by-31 grid, with a single nonzero entry, as shown in the upper left of Figure 6.9. The true solution x is shown in the upper right of the same figure; note that it is everywhere nonzero and gets smaller as we get farther from the center. The bottom left plot in Figure 6.9 shows the solution $x_{J,5}$ after 5 steps of Jacobi's method, starting with an initial solution of all zeros. Note that the solution $x_{J,5}$ is zero more than 5 grid points away from the center, because averaging with neighboring grid points can "propagate information" only one grid point per iteration, and the only nonzero value is initially in the center of the grid. More generally, after k iterations only grid points within k of the center can be nonzero. The bottom right figure shows the best possible solution $x_{Best,5}$ obtainable by any "nearest neighbor" method after 5 steps: it agrees with x on grid points within 5 of the center and is necessarily 0 farther away. We see graphically that the error $x_{Best,5} - x$ is equal to the size of x at the sixth grid point away from the center. This is still a large error; by formalizing this argument, one can show that it would take at least $O(\log n)$ steps on an n-by-n grid to decrease the error by a constant factor less than 1, no matter what "nearest-neighbor" algorithm is used. If we want to do better than $O(\log n)$ steps (and $O(n \log n)$ cost), we need to "propagate information" farther than one grid point per iteration. Multigrid does this by communicating with nearest neighbors on coarser grids, where a nearest neighbor on a coarse grid can be much farther away than a nearest neighbor on a fine grid.

Multigrid uses coarse grids to do *divide-and-conquer* in two related senses. First, it obtains an initial solution for an N-by-N grid by using an $(N/2)$-by-$(N/2)$ grid as an approximation, taking every other grid point from the N-by-N grid. The coarser $(N/2)$-by-$(N/2)$ grid is in turn approximated by an $(N/4)$-by-$(N/4)$ grid, and so on recursively. The second way multigrid

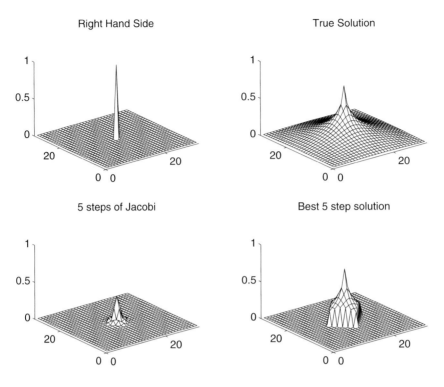

Fig. 6.9. *Limits of averaging neighboring grid points.*

uses divide-and-conquer is in the *frequency domain.* This requires us to think
of the error as a sum of eigenvectors, or sine-curves of different frequencies.
Then, intuitively, the work that we do on a particular grid will attenuate the
error in half of the frequency components not attenuated on coarser grids. In
particular, the work performed on a particular grid—averaging the solution
at each grid point with its neighbors, a variation of Jacobi's method—makes
the solution smoother, which is equivalent to getting rid of the high-frequency
error. We will illustrate these notions further below.

6.9.1. Overview of Multigrid on the Two-Dimensional Poisson's Equation

We begin by stating the algorithm at a high level and then fill in details.
As with block cyclic reduction (section 6.8), it turns out to be convenient to
consider a $(2^k - 1)$-by-$(2^k - 1)$ grid of unknowns rather than the 2^k-by-2^k grid
favored by the FFT (section 6.7). For understanding and implementation, it
is convenient to add the nodes at the boundary, which have the known value
0, to get a $(2^k + 1)$-by-$(2^k + 1)$ grid, as shown in Figures 6.10 and 6.13. We
also let $N_k = 2^k - 1$.

We will let $P^{(i)}$ denote the problem of solving a discrete Poisson equation on
a $(2^i + 1)$-by-$(2^i + 1)$ grid with $(2^i - 1)^2$ unknowns, or equivalently a $(N_i + 2)$-

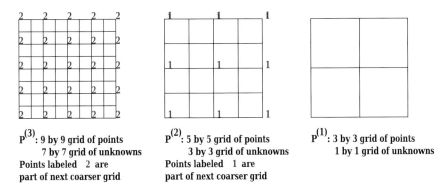

$P^{(3)}$: 9 by 9 grid of points
7 by 7 grid of unknowns
Points labeled 2 are
part of next coarser grid

$P^{(2)}$: 5 by 5 grid of points
3 by 3 grid of unknowns
Points labeled 1 are
part of next coarser grid

$P^{(1)}$: 3 by 3 grid of points
1 by 1 grid of unknowns

Fig. 6.10. *Sequence of grids used by two-dimensional multigrid.*

by-$(N_i + 2)$) grid with N_i^2 unknowns. The problem $P^{(i)}$ is specified by the right-hand side $b^{(i)}$ and implicitly the grid size $2^i - 1$ and the coefficient matrix $T^{(i)} \equiv T_{N_i \times N_i}$. An approximate solution of $P^{(i)}$ will be denoted $x^{(i)}$. Thus, $b^{(i)}$ and $x^{(i)}$ are $(2^i - 1)$-by-$(2^i - 1)$ arrays of values at each grid point. (The zero boundary values are implicit.) We will generate a sequence of related problems $P^{(i)}$, $P^{(i-1)}$, $P^{(i-2)}$, ..., $P^{(1)}$ on increasingly coarse grids, where the solution to $P^{(i-1)}$ is a good approximation to the error in the solution of $P^{(i)}$.

To explain how multigrid works, we need some operators that take a problem on one grid and either improve it or transform it to a related problem on another grid:

- The *solution operator* S takes a problem $P^{(i)}$ and its approximate solution $x^{(i)}$ and computes an improved $x^{(i)}$:

$$\text{improved } x^{(i)} = S(b^{(i)}, x^{(i)}). \tag{6.51}$$

  The improvement is to damp the "high-frequency components" of the error. We will explain what this means below. It is implemented by averaging each grid point value with its nearest neighbors and is a variation of Jacobi's method.

- The *restriction operator* R takes a right-hand side $b^{(i)}$ from problem $P^{(i)}$ and maps it to $b^{(i-1)}$, which is an approximation on the coarser grid:

$$b^{(i-1)} = R(b^{(i)}). \tag{6.52}$$

  Its implementation also requires just a weighted average with nearest neighbors on the grid.

- The *interpolation operator* In takes an approximate solution $x^{(i-1)}$ for $P^{(i-1)}$ and converts it to an approximate solution $x^{(i)}$ for the problem $P^{(i)}$ on the next finer grid:

$$x^{(i)} = In(x^{(i-1)}). \tag{6.53}$$

Its implementation also requires just a weighted average with nearest neighbors on the grid.

Since all three operators are implemented by replacing values at each grid point by some weighted averages of nearest neighbors, each operation costs just $O(1)$ per unknown, or $O(n)$ for n unknowns. This is the key to the low cost of the ultimate algorithm.

Multigrid V-Cycle

This is enough to state the basic algorithm, the *multigrid V-cycle (MGV)*.

ALGORITHM 6.16. *MGV (the lines are numbered for later reference):*

$$\begin{aligned}
&\textit{function } MGV(b^{(i)}, x^{(i)}) \quad &&\textit{... replace an approximate solution } x^{(i)} \\
& &&\textit{... of } P^{(i)} \textit{ with an improved one} \\
&\quad \textit{if } i = 1 &&\textit{... only one unknown} \\
&\qquad \textit{compute the exact solution } x^{(1)} \textit{ of } P^{(1)} \\
&\qquad \textit{return } x^{(1)} \\
&\quad \textit{else}
\end{aligned}$$

$$\begin{array}{lll}
1) & \quad x^{(i)} = S(b^{(i)}, x^{(i)}) & \textit{... improve the solution} \\
2) & \quad r^{(i)} = T^{(i)} \cdot x^{(i)} - b^{(i)} & \textit{... compute the residual} \\
3) & \quad d^{(i)} = In(MGV(4 \cdot R(r^{(i)}), 0)) & \textit{... solve recursively} \\
& & \textit{... on coarser grids} \\
4) & \quad x^{(i)} = x^{(i)} - d^{(i)} & \textit{... correct fine grid solution} \\
5) & \quad x^{(i)} = S(b^{(i)}, x^{(i)}) & \textit{... improve the solution again} \\
& \quad \textit{return } x^{(i)} & \\
& \textit{endif}
\end{array}$$

In words, the algorithm does the following:

1. Starts with a problem on a fine grid $(b^{(i)}, x^{(i)})$.

2. Improves it by damping the high-frequency error: $x^{(i)} = S(b^{(i)}, x^{(i)})$.

3. Computes the residual $r^{(i)}$ of the approximate solution $x^{(i)}$.

4. Approximates the fine grid residual $r^{(i)}$ on the next coarser grid: $R(r^{(i)})$.

5. Solves the coarser problem recursively, with a zero initial guess: $MGV(4 \cdot R(r^{(i)}), 0)$. The factor 4 appears because of the h^2 factor in the right-hand side of Poisson's equation, which changes by a factor of 4 from fine grid to coarse grid.

6. Maps the coarse solution back to the fine grid: $d_i = In(MGV(R(r^{(i)}), 0))$.

7. Subtracts the correction computed on the coarse grid from the fine grid solution: $x^{(i)} = x^{(i)} - d^{(i)}$.

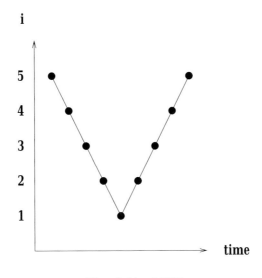

Fig. 6.11. *MGV.*

8. Improves the solution some more: $x^{(i)} = S(b^{(i)}, x^{(i)})$.

We justify the algorithm briefly as follows (we do the details later). Suppose (by induction) that $d^{(i)}$ is the *exact* solution to the equation

$$T^{(i)} \cdot d^{(i)} = r^{(i)} = T^{(i)} \cdot x^{(i)} - b^{(i)}.$$

Rearranging, we get

$$T^{(i)} \cdot (x^{(i)} - d^{(i)}) = b^{(i)}$$

so that $x^{(i)} - d^{(i)}$ is the desired solution.

The algorithm is called a V-cycle, because if we draw it schematically in (grid number i, time) space, with a point for each recursive call to MGV, it looks like Figure 6.11, starting with a call to MGV($b^{(5)}, x^{(5)}$) in the upper left corner. This calls MGV on grid 4, then 3, and so on down to the coarsest grid 1 and then back up to grid 5 again.

Knowing only that the building blocks S, R, and In replace values at grid points by certain weighted averages of their neighbors, we know enough to do an $O(\cdot)$ complexity analysis of MGV. Since each building block does a constant amount of work per grid point, it does a total amount of work proportional to the number of grid points. Thus, each point at grid level i on the "V" in the V-cycle will cost $O((2^i - 1)^2) = O(4^i)$ operations. If the finest grid is at level k with $n = O(4^k)$ unknowns, then the total cost will be given by the geometric sum

$$\sum_{i=1}^{k} O(4^i) = O(4^k) = O(n).$$

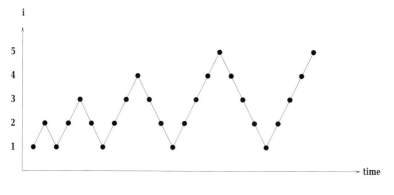

Fig. 6.12. *FMG cycle.*

Full Multigrid

The ultimate multigrid algorithm uses the MGV just described as a building block. It is called *full multigrid (FMG)*.

ALGORITHM 6.17. *FMG:*

> *function $FMG(b^{(k)}, x^{(k)})$ ... return an accurate solution $x^{(k)}$ of $P^{(k)}$*
>   *solve $P^{(1)}$ exactly to get $x^{(1)}$*
>   *for $i = 2$ to k*
> $x^{(i)} = MGV(b^{(i)}, In(x^{(i-1)}))$
>   *end for*

In words, the algorithm does the following:

1. Solves the simplest problem $P^{(1)}$ exactly.

2. Given a solution $x^{(i-1)}$ of the coarse problem $P^{(i-1)}$, maps it to a starting guess $x^{(i)}$ for the next finer problem $P^{(i)}$: $In(x^{(i-1)})$.

3. Solves the finer problem using the MGV with this starting guess: $MGV(b^{(i)}, In(x^{(i-1)}))$.

Now we can do the overall $O(\cdot)$ complexity analysis of FMG. A picture of FMG in (grid number i, time) space is shown in Figure 6.12. There is one "V" in this picture for each call to MGV in the inner loop of FMG. The "V" starting at level i costs $O(4^i)$ as before. Thus the total cost is again given by the geometric sum

$$\sum_{i=1}^{k} O(4^i) = O(4^k) = O(n),$$

which is optimal, since it does a constant amount of work for each of the n unknowns. This explains the entry for multigrid in Table 6.1.

A Matlab implementation of multigrid (for both the one- and the two-dimensional model problems) is available at HOMEPAGE/Matlab/ MG_README.html.

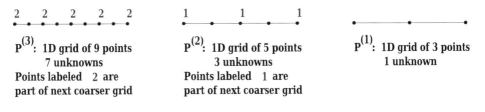

P$^{(3)}$: 1D grid of 9 points
7 unknowns
Points labeled 2 are
part of next coarser grid

P$^{(2)}$: 1D grid of 5 points
3 unknowns
Points labeled 1 are
part of next coarser grid

P$^{(1)}$: 1D grid of 3 points
1 unknown

Fig. 6.13. *Sequence of grids used by one-dimensional multigrid.*

6.9.2. Detailed Description of Multigrid on the One-Dimensional Poisson's Equation

Now we will explain in detail the various operators S, R, and In composing the multigrid algorithm and sketch the convergence proof. We will do this for Poisson's equation in one dimension, since this will capture all the relevant behavior but is simpler to write. In particular, we can now consider a nested set of one-dimensional problems instead of two-dimensional problems, as shown in Figure 6.13.

As before we denote by $P^{(i)}$ the problem to be solved on grid i, namely, $T^{(i)} \cdot x^{(i)} = b^{(i)}$, where as before $N_i = 2^i - 1$ and $T^{(i)} \equiv T_{N_i}$. We begin by describing the solution operator S, which is a form of *weighted Jacobi's method*.

Solution Operator in One Dimension

In this subsection we drop the superscripts on $T^{(i)}$, $x^{(i)}$, and $b^{(i)}$ for simplicity of notation. Let $T = Z\Lambda Z^T$ be the eigendecomposition of T, as defined in Lemma 6.1. The standard Jacobi's method for solving $Tx = b$ is $x_{m+1} = Rx_m + c$, where $R = I - T/2$ and $c = b/2$. We consider *weighted Jacobi's method* $x_{m+1} = R_w x_m + c_w$, where $R_w = I - wT/2$ and $c_w = wb/2$; $w = 1$ corresponds to the standard Jacobi's method. Note that $R_w = Z(I - w\Lambda/2)Z^T$ is the eigendecomposition of R_w. The eigenvalues of R_w determine the convergence of weighted Jacobi in the usual way: Let $e_m = x_m - x$ be the error at the mth iteration of weighted Jacobi convergence so that

$$
\begin{aligned}
e_m &= R_w e_{m-1} \\
&= R_w^m e_0 \\
&= (Z(I - w\Lambda/2)Z^T)^m e_0 \\
&= Z(I - w\Lambda/2)^m Z^T e_0
\end{aligned}
$$

so

$$
Z^T e_m = (I - w\Lambda/2)^m Z^T e_0 \quad \text{or} \quad (Z^T e_m)_j = (I - w\Lambda/2)_{jj}^m (Z^T e_0)_j.
$$

We call $(Z^T e_m)_j$ the *jth frequency component* of the error e_m, since $e_m = Z(Z^T e_m)$ is a sum of columns of Z weighted by the $(Z^T e_m)_j$, i.e., a sum of sinusoids of varying frequencies (see Figure 6.2). The eigenvalues $\lambda_j(R_w) = 1 -$

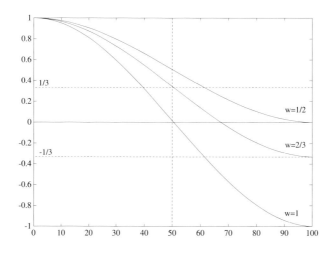

Fig. 6.14. *Graph of the spectrum of R_w for $N = 99$ and $w = 1$ (Jacobi's method), $w = 1/2$, and $w = 2/3$.*

$w\lambda_j/2$ determine how fast each frequency component goes to zero. Figure 6.14 plots $\lambda_j(R_w)$ for $N = 99$ and varying values of the weight w.

When $w = \frac{2}{3}$ and $j > \frac{N}{2}$, i.e., for the upper half of the frequencies λ_j, we have $|\lambda_j(R_w)| \leq \frac{1}{3}$. This means that the upper half of the error components $(Z^T e_m)_j$ are multiplied by $\frac{1}{3}$ or less at every iteration, independently of N. Low-frequency error components are not decreased as much, as we will see in Figure 6.15. So weighted Jacobi convergence with $w = \frac{2}{3}$ is good at decreasing the high-frequency error.

Thus, our solution operator S in equation (6.51) consists of taking one step of weighted Jacobi convergence with $w = \frac{2}{3}$:

$$S(b, x) = R_{2/3} \cdot x + b/3. \tag{6.54}$$

When we want to indicate the grid i on which $R_{2/3}$ operates, we will instead write $R_{2/3}^{(i)}$.

Figure 6.15 shows the effect of taking two steps of S for $i = 6$, where we have $2^i - 1 = 63$ unknowns. There are three rows of pictures, the first row showing the initial solution and error and the following two rows showing the solution x_m and error e_m after successive applications of S. The true solution is a sine curve, shown as a dotted line in the leftmost plot in each row. The approximate solution is shown as a solid line in the same plot. The middle plot shows the error alone, including its two-norm in the label at the bottom. The rightmost plot shows the frequency components of the error $Z^T e_m$. One can see in the rightmost plots that as S is applied, the right (upper) half of the frequency components are damped out. This can also be seen in the middle and left plots, because the approximate solution grows smoother. This is because high-frequency error looks like "rough" error and low-frequency error looks like "smooth" error. Initially, the norm of the vector decreases rapidly, from 1.65 to 1.055, but then decays more gradually, because there is little more error in the high frequencies to damp. Thus, it only makes sense to do a few iterations of S at a time.

Recursive Structure of Multigrid

Using this terminology, we can describe the recursive structure of multigrid as follows. What multigrid does on the finest grid $P^{(k)}$, is to damp the upper half of the frequency components of the error in the solution. This is accomplished by the solution operator S, as just described. On the next coarser grid, with half as many points, multigrid damps the upper half of the remaining frequency components in the error. This is because taking a coarser grid, with half as many points, makes frequencies appear twice as high, as illustrated in the example below.

EXAMPLE 6.16.

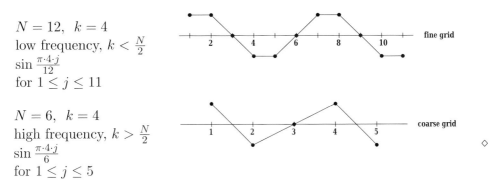

$N = 12, \quad k = 4$
low frequency, $k < \frac{N}{2}$
$\sin \frac{\pi \cdot 4 \cdot j}{12}$
for $1 \le j \le 11$

fine grid

$N = 6, \quad k = 4$
high frequency, $k > \frac{N}{2}$
$\sin \frac{\pi \cdot 4 \cdot j}{6}$
for $1 \le j \le 5$

coarse grid

$\diamond$

On the next coarser grid, the upper half of the remaining frequency components are damped, and so on, until we solve the exact (one-unknown) problem $P^{(1)}$. This is shown schematically in Figure 6.16. The purpose of the restriction and interpolation operators is to change an approximate solution on one grid to one on the next coarser or next finer grid.

Restriction Operator in One Dimension

Now we turn to the restriction operator R, which takes a right-hand side $r^{(i)}$ from problem $P^{(i)}$ and approximates it on the next coarse grid, yielding $r^{(i-1)}$.

The simplest way to compute $r^{(i-1)}$ would be to simply *sample* $r^{(i)}$ at the common grid points of the coarse and fine grids. But it is better to compute $r^{(i-1)}$ at a coarse grid point by averaging values of $r^{(i)}$ on neighboring fine grid points: the value at a coarse grid point is .5 times the value at the corresponding fine grid point, plus .25 times each of the fine grid point neighbors. We call this *averaging*. Both methods are illustrated in Figure 6.17.

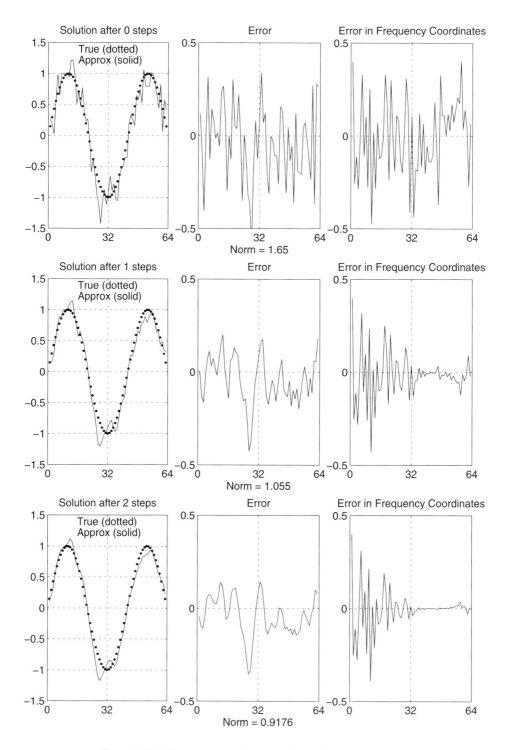

Fig. 6.15. *Illustration of weighted Jacobi convergence.*

Schematic Description of Multigrid

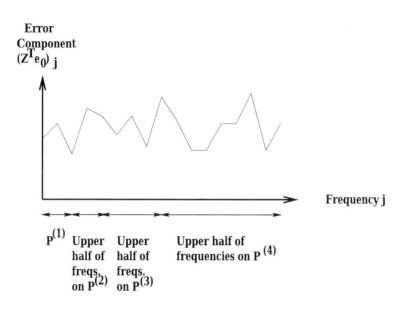

Fig. 6.16. *Schematic description of how multigrid damps error components.*

So altogether, we write the restriction operation as

$$
\begin{aligned}
r^{(i-1)} &= R(r^{(i)}) \\
&\equiv P_i^{i-1} \cdot r^{(i)} \\
&= \begin{bmatrix}
\frac{1}{4} & \frac{1}{2} & \frac{1}{4} & & & & & \\
& & \frac{1}{4} & \frac{1}{2} & \frac{1}{4} & & & \\
& & & & \frac{1}{4} & \frac{1}{2} & \frac{1}{4} & \\
& & & & & \ddots & \ddots & \ddots & \\
& & & & & & \frac{1}{4} & \frac{1}{2} & \frac{1}{4}
\end{bmatrix} \cdot r^{(i)}.
\end{aligned} \qquad (6.55)
$$

The subscript i and superscript $i-1$ on the matrix P_i^{i-1} indicate that it maps from the grid with $2^i - 1$ points to the grid with $2^{i-1} - 1$ points.

In two dimensions, restriction involves averaging with the eight nearest neighbors of each grid points: $\frac{1}{4}$ times the grid cell value itself, plus $\frac{1}{8}$ times the four neighbors to the left, right, top, and bottom, plus $\frac{1}{16}$ times the four remaining neighbors at the upper left, lower left, upper right, and lower right.

Interpolation Operator in One Dimension

The interpolation operator In takes an approximate solution $d^{(i-1)}$ on a coarse grid and maps it to a function $d^{(i)}$ on the next finer grid. The solution $d^{(i-1)}$ is interpolated to the finer grid as shown in Figure 6.18: we do simple linear

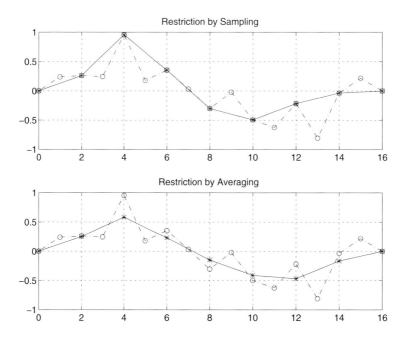

Fig. 6.17. *Restriction from a grid with $2^4 - 1 = 15$ points to a grid with $2^3 - 1 = 7$ points. (0 boundary values also shown.)*

interpolation to fill in the values on the fine grid (using the fact that the boundary values are known to be zero). Mathematically, we write this as

$$
d^{(i)} = In(d^{(i-1)}) \equiv P^{i}_{i-1} \cdot d^{(i-1)} =
\begin{bmatrix}
\frac{1}{2} & & & \\
1 & & & \\
\frac{1}{2} & \frac{1}{2} & & \\
& 1 & \ddots & \\
& \frac{1}{2} & \ddots & \frac{1}{2} \\
& & \ddots & 1 \\
& & & \frac{1}{2}
\end{bmatrix}
\cdot x^{(i-1)}. \qquad (6.56)
$$

The subscript $i - 1$ and superscript i on the matrix P^{i}_{i-1} indicate that it maps from the grid with $2^{i-1} - 1$ points to the grid with $2^i - 1$ points.

Note that $P^{i}_{i-1} = 2 \cdot (P^{i-1}_{i})^T$. In other words, interpolation and smoothing are essentially transposes of one another. This fact will be important in the convergence analysis later.

In two dimensions, interpolation again involves averaging the values at coarse nearest neighbors of a fine grid point (one point if the fine grid point is also a coarse grid point; two neighbors if the fine grid point's nearest coarse neighbors are to the left and right or top and bottom; and four neighbors otherwise).

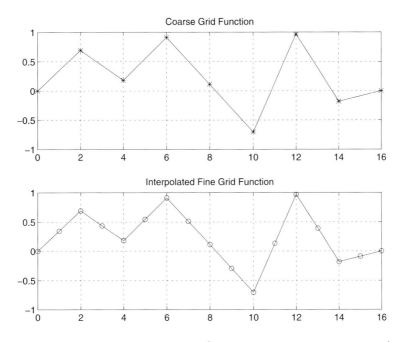

Fig. 6.18. *Interpolation from a grid with $2^3 - 1 = 7$ points to a grid with $2^4 - 1 = 15$ points. (0 boundary values also shown.)*

Putting It All Together

Now we run the algorithm just described for eight iterations on the problem pictured in the top two plots of Figure 6.19; both the true solution x (on the top left) and right-hand side b (on the top right) are shown. The number of unknowns is $2^7 - 1 = 127$. We show how multigrid converges in the bottom three plots. The middle left plot shows the ratio of consecutive residuals $\|r_{m+1}\|/\|r_m\|$, where the subscript m is the number of iterations of multigrid (i.e., calls to FMG, or Algorithm 6.17). These ratios are about .15, indicating that the residual decreases by more than a factor of 6 with each multigrid iteration. This quick convergence is indicated in the middle right plot, which shows a semilogarithmic plot of $\|r_m\|$ versus m; it is a straight line with slope $\log_{10}(.15)$ as expected. Finally, the bottom plot plots all eight error vectors $x_m - x$. We see how they smooth out and become parallel on a semilogarithmic plot, with a constant decrease between adjacent plots of $\log_{10}(.15)$.

Figure 6.20 shows a similar example for a two-dimensional model problem.

Convergence Proof

Finally, we sketch a convergence proof that shows that the overall error in an FMG "V"-cycle is decreased by a constant less than 1, independent of grid size $N_k = 2^k - 1$. This means that the number of FMG V-cycles needed to decrease the error by any factor less than 1 is independent of k, and so the total work

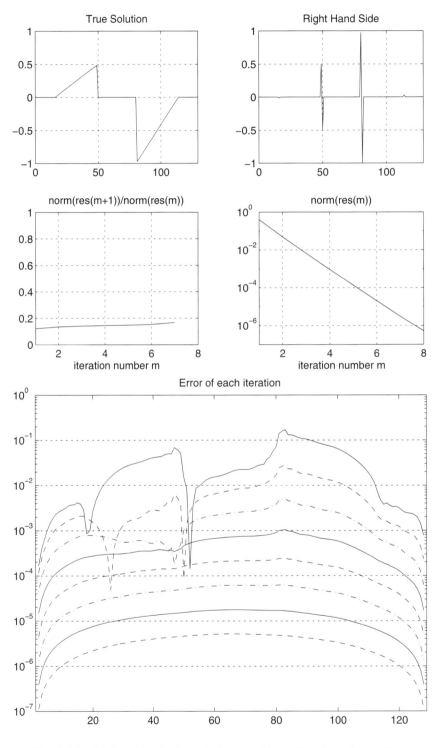

Fig. 6.19. *Multigrid solution of the one-dimensional model problem.*

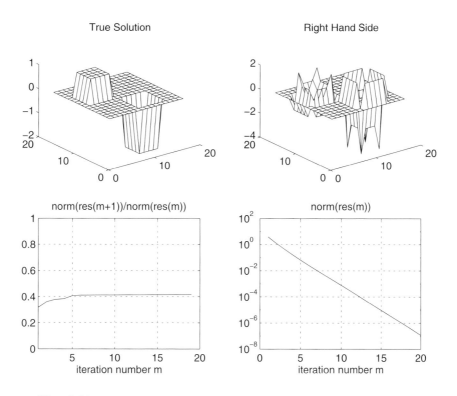

Fig. 6.20. *Multigrid solution of the two-dimensional model problem.*

is proportional to the cost of a single FMG V-cycle, i.e., proportional to the number of unknowns n.

We will simplify the proof by looking at one V-cycle and assuming by induction that the coarse grid problem is solved *exactly* [43]. In reality, the coarse grid problem is not solved quite exactly, but this rough analysis suffices to capture the spirit of the proof: that low-frequency error is attenuated on the coarser grid and high-frequency error is eliminated on the fine grid.

Now let us write all the formulas defining a V-cycle and combine them all to get a single formula of the form "new $e^{(i)} = M \cdot e^{(i)}$," where $e^{(i)} = x^{(i)} - x$ is the error and M is a matrix whose eigenvalues determine the rate of convergence; our goal is to show that they are bounded away from 1, independently of i. The line numbers in the following table refer to Algorithm 6.16.

$$
\begin{array}{rcl}
\text{(a)} \quad x^{(i)} &=& S(b(i), x(i)) = R^{(i)}_{2/3} x^{(i)} + b^{(i)}/3 \\
&& \text{by line 1) and equation (6.54),} \\
\text{(b)} \quad r^{(i)} &=& T^{(i)} \cdot x^{(i)} - b^{(i)} \\
&& \text{by line 2),} \\
d^{(i)} &=& In(MGV(4 \cdot R(r^{(i)}), 0)) \\
&& \text{by line 3)}
\end{array}
$$

$$= In([T^{(i-1)}]^{-1}(4 \cdot R(r^{(i)})))$$

by our assumption that the
coarse grid problem is solved exactly

$$= In([T^{(i-1)}]^{-1}(4 \cdot P_i^{i-1} r^{(i)}))$$

by equation (6.55)

(c) $= P_{i-1}^i([T^{(i-1)})]^{-1}(4 \cdot P_i^{i-1} r^{(i)}))$

by equation (6.56)

(d) $x^{(i)} = x^{(i)} - d^{(i)}$

by line 4)

(e) $x^{(i)} = S(b(i), x(i)) = R_{2/3}^{(i)} x^{(i)} + b^{(i)}/3$

by line 5).

In order to get equations updating the error $e^{(i)}$, we subtract the identity $x = R_{2/3}^{(i)} x + b^{(i)}/3$ from lines (a) and (e) above, $0 = T^{(i)} \cdot x - b^{(i)}$ from line (b), and $x = x$ from line (d) to get

(a) $e^{(i)}$ $= R_{2/3}^{(i)} e^{(i)}$,

(b) $r^{(i)}$ $= T^{(i)} \cdot e^{(i)}$,

(c) $d^{(i)}$ $= P_{i-1}^i([T^{(i-1)}]^{-1}(4 \cdot P_i^{i-1} r^{(i)}))$,

(d) $e^{(i)}$ $= e^{(i)} - d^{(i)}$,

(e) $e^{(i)}$ $= R_{2/3}^{(i)} e^{(i)}$.

Substituting each of the above equations into the next yields the following formula, showing how the error is updated by a V-cycle:

$$\text{new } e^{(i)} = R_{2/3}^{(i)} \left\{ I - P_{i-1}^i \cdot [T^{(i-1)}]^{-1} \cdot (4 \cdot P_i^{i-1} T^{(i)}) \right\} R_{2/3}^{(i)} \cdot e^{(i)}$$

$$\equiv M \cdot e^{(i)}. \tag{6.57}$$

Now we need to compute the eigenvalues of M. We first simplify equation (6.57), using the facts that $P_{i-1}^i = 2 \cdot (P_i^{i-1})^T$ and

$$T^{(i-1)} = 4 \cdot P_i^{i-1} T^{(i)} P_{i-1}^i = 8 \cdot P_i^{i-1} T^{(i)} (P_i^{i-1})^T \tag{6.58}$$

(see Question 6.15). Substituting these into the expression for M in equation (6.57) yields

$$M = R_{2/3}^{(i)} \left\{ I - (P_i^{i-1})^T \cdot [P_i^{i-1} T^{(i)} (P_i^{i-1})^T]^{-1} \cdot (P_i^{i-1} T^{(i)}) \right\} R_{2/3}^{(i)}$$

or, dropping indices to simplify notation,

$$M = R_{2/3} \left\{ I - P^T \cdot [PTP^T]^{-1} \cdot PT \right\} R_{2/3}. \tag{6.59}$$

We continue, using the fact that all the matrices composing M (T, $R_{2/3}$, and P) can be (nearly) diagonalized by the eigenvector matrices $Z = Z^{(i)}$ and

$Z^{(i-1)}$ of $T = T^{(i)}$ and $T^{(i-1)}$, respectively: Recall that $Z = Z^T = Z^{-1}$, $T = Z\Lambda Z$, and $R_{2/3} = Z(I - \Lambda/3)Z \equiv Z\Lambda_R Z$. We leave it to the reader to confirm that $Z^{(i-1)}PZ^{(i)} = \Lambda_P$, where Λ_P is almost diagonal (see Question 6.15):

$$\lambda_{P,jk} = \begin{cases} (+1 + \cos\frac{\pi j}{2^i})/\sqrt{8} & \text{if } k = j, \\ (-1 + \cos\frac{\pi j}{2^i})/\sqrt{8} & \text{if } k = 2^i - j, \\ 0 & \text{otherwise.} \end{cases} \tag{6.60}$$

This lets us write

$$\begin{aligned} ZMZ &= (ZR_{2/3}Z) \\ &\quad \cdot \left\{ I - (ZP^TZ^{(i-1)}) \cdot \left[(Z^{(i-1)}PZ)(ZTZ)(ZP^TZ^{(i-1)})\right]^{-1} \right. \\ &\qquad \left. \cdot (Z^{(i-1)}PZ)(ZTZ) \right\} \cdot (ZR_{2/3}Z) \\ &= \Lambda_R \cdot \left\{ I - \Lambda_P^T \left[\Lambda_P\Lambda\Lambda_P^T\right]^{-1} \Lambda_P\Lambda \right\} \cdot \Lambda_R. \end{aligned}$$

The matrix ZMZ is similar to M since $Z = Z^{-1}$ and so has the same eigenvalues as M. Also, ZMZ is nearly diagonal: it has nonzeros only on its main diagonal and "perdiagonal" (the diagonal from the lower left corner to the upper right corner of the matrix). This lets us compute the eigenvalues of M explicitly.

THEOREM 6.11. *The matrix M has eigenvalues $1/9$ and 0, independent of i. Therefore multigrid converges at a fixed rate independent of the number of unknowns.*

For a proof, see Question 6.15. For a more general analysis, see [268].

For an implementation of this algorithm, see Question 6.16. The Web site [91] contains pointers to an extensive literature, software, and so on.

6.10. Domain Decomposition

Domain decomposition for solving sparse systems of linear equations is a topic of current research. See [49, 116, 205] and especially [232] for recent surveys. We will give only simple examples.

The need for methods beyond those we have discussed arises from of the irregularity and size of real problems and also from the need for algorithms for parallel computers. The fastest methods that we have discussed so far, those based on block cyclic reduction, the FFT, and multigrid, work best (or only) on particularly regular problems such as the model problem, i.e., Poisson's equation discretized with a uniform grid on a rectangle. But the region of solution of a real problem may not be a rectangle but more irregular, representing a physical object like a wing (see Figure 2.12). Figure 2.12 also illustrates that there may be more grid points in regions where the solution is expected to be less smooth than in regions with a smooth solution. Also, we may have more complicated equations than Poisson's equation or even different equations in different regions. Independent of whether the problem is regular, it may be too large to fit in the computer memory and may have to be solved "in pieces." Or we may want to break the problem into pieces that can be solved in parallel on a parallel computer.

Domain decomposition addresses all these issues by showing how to systematically create "hybrid" methods from the simpler methods discussed in previous sections. These simpler methods are applied to smaller and more regular subproblems of the overall problem, after which these partial solutions are "pieced together" to get the overall solution. These subproblems can be solved one at a time if the whole problem does not fit into memory, or in parallel on a parallel computer. We give examples below. There are generally many ways to break a large problem into pieces, many ways to solve the individual pieces, and many ways to piece the solutions together. Domain decomposition theory does not provide a magic way to choose the best way to do this in all cases but rather a set of reasonable possibilities to try. There are some cases (such as problems sufficiently like Poisson's equation) where the theory does yield "optimal methods" (costing $O(1)$ work per unknown).

We divide our discussion into two parts, *nonoverlapping methods* and *overlapping* methods.

6.10.1. Nonoverlapping Methods

This method is also called *substructuring* or a *Schur complement method* in the literature. It has been used for decades, especially in the structural analysis community, to break large problems into smaller ones that fit into computer memory.

For simplicity we will illustrate this method using the usual Poisson's equation with Dirichlet boundary conditions discretized with a 5-point stencil but on an *L-shaped region* rather than a square. This region may be decomposed into two domains: a small square and a large square of twice the side length, where the small square is connected to the bottom of the right side of a larger square. We will design a solver that can exploit our ability to solve problems quickly on squares.

In the figure below, the number of each grid point is shown for a coarse discretization (the number is above and to the left of the corresponding grid point; only grid points interior to the "L" are numbered).

Note that we have numbered first the grid points inside the two subdomains (1 to 4 and 5 to 29) and then the grid points on the boundary (30 and 31). The resulting matrix is

$$\equiv A \equiv \left[\begin{array}{c|c|c} A_{11} & 0 & A_{13} \\ \hline 0 & A_{22} & A_{23} \\ \hline A_{13}^T & A_{23}^T & A_{33} \end{array}\right].$$

Here, $A_{11} = T_{2\times2}$, $A_{22} = T_{5\times5}$, and $A_{33} = T_{2\times1} \equiv T_2 + 2I_2$, where T_N is defined in equation (6.3) and $T_{N\times N}$ is defined in equation (6.14). One of the most important properties of this matrix is that $A_{12} = 0$, since there is no direct coupling between the interior grid points of the two subdomains. The only coupling is through the boundary, which is numbered last (grid points 30 and 31). Thus A_{13} contains the coupling between the small square and the boundary, and A_{23} contains the coupling between the large square and the boundary.

To see how to take advantage of the special structure of A to solve $Ax = b$, write the block LDU decomposition of A as follows:

$$A = \begin{bmatrix} I & 0 & 0 \\ 0 & I & 0 \\ A_{13}^T A_{11}^{-1} & A_{23}^T A_{22}^{-1} & I \end{bmatrix} \cdot \begin{bmatrix} I & 0 & 0 \\ 0 & I & 0 \\ 0 & 0 & S \end{bmatrix} \cdot \begin{bmatrix} A_{11} & 0 & A_{13} \\ 0 & A_{22} & A_{23} \\ 0 & 0 & I \end{bmatrix},$$

where

$$S = A_{33} - A_{13}^T A_{11}^{-1} A_{13} - A_{23}^T A_{22}^{-1} A_{23} \tag{6.61}$$

is called the *Schur complement* of the leading principal submatrix containing A_{11} and A_{22}. Therefore, we may write
$A^{-1} =$

$$\begin{bmatrix} A_{11}^{-1} & 0 & -A_{11}^{-1} A_{13} \\ 0 & A_{22}^{-1} & -A_{22}^{-1} A_{23} \\ 0 & 0 & I \end{bmatrix} \cdot \begin{bmatrix} I & 0 & 0 \\ 0 & I & 0 \\ 0 & 0 & S^{-1} \end{bmatrix} \cdot \begin{bmatrix} I & 0 & 0 \\ 0 & I & 0 \\ -A_{13}^T A_{11}^{-1} & -A_{23}^T A_{22}^{-1} & I \end{bmatrix}.$$

Therefore, to multiply a vector by A^{-1} we need to multiply by the blocks in the entries of this factored form of A^{-1}, namely, A_{13} and A_{23} (and their transposes), A_{11}^{-1} and A_{22}^{-1}, and S^{-1}. Multiplying by A_{13} and A_{23} is cheap because they are very sparse. Multiplying by A_{11}^{-1} and A_{22}^{-1} is also cheap because we chose these subdomains to be solvable by FFT, block cyclic reduction, multigrid, or some other fast method discussed so far. It remains to explain how to multiply by S^{-1}.

Since there are many fewer grid points on the boundary than in the subdomains, A_{33} and S have a much smaller dimension than A_{11} and A_{22}; this effect grows for finer grid spacings. S is symmetric positive definite, as is A, and (in this case) dense. To compute it explicitly one would need to solve with each subdomain once per boundary grid point (from the $A_{11}^{-1} A_{13}$ and $A_{22}^{-1} A_{23}$ terms in (6.61)). This can certainly be done, after which one could factor S using dense Cholesky and proceed to solve the system. But this is expensive, much more so than just multiplying a vector by S, which requires just one solve per subdomain using equation (6.61). This makes a Krylov subspace–based iterative method such as CG look attractive (section 6.6), since these methods require only multiplying a vector by S. The number of matrix-vector multiplications CG requires depends on the condition number of S. What makes

domain decomposition so attractive is that S turns out to be much better conditioned that the original matrix A (a condition number that grows like $O(N)$
instead of $O(N^2)$), and so convergence is fast [116, 205].

More generally, one has $k > 2$ subdomains, separated by boundaries (see
Figure 6.21, where the heavy lines separate subdomains). If we number the
nodes in each subdomain consecutively, followed by the boundary nodes, we
get the matrix

$$
A = \left[
\begin{array}{ccc|c}
A_{1,1} & & 0 & A_{1,k+1} \\
 & \ddots & & \vdots \\
0 & & A_{k,k} & A_{k,k+1} \\
\hline
A_{1,k+1}^T & \cdots & A_{k,k+1}^T & A_{k+1,k+1}
\end{array}
\right], \tag{6.62}
$$

where again we can factor it by factoring each $A_{i,i}$ independently and forming
the Schur complement $S = A_{k+1,k+1} - \sum_{i=1}^{k} A_{i,k+1}^T A_{i,i}^{-1} A_{i,k+1}$.

In this case, when there is more than one boundary segment, S has further
structure that can be exploited to precondition it. For example, by numbering the grid points in the interior of each boundary segment before the grid
points at the intersection of boundary segments, one gets a block structure as
in A. The diagonal blocks of S are complicated but may be approximated by
$T_N^{1/2}$, which may be inverted efficiently using the FFT [36, 37, 38, 39, 40]. To
summarize the state of the art, by choosing the preconditioner for S appropriately, one can make the number of steps of CG independent of the number of
boundary grid points N [231].

6.10.2. Overlapping Methods

The methods in the last section were called *nonoverlapping* because the domains corresponding to the nodes in $A_{i,i}$ were disjoint, leading to the block
diagonal structure in equation (6.62). In this section we permit overlapping
domains, as shown in the figure below. As we will see, this overlap permits us
to design an algorithm comparable in speed with multigrid but applicable to
a wider set of problems.

The rectangle with a dashed boundary in the figure is domain Ω_1, and the
square with a solid boundary is domain Ω_2. We have renumbered the nodes
so that the nodes in Ω_1 are numbered first and the nodes in Ω_2 are numbered

last, with the nodes in the overlap $\Omega_1 \cap \Omega_2$ in the middle.

```
                                        Ω2
        31  26  21  16  13
        30  25  20  15  12
        29  24  19  14  11
        28  23  18  10   8   6   4   2
        27  22  17   9   7   5   3   1
                                        Ω1
```

These domains are shown in the matrix A below, which is the same matrix as in section 6.10.1 but with its rows and columns ordered as shown above:

```
 4 -1 -1
-1  4     -1
-1     4 -1 -1
   -1 -1  4        -1
      -1      4 -1 -1
         -1 -1  4     -1
                -1 || 4 -1 -1
                   -1|-1  4     -1|-1
                     |-1     4 -1              -1
                     |-1 -1  4        -1         -1
                  -1 | 4 -1    -1
                     |-1  4 -1    -1
                     |   -1  4       -1
                  -1|-1        4 -1        -1
                     |-1    -1  4 -1         -1
                     |   -1    -1  4            -1
             -1      |              4 -1        -1
                -1   |             -1  4 -1        -1
                     |   -1           -1  4 -1        -1
                     |      -1           -1  4 -1        -1
                     |         -1           -1  4           -1
                     |               -1         4 -1        -1
                     |                  -1     -1  4 -1        -1
                     |                     -1     -1  4 -1        -1
                     |                        -1     -1  4 -1        -1
                     |                           -1     -1  4           -1
                     |                                 -1     4 -1
                     |                                    -1   -1  4 -1
                     |                                       -1   -1  4 -1
                     |                                          -1   -1  4 -1
                     |                                             -1   -1  4
```

We have indicated the boundaries between domains in the way that we have partitioned the matrix: The single lines divide the matrix into the nodes associated with Ω_1 (1 through 10) and the rest $\Omega \setminus \Omega_1$ (11 through 31). The double lines divide the matrix into the nodes associated with Ω_2 (7 through 31) and the rest $\Omega \setminus \Omega_2$ (1 through 6). The submatrices below are subscripted accordingly:

$$A = \left[\begin{array}{c|c} A_{\Omega_1,\Omega_1} & A_{\Omega_1,\Omega\setminus\Omega_1} \\ \hline A_{\Omega\setminus\Omega_1,\Omega_1} & A_{\Omega\setminus\Omega_1,\Omega\setminus\Omega_1} \end{array}\right] = \left[\begin{array}{c||c} A_{\Omega\setminus\Omega_2,\Omega\setminus\Omega_2} & A_{\Omega\setminus\Omega_2,\Omega_2} \\ \hline A_{\Omega_2,\Omega\setminus\Omega_2} & A_{\Omega_2,\Omega_2} \end{array}\right].$$

We conformally partition vectors such as

$$x = \left[\begin{array}{c} x_{\Omega_1} \\ \hline x_{\Omega\setminus\Omega_1} \end{array}\right] = \left[\begin{array}{c} x(1:10) \\ \hline x(11:31) \end{array}\right]$$

$$= \left[\begin{array}{c} x_{\Omega\setminus\Omega_2} \\ \hline x_{\Omega_2} \end{array}\right] = \left[\begin{array}{c} x(1:6) \\ \hline x(7:31) \end{array}\right].$$

Now we have enough notation to state two basic overlapping domain decomposition algorithms. The simplest one is called the *additive Schwarz method* for historical reasons but could as well be called *overlapping block Jacobi iteration* because of its similarity to (block) Jacobi iteration from sections 6.5 and 6.6.5.

ALGORITHM 6.18. *Additive Schwarz method for updating an approximate solution x_i of $Ax = b$ to get a better solution x_{i+1}:*

$r = b - Ax_i$ */* compute the residual */*
$x_{i+1} = 0$
$x_{i+1,\Omega_1} = x_{i,\Omega_1} + A_{\Omega_1,\Omega_1}^{-1} \cdot r_{\Omega_1}$ */* update the solution on Ω_1 */*
$x_{i+1,\Omega_2} = x_{i+1,\Omega_2} + A_{\Omega_2,\Omega_2}^{-1} \cdot r_{\Omega_2}$ */* update the solution on Ω_2 */*

This algorithm also be written in one line as

$$x_{i+1} = x_i + \left[\begin{array}{c} A_{\Omega_1,\Omega_1}^{-1} \cdot r_{\Omega_1} \\ 0 \end{array}\right] + \left[\begin{array}{c} 0 \\ \hline A_{\Omega_2,\Omega_2}^{-1} \cdot r_{\Omega_2} \end{array}\right].$$

In words, the algorithm works as follows: The update $A_{\Omega_1,\Omega_1}^{-1} r_{\Omega_1}$ corresponds to solving Poisson's equation just on Ω_1, using boundary conditions at nodes 11, 14, 17, 18, and 19, which depend on the previous approximate solution x_i. The update $A_{\Omega_2,\Omega_2}^{-1} r_{\Omega_2}$ is analogous, using boundary conditions at nodes 5 and 6 depending on x_i.

In our case the Ω_i are rectangles, so any one of our earlier fast methods, such as multigrid, could be used to solve $A_{\Omega_i,\Omega_i}^{-1} r_{\Omega_i}$. Since the additive Schwarz method is iterative, it is not necessary to solve the problems on Ω_i exactly.

Indeed, the additive Schwarz method is typically used as a preconditioner for a Krylov subspace method like conjugate gradients (see section 6.6.5). In the notation of section 6.6.5, the preconditioner M is given by

$$M^{-1} = \left[\begin{array}{c|c} A_{\Omega_1,\Omega_1}^{-1} & 0 \\ \hline 0 & 0 \end{array}\right] + \left[\begin{array}{c||c} 0 & 0 \\ \hline 0 & A_{\Omega_2,\Omega_2}^{-1} \end{array}\right].$$

If Ω_1 and Ω_2 did not overlap, then M^{-1} would simplify to

$$
\begin{bmatrix}
A_{\Omega_1,\Omega_1}^{-1} & 0 \\
0 & A_{\Omega_2,\Omega_2}^{-1}
\end{bmatrix}
$$

and we would be doing block Jacobi iteration. But we know that Jacobi's method does not converge particularly quickly, because "information" about the solution from one domain can only move slowly to the other domain across the boundary between them (see the discussion at the beginning of section 6.9). But as long as the overlap is a large enough fraction of the two domains, information will travel quickly enough to guarantee fast convergence. Of course we do not want too large an overlap, because this increases the work significantly. The goal in designing a good domain decomposition method is to choose the domains and the overlaps so as to have fast convergence while doing as little work as possible; we say more on how convergence depends on overlap below.

From the discussion in section 6.5, we know that the Gauss–Seidel method is likely to be more effective than Jacobi's method. This is the case here as well, with the *overlapping block Gauss–Seidel method* (more commonly called the *multiplicative Schwarz method*) often being twice as fast as additive block Jacobi iteration (the additive Schwarz method).

ALGORITHM 6.19. *Multiplicative Schwarz method for updating an approximate solution x_i of $Ax = b$:*

(1) $r_{\Omega_1} = (b - Ax_i)_{\Omega_1}$ /* compute residual of x_i on Ω_1 */
(2) $x_{i+\frac{1}{2},\Omega_1} = x_{i,\Omega_1} + A_{\Omega_1,\Omega_1}^{-1} \cdot r_{\Omega_1}$ /* update solution on Ω_1 */
(2′) $x_{i+\frac{1}{2},\Omega\backslash\Omega_1} = x_{i,\Omega\backslash\Omega_1}$
(3) $r_{\Omega_2} = (b - Ax_{i+\frac{1}{2}})_{\Omega_2}$ /* compute residual of $x_{i+\frac{1}{2}}$ on Ω_2 */
(4) $x_{i+1,\Omega_2} = x_{i+\frac{1}{2},\Omega_2} + A_{\Omega_2,\Omega_2}^{-1} \cdot r_{\Omega_2}$ /* update solution on Ω_2 */
(4′) $x_{i+1,\Omega\backslash\Omega_2} = x_{i+\frac{1}{2},\Omega\backslash\Omega_2}$

Note that lines (2′) *and* (4′) *do not require any data movement, provided that* $x_{i+\frac{1}{2}}$ *and* x_{i+1} *overwrite* x_i.

This algorithm first solves Poisson's equation on Ω_1 using boundary data from x_i, just like Algorithm 6.18. It then solves Poisson's equation on Ω_2, but using boundary data that has just been updated. It may also be used as a preconditioner for a Krylov subspace method.

In practice more domains than just two (Ω_1 and Ω_2) are used. This is done if the domain of solution is more complicated or if there are many independent parallel processors available to solve independent problems $A_{\Omega_i,\Omega_i}^{-1} r_{\Omega_i}$ or just to keep the subproblems $A_{\Omega_i,\Omega_i}^{-1} r_{\Omega_i}$ small and inexpensive to solve.

Here is a summary of the theoretical convergence analysis of these methods for the model problem and similar elliptic partial differential equations. Let h

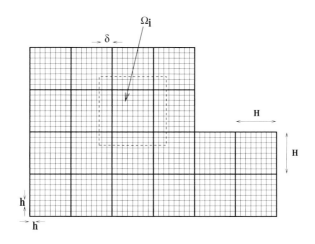

Fig. 6.21. *Coarse and fine discretizations of an L-shaped region.*

be the mesh spacing. The theory predicts how many iterations are necessary to converge as a function of h as h decreases to 0. With two domains, as long as the overlap region $\Omega_1 \cap \Omega_2$ is a nonzero fraction of the total domain $\Omega_1 \cup \Omega_2$, the number of iterations required for convergence is independent of h as h goes to zero. This is an attractive property and is reminiscent of multigrid, which also converged at a rate independent of mesh size h. But the cost of an iteration includes solving subproblems on Ω_1 and Ω_2 *exactly*, which may be comparable in expense to the original problem. So unless the solutions on Ω_1 and Ω_2 are very cheap (as with the L-shaped region above), the cost is still high.

Now suppose we have many domains Ω_i, each of size $H \gg h$. In other words, think of the Ω_i as the regions bounded by a coarse mesh with spacing H, plus some cells beyond the boundary, as shown by the dashed line in Figure 6.21.

Let $\delta < H$ be the amount by which adjacent domains overlap. Now let H, δ, and h all go to zero such that the overlap fraction δ/H remains constant, and $H \gg h$. Then the number of iterations required for convergence grows like $1/H$, i.e., *independently* of the fine mesh spacing h. This is close to, but still not as good as, multigrid, which does a constant number of iterations and $O(1)$ work per unknown.

Attaining the performance of multigrid requires one more idea, which, perhaps not surprisingly, is similar to multigrid. We use an approximation A_H of the problem on the coarse grid with spacing H to get a *coarse grid precon-ditioner* in addition to the fine grid preconditioners $A_{\Omega_i, \Omega_i}^{-1}$. We need three matrices to describe the algorithm. First, let A_H be the matrix for the model problem discretized with coarse mesh spacing H. Second, we need a *restriction operator* R to take a residual on the fine mesh and restrict it to values on the coarse mesh; this is essentially the same as in multigrid (see section 6.9.2). Finally, we need an *interpolation operator* to take values on the coarse mesh and interpolate them to the fine mesh; as in multigrid this also turns out to be R^T.

ALGORITHM 6.20. *Two-level additive Schwarz method for updating an approx-imate solution x_i of $Ax = b$ to get a better solution x_{i+1}:*

$$x_{i+1} = x_i$$
$$\text{for } i = 1 \text{ to the number of domains } \Omega_i$$
$$\quad r_{\Omega_i} = (b - Ax_i)_{\Omega_i}$$
$$\quad x_{i+1,\Omega_i} = x_{i+1,\Omega_i} + A^{-1}_{\Omega_i,\Omega_i} \cdot r_{\Omega_i}$$
$$endfor$$
$$x_{i+1} = x_{i+1} + R^T A_C^{-1} Rr$$

As with Algorithm 6.18, this method is typically used as a preconditioner for a Krylov subspace method.

Convergence theory for this algorithm, which is applicable to more general problems than Poisson's equation, says that as H, δ, and h shrink to 0 with δ/H staying fixed, the number of iterations required to converge is independent of H, h, or δ. This means that as long as the work to solve the subproblems $A^{-1}_{\Omega_i,\Omega_i}$ and A^{-1}_H is proportional to the number of unknowns, the complexity is as good as multigrid.

It is probably evident to the reader that implementing these methods in a real world problem can be complicated. There is software available on-line that implements many of the building blocks described here and also runs on parallel machines. It is called PETSc, for Portable Extensible Toolkit for Scientific computing. PETSc is available at http://www.mcs.anl.gov/petsc/petsc.html and is described briefly in [232].

6.11.   References and Other Topics for Chapter 6

Up-to-date surveys of modern iterative methods are given in [15, 107, 136, 214], and their parallel implementations are also surveyed in [76]. Classical methods such as Jacobi's, Gauss–Seidel, and SOR methods are discussed in detail in [249, 137]. Multigrid methods are discussed in [43, 185, 186, 260, 268] and the references therein; [91] is a Web site with pointers to an extensive bibliography, software, and so on. Domain decomposition are discussed in [49, 116, 205, 232]. Chebyshev and other polynomials are discussed in [240]. The FFT is discussed in any good textbook on computer science algorithms, such as [3] and [248]. A stabilized version of block cyclic reduction is found in [47, 46].

6.12.   Questions for Chapter 6

QUESTION 6.1. *(Easy)* Prove Lemma 6.1.

QUESTION 6.2. *(Easy)* Prove the following formulas for triangular factoriza-tions of T_N.

1. The Cholesky factorization $T_N = B_N^T B_N$ has a upper bidiagonal Cholesky factor B_N with

$$B_N(i, i) = \sqrt{\frac{i+1}{i}} \quad \text{and} \quad B_N(i, i+1) = \sqrt{\frac{i}{i+1}}.$$

2. The result of Gaussian elimination with partial pivoting on T_N is $T_N = L_N U_N$, where the triangular factors are bidiagonal:

$$L_N(i, i) = 1 \quad \text{and} \quad L_N(i+1, i) = -\frac{i}{i+1},$$

$$U_N(i, i) = \frac{i+1}{i} \quad \text{and} \quad U_N(i, i+1) = -1.$$

3. $T_N = D_N D_N^T$, where D_N is the N-by-$(N+1)$ upper bidiagonal matrix with 1 on the main diagonal and -1 on the superdiagonal.

QUESTION 6.3. *(Easy)* Confirm equation (6.13).

QUESTION 6.4. *(Easy)*

1. Prove Lemma 6.2.

2. Prove Lemma 6.3.

3. Prove that the Sylvester equation $AX - XB = C$ is equivalent to $(I_n \otimes A - B^T \otimes I_m) \text{vec}(X) = \text{vec}(C)$.

4. Prove that $\text{vec}(AXB) = (B^T \otimes A) \cdot \text{vec}(X)$.

QUESTION 6.5. *(Medium)* Suppose that $A^{n \times n}$ is diagonalizable, so A has n independent eigenvectors: $Ax_i = \alpha_i x_i$, or $AX = X\Lambda_A$, where $X = [x_1, \ldots, x_n]$ and $\Lambda_A = \text{diag}(\alpha_i)$. Similarly, suppose that $B^{m \times m}$ is diagonalizable, so b has m independent eigenvectors: $By_i = \beta_i y_i$, or $BY = Y\Lambda_B$, where $Y = [y_1, \ldots, y_m]$ and $\Lambda_B = \text{diag}(\beta_j)$. Prove the following results.

1. The mn eigenvalues of $I_m \otimes A + B \otimes I_n$ are $\lambda_{ij} = \alpha_i + \beta_j$, i.e., all possible sums of pairs of eigenvalues of A and B. The corresponding eigenvectors are z_{ij}, where $z_{ij} = x_i \otimes y_j$, whose $(km+l)$th entry is $x_i(k)y_j(l)$. Written another way,

$$(I_m \otimes A + B \otimes I_n)(Y \otimes X) = (Y \otimes X) \cdot (I_m \otimes \Lambda_A + \Lambda_B \otimes I_n). \quad (6.63)$$

2. The Sylvester equation $AX + XB^T = C$ is nonsingular (solvable for X, given any C) if and only if the sum $\alpha_i + \beta_j \neq 0$ for all eigenvalues α_i of A and β_j of B. The same is true for the slightly different Sylvester equation $AX + XB = C$ (see also Question 4.6).

3. The mn eigenvalues of $A \otimes B$ are $\lambda_{ij} = \alpha_i \beta_j$, i.e., all possible products of pairs of eigenvalues of A and B. The corresponding eigenvectors are z_{ij}, where $z_{ij} = x_i \otimes y_j$, whose $(km + l)$th entry is $x_i(k)y_j(l)$. Written another way,

$$(B \otimes A)(Y \otimes X) = (Y \otimes X) \cdot (\Lambda_B \otimes \Lambda_A). \tag{6.64}$$

QUESTION 6.6. *(Easy; Programming)* Write a one-line Matlab program to implement Algorithm 6.2: one step of Jacobi's algorithm for Poisson's equation. Test it by confirming that it converges as fast as predicted in section 6.5.4.

QUESTION 6.7. *(Hard)* Prove Lemma 6.7.

QUESTION 6.8. *(Medium; Programming)* Write a Matlab program to solve the discrete model problem on a square using FFTs. The inputs should be the dimension N and a square N-by-N matrix of values of f_{ij}. The outputs should be an N-by-N matrix of solution v_{ij} and the residual $\|T_{N \times N} v - h^2 f\|_2 / (\|T_{N \times N}\|_2 \cdot \|v\|)$. You should also produce three-dimensional plots of f and v. Use the FFT built into Matlab. Your program should not have to be more than a few lines long if you use all the features of Matlab that you can. Solve it for several problems whose solutions you know and several you do not:

1. $f_{jk} = \sin(j\pi/(N + 1)) \cdot \sin(k\pi/(N + 1))$.

2. $f_{jk} = \sin(j\pi/(N+1)) \cdot \sin(k\pi/(N+1)) + \sin(3j\pi/(N+1)) \cdot \sin(5k\pi/(N+1))$.

3. f has a few sharp spikes (both positive and negative) and is 0 elsewhere. This approximates the electrostatic potential of charged particles located at the spikes and with charges proportional to the heights (positive or negative) of the spikes. If the spikes are all positive, this is also the gravitational potential.

QUESTION 6.9. *(Medium)* Confirm that evaluating the formula in (6.47) by performing the matrix-vector multiplications from right to left is mathematically the same as Algorithm 6.13.

QUESTION 6.10. *(Medium; Hard)*

1. *(Hard)* Let A and H be real symmetric n-by-n matrices that *commute*, i.e., $AH = HA$. Show that there is an orthogonal matrix Q such that $QAQ^T = \text{diag}(\alpha_1, \ldots, \alpha_n)$ and $QHQ^T = \text{diag}(\theta_1, \ldots, \theta_n)$ are both diagonal. In other words, A and H have the same eigenvectors. Hint: First assume A has distinct eigenvalues, and then remove this assumption.

2. *(Medium)* Let

$$
\hat{T} = \begin{bmatrix} \alpha & \theta & & \\ \theta & \ddots & \ddots & \\ & \ddots & \ddots & \theta \\ & & \theta & \alpha \end{bmatrix}
$$

be a symmetric tridiagonal Toeplitz matrix, i.e., a symmetric tridiagonal matrix with constant α along the diagonal and θ along the offdiagonals. Write down *simple* formulas for the eigenvalues and eigenvectors of $\hat{T}$. Hint: Use Lemma 6.1.

3. *(Hard)* Let

$$
T = \begin{bmatrix} A & H & & \\ H & \ddots & \ddots & \\ & \ddots & \ddots & H \\ & & H & A \end{bmatrix}
$$

be an n^2-by-n^2 block tridiagonal matrix, with n copies of A along the diagonal. Let $QAQ^T = \mathrm{diag}(\alpha_1, \ldots, \alpha_n)$ be the eigendecomposition of A, and let $QHQ^T = \mathrm{diag}(\theta_1, \ldots, \theta_n)$ be the eigendecomposition of H as above. Write down *simple* formulas for the n^2 eigenvalues and eigenvectors of T in terms of the α_i, θ_i, and Q. Hint: Use Kronecker products.

4. *(Medium)* Show how to solve $Tx = b$ in $O(n^3)$ time. In contrast, how much bigger are the running times of dense LU factorization and band LU factorization?

5. *(Medium)* Suppose that A and H are (possibly different) symmetric tridiagonal Toeplitz matrices, as defined above. Show how to use the FFT to solve $Tx = b$ in just $O(n^2 \log n)$ time.

QUESTION 6.11. *(Easy)* Suppose that R is upper triangular and nonsingular and that C is upper Hessenberg. Confirm that RCR^{-1} is upper Hessenberg.

QUESTION 6.12. *(Medium)* Confirm that the Krylov subspace $\mathcal{K}_k(A, y_1)$ has dimension k if and only if the Arnoldi algorithm (Algorithm 6.9) or the Lanczos algorithm (Algorithm 6.10) can compute q_k without quitting first.

QUESTION 6.13. *(Medium)* Confirm that when $A^{n \times n}$ is symmetric positive definite and $Q^{n \times k}$ has full column rank, then $T = Q^T A Q$ is also symmetric positive definite. (For this question, Q need not be orthogonal.)

QUESTION 6.14. *(Medium)* Prove Theorem 6.9.

QUESTION 6.15. *(Medium; Hard)*

1. *(Medium)* Confirm equation (6.58).

2. *(Medium)* Confirm equation (6.60).

3. *(Hard)* Prove Theorem 6.11.

QUESTION 6.16. *(Medium; Programming)* A Matlab program implementing multigrid to solve the discrete model problem on a square is available on the class homepage at HOMEPAGE/Matlab/MG_README.html. Start by running the demonstration (type "makemgdemo" and then "testfmgv"). Then, try running testfmg for different right-hand sides (input array b), different numbers of weighted Jacobi iterations before and after each recursive call to the multigrid solver (inputs jac1 and jac2), and different numbers of iterations (input iter). The software will plot the convergence rate (ratio of consecutive residuals); does this depend on the size of b? the frequencies in b? the values of jac1 and jac2? For which values of jac1 and jac2 is the solution most efficient?

QUESTION 6.17. *(Medium; Programming)* Using a fast model problem solver from either Question 6.8 or Question 6.16, use domain decomposition to build a fast solver for Poisson's equation on an L-shaped region, as described in section 6.10. The large square should be 1-by-1 and the small square should be .5-by-.5, attached at the bottom right of the large square. Compute the residual in order to show that your answer is correct.

QUESTION 6.18. *(Hard)* Fill in the entries of a table like Table 6.1, but for solving Poisson's equation in three dimensions instead of two. Assume that the grid of unknowns is $N \times N \times N$, with $n = N^3$. Try to fill in as many entries of columns 2 and 3 as you can.

7

Iterative Methods for Eigenvalue Problems

7.1. Introduction

In this chapter we discuss iterative methods for finding eigenvalues of matrices that are too large to use the direct methods of Chapters 4 and 5. In other words, we seek algorithms that take far less than $O(n^2)$ storage and $O(n^3)$ flops. Since the eigenvectors of most n-by-n matrices would take n^2 storage to represent, this means that we seek algorithms that compute just a few user-selected eigenvalues and eigenvectors of a matrix.

We will depend on the material on Krylov subspace methods developed in section 6.6, the material on symmetric eigenvalue problems in section 5.2, and the material on the power method and inverse iteration in section 5.3. The reader is advised to review these sections.

The simplest eigenvalue problem is to compute just the largest eigenvalue in absolute value, along with its eigenvector. The power method (Algorithm 4.1) is the simplest algorithm suitable for this task: Recall that its inner loop is

$$
\begin{aligned}
y_{i+1} &= A x_i, \\
x_{i+1} &= y_{i+1}/\|y_{i+1}\|_2,
\end{aligned}
$$

where x_i converges to the eigenvector corresponding to the desired eigenvector (provided that there is only one eigenvalue of largest absolute value, and x_1 does not lie in an invariant subspace not containing its eigenvector). Note that the algorithm uses A only to perform matrix-vector multiplication, so all that we need to run the algorithm is a "black-box" that takes x_i as input and returns $A x_i$ as output (see Example 6.13).

A closely related problem is to find the eigenvalue closest to a user-supplied value σ, along with its eigenvector. This is precisely the situation inverse iteration (Algorithm 4.2) was designed to handle. Recall that its inner loop is

$$
\begin{aligned}
y_{i+1} &= (A - \sigma I)^{-1} x_i, \\
x_{i+1} &= y_{i+1}/\|y_{i+1}\|_2,
\end{aligned}
$$

361

i.e., solving a linear system of equations with coefficient matrix $A - \sigma I$. Again x_i converges to the desired eigenvector, provided that there is just one eigenvalue closest to σ (and x_1 satisfies the same property as before). Any of the sparse matrix techniques in Chapter 6 or section 2.7.4 could be used to solve for y_{i+1}, although this is usually much more expensive than simply multiplying by A. When A is symmetric Rayleigh quotient iteration (Algorithm 5.1) can also be used to accelerate convergence (although it is not always guaranteed to converge to the eigenvalue of A closest to σ).

Starting with a given x_1, $k - 1$ iterations of either the power method or inverse iteration produce a sequence of vectors $x_1, x_2, \ldots, x_k$. These vectors span a *Krylov subspace*, as defined in section 6.6.1. In the case of the power method, this Krylov subspace is $\mathcal{K}_k(A, x_1) = \mathrm{span}[x_1, Ax_1, A^2 x_1, \ldots, A^{k-1} x_1]$, and in the case of inverse iteration this Krylov subspace is $\mathcal{K}_k((A - \sigma I)^{-1}, x_1)$. Rather than taking x_k as our approximate eigenvector, it is natural to ask for the "best" approximate eigenvector in $\mathcal{K}_k$, i.e., the best linear combination $\sum_{i=1}^{k} \alpha_i x_i$. We took the same approach for solving $Ax = b$ in section 6.6.2, where we asked for the best approximate solution to $Ax = b$ from $\mathcal{K}_k$. We will see that the best eigenvector (and eigenvalue) approximations from $\mathcal{K}_k$ are much better than x_k alone. Since $\mathcal{K}_k$ has dimension k (in general), we can actually use it to compute k best approximate eigenvalues and eigenvectors. These best approximations are called the *Ritz values* and *Ritz vectors*.

We will concentrate on the symmetric case $A = A^T$. In the last section we will briefly describe the nonsymmetric case.

The rest of this chapter is organized as follows. Section 7.2 discusses the Rayleigh–Ritz method, our basic technique for extracting information about eigenvalues and eigenvectors from a Krylov subspace. Section 7.3 discusses our main algorithm, the Lanczos algorithm, in exact arithmetic. Section 7.4 analyzes the rather different behavior of the Lanczos algorithm in floating point arithmetic, and sections 7.5 and 7.6 describe practical implementations of Lanczos that compute reliable answers despite roundoff. Finally, section 7.7 briefly discusses algorithms for the nonsymmetric eigenproblem.

7.2. The Rayleigh–Ritz Method

Let $Q = [Q_k, Q_u]$ be any n-by-n orthogonal matrix, where Q_k is n-by-k and Q_u is n-by-$(n - k)$. In practice the columns of Q_k will be computed by the Lanczos algorithm (Algorithm 6.10 or Algorithm 7.1 below) and span a Krylov subspace $\mathcal{K}_k$, and the subscript u indicates that Q_u is (mostly) unknown. But for now we do not care where we get Q.

We will use the following notation (which was also used in equation (6.31)):

$$T = Q^T A Q = [Q_k, Q_u]^T A [Q_k, Q_u] = \begin{bmatrix} Q_k^T A Q_k & Q_k^T A Q_u \\ Q_u^T A Q_k & Q_u^T A Q_u \end{bmatrix}$$

$$\equiv \begin{array}{c} k \\ n-k \end{array} \begin{array}{cc} k & n-k \\ \left(\begin{array}{cc} T_k & T_{uk} \\ T_{ku} & T_u \end{array} \right) \end{array}$$

$$= \left[\begin{array}{cc} T_k & T_{ku}^T \\ T_{ku} & T_u \end{array} \right]. \tag{7.1}$$

When $k = 1$, T_k is just the Rayleigh quotient $T_1 = \rho(Q_1, A)$ (see Definition 5.1). So for $k > 1$, T_k is a natural generalization of the Rayleigh quotient.

DEFINITION 7.1. *The* Rayleigh–Ritz procedure *is to approximate the eigenvalues of* A *by the eigenvalues of* $T_k = Q_k^T A Q_k$. *These approximations are called* Ritz values. *Let* $T_k = V \Lambda V^T$ *be the eigendecomposition of* T_k. *The corresponding eigenvector approximations are the columns of* $Q_k V$ *and are called* Ritz vectors.

The Ritz values and Ritz vectors are considered *optimal* approximations to the eigenvalues and eigenvectors of A for several reasons. First, when Q_k and so T_k are known but Q_u and so T_{ku} and T_u are unknown, the Ritz values and vectors are the natural approximations from the known part of the matrix. Second, they satisfy the following generalization of Theorem 5.5. (Theorem 5.5 showed that the Rayleigh quotient was a "best approximation" to a single eigenvalue.) Recall that the columns of Q_k span an invariant subspace of A if and only if $AQ_k = Q_k R$ for some matrix R.

THEOREM 7.1. *The minimum of* $\|AQ_k - Q_k R\|_2$ *over all k-by-k symmetric matrices R is attained by* $R = T_k$, *in which case* $\|AQ_k - Q_k R\|_2 = \|T_{ku}\|_2$. *Let* $T_k = V \Lambda V^T$ *be the eigendecomposition of* T_k. *The minimum of* $\|AP_k - P_k D\|_2$ *over all n-by-k orthogonal matrices P_k where* span(P_k) = span(Q_k) *and over diagonal D is also* $\|T_{ku}\|_2$ *and is attained by* $P_k = Q_k V$ *and* $D = \Lambda$.

In other words, the columns of $Q_k V$ (the Ritz vectors) are the "best" approximate eigenvectors and the diagonal entries of Λ (the Ritz values) are the "best" approximate eigenvalues in the sense of minimizing the residual $\|AP_k - P_k D\|_2$.

Proof. We temporarily drop the subscripts k on T_k and Q_k to simplify notation, so we can write the k-by-k matrix $T = Q^T A Q$. Let $R = T + Z$. We want to show $\|AQ - QR\|_2^2$ is minimized when $Z = 0$. We do this by using a disguised form of the Pythagorean theorem:

$$
\begin{aligned}
\|AQ - QR\|_2^2 &= \lambda_{\max} \left[(AQ - QR)^T (AQ - QR) \right] \\
&\qquad \text{by Part 7 of Lemma 1.7} \\
&= \lambda_{\max} \left[(AQ - Q(T+Z))^T (AQ - Q(T+Z)) \right] \\
&= \lambda_{\max} \left[(AQ - QT)^T (AQ - QT) - (AQ - QT)^T (QZ) \right. \\
&\qquad \left. -(QZ)^T (AQ - QT) + (QZ)^T (QZ) \right]
\end{aligned}
$$

$$
\begin{aligned}
&= \lambda_{\max} \left[(AQ - QT)^T (AQ - QT) - (Q^T AQ - T)Z \right. \\
&\qquad \left. - Z^T (Q^T AQ - T) + Z^T Z \right] \\
&= \lambda_{\max} \left[(AQ - QT)^T (AQ - QT) + Z^T Z \right] \\
&\qquad \text{because } \; Q^T AQ = T \\
&\geq \lambda_{\max} \left[(AQ - QT)^T (AQ - QT) \right] \\
&\qquad \text{by Question 5.5, since } \; Z^T Z \; \text{ is} \\
&\qquad \text{symmetric positive semidefinite} \\
&= \| AQ - QT \|_2^2 \quad \text{by Part 7 of Lemma 1.7.}
\end{aligned}
$$

Restoring subscripts, it is easy to compute the minimum value

$$
\| AQ_k - Q_k T_k \|_2 = \| (Q_k T_k + Q_u T_{ku}) - (Q_k T_k) \|_2 = \| Q_u T_{ku} \|_2 = \| T_{ku} \|_2.
$$

If we replace Q_k with any product $Q_k U$, where U is another orthogonal matrix, then the columns of Q_k and $Q_k U$ span the same space, and

$$
\| AQ_k - Q_k R \|_2 = \| AQ_k U - Q_k R U \|_2 = \| A(Q_k U) - (Q_k U)(U^T R U) \|_2.
$$

These quantities are still minimized when $R = T_k$, and by choosing $U = V$ so that $U^T T_k U$ is diagonal, we solve the second minimization problem in the statement of the theorem. $\square$

This theorem justifies using Ritz values as eigenvalue approximations. When Q_k is computed by the Lanczos algorithm, in which case (see equation (6.31))

$$
T = \left[\begin{array}{c|c} T_k & T_{ku}^T \\ \hline T_{ku} & T_u \end{array} \right] =
\left[\begin{array}{ccccc|cccc}
\alpha_1 & \beta_1 & & & & & & & \\
\beta_1 & \ddots & \ddots & & & & & & \\
 & \ddots & \ddots & \beta_{k-1} & & & & & \\
 & & \beta_{k-1} & \alpha_k & \beta_k & & & & \\
\hline
 & & & \beta_k & \alpha_{k+1} & \beta_{k+1} & & & \\
 & & & & \beta_{k+1} & \ddots & \ddots & & \\
 & & & & & \ddots & \ddots & \beta_{n-1} & \\
 & & & & & & \beta_{n-1} & \alpha_n &
\end{array} \right],
$$

then it is easy to compute all the quantities in Theorem 7.1. This is because there are good algorithms for finding eigenvalues and eigenvectors of the symmetric tridiagonal matrix T_k (see section 5.3) and because the residual norm is simply $\| T_{ku} \|_2 = \beta_k$. (From the Lanczos algorithm we know that β_k is nonnegative.) This simplifies the error bounds on the approximate eigenvalues and eigenvectors in the following theorem.

THEOREM 7.2. *Let T_k, T_{ku}, and Q_k be as in equation (7.1). Let $T_k = V \Lambda V^T$ be the eigendecomposition of T_k, where $V = [v_1, \dots, v_k]$ is orthogonal and $\Lambda = \mathrm{diag}(\theta_1, \dots, \theta_k)$. Then*

1. *There are k eigenvalues $\alpha_1, \ldots, \alpha_k$ of A (not necessarily the largest k) such that $|\theta_i - \alpha_i| \le \|T_{ku}\|_2$ for $i = 1, \ldots, k$. If Q_k is computed by the Lanczos algorithm, then $|\theta_i - \alpha_i| \le \|T_{ku}\|_2 = \beta_k$, where β_k is the single (possibly) nonzero entry in the upper right corner of T_{ku}.*

2. *$\|A(Q_k v_i) - (Q_k v_i)\theta_i\|_2 = \|T_{ku} v_i\|_2$. Thus, the difference between the Ritz value θ_i and some eigenvalue α of A is at most $\|T_{ku} v_i\|_2$, which may be much smaller than $\|T_{ku}\|_2$. If Q_k is computed by the Lanczos algorithm, then $\|T_{ku} v_i\|_2 = \beta_k |v_i(k)|$, where $v_i(k)$ is the kth (bottom) entry of v_i. This formula lets us compute the residual $\|A(Q_k v_i) - (Q_k v_i)\theta_i\|_2$ cheaply, i.e., without multiplying any vector by Q_k or by A.*

3. *Without any further information about the spectrum of T_u, we cannot deduce any useful error bound on the Ritz vector $Q_k v_i$. If we know that the gap between θ_i and any other eigenvalue of T_k or T_u is at least g, then we can bound the angle θ between $Q_k v_i$ and a true eigenvector of A by*

$$\frac{1}{2}\sin 2\theta \le \frac{\|T_{ku}\|_2}{g}. \tag{7.2}$$

If Q_k is computed by the Lanczos algorithm, then the bound simplifies to

$$\frac{1}{2}\sin 2\theta \le \frac{\beta_k}{g}.$$

Proof.

1. The eigenvalues of $\hat{T} = \begin{bmatrix} T_k & 0 \\ 0 & T_u \end{bmatrix}$ include θ_1 through θ_k. Since

$$\|\hat{T} - T\|_2 = \left\| \begin{bmatrix} 0 & T_{ku}^T \\ T_{ku} & 0 \end{bmatrix} \right\|_2 = \|T_{ku}\|_2,$$

Weyl's theorem, Theorem 5.1, tells us that the eigenvalues of $\hat{T}$ and T differ by at most $\|T_{ku}\|_2$. But the eigenvalues of T and A are identical, proving the result.

2. We compute

$$
\begin{aligned}
\|A(Q_k v_i) - (Q_k v_i)\theta_i\|_2 &= \|Q^T A(Q_k v_i) - Q^T(Q_k v_i)\theta_i\|_2 \\
&= \left\| \begin{bmatrix} T_k v_i \\ T_{ku} v_i \end{bmatrix} - \begin{bmatrix} v_i \theta_i \\ 0 \end{bmatrix} \right\|_2 = \left\| \begin{bmatrix} 0 \\ T_{ku} v_i \end{bmatrix} \right\|_2 \\
&\quad \text{since } T_k v_i = \theta_i v_i \\
&= \|T_{ku} v_i\|_2.
\end{aligned}
$$

Then by Theorem 5.5, A has some eigenvalue α satisfying $|\alpha - \theta_i| \le \|T_{ku} v_i\|_2$. If Q_k is computed by the Lanczos algorithm, then $\|T_{ku} v_i\|_2 = \beta_k |v_i(k)|$, because only the top right entry of T_{ku}, namely, β_k, is nonzero.

3. We reuse Example 5.4 to show that we cannot deduce a useful error bound on the Ritz vector without further information about the spectrum of T_u:

$$T = \begin{bmatrix} 1+g & \epsilon \\ \epsilon & 1 \end{bmatrix},$$

where $0 < \epsilon < g$. We let $k = 1$ and $Q_1 = [e_1]$, so $T_1 = 1 + g$ and the approximate eigenvector is simply e_1. But as shown in Example 5.4, the eigenvectors of T are close to $[1, \epsilon/g]^T$ and $[-\epsilon/g, 1]^T$. So without a lower bound on g, i.e., the gap between the eigenvalue of T_k and all the other eigenvalues, including those of T_u, we cannot bound the error in the computed eigenvector. If we do have such a lower bound, we can apply the second bound of Theorem 5.4 to T and $T + E = \mathrm{diag}(T_k, T_u)$ to derive equation (7.2). ⋄

7.3.   The Lanczos Algorithm in Exact Arithmetic

The Lanczos algorithm for finding eigenvalues of a symmetric matrix A combines the Lanczos algorithm for building a Krylov subspace (Algorithm 6.10) with the Rayleigh–Ritz procedure of the last section. In other words, it builds an orthogonal matrix $Q_k = [q_1, \ldots, q_k]$ of orthogonal Lanczos vectors and approximates the eigenvalues of A by the Ritz values (the eigenvalues of the symmetric tridiagonal matrix $T_k = Q_k^T A Q_k$), as in equation (7.1).

ALGORITHM 7.1. *Lanczos algorithm in exact arithmetic for finding eigenvalues and eigenvectors of $A = A^T$:*

> $q_1 = b/\|b\|_2,\ \beta_0 = 0,\ q_0 = 0$
> *for* $j = 1$ *to* k
> $z = Aq_j$
> $\alpha_j = q_j^T z$
> $z = z - \alpha_j q_j - \beta_{j-1} q_{j-1}$
> $\beta_j = \|z\|_2$
>   *if* $\beta_j = 0$, *quit*
> $q_{j+1} = z/\beta_j$
>   *Compute eigenvalues, eigenvectors, and error bounds of T_j*
> *end for*

In this section we explore the convergence of the Lanczos algorithm by describing a numerical example in some detail. This example has been chosen to illustrate both typical convergence behavior, as well as some more problematic behavior, which we call *misconvergence*. Misconvergence can occur because the starting vector q_1 is nearly orthogonal to the eigenvector of the desired eigenvalue or when there are multiple (or very close) eigenvalues.

The title of this section indicates that we have (nearly) eliminated the effects of roundoff error on our example. Of course, the Matlab code (HOME-PAGE/Matlab/LanczosFullReorthog.m) used to produce the example below ran in floating point arithmetic, but we implemented the Lanczos algorithm (in particular the inner loop of Algorithm 7.1) in a particularly careful and expensive way in order to make it mimic the exact result as closely as possible. This careful implementation is called *Lanczos with full reorthogonalization*, as indicated in the titles of the figures below.

In the next section we will explore the same numerical example using the original, inexpensive implementation of Algorithm 7.1, which we call *Lanczos with no reorthogonalization* in order to contrast it with *Lanczos with full reorthogonalization*. (We will also explain the difference in the two implementations.) We will see that the original Lanczos algorithm can behave significantly differently from the more expensive "exact" algorithm. Nevertheless, we will show how to use the less expensive algorithm to compute eigenvalues reliably.

EXAMPLE 7.1. We illustrate the Lanczos algorithm and its error bounds by running a large example, a 1000-by-1000 diagonal matrix A, most of whose eigenvalues were chosen randomly from a normal Gaussian distribution. Figure 7.1 is a plot of the eigenvalues. To make later plots easy to understand, we have also sorted the diagonal entries of A from largest to smallest, so $\lambda_i(A) = a_{ii}$, with corresponding eigenvector e_i, the ith column of the identity matrix. There are a few extreme eigenvalues, and the rest cluster near the center of the spectrum. The starting Lanczos vector q_1 has all equal entries, except for one, as described below.

There is no loss in generality in experimenting with a diagonal matrix, since running the Lanczos algorithm on A with starting vector q_1 is equivalent to running the Lanczos algorithm on $Q^T A Q$ with starting vector $Q^T q_1$ (see Question 7.1).

To illustrate convergence, we will use several plots of the sort shown in Figure 7.2. In this figure the eigenvalues of each T_k are shown plotted in column k, for $k = 1$ to 9 on the top, and for $k = 1$ to 29 on the bottom, with the eigenvalues of A plotted in an extra column at the right. Thus, column k has k pluses, one marking each eigenvalue of T_k. We have also color-coded the eigenvalues as follows: The largest and smallest eigenvalues of each T_k are shown in black, the second largest and second smallest eigenvalues are red, the third largest and third smallest eigenvalues are green, and the fourth largest and fourth smallest eigenvalues are blue. Then these colors recycle into the interior of the spectrum.

To understand convergence, consider the largest eigenvalue of each T_k; these black pluses are on the top of each column. Note that they increase monotonically as k increases; this is a consequence of the Cauchy interlace theorem, since T_k is a submatrix of T_{k+1} (see Question 5.4). In fact, the Cauchy interlace theorem tells us more, that the eigenvalues of T_k *interlace* those of T_{k+1},

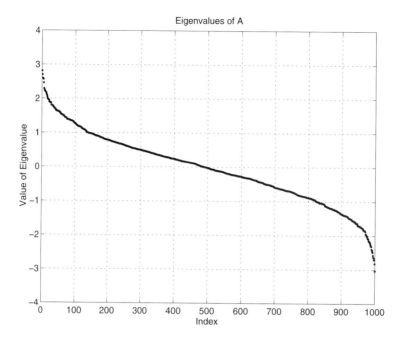

Fig. 7.1. *Eigenvalues of the diagonal matrix A.*

or that $\lambda_i(T_{k+1}) \geq \lambda_i(T_k) \geq \lambda_{i+1}(T_{k+1}) \geq \lambda_{i+1}(T_k)$. In other words, $\lambda_i(T_k)$ increases monotonically with k for any fixed i, not just $i = 1$ (the largest eigenvalue). This is illustrated by the colored sequences of pluses moving right and up in the figure.

A completely analogous phenomenon occurs with the smallest eigenvalues: The bottom black plus sign in each column of Figure 7.2 shows the smallest eigenvalue of each T_k, and these are monotonically decreasing as k increases. Similarly, the ith smallest eigenvalue is also monotonically decreasing. This is also a simple consequence of the Cauchy interlace theorem.

Now we can ask to which eigenvalue of A the eigenvalue $\lambda_i(T_k)$ can converge as k increases. Clearly the largest eigenvalue of T_k, $\lambda_1(T_k)$, ought to converge to the largest eigenvalue of A, $\lambda_1(A)$. Indeed, if the Lanczos algorithm proceeds to step $k = n$ (without quitting early because some $\beta_k = 0$), then T_n and A are similar, and so $\lambda_1(T_n) = \lambda_1(A)$. Similarly, the ith largest eigenvalue $\lambda_i(T_k)$ of T_k must increase monotonically and converge to the ith largest eigenvalue $\lambda_i(A)$ of A (provided that the Lanczos algorithm does not quit early). And the ith smallest eigenvalue $\lambda_{k+1-i}(T_k)$ of T_k must similarly decrease monotonically and converge to the ith smallest eigenvalue $\lambda_{n+1-i}(A)$ of A.

All these converging sequences are represented by sequences of pluses of a common color in Figure 7.2 and other figures in this section. Consider the bottom graph in Figure 7.2: For k larger than about 15, the topmost and bottom-most black pluses form horizontal rows next to the extreme eigenvalues of A, which are plotted in the rightmost column; this demonstrates conver-

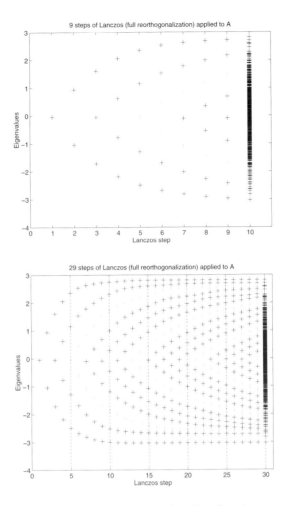

Fig. 7.2. *The Lanczos algorithm applied to A. The first 9 steps are shown on the top, and the first 29 steps are shown on the bottom. Column k shows the eigenvalues of T_k, except that the rightmost columns (column 10 on the top and column 30 on the bottom) show all the eigenvalues of A.*

gence. Similarly, the top sequence of red pluses forms a horizontal row next to the second largest eigenvalue of A in the rightmost column; they converge later than the outermost eigenvalues. A blow-up of this behavior for more Lanczos algorithm steps is shown in the top two graphs of Figure 7.3.

To summarize the above discussion, *extreme eigenvalues, i.e., the largest and smallest ones, converge first, and the interior eigenvalues converge last. Furthermore, convergence is monotonic, with the ith largest (smallest) eigenvalue of T_k increasing (decreasing) to the ith largest (smallest) eigenvalue of A, provided that the Lanczos algorithm does not stop prematurely with some $\beta_k = 0$.*

Now we examine the convergence behavior in more detail, compute the actual errors in the Ritz values, and compare these errors with the error bounds in part 2 of Theorem 7.2. We run the Lanczos algorithm for 99 steps on the same matrix pictured in Figure 7.2 and display the results in Figure 7.3. The top left graph in Figure 7.3 shows only the largest eigenvalues, and the top right graph shows only the smallest eigenvalues.

The middle two graphs in Figure 7.3 show the errors in the four largest computed eigenvalues (on the left) and the four smallest computed eigenvalues (on the right). The colors in the middle graphs match the colors in the top graphs. We measure and plot the errors in three ways:

- The *global errors* (the solid lines) are given by $|\lambda_i(T_k) - \lambda_i(A)|/|\lambda_i(A)|$. We divide by $|\lambda_i(A)|$ in order to normalize all the errors to lie between 1 (no accuracy) and about 10^{-16} (machine epsilon, or full accuracy). As k increases, the global error decreases monotonically, and we expect it to decrease to machine epsilon, unless the Lanczos algorithm quits prematurely.

- The *local errors* (the dotted lines) are given by $\min_j |\lambda_i(T_k) - \lambda_j(A)|/|\lambda_i(A)|$. The local error measures the smallest distance between $\lambda_i(T_k)$ and the *nearest* eigenvalue $\lambda_j(A)$ of A, not just the ultimate value $\lambda_i(A)$. We plot this because sometimes the local error is much smaller than the global error.

- The *error bounds* (the dashed lines) are the quantities $|\beta_k v_i(k)|/|\lambda_i(A)|$ computed by the algorithm (except for the normalization by $|\lambda_i(A)|$, which of course the algorithm does not know!).

The bottom two graphs in Figure 7.3 show the eigenvector components of the Lanczos vectors q_k for the four eigenvectors corresponding to the four largest eigenvalues (on the left) and for the four eigenvectors corresponding to the four smallest eigenvalues (on the right). In other words, they plot $q_k^T e_j = q_k(j)$, where e_j is the jth eigenvector of the diagonal matrix A, for $k = 1$ to 99 and for $j = 1$ to 4 (on the left) and $j = 997$ to 1000 (on the right). The components are plotted on a logarithmic scale, with "+" and "o" to indicate whether the component is positive or negative, respectively. We use these plots to help explain convergence below.

Now we use Figure 7.3 to examine convergence in more detail. The largest eigenvalue of T_k (topmost black pluses in the top left graph of Figure 7.3) begins converging to its final value (about 2.81) right away, is correct to six decimal places after 25 Lanczos steps, and is correct to machine precision by step 50. The global error is shown by the solid black line in the middle left graph. The local error (the dotted black line) is the same as the global error after not too many steps, although it can be "accidentally" much smaller if an eigenvalue $\lambda_i(T_k)$ happens to fall close to some other $\lambda_j(A)$ on its way to $\lambda_i(A)$. The dashed black line in the same graph is the relative error bound computed by the algorithm, which overestimates the true error up to about

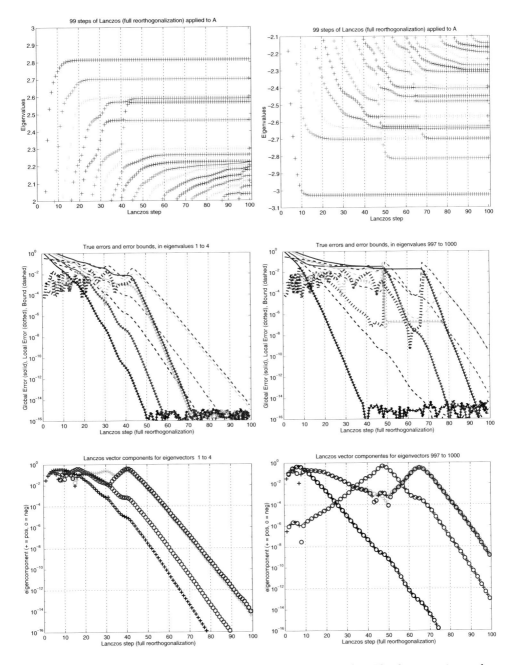

Fig. 7.3. 99 *steps of the Lanczos algorithm applied to A. The largest eigenvalues are shown on the left, and the smallest on the right. The top two graphs show the eigenvalues themselves, the middle two graphs the errors (global = solid, local = dotted, bounds = dashed), and the bottom two graphs show eigencomponents of Lanczos vectors. The colors in a column of three graphs match.*

step 75. Still, the relative error bound correctly indicates that the largest eigenvalue is correct to several decimal digits.

The second through fourth largest eigenvalues (the topmost red, green and blue pluses in the top left graph of Figure 7.3) converge in a similar fashion, with eigenvalue i converging slightly faster than eigenvalue $i+1$. This is typical behavior of the Lanczos algorithm.

The bottom left graph of Figure 7.3 measures convergence in terms of the eigenvector components $q_k^T e_j$. To explain this graph, consider what happens to the Lanczos vectors q_k as the first eigenvalue converges. Convergence means that the corresponding eigenvector e_1 nearly lies in the Krylov subspace spanned by the Lanczos vectors. In particular, since the first eigenvalue has converged after $k = 50$ Lanczos steps, this means that e_1 must very nearly be a linear combination of q_1 through q_{50}. Since the q_k are mutually orthogonal, this means q_k must also be orthogonal to e_1 for $k > 50$. This is borne out by the black curve in the bottom left graph, which has decreased to less than 10^{-7} by step 50. The red curve is the component of e_2 in q_k, and this reaches 10^{-8} by step 60. The green curve (third eigencomponent) and blue curve (fourth eigencomponent) get comparably small a few steps later.

Now we discuss the smallest four eigenvalues, whose behavior is described by the three graphs on the right of Figure 7.3. We have chosen the matrix A and starting vector q_1 to illustrate certain difficulties that can arise in the convergence of the Lanczos algorithm to show that convergence is not always as straightforward as in the case of the four eigenvalues just examined.

In particular, we have chosen $q_1(999)$, the eigencomponent of q_1 in the direction of the second smallest eigenvalue (-2.81), to be about 10^{-7}, which is 10^5 times smaller than all the other components of q_1, which are equal. Also, we have chosen the third and fourth smallest eigenvalues (numbers 998 and 997) to be nearly the same: -2.700001 and -2.7.

The convergence of the smallest eigenvalue of T_k to $\lambda_{1000}(A) \approx -3.03$ is uneventful, similar to the largest eigenvalues. It is correct to 16 digits by step 40.

The *second* smallest eigenvalue of T_k, shown in red, begins by *misconverging* to the *third* smallest eigenvalue of A, near -2.7. Indeed, the dotted red line in the middle right graph of Figure 7.3 shows that $\lambda_{999}(T_k)$ agrees with $\lambda_{998}(A)$ to six decimal places for Lanczos steps $40 < k < 50$. The corresponding error bound (the red dashed line) tells us that $\lambda_{999}(T_k)$ equals *some* eigenvalue of A to three or four decimal places for the same values of k. The reason $\lambda_{999}(T_k)$ misconverges is that the Krylov subspace starts with a very small component of the corresponding Krylov subspace e_{999}, namely, 10^{-7}. This can be seen by the red curve in bottom right graph, which starts at 10^{-7} and takes until step 45 before a large component of e_{999} appears. Only at this point, when the Krylov subspace contains a sufficiently large component of the eigenvector e_{999}, can $\lambda_{999}(T_k)$ start converging again to its final value $\lambda_{999}(A) \approx -2.81$, as shown in the top and middle right graphs. Once this convergence has set in again,

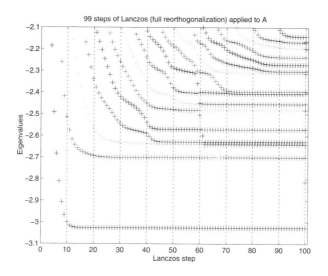

Fig. 7.4. *The Lanczos algorithm applied to* A, *where the starting vector* q_1 *is orthogonal to the eigenvector corresponding to the second smallest eigenvalue* -2.81. *No approximation to this eigenvalue is computed.*

the component of e_{999} starts decreasing again and becomes very small once $\lambda_{999}(T_k)$ has converged to $\lambda_{999}(A)$ sufficiently accurately. (For a quantitative relationship between the convergence rate and the eigencomponent $q_1^T e_{999}$, see the theorem of Kaniel and Saad discussed below.)

Indeed, if q_1 were *exactly* orthogonal to e_{999}, so $q_1^T e_{999} = 0$ rather than just $q_1^T e_{999} = 10^{-7}$, then all later Lanczos vectors would also be orthogonal to e_{999}. This means $\lambda_{999}(T_k)$ would never converge to $\lambda_{999}(A)$. (For a proof, see Question 7.3.) We illustrate this in Figure 7.4, where we have modified q_1 just slightly so that $q_1^T e_{999} = 0$. Note that no approximation to $\lambda_{999}(A) \approx -2.81$ ever appears.

Fortunately, if we choose q_1 at random, it is extremely unlikely to be orthogonal to an eigenvector. We can always rerun the Lanczos algorithm with a different random q_1 to provide more "statistical" evidence that we have not missed any eigenvalues.

Another source of "misconvergence" are (nearly) multiple eigenvalues, such as the the third smallest eigenvalue $\lambda_{998}(A) = -2.700001$ and the fourth smallest eigenvalue $\lambda_{997}(A) = -2.7$. By examining $\lambda_{998}(T_k)$, the bottommost green curve in the top right and middle right graphs of Figure 7.3, we see that during Lanczos steps $50 < k < 75$, $\lambda_{998}(T_k)$ *misconverges* to about -2.7000005, *halfway* between the two closest eigenvalues of A. This is not visible at the resolution provided by the top right graph but is evident from the horizontal segment of the solid green line in the middle right graph during Lanczos steps $50 < k < 75$. At step 76 rapid convergence to the final value $\lambda_{998}(A) = -2.700001$ sets in again.

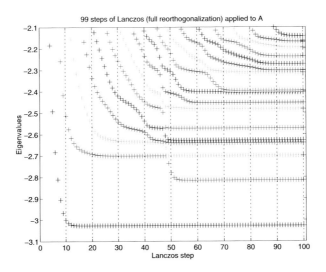

Fig. 7.5. *The Lanczos algorithm applied to A, where the third and fourth smallest eigenvalues are equal. Only one approximation to this double eigenvalue is computed.*

Meanwhile, the fourth smallest eigenvalue $\lambda_{997}(T_k)$, shown in blue, has misconverged to a value near $\lambda_{996}(A) \approx -2.64$; the blue dotted line in the middle right graph indicates that $\lambda_{997}(T_k)$ and $\lambda_{996}(A)$ agree to up to nine decimal places near step $k = 61$. At step $k = 65$ rapid convergence sets in again to the final value $\lambda_{997}(A) = -2.7$. This can also be seen in the bottom right graph, where the eigenvector components of e_{997} and e_{998} grow again during step $50 < k < 65$, after which rapid convergence sets in and they again decrease.

Indeed, if $\lambda_{997}(A)$ were *exactly* a double eigenvalue, we claim that T_k would never have two eigenvalues near that value but only one (in exact arithmetic). (For a proof, see Question 7.3.) We illustrate this in Figure 7.5, where we have modified A just slightly so that it has two eigenvalues exactly equal to -2.7. Note that only one approximation to $\lambda_{998}(A) = \lambda_{997}(A) = -2.7$ ever appears.

Fortunately, there are many applications where it is sufficient to find one copy of each eigenvalue rather than all multiple copies. Also, it is possible to use "block Lanczos" to recover multiple eigenvalues (see the algorithms cited in section 7.6).

Examining other eigenvalues in the top right graph of Figure 7.3, we see that misconvergence is quite common, as indicated by the frequent short horizontal segments of like-colored pluses, which then drop off to the right to the next smaller eigenvalue. For example, the seventh smallest eigenvalue is well-approximated by the fifth (black), sixth (red), and seventh (green) smallest eigenvalues of T_k at various Lanczos steps.

These misconvergence phenomena explain why the computable error bound provided by part 2 of Theorem 7.2 is essential to monitor convergence [198]. If the error bound is small, the computed eigenvalue is indeed a good approximation to some eigenvalue, even if one is "missing." ⋄

There is another error bound, due to Kaniel and Saad, that sheds light on why misconvergence occurs. This error bound depends on the angle between the starting vector q_1 and the desired eigenvectors, the Ritz values, and the desired eigenvalues. In other words, it depends on quantities unknown during the computation, so it is not of practical use. But it shows that if q_1 is nearly orthogonal to the desired eigenvector, or if the desired eigenvalue is nearly multiple, then we can expect slow convergence. See [197, sect. 12-4] for details.

7.4. The Lanczos Algorithm in Floating Point Arithmetic

The example in the last section described the behavior of the "ideal" Lanczos algorithm, essentially without roundoff. We call the corresponding careful but expensive implementation of Algorithm 6.10 *Lanczos with full reorthogonalization* to contrast it with the original inexpensive implementation, which we call *Lanczos with no reorthogonalization* (HOMEPAGE/Matlab/LanczosNoReorthog.m). Both algorithms are shown below.

ALGORITHM 7.2. *Lanczos algorithm with full or no reorthogonalization for finding eigenvalues and eigenvectors of $A = A^T$:*

$q_1 = b/\|b\|_2$, $\beta_0 = 0$, $q_0 = 0$
for $j = 1$ *to* k
    $z = Aq_j$
    $\alpha_j = q_j^T z$
    $\begin{cases} z = z - \sum_{i=1}^{j-1}(z^T q_i)q_i, \quad z = z - \sum_{i=1}^{j-1}(z^T q_i)q_i & \textit{full reorthogonalization} \\ z = z - \alpha_j q_j - \beta_{j-1} q_{j-1} & \textit{no reorthogonalization} \end{cases}$
    $\beta_j = \|z\|_2$
    if $\beta_j = 0$, *quit*
    $q_{j+1} = z/\beta_j$
    Compute eigenvalues, eigenvectors, and error bounds of T_k
end for

Full reorthogonalization corresponds to applying the Gram–Schmidt orthogonalization process "$z = z - \sum_{i=1}^{j-1}(z^T q_i)q_i$" *twice* in order to almost surely make z orthogonal to q_1 through q_{j-1}. (See Algorithm 3.1 as well as [197, sect. 6-9] and [171, chap. 7] for discussions of when "twice is enough.") In exact arithmetic, we showed in section 6.6.1 that z is orthogonal to q_1 through q_{j-1} without reorthogonalization. Unfortunately, we will see that roundoff destroys this orthogonality property, upon which all of our analysis has depended so far.

This loss of orthogonality does not cause the algorithm to behave completely unpredictably. Indeed, we will see that the price we pay is to get *multiple copies* of converged Ritz values. In other words, instead of T_k having one eigenvalue nearly equal to $\lambda_i(A)$ for k large, it may have many eigenvalues nearly equal to $\lambda_i(A)$. This is not a disaster if one is not concerned about computing multiplicities of eigenvalues and does not mind the resulting delayed convergence of interior eigenvalues. See [57] for a detailed description of a Lanczos implementation that operates in this fashion, and NETLIB/lanczos for the software itself. (This software has heuristics for estimating multiplicities of eigenvalues.)

But if accurate multiplicities are important, then one needs to keep the Lanczos vectors (nearly) orthogonal. So one could use the Lanczos algorithm with full reorthogonalization, as we did in the last section. But one can easily confirm that this costs $O(k^2 n)$ flops instead of $O(kn)$ flops for k steps, and $O(kn)$ space instead of $O(n)$ space, which may be too high a price to pay.

Fortunately, there is a middle ground between no reorthogonalization and full reorthogonalization, which nearly gets the best of both worlds. It turns out that the q_k lose their orthogonality in a very systematic way by developing large components in the directions of already converged Ritz vectors. (This is what leads to multiple copies of converged Ritz values.) This systematic loss of orthogonality is illustrated by the next example and explained by Paige's theorem below. We will see that by monitoring the computed error bounds, we can conservatively predict which q_k will have large components of which Ritz vectors. Then we can *selectively orthogonalize* q_k against just those few prior Ritz vectors, rather than against all the earlier q_is at each step, as with full reorthogonalization. This keeps the Lanczos vectors (nearly) orthogonal for very little extra work. The next section discusses selective orthogonalization in detail.

EXAMPLE 7.2. Figure 7.7 shows the convergence behavior of 149 steps of Lanczos on the matrix in Example 7.1. The graphs on the right are with full reorthogonalization, and the graphs on the left are with no reorthogonalization. These graphs are similar to those in Figure 7.3, except that the global error is omitted, since this clutters the middle graphs.

Figure 7.6 plots the smallest singular value $\sigma_{\min}(Q_k)$ versus Lanczos step k. In exact arithmetic, Q_k is orthogonal and so $\sigma_{\min}(Q_k) = 1$. With roundoff, Q_k loses orthogonality starting at around step $k = 70$, and $\sigma_{\min}(Q_k)$ drops to .01 by step $k = 80$, which is where the top two graphs in Figure 7.7 begin to diverge visually.

In particular, starting at step $k = 80$ in the top left graph of Figure 7.7, the second smallest (red) eigenvalue $\lambda_2(T_k)$, which had converged to $\lambda_2(A) \approx 2.7$ to almost 16 digits, leaps up to $\lambda_1(A) \approx 2.81$ in just a few steps, yielding a "second copy" of $\lambda_1(A)$ along with $\lambda_1(T_k)$ (in black). (This may be hard to see, since the red pluses overwrite and so obscure the black pluses.) This transition can be seen in the leap in the dashed red error bound in the middle left graph. Also, this transition was "foreshadowed" by the increasing component of e_1

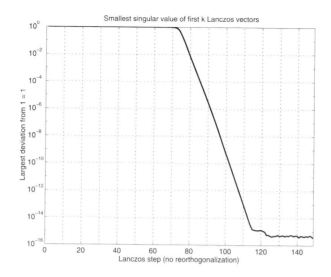

Fig. 7.6. *Lanczos algorithm without reorthogonalization applied to A. The smallest singular value $\sigma_{\min}(Q_k)$ of the Lanczos vector matrix Q_k is shown for $k = 1$ to 149. In the absence of roundoff, Q_k is orthogonal, and so all singular values should be one. With roundoff, Q_k becomes rank deficient.*

in the bottom left graph, where the black curve starts rising again at step $k = 50$ rather than continuing to decrease to machine epsilon, as it does with full reorthogonalization in the bottom right graph. Both of these indicate that the algorithm is diverging from its exact path (and that some selective orthogonalization is called for). After the second copy of $\lambda_1(A)$ has converged, the component of e_1 in the Lanczos vectors starts dropping again, starting a little after step $k = 80$.

Similarly, starting at about step $k = 95$, a second copy of $\lambda_2(A)$ appears when the blue curve ($\lambda_4(T_k)$) in the upper left graph moves from about $\lambda_3(A) \approx 2.6$ to $\lambda_2(A) \approx 2.7$. At this point we have two copies of $\lambda_1(A) \approx 2.81$ and two copies of $\lambda_2(A)$. This is a bit hard to see on the graphs, since the pluses of one color obscure the pluses of the other color (red overwrites black, and blue overwrites green). This transition is indicated by the dashed blue error bound for $\lambda_4(T_k)$ in the middle left graph rising sharply near $k = 95$ and is foreshadowed by the rising red curve in the bottom left graph, which indicates that the component of e_2 in the Lanczos vectors is rising. This component peaks near $k = 95$ and starts dropping again.

Finally, around step $k = 145$, a *third* copy of $\lambda_1(A)$ appears, again indicated and foreshadowed by changes in the two bottom left graphs. If we were to continue the Lanczos process, we would periodically get additional copies of many other converged Ritz values. $\diamond$

The next theorem provides an explanation for the behavior seen in the above example, and hints at a practical criterion for selectively orthogonalizing Lanczos vectors. In order not to be overwhelmed by taking all possible roundoff

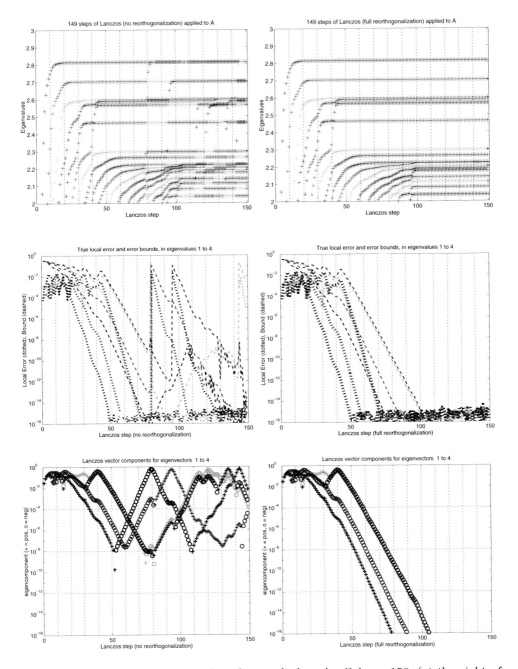

Fig. 7.7. 149 *steps of Lanczos algorithm applied to A. Column* 150 *(at the right of the top graphs) shows the eigenvalues of A. In the left graphs, no reorthogonalization is done. In the right graphs, full reorthogonalization is done.*

errors into account, we will draw on others' experience to identify those few rounding errors that are important, and simply ignore the rest [197, sect. 13-4]. This lets us summarize the Lanczos algorithm with no reorthogonalization in one line:

$$\beta_j q_{j+1} + f_j = Aq_j - \alpha_j q_j - \beta_{j-1} q_{j-1}. \tag{7.3}$$

In this equation the variables represent the values actually stored in the machine, except for f_j, which represents the roundoff error incurred by evaluating the right-hand side and then computing β_j and q_{j+1}. The norm $\|f_j\|_2$ is bounded by $O(\varepsilon\|A\|)$, where ε is machine epsilon, which is all we need to know about f_j. In addition, we will write $T_k = V\Lambda V^T$ *exactly*, since we know that the roundoff errors occurring in this eigendecomposition are not important. Thus, Q_k is not necessarily an orthogonal matrix, but V is.

THEOREM 7.3. *Paige. We use the notation and assumptions of the last paragraph. We also let* $Q_k = [q_1, \ldots, q_k]$, $V = [v_1, \ldots, v_k]$, *and* $\Lambda = \mathrm{diag}(\theta_1, \ldots, \theta_k)$. *We continue to call the columns* $y_{k,i} = Q_k v_i$ *of* $Q_k V$ *the Ritz vectors and the* θ_i *the Ritz values. Then*

$$y_{k,i}^T q_{k+1} = \frac{O(\varepsilon\|A\|)}{\beta_k |v_i(k)|}.$$

In other words the component $y_{k,i}^T q_{k+1}$ of the computed Lanczos vector q_{k+1} in the direction of the Ritz vector $y_{k,i} = Q_k v_i$ is proportional to the reciprocal of $\beta_k |v_i(k)|$, which is the error bound on the corresponding Ritz value θ_i (see Part 2 of Theorem 7.2). Thus, when the Ritz value θ_i converges and its error bound $\beta_k |v_i(k)|$ goes to zero, the Lanczos vector q_{k+1} acquires a large component in the direction of Ritz vector $y_{k,i}$. Thus, the Ritz vectors become linearly dependent, as seen in Example 7.2. Indeed, Figure 7.8 plots both the error bound $|\beta_k v_i(k)|/|\lambda_i(A)| \approx |\beta_k v_i(k)|/\|A\|$ and the Ritz vector component $y_{k,i}^T q_{k+1}$ for the largest Ritz value ($i = 1$, the top graph) and for the second largest Ritz value ($i = 2$, the bottom graph) of our 1000-by-1000 diagonal example. According to Paige's theorem, the product of these two quantities should be $O(\varepsilon)$. Indeed it is, as can be seen by the symmetry of the curves about the middle line $\sqrt{\varepsilon}$ of these semilogarithmic graphs.

Proof of Paige's theorem. We start with equation (7.3) for $j = 1$ to $j = k$, and write these k equations as the single equation

$$\begin{aligned} AQ_k &= Q_k T_k + [0, \ldots, 0, \beta_k q_{k+1}] + F_k \\ &= Q_k T_k + \beta_k q_{k+1} e_k^T + F_k, \end{aligned}$$

where e_k^T is the k-dimensional row vector $[0, \ldots, 0, 1]$ and $F_k = [f_1, \ldots, f_k]$ is the matrix of roundoff errors. We simplify notation by dropping the subscript k to get $AQ = QT + \beta q e^T + F$. Multiply on the left by Q^T to get $Q^T AQ =$

$Q^T Q T + \beta Q^T q e^T + Q^T F$. Since $Q^T A Q$ is symmetric, we get that $Q^T Q T + \beta Q^T q e^T + Q^T F$ equals its transpose or, rearranging this equality,

$$0 = (Q^T Q T - T Q^T Q) + \beta(Q^T q e^T - e q^T Q) + (Q^T F - F^T Q). \qquad (7.4)$$

If θ and v are a Ritz value and Ritz vector, respectively, so that $Tv = \theta v$, then note that

$$v^T \beta(e q^T Q)v = [\beta v(k)] \cdot [q^T(Qv)] \qquad (7.5)$$

is the product of error bound $\beta v(k)$ and the Ritz vector component $q^T(Qv) = q^T y$, which Paige's theorem says should be $O(\varepsilon \|A\|)$. Our goal is now to manipulate equation (7.4) to get an expression for $e q^T Q$ alone, and then use equation (7.5).

To this end, we now invoke more simplifying assumptions about roundoff: Since each column of Q is gotten by dividing a vector z by its norm, the diagonal of $Q^T Q$ is equal to 1 to full machine precision; we will suppose that it is exactly 1. Furthermore, the vector $z' = z - \alpha_j q_j = z - (q_j^T z) q_j$ computed by the Lanczos algorithm is constructed to be orthogonal to q_j, so it is also true that q_{j+1} and q_j are orthogonal to nearly full machine precision. Thus $q_{j+1}^T q_j = (Q^T Q)_{j+1,j} = O(\varepsilon)$; we will simply assume $(Q^T Q)_{j+1,j} = 0$. Now write $Q^T Q = I + C + C^T$, where C is lower triangular. Because of our assumptions about roundoff, C is in fact nonzero only on the second subdiagonal and below. This means

$$Q^T Q T - T Q^T Q = (CT - TC) + (C^T T - TC^T),$$

where we can use the zero structures of C and T to easily show that $CT - TC$ is strictly lower triangular and $C^T T - TC^T$ is strictly upper triangular. Also, since e is nonzero only in its last entry, $e q^T Q$ is nonzero only in the last row. Furthermore, the structure of $Q^T Q$ just described implies that the last entry of the last row of $e q^T Q$ is zero. So in particular, $e q^T Q$ is also strictly lower triangular and $Q^T q e^T$ is strictly upper triangular. Applying the fact that $e q^T Q$ and $CT - TC$ are both strictly lower triangular to equation (7.4) yields

$$0 = (CT - TC) - \beta e q^T Q + L, \qquad (7.6)$$

where L is the strict lower triangle of $Q^T F - F^T Q$. Multiplying equation (7.6) on the left by v^T and on the right by v, using equation (7.5) and the fact that $v^T(CT - TC)v = v^T Cv\theta - \theta v^T Cv = 0$, yields

$$v^T \beta(e q^T Q)v = [\beta v(k)] \cdot [q^T(Qv)] = v^T L v.$$

Since $|v^T L v| \leq \|L\| = O(\|Q^T F - F^T Q\|) = O(\|F\|) = O(\varepsilon \|A\|)$, we get

$$[\beta v(k)] \cdot [q^T(Qv)] = O(\varepsilon \|A\|),$$

which is equivalent to Paige's theorem. $\quad\square$

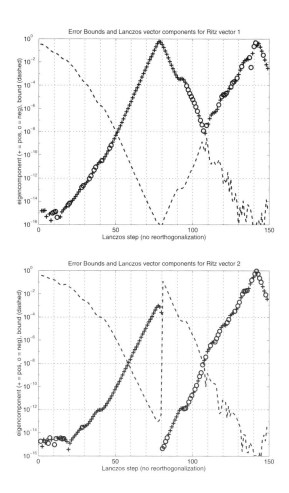

Fig. 7.8. *Lanczos with no reorthogonalization applied to A. The first* 149 *steps are shown for the largest eigenvalue (in black, at top) and for the second largest eigenvalue (in red, at bottom). The dashed lines are error bounds as before. The lines marked by pluses and o's show* $y_{k,i}^T q_{k+1}$, *the component of Lancos vector* $k+1$ *in the direction of the Ritz vector for the largest Ritz value* $(i = 1,$ *at top) or for the second largest Ritz value* $(i = 2,$ *at bottom).*

7.5.   The Lanczos Algorithm with Selective Orthogonalization

We discuss a variation of the Lanczos algorithm which has (nearly) the high accuracy of the Lanczos algorithm with full reorthogonalization but (nearly) the low cost of the Lanczos algorithm with no reorthogonalization. This algorithm is called the Lanczos algorithm with *selective orthogonalization*. As discussed in the last section, our goal is to keep the computed Lanczos vectors q_k as nearly orthogonal as possible (for high accuracy) by orthogonalizing them against as few other vectors as possible at each step (for low cost). Paige's theorem (Theorem 7.3 in the last section) tells us that the q_k lose orthogonality because they acquire large components in the direction of Ritz vectors $y_{i,k} = Q_k v_i$ whose Ritz values θ_i have converged, as measured by the error bound $\beta_k |v_i(k)|$ becoming small. This phenomenon was illustrated in Example 7.2.

Thus, the simplest version of selective orthogonalization simply monitors the error bound $\beta_k |v_i(k)|$ at each step, and when it becomes small enough, the vector z in the inner loop of the Lanczos algorithm is orthogonalized against $y_{i,k}$: $z = z - (y_{i,k}^T z) y_{i,k}$. We consider $\beta_k |v_i(k)|$ to be small when it is less than $\sqrt{\varepsilon} \|A\|$, since Paige's theorem tells us that the vector component $|y_{i,k}^T q_{k+1}| = |y_{i,k}^T z / \|z\|_2|$ is then likely to exceed $\sqrt{\varepsilon}$. (In practice we may replace $\|A\|$ by $\|T_k\|$, since $\|T_k\|$ is known and $\|A\|$ may not be.) This leads to the following algorithm.

ALGORITHM 7.3. *The Lanczos algorithm with selective orthogonalization for finding eigenvalues and eigenvectors of $A = A^T$:*

> $q_1 = b/\|b\|_2$, $\beta_0 = 0$, $q_0 = 0$
> for $j = 1$ to k
> $\qquad z = A q_j$
> $\qquad \alpha_j = q_j^T z$
> $\qquad z = z - \alpha_j q_j - \beta_{j-1} q_{j-1}$
> $\qquad$ /* Selectively orthogonalize against converged Ritz vectors */
> $\qquad$ for all $i \leq k$ such that $\beta_k |v_i(k)| \leq \sqrt{\varepsilon} \|T_k\|$
> $\qquad\qquad z = z - (y_{i,k}^T z) y_{i,k}$
> $\qquad$ end for
> $\qquad \beta_j = \|z\|_2$
> $\qquad$ if $\beta_j = 0$, quit
> $\qquad q_{j+1} = z/\beta_j$
> $\qquad$ Compute eigenvalues, eigenvectors, and error bounds of T_k
> end for

The following example shows what will happen to our earlier 1000-by-1000 diagonal matrix when this algorithm is used (HOMEPAGE/Matlab/LanczosSelectOrthog.m).

EXAMPLE 7.3. The behavior of the Lanczos algorithm with selective orthogonalization is visually indistinguishable from the behavior of the Lanczos algorithm with full orthogonalization shown in the three graphs on the right of Figure 7.7. In other words, selective orthogonalization provided as much accuracy as full orthogonalization.

The smallest singular values of all the Q_k were greater than $1 - 10^{-8}$, which means that selective orthogonalization did keep the Lanczos vectors orthogonal to about half precision, as desired.

Figure 7.9 shows the Ritz values corresponding to the Ritz vectors selected for reorthogonalization. Since the selected Ritz vectors correspond to converged Ritz values and the largest and smallest Ritz values converge first, there are two graphs: the large converged Ritz values are at the top, and the small converged Ritz values are at the bottom. The top graph matches the Ritz values shown in the upper right graph in Figure 7.7 that have converged to at least half precision. All together, 1485 Ritz vectors were selected for orthogonalization of a total possible $149*150/2 = 11175$. Thus, selective orthogonalization did only $1485/11175 \approx 13\%$ as much work reorthogonalizing to keep the Lanczos vectors (nearly) orthogonal as full reorthogonalization.

Figure 7.10 shows how the Lanczos algorithm with selective reorthogonalization keeps the Lanczos vectors orthogonal just to the Ritz vectors for the largest two Ritz values. The graph at the top is a superposition of the two graphs in Figure 7.8, which show the error bounds and Ritz vectors components for the Lanczos algorithm with no reorthogonalization. The graph at the bottom is the corresponding graph for the Lanczos algorithm with selective orthogonalization. Note that at step $k = 50$, the error bound for the largest eigenvalue (the dashed black line) has reached the threshold of $\sqrt{\varepsilon}$. The Ritz vector is selected for orthogonalization (as shown by the top black pluses in the top of Figure 7.9), and the component in this Ritz vector direction disappears from the bottom graph of Figure 7.10. A few steps later, at $k = 58$, the error bound for the second largest Ritz value reaches $\sqrt{\varepsilon}$, and it too is selected for orthogonalization. The error bounds in the bottom graph continue to decrease to machine epsilon ε and stay there, whereas the error bounds in the top graph eventually grow again. ◇

7.6. Beyond Selective Orthogonalization

Selective orthogonalization is not the end of the story, because the symmetric Lanczos algorithm can be made even less expensive. It turns out that once a Lanczos vector has been orthogonalized against a particular Ritz vector y, it takes many steps before the Lanczos vector again requires orthogonalization against y. So much of the orthogonalization work in Algorithm 7.3 can be eliminated. Indeed, there is a simple and inexpensive recurrence for deciding when to reorthogonalize [224, 192]. Another enhancement is to use the error bounds to efficiently distinguish between converged and "misconverged" eigen-

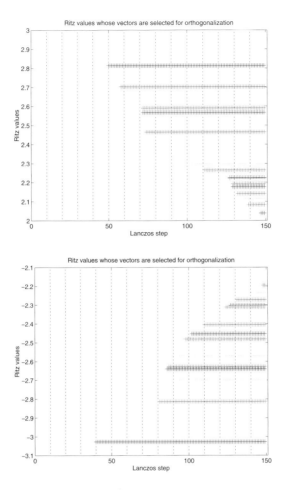

Fig. 7.9. *The Lanczos algorithm with selective orthogonalization applied to A. The Ritz values whose Ritz vectors are selected for orthogonalization are shown.*

values [198]. A state-of-the-art implementation of the Lanczos algorithm is described in [125]. A different software implementation is available in ARPACK (NETLIB/scalapack/readme.arpack [171, 233]).

If we apply the Lanczos algorithm to the shifted and inverted matrix $(A - \sigma I)^{-1}$, then we expect the eigenvalues closest to σ to converge first. There are other methods to "precondition" a matrix A to converge to certain eigenvalues more quickly. For example, Davidson's method [60] is used in quantum chemistry problems, where A is strongly diagonally dominant. It is also possible to combine Davidson's method with Jacobi's method [229].

7.7.   Iterative Algorithms for the Nonsymmetric Eigenproblem

When A is nonsymmetric, the Lanczos algorithm described above is no longer applicable. There are two alternatives.

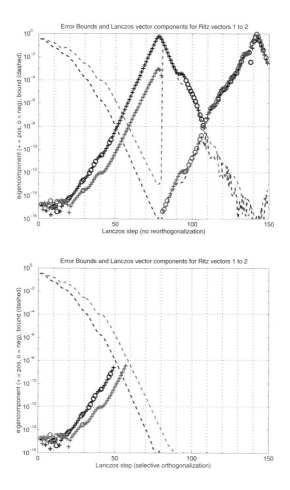

Fig. 7.10. *The Lanczos algorithm with selective orthogonalization applied to A. The top graph shows the first 149 steps of the Lanczos algorithm with no reorthogonalization, and the bottom graph shows the Lanczos algorithm with selective orthogonalization. The largest eigenvalue is shown in black, and the second largest eigenvalue is shown in red. The dashed lines are error bounds as before. The lines marked by pluses and o's show $y_{k,i}^T q_{k+1}$, the component of Lancos vector $k+1$ in the direction of the Ritz vector for the largest Ritz value ($i = 1$, in black) or for the second largest Ritz value ($i = 2$, in red). Note that selective orthogonalization eliminates these components after the first selective orthogonalizations at steps 50 ($i = 1$) and 58 ($i = 2$).*

The first alternative is to use the *Arnoldi algorithm* (Algorithm 6.9). Recall that the Arnoldi algorithm computes an orthogonal basis Q_k of a Krylov subspace $\mathcal{K}_k(A, q_1)$ such that $Q_k^T A Q_k = H_k$ is upper Hessenberg rather than symmetric tridiagonal. The analogue of the Rayleigh–Ritz procedure is again to approximate the eigenvalues of A by the eigenvalues of H_k. Since A is nonsymmetric, its eigenvalues may be complex and/or badly conditioned, so many of the attractive error bounds and monotonic convergence properties enjoyed by the Lanczos algorithm and described in section 7.3 no longer hold. Nonetheless, effective algorithms and implementations exist. Good references include [154, 171, 212, 216, 217, 233] and the book [213]. The latest software is described in [171, 233] and may be found in NETLIB/scalapack/readme.arpack. The Matlab command `eigs` (for "sparse eigenvalues") uses this software.

A second alternative is to use the *nonsymmetric Lanczos algorithm*. This algorithm attempts to reduce A to nonsymmetric tridiagonal form by a nonorthogonal similarity. The hope is that it will be easier to find the eigenvalues of a (sparse!) nonsymmetric tridiagonal matrix than the Hessenberg matrix produced by the Arnoldi algorithm. Unfortunately, the similarity transformations can be quite ill-conditioned, which means that the eigenvalues of the tridiagonal and of the original matrix may greatly differ. In fact, it is not always possible to find an appropriate similarity because of a phenomenon known as "breakdown" [42, 134, 135, 199]. Attempts to repair breakdown by a process called "look-ahead" have been proposed, implemented, and analyzed in [16, 18, 55, 56, 64, 108, 202, 265, 266].

Finally, it is possible to apply subspace iteration (Algorithm 4.3) [19], Davidson's algorithm [216], or the Jacobi–Davidson algorithm [230] to the sparse nonsymmetric eigenproblem.

7.8. References and Other Topics for Chapter 7

In addition to the references in sections 7.6 and 7.7, there are a number of good surveys available on algorithms for sparse eigenvalues problems: see [17, 51, 125, 163, 197, 213, 262]. Parallel implementations are also discussed in [76].

In section 6.2 we discussed the existence of on-line help to choose from among the variety of iterative methods available for solving $Ax = b$. A similar project is underway for eigenproblems and will be incorporated in a future edition of this book.

7.9. Questions for Chapter 7

QUESTION 7.1. *(Easy)* Confirm that running the Arnoldi algorithm (Algorithm 6.9) or the Lanczos algorithm (Algorithm 6.10) on A with starting vector q yields the identical tridiagonal matrices T_k (or Hessenberg matrices H_k) as running on $Q^T A Q$ with starting vector $Q^T q$.

QUESTION 7.2. *(Medium)* Let λ_i be a simple eigenvalue of A. Confirm that if q_1 is orthogonal to the corresponding eigenvector of A, then the eigenvalues of the tridiagonal matrices T_k computed by the Lanczos algorithm in exact arithmetic cannot converge to λ_i in the sense that the largest T_k computed cannot have λ_i as an eigenvalue. Show, by means of a 3-by-3 example, that an eigenvalue of some other T_k can equal λ_i "accidentally."

QUESTION 7.3. *(Medium)* Confirm that no symmetric tridiagonal matrix T_k computed by the Lanczos algorithm can have an exactly multiple eigenvalue. Show that if A has a multiple eigenvalue, then the Lanczos algorithm applied to A must break down before the last step.

Bibliography

[1] R. Agarwal, F. Gustavson, and M. Zubair. Exploiting functional parallelism of POWER2 to design high performance numerical algorithms. *IBM J. Res. Development*, 38:563–576, 1994.

[2] L. Ahlfors. *Complex Analysis*. McGraw-Hill, New York, 1966.

[3] A. Aho, J. Hopcroft, and J. Ullman. *The Design and Analysis of Computer Algorithms*. Addison-Wesley, Reading, MA, 1974.

[4] G. Alefeld and J. Herzberger. *Introduction to Interval Computations*. Academic Press, New York, 1983.

[5] P. R. Amestoy and I. S. Duff. Vectorization of a multiprocessor multifrontal code. *International Journal of Supercomputer Applications*, 3:41–59, 1989.

[6] P. R. Amestoy. Factorization of large unsymmetric sparse matrices based on a multifrontal approach in a multiprocessor environment. Technical Report TH/PA/91/2, CERFACS, Toulouse, France, February 1991. Ph.D. thesis.

[7] A. Anda and H. Park. Fast plane rotations with dynamic scaling. *SIAM J. Matrix Anal. Appl.*, 15:162–174, 1994.

[8] A. Anda and H. Park. Self scaling fast rotations for stiff least squares problems. *Linear Algebra Appl.*, 234:137–162, 1996.

[9] A. Anderson, D. Culler, D. Patterson, and the NOW Team. A case for networks of workstations: NOW. *IEEE Micro*, 15(1):54–64, February 1995.

[10] E. Anderson, Z. Bai, C. Bischof, J. Demmel, J. Dongarra, J. Du Croz, A. Greenbaum, S. Hammarling, A. McKenney, S. Ostrouchov, and D. Sorensen. *LAPACK Users' Guide (2nd edition)*. SIAM, Philadelphia, PA, 1995.

[11] ANSI/IEEE, New York. *IEEE Standard for Binary Floating Point Arithmetic, Std 754-1985 edition*, 1985.

[12] ANSI/IEEE, New York. *IEEE Standard for Radix Independent Floating Point Arithmetic, Std 854-1987 edition*, 1987.

[13] P. Arbenz and G. Golub. On the spectral decomposition of Hermitian matrices modified by row rank perturbations with applications. *SIAM J. Matrix Anal. Appl.*, 9:40–58, 1988.

[14] M. Arioli, J. Demmel, and I. S. Duff. Solving sparse linear systems with sparse backward error. *SIAM J. Matrix Anal. Appl.*, 10:165–190, 1989.

[15] O. Axelsson. *Iterative Solution Methods.* Cambridge University Press, Cambridge, UK, 1994.

[16] Z. Bai. Error analysis of the Lanczos algorithm for the nonsymmetric eigenvalue problem. *Math. Comp.*, 62:209–226, 1994.

[17] Z. Bai. Progress in the numerical solution of the nonsymmetric eigenvalue problem. *J. Numer. Linear Algebra Appl.*, 2:219–234, 1995.

[18] Z. Bai, D. Day, and Q. Ye. ABLE: An adaptive block Lanczos method for non-Hermitian eigenvalue problems. Mathematics Dept. Report 95-04, University of Kentucky, May 1995. Submitted to *SIAM J. Matrix Anal. Appl.*

[19] Z. Bai and G. W. Stewart. SRRIT: A Fortran subroutine to calculate the dominant invariant subspace of a nonsymmetric matrix. Computer Science Dept. Report TR 2908, University of Maryland, April 1992. Available as pub/reports for reports and pub/srrit for programs via anonymous ftp from thales.cs.umd.edu.

[20] D. H. Bailey. Multiprecision translation and execution of Fortran programs. *ACM Trans. Math. Software*, 19:288–319, 1993.

[21] D. H. Bailey. A Fortran-90 based multiprecision system. *ACM Trans. Math. Software*, 21:379–387, 1995.

[22] D. H. Bailey, K. Lee, and H. D. Simon. Using Strassen's algorithm to accelerate the solution of linear systems. *J. Supercomputing*, 4:97–371, 1991.

[23] J. Barnes and P. Hut. A hierarchical $O(n \log n)$ force calculation algorithm. *Nature*, 324:446–449, 1986.

[24] R. Barrett, M. Berry, T. Chan, J. Demmel, J. Donato, J. Dongarra, V. Eijkhout, V. Pozo, C. Romine, and H. van der Vorst. *Templates for the Solution of Linear Systems: Building Blocks for Iterative Methods.* SIAM, Philadelphia, PA, 1994. Also available electronically at http://www.netlib.org/templates.

[25] S. Batterson. Convergence of the shifted QR algorithm on 3 by 3 normal matrices. *Numer. Math.*, 58:341–352, 1990.

[26] F. L. Bauer. Genauigkeitsfragen bei der Lösung linearer Gleichungssysteme. *Z. Angew. Math. Mech.*, 46:409–421, 1966.

[27] T. Beelen and P. Van Dooren. An improved algorithm for the computation of Kronecker's canonical form of a singular pencil. *Linear Algebra Appl.*, 105:9–65, 1988.

[28] C. Bischof. Incremental condition estimation. *SIAM J. Matrix Anal. Appl.*, 11:312–322, 1990.

[29] C. Bischof, A. Carle, G. Corliss, A. Griewank, and P. Hovland. ADIFOR: Generating derivative codes from Fortran programs. *Scientific Programming*, 1:11–29, 1992. Software available at http://www.mcs.anl.gov/adifor/.

[30] C. Bischof and G. Quintana-Orti. Computing rank-revealing QR factorizations of dense matrices. Argonne Preprint ANL-MCS-P559-0196, Argonne National Laboratory, Argonne, IL, 1996.

[31] Å. Björck. *Solution of Equations in $\mathbb{R}^n$*, volume 1 of *Handbook of Numerical Analysis*, chapter Least Squares Methods. Elsevier/North Holland, Amsterdam, 1987.

[32] Å. Björck. Least squares methods. Mathematics Department Report, Linköping University, 1991.

[33] Å. Björck. *Numerical Methods for Least Squares Problems*. SIAM, Philadelphia, PA, 1996.

[34] L. S. Blackford, J. Choi, A. Cleary, E. D'Azevedo, J. Demmel, I. Dhillon, J. Dongarra, S. Hammarling, G. Henry, A. Petitet, K. Stanley, D. Walker, and R. C. Whaley. *ScaLAPACK Users' Guide*. Software, Environments, and Tools 4. SIAM, Philadelphia, PA, 1997.

[35] J. Blue. A portable FORTRAN program to find the Euclidean norm of a vector. *ACM Trans. Math. Software*, 4:15–23, 1978.

[36] J. H. Bramble, J. E. Pasciak, and A. H. Schatz. The construction of preconditioners for elliptic problems by substructuring, I. *Math. Comp.*, 47:103–134, 1986.

[37] J. H. Bramble, J. E. Pasciak, and A. H. Schatz. An iterative method for elliptic problems on regions partitioned into substructures. *Math. Comp.*, 46:361–369, 1986.

[38] J. H. Bramble, J. E. Pasciak, and A. H. Schatz. The construction of preconditioners for elliptic problems by substructuring, II. *Math. Comp.*, 49:1–16, 1987.

[39] J. H. Bramble, J. E. Pasciak, and A. H. Schatz. The construction of pre-conditioners for elliptic problems by substructuring, III. *Math. Comp.*, 51:415–430, 1988.

[40] J. H. Bramble, J. E. Pasciak, and A. H. Schatz. The construction of preconditioners for elliptic problems by substructuring, IV. *Math. Comp.*, 53:1–24, 1989.

[41] K. Brenan, S. Campbell, and L. Petzold. *Numerical Solution of Initial-Value Problems in Differential-Algebraic Equations*. North Holland, New York, 1989.

[42] C. Brezinski, M. Redivo Zaglia, and H. Sadok. Avoiding breakdown and near-breakdown in Lanczos type algorithms. *Numer. Algorithms*, 1:261–284, 1991.

[43] W. Briggs. *A Multigrid Tutorial*. SIAM, Philadelphia, PA, 1987.

[44] J. Bunch and L. Kaufman. Some stable methods for calculating inertia and solving symmetric linear systems. *Math. Comp.*, 31:163–179, 1977.

[45] J. Bunch, P. Nielsen, and D. Sorensen. Rank-one modification of the symmetric eigenproblem. *Numer. Math.*, 31:31–48, 1978.

[46] B. Buzbee, F. Dorr, J. George, and G. Golub. The direct solution of the discrete Poisson equation on irregular regions. *SIAM J. Numer. Anal.*, 8:722–736, 1971.

[47] B. Buzbee, G. Golub, and C. Nielsen. On direct methods for solving Poisson's equation. *SIAM J. Numer. Anal.*, 7:627–656, 1970.

[48] T. Chan. Rank revealing QR factorizations. *Linear Algebra Appl.*, 88/89:67–82, 1987.

[49] T. Chan and T. Mathew. Domain decomposition algorithms. In A. Iserles, editor, *Acta Numerica, Volume* 3. Cambridge University Press, Cambride, UK, 1994.

[50] S. Chandrasekaran and I. Ipsen. On rank-revealing factorisations. *SIAM J. Matrix Anal. Appl.*, 15:592–622, 1994.

[51] F. Chatelin. *Eigenvalues of Matrices*. Wiley, Chichester, England, 1993. English translation of the original 1988 French edition.

[52] F. Chaitin-Chatelin and V. Fraysse. *Lectures on Finite Precision Computations*. SIAM, Philadelphia, PA, 1996.

[53] J. Choi, J. Demmel, I. Dhillon, J. Dongarra, S. Ostrouchov, A. Petitet, K. Stanley, D. Walker, and R. C. Whaley. ScaLAPACK: A portable linear algebra library for distributed memory computers—Design issues and performance. Computer Science Dept. Technical Report CS-95-283, University of Tennessee, Knoxville, TN, March 1995. (LAPACK Working Note 95.)

[54] J. Coonen. Underflow and the denormalized numbers. *Computer*, 14:75–87, 1981.

[55] J. Cullum, W. Kerner, and R. Willoughby. A generalized nonsymmetric Lanczos procedure. *Comput. Phys. Comm.*, 53:19–48, 1989.

[56] J. Cullum and R. Willoughby. A practical procedure for computing eigenvalues of large sparse nonsymmetric matrices. In J. Cullum and R. Willoughby, editors, *Large Scale Eigenvalue Problems*. North Holland, Amsterdam, 1986. Mathematics Studies Series Vol. 127, Proceedings of the IBM Institute Workshop on Large Scale Eigenvalue Problems, July 8–12, 1985, Oberlech, Austria.

[57] J. Cullum and R. A. Willoughby. *Lanczos Algorithms for Large Symmetric Eigenvalue Computations*. Birkhaüser, Basel, 1985. Vol. 1, Theory, Vol. 2, Program.

[58] J. J. M. Cuppen. The singular value decomposition in product form. *SIAM J. Sci. Statist. Comput.*, 4:216–221, 1983.

[59] J. J. M. Cuppen. A divide and conquer method for the symmetric tridiagonal eigenproblem. *Numer. Math.*, 36:177–195, 1981.

[60] E. Davidson. The iteration calculation of a few of the lowest eigenvalues and corresponding eigenvectors of large real symmetric matrices. *J. Comp. Phys.*, 17:87–94, 1975.

[61] P. Davis. *Interpolation and Approximation*. Dover, New York, 1975.

[62] T. A. Davis and I. S. Duff. An unsymmetric-pattern multifrontal method for sparse LU factorization. Technical Report RAL 93-036, Rutherford Appleton Laboratory, Chilton, Didcot, Oxfordshire, UK, 1994.

[63] T. A. Davis and I. S. Duff. A combined unifrontal/multifrontal method for unsymmetric sparse matrices. Technical Report TR-95-020, Computer and Information Sciences Department, University of Florida, 1995.

[64] D. Day. *Semi-duality in the two-sided Lanczos algorithm*. Ph.D. thesis, University of California, Berkeley, CA, 1993.

[65] D. Day. How the QR algorithm fails to converge and how to fix it. Technical Report 96-0913J, Sandia National Laboratory, Albuquerque, NM, April 1996.

[66] A. Deichmoller. *Über die Berechnung verallgemeinerter singulärer Werte mittles Jacobi-ähnlicher Verfahren*. Ph.D. thesis, Fernuniversität-Hagen, Hagen, Germany, 1991.

[67] P. Deift, J. Demmel, L.-C. Li, and C. Tomei. The bidiagonal singular values decomposition and Hamiltonian mechanics. *SIAM J. Numer. Anal.*, 28:1463–1516, 1991. (LAPACK Working Note 11.)

[68] P. Deift, T. Nanda, and C. Tomei. ODEs and the symmetric eigenvalue problem. *SIAM J. Numer. Anal.*, 20:1–22, 1983.

[69] J. Demmel. The condition number of equivalence transformations that block diagonalize matrix pencils. *SIAM J. Numer. Anal.*, 20:599–610, 1983.

[70] J. Demmel. Underflow and the reliability of numerical software. *SIAM J. Sci. Statist. Comput.*, 5:887–919, 1984.

[71] J. Demmel. On condition numbers and the distance to the nearest ill-posed problem. *Numer. Math.*, 51:251–289, 1987.

[72] J. Demmel. The componentwise distance to the nearest singular matrix. *SIAM J. Matrix Anal. Appl.*, 13:10–19, 1992.

[73] J. Demmel, I. Dhillon, and H. Ren. On the correctness of some bisection-like parallel eigenvalue algorithms in floating point arithmetic. *Electronic Trans. Numer. Anal.*, 3:116–140, December 1995. (LAPACK Working Note 70.)

[74] J. Demmel and W. Gragg. On computing accurate singular values and eigenvalues of acyclic matrices. *Linear Algebra Appl.*, 185:203–218, 1993.

[75] J. Demmel, M. Gu, S. Eisenstat, I. Slapničar, K. Veselić, and Z. Drmač. Computing the singular value decomposition with high relative accuracy. Technical Report CSD-97-934, Computer Science Division, University of California, Berkeley, CA, February 1997. LAPACK Working Note 119. Submitted to *Linear Algebra Appl.*

[76] J. Demmel, M. Heath, and H. van der Vorst. Parallel numerical linear algebra. In A. Iserles, editor, *Acta Numerica, Volume* 2. Cambridge University Press, Cambridge, UK, 1993.

[77] J. Demmel and N. J. Higham. Stability of block algorithms with fast Level 3 BLAS. *ACM Trans. Math. Software*, 18:274–291, 1992.

[78] J. Demmel and B. Kågström. Accurate solutions of ill-posed problems in control theory. *SIAM J. Matrix Anal. Appl.*, 9:126–145, 1988.

[79] J. Demmel and B. Kågström. The generalized Schur decomposition of an arbitrary pencil $A - \lambda B$: Robust software with error bounds and applications. Parts I and II. *ACM Trans. Math. Software*, 19:160–201, June 1993.

[80] J. Demmel and W. Kahan. Accurate singular values of bidiagonal matrices. *SIAM J. Sci. Statist. Comput.*, 11:873–912, 1990.

[81] J. Demmel and X. Li. Faster numerical algorithms via exception handling. *IEEE Trans. Comput.*, 43:983–992, 1994. (LAPACK Working Note 59.)

[82] J. Demmel and K. Veselić. Jacobi's method is more accurate than QR. *SIAM J. Matrix Anal. Appl.*, 13:1204–1246, 1992. (LAPACK Working Note 15.)

[83] I. S. Dhillon. *A New $O(n^2)$ Algorithm for the Symmetric Tridiagonal Eigenvalue/Eigenvector Problem*. Ph.D. thesis, Computer Science Division, University of California, Berkeley, May 1997.

[84] P. Dierckx. *Curve and Surface Fitting with Splines*. Oxford University Press, Oxford, UK, 1993.

[85] J. Dongarra. Performance of various computers using standard linear equations software. Computer Science Dept. Technical Report, University of Tennessee, Knoxville, April 1996. Up-to-date version available at NETLIB/benchmark.

[86] J. Dongarra, J. Du Croz, I. Duff, and S. Hammarling. Algorithm 679: A set of Level 3 Basic Linear Algebra Subprograms. *ACM Trans. Math. Software*, 16:18–28, 1990.

[87] J. Dongarra, J. Du Croz, I. Duff, and S. Hammarling. A set of Level 3 Basic Linear Algebra Subprograms. *ACM Trans. Math. Software*, 16:1–17, 1990.

[88] J. Dongarra, J. Du Croz, S. Hammarling, and R. J. Hanson. Algorithm 656: An extended set of FORTRAN Basic Linear Algebra Subroutines. *ACM Trans. Math. Software*, 14:18–32, 1988.

[89] J. Dongarra, J. Du Croz, S. Hammarling, and R. J. Hanson. An extended set of FORTRAN Basic Linear Algebra Subroutines. *ACM Trans. Math. Software*, 14:1–17, 1988.

[90] J. Dongarra and D. Sorensen. A fully parallel algorithm for the symmetric eigenproblem. *SIAM J. Sci. Statist. Comput.*, 8:139–154, 1987.

[91] C. Douglas. MGNET: Multi-Grid net. http://NA.CS.Yale.EDU/mgnet/ www/mgnet.html.

[92] Z. Drmač. *Computing the Singular and the Generalized Singular Values.* Ph.D. thesis, Fernuniversität-Hagen, Hagen, Germany, 1994.

[93] I. S. Duff, A. M. Erisman, and J. K. Reid. *Direct Methods for Sparse Matrices.* Oxford University Press, London, 1986.

[94] I. S. Duff. Sparse numerical linear algebra: Direct methods and pre- conditioning. Technical Report RAL-TR-96-047, Rutherford Appleton Laboratory, Chilton, Didcot, Oxfordshire, UK, 1996.

[95] I. S. Duff and J. K. Reid. MA47, a Fortran code for direct solution of indefinite sparse symmetric linear systems. Technical Report RAL-95-001, Rutherford Appleton Laboratory, Chilton, Didcot, Oxfordshire, UK, 1995.

[96] I. S. Duff and J. K. Reid. The design of MA48, a code for the direct solution of sparse unsymmetric linear systems of equations. *ACM Trans. Math. Software*, 22:187–226, 1996.

[97] I. S. Duff and J. K. Reid. The multifrontal solution of indefinite sparse symmetric linear equations. *ACM Trans. Math. Software*, 9:302–325, 1983.

[98] I. S. Duff and J. A. Scott. The design of a new frontal code for solving sparse unsymmetric systems. *ACM Trans. Math. Software*, 22:30–45, 1996.

[99] A. Edelman. The complete pivoting conjecture for Gaussian elimination is false. *The Mathematica Journal*, 2:58–61, 1992.

[100] A. Edelman and H. Murakami. Polynomial roots from companion ma- trices. *Math. Comp.*, 64:763–776, 1995.

[101] S. Eisenstat and I. Ipsen. Relative perturbation techniques for singular value problems. *SIAM J. Numer. Anal.*, 32:1972–1988, 1995.

[102] V. Faber and T. Manteuffel. Necessary and sufficient conditions for the existence of a conjugate gradient method. *SIAM J. Numer. Anal.*, 21:315–339, 1984.

[103] D. M. Fenwick, D. J. Foley, W. B. Gist, S. R. VanDoren, and D. Wissel. The AlphaServer 8000 series: High-end server platform development. *Digital Technical Journal*, 7:43–65, 1995.

[104] K. Fernando and B. Parlett. Accurate singular values and differential qd algorithms. *Numer. Math.*, 67:191–229, 1994.

[105] V. Fernando, B. Parlett, and I. Dhillon. A way to find the most redundant equation in a tridiagonal system. Berkeley Mathematics Dept. Preprint, 1995.

[106] H. Flaschka. *Dynamical Systems, Theory and Applications*, volume 38 of Lecture Notes in Physics, chapter Discrete and periodic solutions of some aspects of the inverse method. Springer-Verlag, New York, 1975.

[107] R. Freund, G. Golub, and N. Nachtigal. Iterative solution of linear systems. In A. Iserles, editor, *Acta Numerica* 1992, pages 57–100. Cambridge University Press, Cambridge, UK, 1992.

[108] R. Freund, M. Gutknecht, and N. Nachtigal. An implementation of the look-ahead Lanczos algorithm for non-Hermitian matrices. *SIAM J. Sci. Comput.*, 14:137–158, 1993.

[109] X. Sun, G. Quintana-Orti, and C. Bischof. A blas-3 version of the QR factorization with column pivoting. Argonne Preprint MCS-P551-1295, Argonne National Laboratory, Argonne, IL, 1995.

[110] F. Gantmacher. *The Theory of Matrices, vol.* II *(translation)*. Chelsea, New York, 1959.

[111] M. Garey and D. Johnson. *Computers and Intractability*. W. H. Freeman, San Francisco, 1979.

[112] A. George. Nested dissection of a regular finite element mesh. *SIAM J. Numer. Anal.*, 10:345–363, 1973.

[113] A. George, M. Heath, J. Liu, and E. Ng. Solution of sparse positive definite systems on a shared memory multiprocessor. *Internat. J. Parallel Programming*, 15:309–325, 1986.

[114] A. George and J. Liu. *Computer Solution of Large Sparse Positive Definite Systems*. Prentice-Hall, Englewood Cliffs, NJ, 1981.

[115] A. George and E. Ng. Parallel sparse Gaussian elimination with partial pivoting. *Ann. Oper. Res.*, 22:219–240, 1990.

[116] R. Glowinski, G. Golub, G. Meurant, and J. Periaux, editors. *Domain Decomposition Methods for Partial Differential Equations*, SIAM, Philadelphia, PA, 1988. Proceedings of the First International Symposium on Domain Decomposition Methods for Partial Differential Equations, Paris, France, January 1987.

[117] S. Goedecker. Remark on algorithms to find roots of polynomials. *SIAM J. Sci. Statist. Comp.*, 15:1059–1063, 1994.

[118] I. Gohberg, P. Lancaster, and L. Rodman. *Matrix Polynomials*. Academic Press, New York, 1982.

[119] D. Goldberg. What every computer scientist should know about floating point arithmetic. *ACM Computing Surveys*, 23:5–48, 1991.

[120] G. Golub and W. Kahan. Calculating the singular values and pseudo-inverse of a matrix. *SIAM J. Numer. Anal. (Series B)*, 2:205–224, 1965.

[121] G. Golub and C. Van Loan. *Matrix Computations*. Johns Hopkins University Press, Baltimore, MD, 3rd edition, 1996.

[122] N. Gould. On growth in Gaussian elimination with complete pivoting. *SIAM J. Matrix Anal. Appl.*, 12:354–361, 1991. See also editor's note in *SIAM J. Matrix Anal. Appl.*, 12(3), 1991.

[123] A. Greenbaum and Z. Strakos. Predicting the behavior of finite precision Lanczos and conjugate gradient computations. *SIAM J. Matrix Anal. Appl.*, 13:121–137, 1992.

[124] L. Greengard and V. Rokhlin. A fast algorithm for particle simulations. *J. Comput. Phys.*, 73:325–348, 1987.

[125] R. Grimes, J. Lewis, and H. Simon. A shifted block Lanczos algorithm for solving sparse symmetric generalized eigenproblems. *SIAM J. Matrix Anal. Appl.*, 15:228–272, 1994.

[126] M. Gu. *Numerical Linear Algebra Computations*. Ph.D. thesis, Dept. of Computer Science, Yale University, November 1993.

[127] M. Gu and S. Eisenstat. A stable algorithm for the rank-1 modification of the symmetric eigenproblem. Computer Science Dept. Report YALEU/DCS/RR-916, Yale University, September 1992.

[128] M. Gu and S. Eisenstat. An efficient algorithm for computing a rank-revealing QR decomposition. Computer Science Dept. Report YALEU/DCS/RR-967, Yale University, June 1993.

[129] M. Gu and S. C. Eisenstat. A stable and efficient algorithm for the rank-1 modification of the symmetric eigenproblem. *SIAM J. Matrix Anal. Appl.*, 15:1266–1276, 1994. Yale Technical Report YALEU/DCS/RR-916, September 1992.

[130] M. Gu and S. C. Eisenstat. A divide-and-conquer algorithm for the bidiagonal SVD. *SIAM J. Matrix Anal. Appl.*, 16:79–92, 1995.

[131] M. Gu and S. C. Eisenstat. A divide-and-conquer algorithm for the symmetric tridiagonal eigenproblem. *SIAM J. Matrix Anal. Appl.*, 16:172–191, 1995.

[132] A. Gupta and V. Kumar. Optimally scalable parallel sparse Cholesky factorization. In *Proceedings of the Seventh SIAM Conference on Parallel Processing for Scientific Computing*, pages 442–447. SIAM, Philadelphia, PA, 1995.

[133] A. Gupta, E. Rothberg, E. Ng, and B. W. Peyton. Parallel sparse Cholesky factorization algorithms for shared-memory multiprocessor systems. In R. Vichnevetsky, D. Knight, and G. Richter, editors, *Advances in Computer Methods for Partial Differential Equations*—VII. IMACS, 1992.

[134] M. Gutknecht. A completed theory of the unsymmetric Lanczos process and related algorithms, Part I. *SIAM J. Matrix Anal. Appl.*, 13:594–639, 1992.

[135] M. Gutknecht. A completed theory of the unsymmetric Lanczos process and related algorithms, Part II. *SIAM J. Matrix Anal. Appl.*, 15:15–58, 1994.

[136] W. Hackbusch. *Iterative Solution of Large Sparse Linear Systems of Equations*. Springer-Verlag, Berlin, 1994.

[137] L. A. Hageman and D. M. Young. *Applied Iterative Methods*. Academic Press, New York, 1981.

[138] W. W. Hager. Condition estimators. *SIAM J. Sci. Statist. Comput.*, 5:311–316, 1984.

[139] P. Halmos. *Finite Dimensional Vector Spaces*. Van Nostrand, New York, 1958.

[140] E. R. Hansen. *Global Optimization Using Interval Analysis*. Marcel Dekker, New York, 1992.

[141] P. C. Hansen. The truncated SVD as a method for regularization. *BIT*, 27:534–553, 1987.

[142] P. C. Hansen. Truncated singular value decomposition solutions to discrete ill-posed problems ill-determined numerical rank. *SIAM J. Sci. Statist. Comput.*, 11:503–518, 1990.

[143] M. T. Heath and P. Raghavan. Performance of a fully parallel sparse solver. In *Proceedings of the Scalable High-Performance Computing Conference*, pages 334–341, IEEE, Los Alamitos, CA, 1994.

[144] M. Hénon. Integrals of the Toda lattice. *Phys. Rev. B*, 9:1421–1423, 1974.

[145] M. R. Hestenes and E. Stiefel. Methods of conjugate gradients for solving linear systems. *J. Res. Natl. Bur. Stand.*, 49:409–436, 1954.

[146] N. J. Higham. A survey of condition number estimation for triangular matrices. *SIAM Rev.*, 29:575–596, 1987.

[147] N. J. Higham. FORTRAN codes for estimating the one-norm of a real or complex matrix, with applications to condition estimation. *ACM Trans. Math. Software*, 14:381–396, 1988.

[148] N. J. Higham. Experience with a matrix norm estimator. *SIAM J. Sci. Statist. Comput.*, 11:804–809, 1990.

[149] N. J. Higham. *Accuracy and Stability of Numerical Algorithms.* SIAM, Philadelphia, PA, 1996.

[150] P. Hong and C. T. Pan. The rank revealing QR and SVD. *Math. Comp.*, 58:575–232, 1992.

[151] X. Hong and H. T. Kung. I/O complexity: The red blue pebble game. In *Proceedings of the* 13*th Symposium on the Theory of Computing*, pages 326–334. ACM, New York, 1981.

[152] A. K. Jain. *Fundamentals of Digital Image Processing.* Prentice-Hall, Englewood Cliffs, NJ, 1989.

[153] E. Jessup and D. Sorensen. A divide and conquer algorithm for computing the singular value decomposition of a matrix. In *Proceedings of the Third SIAM Conference on Parallel Processing for Scientific Computing*, pages 61–66, SIAM, Philadelphia, PA, 1989.

[154] Z. Jia. *Some Numerical Methods for Large Unsymmetric Eigenproblems.* Ph.D. thesis, Universität Bielefeld, Bielefeld, Germany, 1994.

[155] W.-D. Webber, J. P. Singh, and A. Gupta. Splash: Stanford parallel applications for shared-memory. *Computer Architecture News*, 20:5–44, 1992.

[156] W. Kahan. Accurate eigenvalues of a symmetric tridiagonal matrix. Computer Science Dept. Technical Report CS41, Stanford University, Stanford, CA, July 1966 (revised June 1968).

[157] W. Kahan. A survey of error analysis. In *Information Processing* 71, pages 1214–1239, North Holland, Amsterdam, 1972.

[158] W. Kahan. The baleful effect of computer benchmarks upon applied mathematics, physics and chemistry. http://HTTP.CS.Berkeley.EDU/~wkahan/ieee754status/baleful.ps, 1995.

[159] W. Kahan. Lecture notes on the status of IEEE standard 754 for binary floating point arithmetic. http://HTTP.CS.Berkeley.EDU/~wkahan/ieee754status/ieee754.ps, 1995.

[160] T. Kailath and A. H. Sayed. Displacement structure: Theory and applications. *SIAM Rev.*, 37:297–386, 1995.

[161] T. Kato. *Perturbation Theory for Linear Operators*. Springer-Verlag, Berlin, 2nd edition, 1980.

[162] R. B. Kearfott. *Rigorous Global Search: Continuous Problems*. Kluwer, Dordrecht, the Netherlands, 1996. See also http://interval.usl.edu/euromath.html.

[163] W. Kerner. Large-scale complex eigenvalue problems. *J. Comput. Phys.*, 85:1–85, 1989.

[164] G. Kolata. Geodesy: Dealing with an enormous computer task. *Science*, 200:421–422, 1978.

[165] S. Krishnan, A. Narkhede, and D. Manocha. BOOLE: A system to compute Boolean combinations of sculptured solids. Computer Science Dept. Technical Report TR95-008, University of North Carolina, Chapel Hill, 1995. http://www.cs.unc.edu/~geom/geom.html.

[166] M. Kruskal. *Dynamical Systems, Theory and Applications*, volume 38 of Lecture Notes in Physics, chapter Nonlinear Wave Equations. Springer-Verlag, New York, 1975.

[167] K. Kundert. Sparse matrix techniques. In A. Ruehli, editor, *Circuit Analysis, Simulation and Design*. North Holland, Amsterdam, 1986.

[168] C. Lawson and R. Hanson. *Solving Least Squares Problems*. Prentice-Hall, Englewood Cliffs, NJ, 1974.

[169] C. Lawson, R. Hanson, D. Kincaid, and F. Krogh. Basic Linear Algebra Subprograms for Fortran usage. *ACM Trans. Math. Software*, 5:308–323, 1979.

[170] P. Lax. Integrals of nonlinear equations of evolution and solitary waves. *Comm. Pure Appl. Math.*, 21:467–490, 1968.

[171] R. Lehoucq. *Analysis and Implementation of an Implicitly Restarted Arnoldi Iteration*. Ph.D. thesis, Rice University, Houston, TX, 1995.

[172] R.-C. Li. Solving secular equations stably and efficiently. Computer Science Dept. Technical Report CS-94-260, University of Tennessee, Knoxville, TN, November 1994. (LAPACK Working Note 89.)

[173] T.-Y. Li and Z. Zeng. Homotopy-determinant algorithm for solving non-symmetric eigenvalue problems. *Math. Comp.*, 59:483–502, 1992.

[174] T.-Y. Li and Z. Zeng. Laguerre's iteration in solving the symmetric tridiagonal eigenproblem—a revisit. Michigan State University Preprint, 1992.

[175] T.-Y. Li, Z. Zeng, and L. Cong. Solving eigenvalue problems of nonsymmetric matrices with real homotopies. *SIAM J. Numer. Anal.*, 29:229–248, 1992.

[176] T.-Y. Li, H. Zhang, and X.-H. Sun. Parallel homotopy algorithm for symmetric tridiagonal eigenvalue problem. *SIAM J. Sci. Statist. Comput.*, 12:469–487, 1991.

[177] X. Li. *Sparse Gaussian Elimination on High Performance Computers*. Ph.D. thesis, Computer Science Division, Department of Electrical Engineering and Computer Science, University of California, Berkeley, September 1996.

[178] S.-S. Lo, B. Phillipe, and A. Sameh. A multiprocessor algorithm for the symmetric eigenproblem. *SIAM J. Sci. Statist. Comput.*, 8:155–165, 1987.

[179] K. Löwner. Über monotone matrixfunctionen. *Math. Z.*, 38:177–216, 1934.

[180] R. Lucas, W. Blank, and J. Tieman. A parallel solution method for large sparse systems of equations. *IEEE Trans. Computer Aided Design*, CAD-6:981–991, 1987.

[181] D. Manocha and J. Demmel. Algorithms for intersecting parametric and algebraic curves i: simple intersections. *ACM Transactions on Graphics*, 13:73–100, 1994.

[182] D. Manocha and J. Demmel. Algorithms for intersecting parametric and algebraic curves ii: Higher order intersections. *Computer Vision, Graphics and Image Processing: Graphical Models and Image Processing*, 57:80–100, 1995.

[183] R. Mathias. Accurate eigensystem computations by Jacobi methods. *SIAM J. Matrix Anal. Appl.*, 16:977–1003, 1996.

[184] The MathWorks, Inc., Natick, MA. *MATLAB Reference Guide*, 1992.

[185] S. McCormick, editor. *Multigrid Methods*, volume 3 of SIAM Frontiers in Applied Mathematics. SIAM, Philadelphia, PA, 1987.

[186] S. McCormick. *Multilevel Adaptive Methods for Partial Differential Equations*, volume 6 of SIAM Frontiers in Applied Mathematics. SIAM, Philadelphia, PA, 1989.

[187] J. Moser. *Dynamical Systems, Theory and Applications*, volume 38 of Lecture Notes in Physics, chapter Finitely many mass points on the line under the influence of an exponential potential—an integrable system. Springer-Verlag, New York, 1975.

[188] J. Moser, editor. *Dynamical Systems, Theory and Applications*, volume 38 of Lecture Notes in Physics. Springer-Verlag, New York, 1975.

[189] A. Netravali and B. Haskell. *Digital Pictures*. Plenum Press, New York, 1988.

[190] A. Neumaier. *Interval Methods for Systems of Equations*. Cambridge University Press, Cambridge, UK, 1990.

[191] E. G. Ng and B. W. Peyton. Block sparse Cholesky algorithms on advanced uniprocessor computers. *SIAM J. Sci. Statist. Comp.*, 14:1034–1056, 1993.

[192] B. Nour-Omid, B. Parlett, and A. Liu. How to maintain semi-orthogonality among Lanczos vectors. CPAM Technical Report 420, University of California, Berkeley, CA, 1988.

[193] W. Oettli and W. Prager. Compatibility of approximate solution of linear equations with given error bounds for coefficients and right hand sides. *Numer. Math.*, 6:405–409, 1964.

[194] C. C. Paige and M. A. Saunders. Solution of sparse indefinite systems of linear equations. *SIAM J. Numer. Anal.*, 12:617–629, 1975.

[195] V. Pan. How can we speed up matrix multiplication. *SIAM Rev.*, 26:393–416, 1984.

[196] V. Pan and P. Tang. Bounds on singular values revealed by QR factorization. Technical Report MCS-P332-1092, Mathematics and Computer Science Division, Argonne National Laboratory, Argonne, IL, 1992.

[197] B. Parlett. *The Symmetric Eigenvalue Problem*. Prentice Hall, Englewood Cliffs, NJ, 1980.

[198] B. Parlett. Misconvergence in the Lanczos algorithm. In M. G. Cox and S. Hammarling, editors, *Reliable Numerical Computation*, chapter 1. Clarendon Press, Oxford, UK, 1990.

[199] B. Parlett. Reduction to tridiagonal form and minimal realizations. *SIAM J. Matrix Anal. Appl.*, 13:567–593, 1992.

[200] B. Parlett. *Acta Numerica*, The new qd algorithms, pages 459–491. Cambridge University Press, Cambridge, UK, 1995.

[201] B. Parlett. The construction of orthogonal eigenvectors for tight clusters by use of submatrices. Center for Pure and Applied Mathematics PAM-664, University of California, Berkeley, CA, January 1996. Submitted to *SIAM J. Matrix Anal. Appl.*

[202] B. N. Parlett, D. R. Taylor, and Z. A. Liu. A look-ahead Lanczos algorithm for unsymmetric matrices. *Math. Comp.*, 44:105–124, 1985.

[203] B. N. Parlett and I. S. Dhillon. Fernando's solution to Wilkinson's problem: An application of double factorization. *Linear Algebra Appl.*, 1997. To appear.

[204] D. Priest. Algorithms for arbitrary precision floating point arithmetic. In P. Kornerup and D. Matula, editors, *Proceedings of the 10th Symposium on Computer Arithmetic*, pages 132–145, Grenoble, France, June 26–28, 1991. IEEE Computer Society Press, Los Alamitos, CA.

[205] A. Quarteroni, editor. *Domain Decomposition Methods*, AMS, Providence, RI, 1993. Proceedings of the Sixth International Symposium on Domain Decomposition Methods, Como, Italy, 1992.

[206] H. Ren. *On Error Analysis and Implementation of Some Eigenvalue and Singular Value Algorithms.* Ph.D. thesis, University of California at Berkeley, 1996.

[207] E. Rothberg and R. Schreiber. Improved load distribution in parallel sparse Cholesky factorization. In *Supercomputing*, pages 783–792, November 1994.

[208] S. Rump. Bounds for the componentwise distance to the nearest singular matrix. *SIAM J. Matrix Anal. Appl.*, 18:83–103, 1997.

[209] H. Rutishauser. *Lectures on Numerical Mathematics.* Birkhäuser, Basel, 1990.

[210] J. Rutter. A serial implementation of Cuppen's divide and conquer algorithm for the symmetric eigenvalue problem. Mathematics Dept. Master's Thesis, University of California, 1994. Available by anonymous ftp from tr-ftp.cs.berkeley.edu, directory pub/tech-reports/csd/csd-94-799, file all.ps.

[211] Y. Saad. Krylov subspace methods for solving large unsymmetric linear system. *Math. Comp.*, 37:105–126, 1981.

[212] Y. Saad. Numerical solution of large nonsymmetric eigenvalue problems. *Comput. Phys. Comm.*, 53:71–90, 1989.

[213] Y. Saad. *Numerical Methods for Large Eigenvalue Problems*. Manchester University Press, Manchester, UK, 1992.

[214] Y. Saad. *Iterative Methods for Sparse Linear Systems*. PWS Publishing Co., Boston, 1996.

[215] Y. Saad and M. H. Schultz. GMRES: A generalized minimal residual algorithm for solving nonsymmetric linear systems. *SIAM J. Sci. Statist. Comput.*, 7:856–869, 1986.

[216] M. Sadkane. Block-Arnoldi and Davidson methods for unsymmetric large eigenvalue problems. *Numer. Math.*, 64:195–211, 1993.

[217] M. Sadkane. A block Arnoldi-Chebyshev method for computing the leading eigenpairs of large sparse unsymmetric matrices. *Numer. Math.*, 64:181–193, 1993.

[218] J. R. Shewchuk. Adaptive Precision Floating-Point Arithmetic and Fast Robust Geometric Predicates. Technical Report CMU-CS-96-140, School of Computer Science, Carnegie Mellon University, Pittsburgh, PA, May 1996.

[219] M. Shub and S. Smale. Complexity of Bezout's theorem I: Geometric aspects. *J. Amer. Math. Soc.*, 6:459–501, 1993.

[220] M. Shub and S. Smale. Complexity of Bezout's theorem II: Volumes and probabilities. In F. Eyssette and A. Galligo, editors, *Progress in Mathematics, Vol.* 109—*Computational Algebraic Geometry.* Birkhäuser, Basel, 1993.

[221] M. Shub and S. Smale. Complexity of Bezout's theorem III: Condition number and packing. *J. Complexity*, 9:4–14, 1993.

[222] M. Shub and S. Smale. Complexity of Bezout's theorem IV: Probability of success; extensions. Mathematics Department Preprint, University of California, 1993.

[223] SGI Power Challenge. Technical Report, Silicon Graphics, 1995.

[224] H. Simon. The Lanczos algorithm with partial reorthogonalization. *Math. Comp.*, 42:115–142, 1984.

[225] R. D. Skeel. Scaling for numerical stability in Gaussian elimination. *Journal of the ACM*, 26:494–526, 1979.

[226] R. D. Skeel. Iterative refinement implies numerical stability for Gaussian elimination. *Math. Comp.*, 35:817–832, 1980.

[227] R. D. Skeel. Effect of equilibration on residual size for partial pivoting. *SIAM J. Numer. Anal.*, 18:449–454, 1981.

[228] I. Slapničar. *Accurate Symmetric Eigenreduction by a Jacobi Method.* Ph.D. thesis, Fernuniversität-Hagen, Hagen, Germany, 1992.

[229] G. Sleijpen and H. van der Vorst. A Jacobi-Davidson iteration method for linear eigenvalue problems. Dept. of Mathematics Report 856, University of Utrecht, 1994.

[230] G. Sleijpen, A. Booten, D. Fokkema, and H. van der Vorst. Jacobi-Davidson type methods for generalized eigenproblems and polynomial eigenproblems, Part I. Dept. of Mathematics Report 923, University of Utrecht, 1995.

[231] B. Smith. Domain decomposition algorithms for partial differential equations of linear elasticity. Technical Report 517, Department of Computer Science, Courant Institute, September 1990. Ph.D. thesis.

[232] B. Smith, P. Bjorstad, and W. Gropp. *Domain decomposition: Parallel multilevel methods for elliptic partial differential equations.* Cambridge University Press, Cambridge, UK, 1996. Corresponding PETSc software available at http://www.mcs.anl.gov/petsc/petsc.html.

[233] D. Sorensen. Implicit application of polynomial filters in a k-step Arnoldi method. *SIAM J. Matrix Anal. Appl.*, 13:357–385, 1992.

[234] D. Sorensen and P. Tang. On the orthogonality of eigenvectors computed by divide-and-conquer techniques. *SIAM J. Numer. Anal.*, 28:1752–1775, 1991.

[235] G. W. Stewart. *Introduction to Matrix Computations.* Academic Press, New York, 1973.

[236] G. W. Stewart. Rank degeneracy. *SIAM J. Sci. Statist. Comput.*, 5:403–413, 1984.

[237] G. W. Stewart and J.-G. Sun. *Matrix Perturbation Theory.* Academic Press, New York, 1990.

[238] SPARCcenter 2000 architecture and implementation. Sun Microsystems, Inc., November 1993. Technical White Paper.

[239] W. Symes. The QR algorithm for the finite nonperiodic Toda lattice. *Phys. D*, 4:275–280, 1982.

[240] G. Szegö. *Orthogonal Polynomials.* AMS, Providence, RI, 1967.

[241] K.-C. Toh and L. N. Trefethen. Pseudozeros of polynomials and pseudospectra of companion matrices. *Numer. Math.*, 68:403–425, 1994.

[242] L. Trefethen and R. Schreiber. Average case analysis of Gaussian elimination. *SIAM J. Matrix Anal. Appl.*, 11:335–360, 1990.

[243] L. N. Trefethen and D. Bau. *Numerical Linear Algebra*. SIAM, Philadelphia, PA, 1997.

[244] A. Van Der Sluis. Condition numbers and equilibration of matrices. *Numer. Math.*, 14:14–23, 1969.

[245] A. F. van der Stappen, R. H. Bisseling, and J. G. G. van der Vorst. Parallel sparse LU decomposition on a mesh network of transputers. *SIAM J. Matrix Anal. Appl.*, 14:853–879, 1993.

[246] P. Van Dooren. The computation of Kronecker's canonical form of a singular pencil. *Linear Algebra Appl.*, 27:103–141, 1979.

[247] P. Van Dooren. The generalized eigenstructure problem in linear system theory. *IEEE Trans. Automat. Control*, AC-26:111–128, 1981.

[248] C. V. Van Loan. *Computational Frameworks for the Fast Fourier Transform*. SIAM, Philadelphia, 1992.

[249] R. S. Varga. *Matrix Iterative Analysis*. Prentice-Hall, Englewood Cliffs, NJ, 1962.

[250] K. Veselić and I. Slapničar. Floating point perturbations of Hermitian matrices. *Linear Algebra Appl.*, 195:81–116, 1993.

[251] V. Voevodin. The problem of non-self-adjoint generalization of the conjugate gradient method is closed. *Comput. Math. Math. Phys.*, 23:143–144, 1983.

[252] D. Watkins. *Fundamentals of Matrix Computations*. Wiley, Chichester, UK, 1991.

[253] The Cray C90 series. http://www.cray.com/PUBLIC/product-info/C90/. Cray Research, Inc.

[254] The Cray J90 series. http://www.cray.com/PUBLIC/product-info/J90/. Cray Research, Inc.

[255] The Cray T3E series. http://www.cray.com/PUBLIC/product-info/T3E/. Cray Research, Inc.

[256] The IBM SP-2. http://www.rs6000.ibm.com/software/sp_products/sp2.html. IBM.

[257] The Intel Paragon. http://www.ssd.intel.com/homepage.html. Intel.

[258] P.-Å. Wedin. Perturbation theory for pseudoinverses. *BIT*, 13:217–232, 1973.

[259] S. Weisberg. *Applied Linear Regression*. Wiley, Chichester, UK, 2nd edition, 1985.

[260] P. Wesseling. *An Introduction to Multigrid Methods*. Wiley, Chichester, UK, 1992.

[261] J. H. Wilkinson. *Rounding Errors in Algebraic Processes*. Prentice Hall, Englewood Cliffs, NJ, 1963.

[262] J. H. Wilkinson. *The Algebraic Eigenvalue Problem*. Oxford University Press, Oxford, UK, 1965.

[263] S. Winograd and D. Coppersmith. Matrix multiplication via arithmetic progressions. In *Proceedings of the Nineteenth Annual ACM Symposium on the Theory of Computing*, pages 1–6. ACM, New York, 1987.

[264] M. Wolfe. *High Performance Compilers for Parallel Computing*. Addison-Wesley, Reading, MA, 1996.

[265] Q. Ye. A convergence analysis for nonsymmetric Lanczos algorithms. *Math. Comp.*, 56:677–691, 1991.

[266] Q. Ye. A breakdown-free variation of the nonsymmetric Lanczos algorithm. *Math. Comp.*, 62:179–207, 1994.

[267] D. Young. *Iterative Solution of Large Linear Systems*. Academic Press, New York, 1971.

[268] H. Yserentant. Old and new convergence proofs for multigrid methods. In A. Iserles, editor, *Acta Numerica 1993*, pages 285–326. Cambridge University Press, Cambrigde, UK, 1993.

[269] Z. Zeng. *Homotopy-Determinant Algorithm for Solving Matrix Eigenvalue Problems and Its Parallelizations*. Ph.D. thesis, Michigan State University, East Lansing, MI, 1991.

[270] Z. Zlatev, J. Waśniewski, P. C. Hansen, and Tz. Ostromsky. PARASPAR: a package for the solution of large linear algebraic equations on parallel computers with shared memory. Technical Report 95-10, Technical University of Denmark, Lyngby, September 1995.

[271] Z. Zlatev. *Computational Methods for General Sparse Matrices*. Kluwer Academic, Dordrecht, Boston, 1991.

Index

Arnoldi's algorithm, 119, 303, 304, 320, 359, 386
ARPACK, 384

backward error, *see* backward stability
backward stability, 5
  bisection, 230, 246
  Cholesky, 79, 84, 253, 263
  convergence criterion, 164
  direct versus iterative methods for $Ax = b$, 31
  eigenvalue problem, 123
  Gaussian elimination, 41
  GEPP, 41, 46, 49
  Gram–Schmidt, 108, 134
  instability of Cramer's rule, 95
  Jacobi's method for $Ax = \lambda x$, 242
  Jacobi's method for the SVD, 263
  Jordan canonical form, 146
  Lanczos algorithm, 305, 321
  linear equations, 44, 49
  normal equations, 118
  orthogonal transformations, 124
  polynomial evaluation, 16
  QR decomposition, 118, 119, 123
  secular equation, 224
  single precision iterative refinement, 62
  Strassen's method, 72, 93
  substitution, 25
  SVD, 118, 119, 123, 128
band matrices
  linear equations, 76, 79–83, 85, 86
  symmetric eigenproblem, 186
Bauer–Fike theorem, 150

biconjugate gradients, 321
bidiagonal form, 131, 240, 308, 357
  condition number, 95
  dqds algorithm, 242
  LR iteration, 242
  perturbation theory, 207, 242, 244, 245, 262
  qds algorithm, 242
  QR iteration, 241, 242
  reduction, 166, 237, 253
  SVD, 245, 260
Bisection
  finding zeros of polynomials, 9, 30
  SVD, 240–242, 246, 249
  symmetric eigenproblem, 201, 210, 211, 228, 235, 240, 260
bisection
  symmetric eigenproblem, 119
BLAS (Basic Linear Algebra Subroutines), 28, 66–75, 90, 93
  in Cholesky, 78, 98
  in Hessenberg reduction, 166
  in Householder transformations, 137
  in nonsymmetric eigenproblem, 185, 186
  in QR decomposition, 121
  in sparse Gaussian elimination, 91
block algorithms
  Cholesky, 66, 78, 98
  Gaussian elimination, 72–75
  Hessenberg reduction, 166
  Householder reflection, 137
  matrix multiplication, 67
  nonsymmetric eigenproblem, 185, 186

QR decomposition, 121, 137
sparse Gaussian elimination, 90
block cyclic reduction, 266, 327–330,
    332, 356
    model problem, 277
boundary value problem
    Dirichlet, 267
    eigenproblem, 270
    L-shaped region, 348
    one-dimensional heat equation,
        81
    Poisson's equation, 267, 324, 348
    Toda lattice, 255
bulge chasing, 169, 171, 213

canonical form, 139, 140, 145
    generalized Schur for real regu-
        lar pencils, 179, 185
    generalized Schur for regular pen-
        cils, 178, 181, 185
    generalized Schur for singular
        pencils, 181, 186
    Jordan, 3, 19, 140, 141, 145,
        146, 150, 175, 176, 178, 180,
        184, 185, 188, 280
    Kronecker, 180–182, 186, 187
    polynomial, 19
    real Schur, 147, 163, 184, 212
    Schur, 4, 140, 146–148, 152, 158,
        160, 161, 163, 175, 178, 181,
        184–186, 188
    Weierstrass, 173, 176, 178, 180,
        181, 185–187
CAPSS, 91
Cauchy interlace theorem, 261, 367
Cauchy matrices, 92
Cayley transform, 264
Cayley–Hamilton theorem, 295
CG, *see* conjugate gradients
CGS, *see* Gram–Schmidt orthog-
    onalization process (classi-
    cal); conjugate gradients
    squared
characteristic polynomial, 140, 149,

    295
companion matrix, 301
    of $A - \lambda B$, 174
    of $R_{SOR(\omega)}$, 290
    of a matrix polynomial, 183
    secular equation, 218, 224, 231
Chebyshev acceleration, 279, 294–
    299, 331
    model problem, 277
Chebyshev polynomial, 296, 313, 330,
    356, 358
Cholesky, 2, 76–79, 253
    band, 2, 81, 82, 277
    block algorithm, 66, 98
    condition number, 95
    conjugate gradients, 308
    definite pencils, 179
    incomplete (as preconditioner),
        318
    LINPACK, 64
    LR iteration, 243, 263
    mass-spring system, 180
    model problem, 277
    normal equations, 107
    of T_N, 270, 357
    on a Cray YMP, 63
    sparse, 84, 85, 277
    symmetric eigenproblem, 253, 263
    tridiagonal, 82, 330
CLAPACK, 63, 93, 96
companion matrix, 184, 301
    block, 184
computational geometry, 139, 175,
    184, 187, 192
condition number, 2, 4, 5
    convergence of iterative meth-
        ods, 285, 312, 314, 316, 319,
        351
    distance to ill-posedness, 17, 19,
        24, 33, 93, 152
    equilibration, 63
    estimation, 50
    infinite, 17, 148

iterative refinement of linear systems, 60

least squares, 101, 102, 105, 108, 117, 125, 126, 128, 129, 134

linear equations, 32–38, 46, 50, 94, 96, 105, 124, 132, 146

nonsymmetric eigenproblem, 32, 148–153, 189, 190

Poisson's equation, 269

polynomial evaluation, 15, 17, 19, 25

polynomial roots, 29

preconditioning, 316

rank-deficient least squares, 101, 125, 126, 128, 129

relative, for $Ax = b$, 35, 54, 62

symmetric eigenproblem, 197

conjugate gradients, 266, 278, 301, 306–319, 350

convergence, 305, 312, 351

model problem, 277

preconditioning, 316, 350, 353

conjugate gradients squared, 321

conjugate gradients stabilized, 321

conjugate transpose, 1

conservation law, 255

consistently ordered, 293

controllable subspace, 182, 187

convolution, 323, 325

Courant–Fischer minimax theorem, 198, 199, 201, 261

Cray, 13, 14

2, 226

C90/J90, 13, 63, 90, 226

extended precision, 27

roundoff error, 13, 25, 27, 224, 226

square root, 27

T3 series, 13, 63, 90

YMP, 63, 65

DAEs, *see* differential algebraic equations

DEC

symmetric multiprocessor, 63, 90

workstations, 10, 13, 14

deflation, 221

during QR iteration, 214

in secular equation, 221, 236, 262

diagonal dominance, 98, 384

convergence of Jacobi and Gauss–Seidel, 286–294

weak, 289

differential algebraic equations, 175, 178, 185, 186

divide-and-conquer, 13, 195, 211, 212, 216–228, 231, 235

SVD, 133, 240, 241

domain decomposition, 266, 285, 317, 319, 347–356, 360

dqds algorithm, 195, 242

eigenvalue, 140

generalized nonsymmetric eigenproblem, 174

algorithms, 173–184

nonsymmetric eigenproblem

algorithms, 153–173, 184

perturbation theory, 148–153

symmetric eigenproblem

algorithms, 210–237

perturbation theory, 197–210

eigenvector, 140

generalized nonsymmetric eigenproblem, 175

algorithms, 173–184

nonsymmetric eigenproblem

algorithms, 153–173, 184

of Schur form, 148

symmetric eigenproblem

algorithms, 210–237

perturbation theory, 197–210

EISPACK, 63

equilibration, 37, 62

equivalence transformation, 175

fast Fourier transform, 266, 278, 319,

321–327, 332, 347, 350, 351, 356, 358–360
    model problem, 277
FFT, *see* fast Fourier transform
floating point arithmetic, 3, 5, 9, 24
    ∞, 12, 28, 230
    complex numbers, 12, 26
    cost of comparison, 50
    cost of division, square root, 244
    cost versus memory operations, 65
    Cray, 13, 27, 226
    exception handling, 12, 28, 230
    extended precision, 14, 27, 45, 62, 224
    IEEE standard, 10, 241
    interval arithmetic, 14, 45
    Lanczos algorithm, 375
    machine epsilon, machine precision, macheps, 12
    NaN (Not a Number), 12
    normalized numbers, 9
    overflow, 11
    roundoff error, 11
    subnormal numbers, 12
    underflow, 11
flops, 5

Gauss–Seidel, 266, 278, 279, 282–283, 285–294, 356
    in domain decomposition, 354
    model problem, 277
Gaussian elimination, 31, 38–44
    band matrices, 79–83
    block algorithm, 31, 63–76
    error bounds, 31, 44–60
    GECP, 46, 50, 55, 56, 96
    GEPP, 46, 49, 55, 56, 94, 96, 132
    iterative refinement, 31, 60–63
    pivoting, 45
    sparse matrices, 83–90
    symmetric matrices, 79

    symmetric positive definite matrices, 76–79
Gershgorin's Theorem, 98
Gershgorin's theorem, 82, 83, 150
Givens rotation, 119, 121–123
    error analysis, 123
    in GMRES, 320
    in Jacobi's method, 232, 250
    in QR decomposition, 121, 135
    in QR iteration, 168, 169
GMRES, 306, 320
    restarted, 320
Gram–Schmidt orthogonalization process, 107, 375
    Arnoldi's algorithm, 303, 320
    classical, 107, 119, 134
    modified, 107, 119, 134, 231
    QR decomposition, 107, 119
    stability, 108, 118, 134
graph
    bipartite, 286, 291
    directed, 288
    strongly connected, 289
guptri (generalized upper triangular form), 187

Hessenberg form, 164, 184, 213, 301, 359
    double shift QR iteration, 170, 173
    implicit Q theorem, 168
    in Arnoldi's algorithm, 302, 303, 386
    in GMRES, 320
    QR iteration, 166–173, 184
    reduction, 164–166, 212, 302, 386
    single shift QR iteration, 169
    unreduced, 166
Hilbert matrix, 92
Householder reflection, 119–123, 135
    block algorithm, 133, 137, 166
    error analysis, 123
    in bidiagonal reduction, 166, 252
    in double shift QR iteration, 170

in Hessenberg reduction, 212
in QR decomposition, 119, 134,
   135, 157
in QR decomposition with pivoting, 132
in tridiagonal reduction, 213
HP workstations, 10

IBM
   370, 9
   RS6000, 6, 14, 27, 70, 71, 133,
      185, 236
   SP-2, 63, 90
   workstations, 10
ill-posedness, 17, 24, 33, 34, 93, 148
implicit Q theorem, 168
impulse response, 178
incomplete Cholesky, 318
incomplete LU decomposition, 319
inertia, 202, 208, 228, 246
Intel
   8086/8087, 14
   Paragon, 63, 75, 90
   Pentium, 14, 62
invariant subspace, 145–147, 153, 154,
      156–158, 189, 207
inverse iteration, 155, 162
   SVD, 241
   symmetric eigenproblem, 119, 211,
      214, 215, 228–232, 235, 236,
      240, 260, 361
inverse power method, *see* inverse
   iteration
irreducibility, 286, 288–290
iterative methods
   for $Ax = \lambda x$, 361–387
   for $Ax = b$, 265–360
      convergence rate, 281
      splitting, 279

Jacobi's method (for $Ax = \lambda x$), 195,
      210, 212, 232–235, 237, 260,
      263
Jacobi's method (for $Ax = b$), 278,
      279, 281–282, 285–294, 356

in domain decomposition, 354
   model problem, 277
Jacobi's method (for the SVD), 242,
      248–254, 262, 263
Jordan canonical form, 3, 19, 140,
      141, 145, 146, 150, 175, 176,
      178, 180, 184, 185, 188, 280
   instability, 146, 178
   solving differential equations, 176

Korteweg–de Vries equation, 259
Kronecker canonical form, 180–182,
      186, 187
   solving differential equations, 181
Kronecker product, 274, 357
Krylov subspace, 266, 278, 299–321,
      350, 353, 359, 361–387

Lanczos algorithm, 119, 304, 305,
      307, 309, 320, 359, 362–387
   nonsymmetric, 320, 386
LAPACK, 6, 63, 93, 94, 153
   dlamch, 14
   sbdsdc, 241
   sbdsqr, 241, 242
   sgebrd, 167
   sgeequ, 63
   sgees(x), 153
   sgeesx, 185
   sgees, 185
   sgeev(x), 153
   sgeevx, 185
   sgeev, 185
   sgehrd, 166
   sgelqf, 132
   sgelss, 133
   sgels, 121
   sgeqlf, 132
   sgeqpf, 132, 133
   sgeqrf, 137
   sgerfs, 63
   sgerqf, 132
   sgesvx, 35, 54, 55, 58, 62, 63,
      96
   sgesv, 96

sgetf2, 75, 96
sgetrf, 75, 96
sggesx, 186
sgges, 179, 186
sggevx, 186
sggev, 186
sgglse, 138
slacon, 54
slaed3, 226
slaed4, 222, 223
slahqr, 164
slamch, 14
slatms, 97
spotrf, 78
sptsv, 83
ssbsv, 81
sspsv, 81
sstebz, 231, 236
sstein, 231
ssteqr, 214
ssterf, 214
sstevd, 211, 217
sstev, 211
ssyevd, 217, 236
ssyevx, 212
ssyev, 211, 214
ssygv, 179, 186
ssysv, 79
ssytrd, 166
strevc, 148
strsen, 153
strsna, 153
LAPACK++, 63
LAPACK90, 63
Laplace's equation, 265
least squares, 101–138
    condition number, 117–118, 125, 126, 128, 134
    in GMRES, 320
    normal equations, 105–107
    overdetermined, 2, 101
    performance, 132–133
    perturbation theory, 117–118
    pseudoinverse, 127

QR decomposition, 105, 107–109, 114, 121
    rank-deficient, 125–132
        failure to recognize, 132
        pseudoinverse, 127
    roundoff error, 123–124
    software, 121
    SVD, 105, 109–117
    underdetermined, 2, 101, 136
    weighted, 135
linear equations
    Arnoldi's method, 320
    band matrices, 76, 79–83, 85, 86
    block algorithm, 63–76
    block cyclic reduction, 327–330
    Cauchy matrices, 92
    Chebyshev acceleration, 279, 294–299
    Cholesky, 76–79, 277
    condition estimation, 50
    condition number, 32–38
    conjugate gradients, 307–321
    direct methods, 31–99
    distance to ill-posedness, 33
    domain decomposition, 319, 347–356
    error bounds, 44–60
    fast Fourier transform, 321–327
    FFT, see   fast Fourier transform
    Gauss–Seidel, 279, 282–283, 285–294
    Gaussian elimination, 38–44
        with complete pivoting (GECP), 41, 50
        with partial pivoting (GEPP), 41, 49, 94
    iterative methods, 265–360
    iterative refinement, 60–63
    Jacobi's method (for $Ax = b$), 279, 281–282, 285–294
    Krylov subspace methods, 299–321

LAPACK, 96
multigrid, 331–347
perturbation theory, 32–38
pivoting, 44
relative condition number, 35–38
relative perturbation theory, 35–38
sparse Cholesky, 83–90
sparse Gaussian elimination, 83–90
sparse matrices, 83–90
SSOR, *see* symmetric successive overrelaxation
successive overrelaxation, 279, 283–294
symmetric matrices, 79
symmetric positive definite, 76–79
symmetric successive overrelaxation, 279, 294–299
Toeplitz matrices, 93
Vandermonde matrices, 92
LINPACK, 63, 65
    spofa, 63
    benchmark, 75, 94
LR iteration, 242, 263
Lyapunov equation, 188

machine epsilon, machine precision, macheps, 12
mass matrix, 143, 180, 254
mass-spring system, 142, 175, 179, 183, 184, 196, 209, 254
Matlab, 6, 59
    cond, 54
    eig, 179, 185, 186, 211
    fft, 327
    hess, 166
    pinv, 117
    polyfit, 102
    rcond, 54
    roots, 184
    schur, 185

speig, 386
bisect.m, 30
clown, 114
eigscat.m, 150, 190
FFT, 358
homework, 29, 30, 98, 134, 138, 190–192, 358, 360
iterative methods for $Ax = b$, 266, 301
Jacobi's method for $Ax = b$, 282, 358
Lanczos method for $Ax = \lambda x$, 367, 375, 382
least squares, 121, 129
massspring.m, 144, 197
multigrid, 336, 360
notation, 1, 41, 42, 98, 99, 251, 326
pivot.m, 50, 55, 62
Poisson's equation, 275, 358
polyplot.m, 29
qrplt.m, 161, 191
QRStability.m, 134
RankDeficient.m, 129
RayleighContour.m, 201
sparse matrices, 90
matrix pencils, 173
    regular, 174
    singular, 174
memory hierarchy, 64
MGS, *see* Gram–Schmidt orthogonalization process, modified
minimum residual algorithm, 319
MINRES, *see* minimum residual algorithm
model problem, 265–276, 285–286, 299, 314, 319, 323, 324, 327, 331, 347, 360
    diagonal dominance, 288, 290
    irreducibility, 290
    red-black ordering, 291
    strong connectivity, 289
    summary of methods, 277–279

symmetric positive definite, 291
Moore–Penrose pseudoinverse, *see* pseudoinverse
multigrid, 331–347, 356, 360
  model problem, 277

NETLIB, 93
Newton's method, 60, 219, 221, 231, 300
nonsymmetric eigenproblem, 139
  algorithms, 153–173
  condition number, 148
  eigenvalue, 140
  eigenvector, 140
  equivalence transformation, 175
  generalized, 173–184
    algorithms, 184
  ill-posedness, 148
  invariant subspace, 145
  inverse iteration, 155
  inverse power method, *see* inverse iteration
  matrix pencils, 173
  nonlinear, 183
  orthogonal iteration, 156
  perturbation theory, 148
  power method, 154
  QR iteration, 159
  regular pencil, 174
  Schur canonical form, 146
  similarity transformation, 141
  simultaneous iteration, *see* orthogonal iteration
  singular pencil, 174
  software, 153
  subspace iteration, *see* orthogonal iteration
  Weierstrass canonical form, 176
normal equations, 105, 106, 118, 135, 136, 319
  backward stability, 118
norms, 19
notation, 1
null space, 111

ODEs, *see* ordinary differential equations
ordinary differential equations, 175, 178, 184–186
  impulse response, 178
  overdetermined, 182
  underdetermined, 181
  with algebraic constraints, 178
orthogonal iteration, 156
orthogonal matrices, 22, 77, 118, 126, 131, 161
  backward stability, 124
  error analysis, 123
  Givens rotation, 119
  Householder reflection, 119
  implicit Q theorem, 168
  in bidiagonal reduction, 167
  in definite pencils, 179
  in generalized real Schur form, 179
  in Hessenberg reduction, 164
  in orthogonal iteration, 157
  in Schur form, 147
  in symmetric QR iteration, 213
  in Toda flow, 256
  Jacobi rotations, 232

PARPRE, 319
PCs, 10
pencils, *see* matrix pencils
perfect shuffle, 240, 262
perturbation theory, 2, 4, 7, 17
  generalized nonsymmetric eigenproblem, 181
  least squares, 101, 117, 125
  linear equations, 31, 32, 44, 49
  nonsymmetric eigenproblem, 83, 139, 142, 148, 181, 187, 190
  polynomial roots, 29
  rank-deficient least squares, 125
  relative, for $Ax = \lambda x$, 195, 198, 207–210, 212, 241, 242, 244–247, 249, 260, 262

relative, for $Ax = b$, 32, 35–38,
    62
relative, for SVD, 207–210, 245–
    248, 250
singular pencils, 181
symmetric eigenproblem, 195, 197,
    207, 260, 262, 365
pivoting, 41
    average pivot growth, 93
    band matrices, 80
    by column in QR decomposi-
        tion, 130
    Cholesky, 78
    Gaussian elimination with com-
        plete pivoting (GECP), 50
    Gaussian elimination with par-
        tial pivoting (GEPP), 49,
        132
    growth factor, 49, 60
Poisson's equation, 266–279
    in one dimension, 267–270
    in two dimensions, 270–279
    see also model problem, 265
polynomial
    characteristic, see characteris-
        tic polynomial
    convolution, 325
    evaluation, 34, 92
        at roots of unity, 326
        backward stability, 16
        condition number, 15, 17, 25
        roundoff error, 15, 46
        with Horner's rule, 7, 15
    fitting, 101, 138
    interpolation, 92
        at roots of unity, 326
    multiplication, 325
    zero finding
        bisection, 9
        computational geometry, 192
        condition number, 29
power method, 154
preconditioning, 316, 351, 353–356,
    384

projection, 189
pseudoinverse, 114, 127, 136
pseudospectrum, 191

qds algorithm, 242
QMR, see quasi-minimum residu-
    als
QR algorithm, see QR iteration
QR decomposition, 105, 107, 131,
    147
    backward stability, 118, 119
    block algorithm, 137
    column pivoting, 130
    in orthogonal iteration, 157
    in QR flow, 257
    in QR iteration, 163, 171
    rank-revealing, 132, 134
    underdetermined least squares,
        136
QR iteration, 159, 191, 210
    backward stability, 119
    bidiagonal, 241
    convergence failure, 173
    Hessenberg, 164, 166, 184, 212
    implicit shifts, 167–173
    tridiagonal, 211, 212, 235
        convergence, 214
quasi-minimum residuals, 321
quasi-triangular matrix, 147

range space, 111
Rayleigh quotient, 198, 205
    iteration, 211, 214, 262, 362
Rayleigh–Ritz method, 205, 261, 362
red-black ordering, 283, 291
relative perturbation theory
    for $Ax = \lambda x$, 207–210
    for $Ax = b$, 35–38
    for SVD, 207–210, 245–248
roundoff error, 3, 5, 10, 11, 300
    Bisection, 30
    bisection, 230
    block cyclic reduction, 330
    conjugate gradients (CG), 316
    Cray, 13, 27

dot product, 26
Gaussian elimination, 26, 44, 59
geometric modeling, 193
in logarithm, 25
inverse iteration, 231
iterative refinement, 60
Jacobi's method for $Ax = \lambda x$, 253
Jacobi's method for the SVD, 250
Jordan canonical form, 146
Lanczos algorithm, 305, 362, 367, 375, 376, 379
matrix multiplication, 26
orthogonal iteration, 157
orthogonal transformations, 101, 123
polynomial evaluation, 15
polynomial root finding, 30
QR iteration, 164
rank-deficient least squares, 125, 128
rank-revealing QR decomposition, 131
simulating quadruple precision, 27
substitution, forward or back, 26
SVD, 241, 247
symmetric eigenproblem, 191

ScaLAPACK, 63, 75
ARPACK, 384
PARPRE, 319
Schur canonical form, 4, 140, 146–148, 152, 158, 160, 161, 163, 175, 178, 181, 184–186
block diagonalization, 188
computing eigenvectors, 148
computing matrix functions, 188
for real matrices, 147, 163, 184, 212
generalized for real regular pencils, 179, 185
generalized for regular pencils, 178, 181, 185
generalized for singular pencils, 181, 186
solving Sylvester or Lyapunov equations, 188
Schur complement, 98, 99, 350
secular equation, 218
SGI symmetric multiprocessor, 63, 90, 91
shifting, 155
convergence failure, 173
exceptional shift, 173
Francis shift, 173
in double shift Hessenberg QR iteration, 164, 170, 173
in QR iteration, 161, 173
in single shift Hessenberg QR iteration, 169
in tridiagonal QR iteration, 213
Rayleigh quotient shift, 214
Wilkinson shift, 213
zero shift, 241
similarity transformation, 141
best conditioned, 153, 187
simultaneous iteration, see orthogonal iteration
singular value, 109
algorithms, 237–254
singular value decomposition, see SVD

singular vector, 109
algorithms, 237–254
SOR, see successive overrelaxation
sparse matrices
direct methods for $Ax = b$, 83–90
iterative methods for $Ax = \lambda x$, 361–387
iterative methods for $Ax = b$, 265–360
spectral projection, 189
splitting, 279
SSOR, see symmetric successive over-

relaxation

stiffness matrix, 143, 180, 254

Strassen's method, 70

strong connectivity, 289

subspace iteration, *see* orthogonal
        iteration

substitution (forward or backward),
        3, 38, 44, 48, 94, 178, 188
    error analysis, 25

successive overrelaxation, 279, 283–
        294, 356
    model problem, 277

Sun
    symmetric multiprocessor, 63,
        90
    workstations, 10, 14

SVD, 105, 109–117, 134, 136, 174,
        195
    algorithms, 237–254, 260
    backward stability, 118, 119, 128
    high relative accuracy, 245–254
    reduction to bidiagonal form, 166,
        237
    relative perturbation theory, 207–
        210
    underdetermined least squares,
        136

Sylvester equation $AX - XB = C$,
        188, 357

Sylvester's inertia theorem, 202

symmetric eigenproblem, 195
    algorithms, 210
    bisection, 211, 260
    condition numbers, 197, 207
    Courant–Fischer minimax the-
        orem, 199, 261
    definite pencil, 179
    divide-and-conquer, 13, 211, 216,
        260
    inverse iteration, 211
    Jacobi's method, 212, 232, 260
    perturbation theory, 197
    Rayleigh quotient, 198
    Rayleigh quotient iteration, 211,

        214
    relative perturbation theory, 207
    Sylvester's inertia theorem, 202
    tridiagonal QR iteration, 211,
        212

symmetric successive overrelaxation,
        279, 294–299
    model problem, 277

SYMMLQ, 319

templates for $Ax = b$, 266, 279, 301

Toda flow, 255, 260

Toda lattice, 255

Toeplitz matrices, 93

transpose, 1

tridiagonal form, 119, 166, 180, 232,
        235–237, 243, 246, 255, 307,
        330
    bisection, 228–232
    block, 293, 358
    divide-and-conquer, 216
    in block cyclic reduction, 330
    in boundary value problems, 82
    inverse iteration, 228–232
    nonsymmetric, 320
    QR iteration, 211, 212
    reduction, 164, 166, 197, 213,
        236, 253
        using Lanczos, 302, 304, 320,
        364, 386
    relation to bidiagonal form, 240

unitary matrices, 22

Vandermonde matrices, 92

vec($\cdot$), 274

Weierstrass canonical form, 173, 176,
        178, 180, 181, 185–187
    solving differential equations, 176